T0201728

Geophysical Monograph Series

Including
IUGG Volumes
Maurice Ewing Volumes
Mineral Physics Volumes

Geophysical Monograph Series

Geophysical Monograph 168

Earth's Deep Water Cycle

Steven D. Jacobsen

Suzan van der Lee

Editors

American Geophysical Union
Washington, DC

Published under the aegis of the AGU Books Board

Jean-Louis Bougeret, Chair; Gray E. Bebout, Cassandra G. Fesen, Carl T. Friedrichs, Ralf R. Haese, W. Berry Lyons, Kenneth R. Minschwaner, Andrew Nyblade, Darrell Strobel, and Chunzai Wang, members.

Library of Congress Cataloging-in-Publication Data

Earth's deep water cycle / Steven D. Jacobsen, Suzan van der Lee, editors.
    p. cm. -- (Geophysical monograph series ; 168)
 Includes bibliographical references.
 ISBN-13: 978-0-87590-433-7
 ISBN-10: 0-87590-433-5
 1. Groundwater. 2. Earth--Mantle. 3. Hydrologic cycle. 4. Geodynamics.
 5. Geology, Structural. I. Jacobsen, Steven D. II. Lee, Suzan Frederike Maria van der, 1966- GB1005.E27 2006
 551.48--dc22
                2006031417

 ISBN-13: 978-0-87590-433-7
 ISBN-10: 0-87590-433-5

ISSN 0065-8448

Copyright 2006 by the American Geophysical Union
2000 Florida Avenue, N.W.
Washington, DC 20009

Figures, tables and short excerpts may be reprinted in scientific books and journals if the source is properly cited.

Authorization to photocopy items for internal or personal use, or the internal or personal use of specific clients, is granted by the American Geophyscial Union for libraries and other users registered with the Copyright Clearance Center (CCC) Transactional Reporting Service, provided that the base fee of $1.50 per copy plus $0.35 per page is paid directly to CCC, 222 Rosewood Dr., Danvers, MA 01923. 0065-8448/06/$01.50+0.35.

This consent does not extend to other kinds of copying, such as copying for creating new collective works or for resale. The reproduction of multiple copies and the use of full articles or the use of extracts, including figures and tables, for commercial purposes requires permission from the American Geophysical Union.

Printed in the United States of America.

CONTENTS

IV. Observational Constraints on Water in the Deep Mantle

V. Models of a Deep Water Cycle

PREFACE

Earth's bulk water content likely exceeds that of all other terrestrial planets combined. On Earth, plate tectonics is responsible for the recycling of water between the crust and uppermost mantle. Water descends in hydrated oceanic crust, mostly in the form of hydrous minerals. Much of this water returns to the surface in the form of back-arc volcanism. But to what extent, if at all, does water recirculate deeper into the mantle below a few hundred kilometers? What are the observed fluxes, and how do they compare with potential fluxes based on the capacity of deep nominally anhydrous minerals to store water? Does some amount of Earth's primordial hydrogen or "water" remain trapped at deeper levels in the mantle? Is it possible to detect water in the deep mantle seismically or from electrical conductivity profiles? And to what extent may water be necessary for plate tectonics in the first place? These and other questions surrounding Earth's potential deep water cycle are covered in this volume through interdisciplinary studies from mineral physics, geochemistry, and geophysics. Indeed, through laboratory synthesis and physical properties measurements, seismology, and geodynamic modeling, we are beginning to understand what parameters will allow us to obtain a broader appreciation of the dynamic role that water plays in our planet.

Liquid water covers 70% of Earth's surface, but constitutes only about 0.025 wt% of the planet's mass – far less than Earth is thought to have contained during accretion and before core formation. Hydrogen loss during accretion must have been extensive, and there is compelling evidence that degassing of the mantle with greater than ninety percent efficiency has led to the formation of Earth's oceans [see e.g. *Rüpke et al.* of this volume]. However, the mass fraction of liquid water on Earth (0.025 wt% H_2O) is still on the order of ten times less than the water content of mid-ocean ridge magmas (0.2-0.3 wt% H_2O), and about half of what model "enriched mantle" sources contain (~0.04 wt% H_2O). Perhaps rather than asking from where did Earth's water originate, we should be asking to where has Earth's water gone?

It is possible that the majority of Earth's "missing water" resides in the deep mantle. Trace amounts of hydrogen have been measured in nearly all natural upper mantle-derived rocks, and evidence for dehydration and flux melting at shallow depths is well known, but the story below the asthenosphere becomes more elusive. Chapters in this volume will focus on the deeper upper mantle and transition zone at 410-660 km depth. Because of the large volume of rock associated with this part of the mantle, a small fraction of hydrogen goes a long way. Just a thousand ppm H_2O by weight, or 0.1 wt% H_2O, if spread uniformly throughout the transition zone, would roughly equate in mass to the Atlantic Ocean.

The present volume is divided into five sections beginning with (I) two overview chapters where mineral physics and geochemical constraints are reviewed by *Smyth and Jacobsen*, followed by a review of seismological studies and constraints on water in the deep Earth by *Van der Lee and Wiens*. Subsequent chapters are grouped into sections dealing with (II) water storage and stability of OH-bearing phases, (III) properties of a deep hydrous mantle, (IV) observational studies from seismology and electrical conductivity, and (V) global models of the deep-Earth water cycle.

In section II, *Komabayashi* presents a complete phase diagram of hydrous peridotite in the $MgO-SiO_2-H_2O$ system extending to lower mantle conditions. His thorough treatment of phase stability is based upon the latest high-pressure experimental results and new thermodynamic calculations. The section on storage capacity and stability continues with a study of natural samples from the upper mantle; *Mosenfelder et al.* examine the OH-signatures in mantle-derived olivine using infrared spectroscopy and high-resolution microscopy to evaluate some key differences between natural samples and those synthesized in laboratory experiments. Going deeper, *Bolfan-Casanova et al.* report on the storage capacity and hydration mechanisms in phases of the transition zone and lower mantle including wadsleyite, ringwoodite, silicate perovskite, and magnesiowüstite. Section II concludes with a chapter on Raman spectroscopic studies of OH-bearing minerals by *Kleppe and Jephcoat*. Their chapter features a new graphical database of Raman spectra for the most important OH-bearing and hydrous phases.

Section III deals mainly with physical properties of hydrated mantle minerals. Experimental studies on the physical properties of hydrous and OH-bearing phases provides a key link between mineral physics and the observational studies and models to follow in subsequent sections. *Litasov et al.* report experimental results on the influence of water on the phase transformations related to the major seismic discontinuities at 410 and 660-km depth. *Karato* presents a critical review of the effects of hydrogen on transport properties including electrical conductivity and plastic deformation. *Jacobsen and Smyth* have measured compressional and shear wave velocities in hydrous ringwoodite ultrasonically at high pressures and model hydrous velocities in the transition zone. *Inoue et al.* report new *in situ* high-pressure and high-temperature studies on the stability and equation of state of superhydrous phase B, a dense hydrous

Earth's Deep Water Cycle
Geophysical Monograph Series 168
Copyright 2006 by the American Geophysical Union.
10.1029/168GM01

magnesium silicate that may play a role in transporting large amounts of water along cooler subduction geotherms. Section III concludes with a chapter on the properties of pure H_2O – *Lin et al.* examine the phase diagram of H_2O as it pertains to planetary interiors and review various lines of evidence for the newly discovered triple point beyond which solid-H_2O becomes "superionic", with freely moving H^+ (protons).

Section IV highlights current methods to infer the presence and amount of hydrogen in the mantle. *Koyama et al.* interpret a region of high electrical conductivity and high seismic P wave velocities in the transition zone beneath the Mariana Islands as ~0.3 wt% more hydrous than the surrounding mantle. Next, *Courtier and Revenaugh* examine transition-zone discontinuities through *ScS* reverberations and find that observed depths and impedance contrasts are consistent with a locally hydrous transition zone beneath the eastern US. In the western US, *Song and Helmberger* study triplicated S waves and infer a possibly water-triggered low-velocity layer atop the 410-km discontinuity that has a sharp western edge. To complete the variety of seismic observations used to detect "deep protons", *Braunmiller et al.* turn to S waves converted from P waves at transition zone discontinuities and find that the transition zone beneath the Andes could be either saturated with water or dry as a desert. *Shito et al.* combine various seismic observations, specifically anomalies in P and S velocities and S-wave attenuation, to systematically map the possible distribution of water in the upper mantle. They find up to 0.1 wt% H_2O above the transition zone beneath the Philippine Sea. *Suetsugu et al.* combine P-velocity anomalies with relief on the 660-km discontinuity and deduce 1-1.5 wt% H_2O at the base of the transition zone below the Philippine Sea and western Japan. Finally, *Lawrence and Wysession* study a large attenuation anomaly imaged in the top of the lower mantle beneath China that may be related to deep water cycling. Observational inferences for water content in the mantle require special attention to associated uncertainties both in material properties and data processing, but the emerging consensus suggests that there is a detectable level of hydration heterogeneously distributed in the Earth's mantle.

The volume concludes with several models of a deep-Earth water cycle in section V. *Rupke et al.* provide compelling arguments that Earth's mantle has efficiently degassed, with the majority of the planet's water locked in the hydrosphere and shallowest upper mantle. In the case that some water resists dehydration and cycles to deeper depths of the transition zone, properties of the 410-km discontinuity will likely hold key evidence one way or the other. *Hirschmann et al.* model the influence of water on various properties of the 410-discontinuity, which will help to direct seismologists towards observable

features related to water, or the absence of it. Finally, *Karato et al.* present an updated "transition zone water filter model" that seeks to explain how water and associated melts in the transition zone may explain paradoxical geochemical trends in surface volcanism.

In its entirety, we hope this volume portrays the nature, current directions, and future needs in the study of Earth's deep water cycle. Earth is the only planet we know of that looses internal heat through plate tectonics. Large spatial and temporal variations in viscosity are required in a planet's mantle for plate tectonics to operate. Cycling of water between the deep upper mantle and the surface could provide viscosity contrasts; thus, water content could be the second most important factor, next to temperature, influencing the viscosity of Earth's mantle and the fate of plate tectonics. In this larger context, the study of Earth's deep water cycle is central to understanding the evolution of the planet.

At the same time, this volume comes at a crossroads in the study of water in the deep Earth. Up until about ten years ago the study of OH in nominally anhydrous minerals was more or less restricted to a few papers in the mineralogical literature. Since then, we have come to recognize the strong possibility that water is playing a major role in mantle dynamics. Despite recent progress on many fronts, as exemplified in this volume, considerable work remains to be done. The effects of hydration on elastic moduli and their derivatives to temperature, seismic attenuation, and the solubility of water in the lower mantle stand out as areas especially in need of exploration. Equally vital is the need to reduce ambiguities associated with the interpretation of geophysical data used to remotely sense water in the deep upper mantle.

Numerous people have contributed to the development and production of this volume. We wish to thank AGU Acquisitions Editor Allan Graubard, Program Coordinator Maxine Aldred, Production Coordinator Colleen Matan, and Dawn Seigler, who orchestrated the review and editing process. We also wish to thank the initial review committee, who skeptically accepted the ambitious plan of a monograph on Earth's Deep Water Cycle from the two proposing editors. This volume would not have been possible without the contributing authors, whose hard work and scientific insight constitute the body of this work. Finally, we are indebted to the numerous reviewers, many of whom are acknowledged by name at the end of each chapter, for their crucial contribution to this volume.

Steven D. Jacobsen
Suzan van der Lee
Editors

Nominally Anhydrous Minerals and Earth's Deep Water Cycle

Joseph R. Smyth

Department of Geological Sciences, University of Colorado, Boulder, CO 80309 USA

Steven D. Jacobsen

Department of Geological Sciences, Northwestern University, Evanston, IL 60208 USA

Deep reservoirs of water incorporated as hydroxyl into solid silicate minerals of the Earth's interior may contain the majority of the planet's hydrogen and have acted as buffers to maintain ocean volume and continental freeboard over geologic time. Two tenths of one weight percent H_2O in subducted oceanic crustal material and subsequently released to the hydrosphere from mid-ocean ridge basalt is sufficient to recycle the total ocean volume once over 4.5 billion years. It is possible that actual fluxes are several times this amount. The nominally anhydrous minerals of the transition zone (410–660 km depth) may serve as a large internal reservoir. New and recent data on molar volumes and elastic properties indicate that hydration has a larger effect on shear velocities than does temperature within their respective uncertainties. Based on these new data, seismic velocities in this region are consistent with significant hydration (one-half percent or more H_2O by weight in a pyrolite-composition mantle). The data indicate that lateral velocity variations in the Transition Zone (TZ) may reflect variations in hydration rather than variations in temperature, at least in regions distant from subduction zones.

1. INTRODUCTION

Earth is unique among the Terrestrial planets in having liquid water on its surface. Water controls the entire character of the biology and geology of the planet. Without liquid water, carbon-based life is highly unlikely if not impossible. Geologically, water controls igneous activity by lowering the melting points of rocks by hundreds of degrees Celsius. It fluxes melting below divergent plate boundaries [*Asimow and Langmuir*, 2003] and controls major-element partitioning [*Inoue*, 1994]. Water controls tectonics by affecting the strength and deformation mechanisms of minerals [*Kavner,*

2003] and thus the rheology of rocks [*Karato*, 1998; *Mei and Kohlstedt*, 2000]. It also controls weathering and all low temperature geochemical reactions that generate sediments from rocks. Water transports the sediments in streams and glaciers, and deposits them in rivers, lakes, and oceans. Although oceans cover more than seventy percent of the surface, they constitute only 0.025 percent of the planet's mass. The presence of quartz pebble conglomerates of earliest Archean age tells us that there has been moving water on the surface, hence oceans and dry land, almost as far back as we can see in geologic time. Oxygen isotope ratios in ancient detrital zircons indicate the presence of liquid water as early as 4.4 GY ago [*Wilde et al.*, 2001; *Mojzsis et al.*, 2001]. Although Earth's oceans are unique in our solar system, how common might Earth-like water planets be in our galaxy or in other galaxies?

Earth's Deep Water Cycle
Geophysical Monograph Series 168
Copyright 2006 by the American Geophysical Union.
10.1029/168GM02

Despite its importance to understanding the biology and geology of the planet, hydrogen is the most poorly constrained major-element compositional variable in the bulk Earth. The total complement of Earth's hydrogen is unknown within an order of magnitude [*Abe et al.*, 2000]. We do not have good constraints on the amount of H that may be incorporated in hydrated ocean crust and lithospheric mantle [*Dixon et al.*, 2002] or in interior reservoirs, particularly in the transition zone [*Richard et al.*, 2002], but also in the deep mantle and core [*Williams and Hemley*, 2001]. Further, the origin of the hydrogen on the planet is a subject of considerable debate. Is it endogenous, part of the planet's original accretion of material [*Drake and Righter*, 2001], or exogenous, added after accretion and differentiation from relatively D-enriched comets [*Robert*, 2001; *Delsemme*, 1999]? Has ocean volume been relatively constant over geologic time as is widely assumed, or has it fluctuated significantly? Is there a large reservoir of hydrogen incorporated in solid silicate minerals in the interior [*Smyth*, 1987; *Bercovici and Karato*, 2003], or is the interior relatively dry [*Dixon and Clague*, 2001; *Wood*, 1995]? What are the fluxes of water into and out of the uppermost (<200 km) [*Iwamori*, 1998; *Schmidt and Poli*, 1998] and deeper mantle (>200km) [*Schmidt*, 1996; *Kawamoto et al.*, 1996]? Such questions can be and have been addressed through various Earth Science sub disciplines, but as yet there is relatively little sense of consensus.

Hirschmann et al. [2005] have estimated the H storage capacities of the nominally anhydrous minerals in the various regions of the upper mantle, transition zone, and lower mantle. Recent results on the hydration of olivine indicating H_2O contents approaching one percent by weight [*Smyth et al.*, 2006; *Mosenfelder et al.*, 2006] at temperatures and pressures near the 410 km discontinuity, suggest that there may be a significant flux of water into the transition zone and perhaps the lower mantle as hydroxyl in the nominally anhydrous silicates of the region. An assessment of the hydration of the upper mantle and transition zone will depend on measurements of the molar volume, bulk moduli and full elastic properties of these phases as a function of water content. Such data are far from complete, but preliminary findings indicate that hydration may have a larger effect on these properties than does temperature within the uncertainties of each. The objective of this paper is to review the effects of hydration on the physical properties of these phases and provide a preliminary assessment of the possibility of a deep water cycle within the mantle. To estimate the significance of a deep water cycle, we begin with a review of the possible fluxes and potential reservoirs in the nominally anhydrous minerals of the upper mantle and transition zone.

2. FLUXES AND RESERVOIRS

The current flux of crustal material into the deep mantle (> 200 km depth) with subduction must be approximately equal to the amount of new ocean crust generated from mantle sources at mid-ocean ridges plus small amounts of ocean island basalts. If the total length of mid-ocean ridge is 70,000 km, the average spreading rate 10 cm per year [*Gordon*, 1995], and the average ocean crust thickness is 7 km [*Parsons*, 1982], then approximately 49 km^3 of crust is created and consumed per year. Alternatively, if the total area of oceanic crust on the Earth is 3.6 x 10^8 km^2, and it has a mean age of 80 million years [*Parsons*, 1982], we estimate 32 km^3 per year. If this material were 0.1 percent water by weight, it would be 0.3 percent water by volume for a net flux of about 0.1 to 0.15 km^3 per year of liquid-water equivalent. If subduction rates have been constant over the past 4.5 billion years, the integrated volume is equal to about one-half the entire current ocean volume. In other words, 0.2 percent water by weight (2000 ppmw) in deeply subducted crust (i.e., just the crustal portion of the slab) and returning in mid-ocean ridge basalt is roughly sufficient to recycle the ocean once in 4.5 billion years, assuming current subduction and spreading rates. It is likely that subduction rates were higher in the past, so it is probable that geologically significant amounts of H have been carried to the interior over geologic time. Also, hydration is likely to extend well into the oceanic lithosphere, and Wen and Anderson [1995] estimate an annual flux of lithosphere of 265 km^3. Furthermore, water released by dehydration of the subducting slab below 200 km depth, will likely be captured by the hot olivine in the overlying mantle wedge. Hydration greatly reduces the strength and effective viscosity of olivine [*Karato et al.*, 1986] allowing it to lubricate the downward movement of the cold slab. Mid-ocean ridge basalt (MORB) liquids contain anywhere from 0.1 to 1.5 weight percent H_2O, but are generally thought to average 0.25 percent by weight [*Dixon et al.*, 2002]. Eclogites from kimberlites and ultra-high pressure metamorphic zones that are thought to represent subducted crustal material, contain comparable amounts of water [*Rossman and Smyth*, 1990; *Smyth et al.*, 1991; *Katayama and Nakashima*, 2003]. Thus, if the total ocean volume has not been increasing over geologic time, it probably represents some equilibrium exchange volume with an internal reservoir contained as hydroxyl in silicate minerals of the mantle at depths greater than about 200 km.

Hydrated oceanic crust ranges up to nearly 8 percent and averages 1.0 to 1.5 percent by weight H_2O [*Alt*, 2004]. Major hydrous minerals in the hydrated crust include iron oxy-hydroxides, celadonite, and saponite at low temperature and talc, chlorite, lawsonite, and amphiboles at higher temperatures. Below the basalts and gabbroic sheeted dikes, lie cumulate gabbros and ultramafic rocks. Most of nominally hydrous minerals of the crustal portion of the slab break down before the slab reaches the 410 km discontinuity, and much of this water is returned via arc volcanism. Talc and chlorite dehydrate at relatively low pressures (< 3GPa), whereas lawsonite and phlogopite are stable to about 7 GPa on a cold geotherm [*Kawamoto et al.,* 1996, *Schmidt and Poli,* 1998]. Talc breaks down to 10Å-phase and eventually to phase E or will dehydrate to enstatite. Phengite however is stable to 11 GPa (330 km) [*Schmidt,* 1996] but requires the presence of potassium. As mentioned earlier, the nominally anhydrous minerals of eclogite can contain 1000 to 2000 ppmw H_2O.

Below the crust, the lithospheric mantle is composed of harzbergites and peridotites. Hydrothermal alteration of the ultramafic rocks produces mainly serpentine and talc. Serpentine and chlorite contain roughly ten percent water by weight or nearly 30 percent by volume, so a fully hydrated oceanic lithosphere could sequester more than one entire ocean volume of water. If hydration of the lithospheric mantle is significant, the amount of water being subducted may be larger. The major hydrous minerals of ultramafic rocks include serpentine, talc, chlorite, and amphibole. Talc and chlorite break down at shallow depths (< 100 km), but serpentine and amphibole are stable to 200 km or more in a cold subduction geotherm [*Kawamoto et al.,* 1996]. Serpentine breaks down to phase A and eventually to clinohumite plus olivine.

The breakdown of these minerals can release potentially very large amounts of aqueous fluids, which can cause flux melting in the overlying hot mantle wedge. Dixon et al. [2002] have argued that dehydration is highly efficient with 99 percent of water from igneous rocks and 97% from sediments being returned, on the basis of the incompatible behavior of hydrogen in these systems. This estimate ignores the H contents of nominally anhydrous minerals and the pressure effect on the compatibility of H, and so may be a high estimate of the dehydration efficiency. The amount of residual water remaining in the slab would be significant given the volumes of subducted material. About 2000 ppmw in the crustal portion of the slab alone (~90% efficiency) would be sufficient to recycle the current ocean volume once in 4.5 billion years at current subduction rates. The more rapid the subduction, the cooler the slab and the deeper the nominally hydrous phases are stable. The amount of H soluble in olivine increases at least a hundredfold relative to ambient conditions at pressures above 10 GPa (300 km) [*Kohlstedt et al.,* 1996; *Locke et al.,* 2002, *Mosenfelder et al.,* 2006; *Smyth et al.,* 2006], meaning that

at high pressure, hydrogen is a more compatible element than is commonly believed. If there is significant hydration of the lithospheric mantle below the crust, the flux would be larger, as it would be if there were significant capture of H by olivine in the hot mantle wedge overlying the slab. If subduction rates in the early Earth were much higher than at present, H fluxes would have been higher in the past.

In the Transition Zone, water can be held in the wadsleyite and ringwoodite phases. Wadsleyite is known to incorporate up to 3.0 wt%, and ringwoodite 2.8 wt% H_2O [*Kohlstedt et al.,* 1996]. A mantle of 75 percent olivine-stoichiometry (i.e., pyrolite) would then be able to incorporate some eight times the ocean volume in the Transition Zone before the nominally anhydrous phases become saturated. The amount of H_2O in both wadsleyite and ringwoodite decreases with temperature above 1100°C, but there is still a nearly percent H_2O at 1400°C [*Demouchy et al.,* 2005; *Smyth et al.,* 2003]. The experimental evidence is now indisputable that these silicates are capable of incorporating globally significant amounts of water, but the actual H-content (or hydration state) of the Transition Zone is unknown. In order to evaluate how much water might actually be present, careful laboratory measurements of the effects of hydration on transformation pressures, and molar volume and elastic properties of these minerals will be required. These data can then be compared with seismic models of the interior. Such measurements are in progress, but are by no means complete.

3. VOLUMES OF HYDRATION

Hydration has a significant effect on the molar volume of the nominally anhydrous phases of the upper mantle and transition zone. Smyth et al. [2006] report that 0.5 percent by weight H_2O in olivine has about the same effect on density as raising the temperature by 240°C. Smyth et al. [2006] give unit cell volume data for forsterite as a function of H_2O contents, based on the olivine-specific calibration of Bell et al. [2003]. Converting their data to molar volumes for comparison with wadsleyite and ringwoodite and fitting their data to a linear relation we obtain

$$MV= 14.67241 (\pm0.00078) + 2.07(\pm0.15) \times 10^{-6} * c_{H2O} \quad [1]$$

with a correlation coefficient, R, of 0.86, where MV is molar volume in cm^3 and c_{H2O} is the ppm by weight H_2O as determined from FTIR using the calibration of Bell et al. [2003] for polarized infrared spectra. This calibration give H_2O contents nearly three times those of the general (unpolarized) calibrations of Libowitzky and Rossman [1997] or Patterson [1982].

Jacobsen et al. (2005) and Holl et al. (2003) give cell volume data for wadsleyite as a function of H_2O content. Fitting these data to a linear relation, we obtain

$$MV = 13.61674\ (\pm0.00038) + 1.43\ (\pm0.05) \times 10^{-6} * c_{H2O}\ [\ 2\ ]$$

with a correlation coefficient, R, of 0.96. H_2O contents were determined based on the general infrared calibration Libowitzky and Rossman [1997] which is close to that of Patterson [1981].

Similarly, Smyth et al. [2003] give volume data for pure magnesian ringwoodite as a function of H_2O content. From these data we obtain:

$$MV = 13.2478\ (\pm0.0018) + 4.75\ (\pm0.35) \times 10^{-6} * c_{H2O}\ [\ 3\ ]$$

with a correlation coefficient, R, of 0.99. These data indicate similar relative volume expansion with hydration for olivine and wadsleyite and a larger expansion for ringwoodite. If the cell volume data of Kudoh et al. [1998] for a sample with estimated 20000 ppmw H_2O is included is included, the volume effect is larger, as are the errors in the regression. This relation is based on H_2O determination by infrared spectroscopy and therefore dependent on the calibration used. The relative

volume expansions as a function of H_2O content at ambient conditions for forsterite, wadsleyite, and ringwoodite are plotted in Figure 1. The H contents of wadsleyite and ringwoodite used for these volume effect estimates are based on the general calibrations and may be subject to revision if phase-specific calibrations are subsequently developed for wadsleyite and ringwoodite. These volumes should not be used to estimate the buoyancy of hydrous materials in the Earth, because we do not yet have experimental measurements of the effect of hydration on thermal expansion and compressibility, but work is proceeding rapidly in this area as outlined below.

4. ELASTIC PROPERTIES

Bulk modulus measurements on hydrous ringwoodite [*Inoue et al.* 1998; *Smyth et al.* 2004], hydrous wadsleyite [*Crichton et a*l., 1999; *Smyth et al.*, 2005b; *Holl et al.*, 2003], and olivine [*Smyth et al.*, 2005a; 2006] indicate significant softening with hydration. Unit cell volume measurements as a function of pressure yield an isothermal bulk modulus measurement, K_T, as well as a value for K′ (*dK/dP*), if the volume measurements are of sufficient precision. In general, the effect of hydration on these phases is to decrease bulk

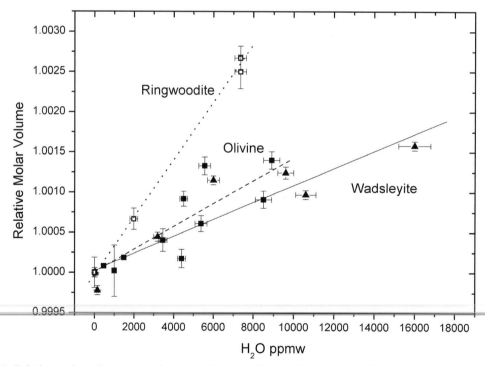

Figure 1. Relative molar volume expansions at ambient conditions with hydration for olivine, wadsleyite, and ringwoodite. The larger expansion effect observed for ringwoodite may be due to a larger role of tetrahedral vacancy in the hydration of ringwoodite relative to wadsleyite and olivine. The flexure in the expansion curve of wadsleyite correlates with the decrease in space group symmetry from *Imma* to *I2/m*.

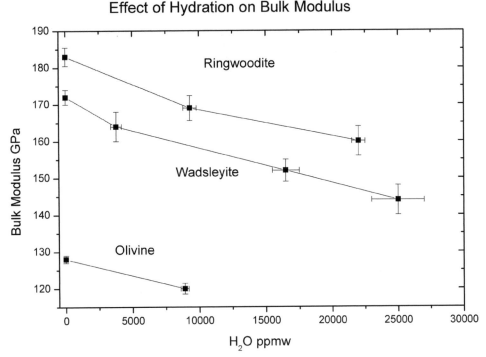

Figure 2. Effect of hydration on isothermal bulk modulus at ambient conditions for olivine, wadsleyite, and ringwoodite.

modulus and increase K′. We have plotted the isothermal bulk modulus values for olivine, wadsleyite, and ringwoodite as a function of H_2O content in Figure 2. For anhydrous samples we use values of 128 GPa for olivine [*Will et al.*, 1986; *Kudoh and Takéuchi* 1985, *Downs et al.*, 1996], 172 GPa for wadsleyite [*Hazen et al.*, 2000], and 183 GPa for ringwoodite [*Zerr et al.*, 1993]. A preliminary value of the bulk modulus for hydrous olivine is 120 GPa from *Smyth et al.* [2005a], and for hydrous wadsleyite we use 162 GPa (~1 wt% H_2O) and 155 GPa (~2.5 wt% H_2O) from *Holl et al.* [2003] and *Yusa and Inoue* [1997]. For ringwoodite, values are 175 GPa (~1 wt% H_2O) and 165 GPa (~2.8 wt% H_2O) from the studies of *Manghnani et al.* [2005] and *Wang et al.* [2003]. The data indicate similar elastic softening of all three phases. Bulk modulus measurements alone do not yield values for V_p and V_S, however full elastic property measurements for hydrous ringwoodite are available and may give some indication of the effects for olivine and wadsleyite.

Elastic property measurements on hydrous ringwoodite at ambient conditions [*Wang et al.*, 2003; *Jacobsen et al.*, 2004a], and at elevated pressure [see *Jacobsen and Smyth, this volume*] also indicate significant reduction of both bulk and shear moduli with hydration. The high-pressure acoustic measurements on hydrous ringwoodite were made using a novel GHz-ultrasonic method [*Jacobsen et al.*, 2004b], where measured velocities (V_p and V_S) in various crystal-

lographic orientations were used to determine the complete elastic tensor of the single crystals grown at high pressure. Aggregate moduli were fit to the single-crystal elastic data up to 10 GPa, resulting in $K_{S0} = 177(4)$ GPa, $K_S' = 5.3(4)$, $G_0 = 103(1)$ GPa and $G' = 2.0(2)$ for the ringwoodite of Fo_{90} composition and containing ~1 wt% H_2O [see *Jacobsen and Smyth, this volume*]. In agreement with static compression measurements [*Manghnani et al.*, 2005], the bulk modulus of ringwoodite is reduced by about 6% upon hydration. The shear modulus is reduced by about 12%, indicating that water has a greater affect of shear properties. At ambient conditions, the magnitude of reduction is large; adding 1 wt% H_2O into ringwoodite has the same effect on V_p and V_S as raising the temperature by about 600°C and 1000°C, respectively. Due to elevated pressure derivatives, the difference between anhydrous and hydrous velocities diminishes, but temperature derivatives of the moduli are still needed to make a more accurate evaluation of the effects of water on velocities in the transition zone.

To illustrate the magnitude of effects of ~1 wt% H_2O hydration on velocity we have calculated monomineralic P and S velocities for hydrous and anhydrous ringwoodite and wadsleyite using finite strain equations of state [*Davies and Dziewonski* 1975] along an adiabat with 1673 K foot temperature (Plate 1). For comparison, we also show several global velocity models; PEMC [*Dziewonski et al.*, 1975], PREM

[*Dziewonski and Anderson* 1981], and IASPEI91 [*Kennett and Engdahl,* 1991]. For anhydrous phases we use K_S values of 129, 172, and 188 GPa with dKs/dP of 4.3 for all three phases [*Duffy et al.,* 1995; *Li et al.,* 1998; *Sinogeikin et al.,* 2003; and *Li,* 2003]. We use shear moduli (G) of 82, 106, and 120 GPa for dry olivine, wadsleyite, and ringwoodite with dG/dP of 1.4 [*Duffy et al.,* 1995; *Li et al.,* 1998; *Sinogeikin et al.,* 2003; *Li,* 2003]. The parameters for hydrous ringwoodite are from our measurements, and those of hydrous wadsleyite are scaled from our measurement on ringwoodite. We assume pressure derivatives for the hydrous phases are increased to 5.3 for K' and 1.8 for G'. For comparison, we also plot velocities for majorite garnet using $\rho_0 = 3.53$ kg/m^3, $K_{S0} = 165$ GPa and $G_0 = 88$ GPa [*Gwanmesia et al.,* 1998; *Sinogeikin et al.,* 1997] with $dK_S/dT = -0.020$ and $dG/dT = -0.010$ [*Suzuki and Anderson* 1983] and assuming pressure derivatives of $K' = 4.3$ and $G' = 1.4$. Given the absence of temperature derivatives of the moduli for the hydrous phases, in this simple model we are required to assume anhydrous values. Temperature dependence of the bulk modulus (dK_S/dT) was taken to be -0.016 GPa/K for dry olivine [*Liu et al.,* 2005], -0.019 GPa/K for dry wadsleyite [*Li et al.,* 1998; *Katsura et al.,* 2001], and -0.021 GPa/K for dry ringwoodite [*Sinogeikin et al.,* 2003; *Mayama et al.,* 2005]. Temperature dependence of the shear modulus (dG/dT) was taken to be -0.013 GPa/K for dry olivine [*Liu et al.,* 2005], -0.017 GPa/K for dry wadsleyite [*Li et al.,* 1998; *Katsura et al.,* 2001], and -0.015 GPa/K for dry ringwoodite [*Sinogeikin et al.,* 2003; *Mayama et al.,* 2005]. A linear coefficient of volume thermal expansion $\alpha_V = 27 \times 10^{-6}$ K^{-1} was assumed for all phases. The set of thermoelastic parameters used in this illustrative model are given in Table 1.

Whereas elevated pressure derivatives act to bring hydrous and anhydrous P-wave velocities to within their mutual uncertainty in the transition zone, S-wave velocities for the hydrous phases remain several percent lower than anhydrous S-wave velocities. Temperature derivatives of the elastic moduli for the hydrous phases will be needed in order to improve estimates of hydrous velocities in the transition zone. Considering the current model, S-wave velocities for both hydrous and anhydrous velocities for wadsleyite match

the seismic velocity models equally well, whereas in the deeper part of the TZ where ringwoodite is the stable phase, hydrous ringwoodite, or a mixture of anhydrous ringwoodite and majorite best fit the velocity models. The key feature of Plate 1 is that both reduced S-wave velocities, and elevated V_P/V_S ratios are generally indicative of hydration.

5. DISPLACEMENT OF TRANSITIONS

There are several possible ways to detect hydration (or lateral variations of water) in Earth's Transition Zone, and, as outlined above, there has been much recent progress in quantifying the physical properties of hydrous high-pressure phases to make this possible. The seismic discontinuities at 410, 525, and 660 km depths correspond to phase transitions in the Mg_2SiO_4 phases (olivine, wadsleyite, ringwoodite) composing 50 to 80 percent of the upper mantle and TZ [*Duffy et al.,* 1995; *Duffy and Anderson,* 1989]. Water can affect not only the seismic velocities, but also the depths of transitions, and the width or sharpness of the transitions in these phases [*Wood,* 1995; *Smyth and Frost,* 2002; *Chen et al.,* 2002; *Hirchmann et al.,* 2005, *also see Hirschmann et al., this volume*]. These are the major seismic observables that can be evaluated in the laboratory. The absolute thickness of the TZ (between 410 and 660 km) can be determined more accurately than the absolute depth of the discontinuities using the relative arrival times of P-to-S converted phases [*Gilbert et al.,* 2003]. Because water is more soluble in wadsleyite and ringwoodite than in either olivine above the TZ or perovskite plus ferropericlase below the TZ, we expect H to expand the wadsleyite and ringwoodite stability fields relative to olivine at 410 and perovskite plus ferropericlase at 660 km [*Wood,* 1995]. The olivine to wadsleyite transition is exothermic with a positive Clapeyron slope (dP/dT), so hydration and decreased temperature have similar effects; moving the olivine-wadsleyite transition (410 km) to shallower depths, whereas the negative Clapeyron slope of the ringwoodite-perovskite transition (660 km) would move it to deeper depths. Smyth and Frost [2002] indeed observed a decrease in the pressure of the olivine-wadsleyite transition of about 1.0 GPa (~30 km) with the addition of 2 wt%

Table 1. Thermoelastic Properties of Hydrous and Anhydrous (10,000 ppmw H$_2$O) Polymorphs of (Mg$_{0.9}$Fe$_{0.1}$)$_2$SiO$_4$

	ρ_{STP} (g/cm^3)	Ks (GPa)	dKs/dP	dKs/dT (GPa K^{-1})	G (GPa)	dG/dP	dG/dT (GPa K^{-1})
Olivine	3.36	129	4.3	-0.016	82	1.4	-0.013
Wadsleyite (dry)	3.57	172	4.3	-0.019	106	1.4	-0.017
Wadsleyite (wet)	3.51	155	5.3	-0.019	93	1.8	-0.017
Ringwoodite (dry)	3.70	188	4.3	-0.021	120	1.4	-0.015
Ringwoodite (wet)	3.65	177	5.3	-0.021	105	1.8	-0.015

H_2O into the system. However, hydration and decreased temperature have the opposite effect on the seismic velocities of the intervening transition zone; decreased temperature increases P and S-wave velocities, whereas hydration decreases these velocities. Therefore, a slow (red) and thick TZ would suggest the presence of water (Plate 2). In order to further distinguish the effects of temperature and hydration, it is important to evaluate the relative magnitudes of these effects on P and S velocities. Results of Jacobsen and Smyth [*this volume*] indicate that hydration has a larger effect on Vs than on Vp in ringwoodite. A systematic search for regions of anti-correlation of the depths to the 410 and 660 km discontinuities and estimation of intervening seismic velocities, and in particular elevated V_P/V_S ratios, might reveal region of significant hydration in the transition zone. Such a search is beyond the scope of the current study.

6. DISCUSSION

The pyrolite model of mantle composition and mineralogy was developed by Ringwood [1976] to be a single composition throughout the upper and lower mantle that is roughly consistent with a bulk Earth of chondritic composition, with phase transformations observed in the laboratory, with velocities and densities derived from seismology, with high pressure inclusions seen in basalts and kimberlites, and to yield basalt by partial melting. The model has more than 60% normative olivine. Are observed velocities in the TZ consistent with significant hydration (i.e. >0.1 wt % H_2O) of this composition? The velocity jump at 410 km observed seismically is too small to be consistent with an olivine to wadsleyite transition in a dry, olivine-rich (pyrolite) mantle [*Duffy and Anderson*, 1989; *Anderson*, 1989; *Anderson and Bass*, 1986]. The small velocity increase at 410 km led these authors to suggest that the mantle in this region must have less than 50% olivine. However, the current ultrasonic results together with static compression measurements indicate a dramatic decrease in elastic stiffness of wadsleyite and ringwoodite with hydration [*Crichton et al.*, 1999; *Duffy et al*, 1995; *Yusa and Inoue*, 1997; *Smyth et al.*, 2004]. This means that the observed velocity jump at 410 km would be more consistent with a hydrous than an anhydrous mantle for an olivine-rich stoichiometry, i.e. pyrolite, with more than 60% normative olivine.

Several authors have pointed out that the observed velocities in the transition zone are inconsistent with a dry pyrolite composition mantle [*Duffy et al.*, 1995; *Duffy and Anderson*, 1989; *Anderson*, 1989; *Anderson and Bass*, 1986]. However, it can be seen from Plate 1 that the very large effect of hydration on the P and S velocities in ringwoodite and inferred for wadsleyite, means that PREM model velocities are consistent with a pyrolite-composition Transition Zone with one-half to one weight percent H_2O. Because of its very high diffusive mobility and fractionation into melt phases, hydrogen could be effectively decoupled from other compositional variables, that is, exhibit open system behavior. Wood (1995) argued that the sharpness of the 410 km discontinuity precludes water contents above 0.025% by weight. However, gravity-driven diffusion of H would likely sharpen the olivine-wadsleyite boundary [*Smyth and Frost*, 2002] in regions without strong vertical convection, and the presence of a small amount of partial melt [*Bercovici and Karato*, 2003] or a mobile aqueous fluid, could sharpen even a convecting boundary. It is therefore unlikely that the observed width of the 410 km discontinuity will provide a strong constraint on the amount of H present. Although hydrogen diffusion rates for wadsleyite and ringwoodite remain unknown, they must be less than the diffusion rate for heat and so simple diffusion cannot compete with convection for redistribution of H on a global scale [*Richard et al.*, 2002].

In summary, the observation of the very strong effect of hydration on P and S velocities in ringwoodite, and inferred from isothermal compression of wadsleyite, indicate that hydration is likely to have a larger effect on seismic velocities than temperature within the uncertainties in each in the Transition Zone. This means that tomographic images of the transition zone in regions distant from active subduction, blue coloration (fast) is more likely to indicate dry than cool, and red (slow) is more likely to indicate wet, than hot. Seismic data are consistent with a transition zone hydration between 0.5 and 1.0 weight percent H_2O for a pyrolite-composition model. This estimate is based on some simple assumptions about the effect of temperature and pressure on the elastic properties of the major hydrous phases (Table 1) which are not yet well constrained by experimental data. Hydrogen abundance at the higher end of this estimate would be sufficient to generate the small amount of partial melt near 410 km, postulated [*Bercovici and Karato*, 2003] and recently reported to be observed below the western US [*Song et al.*, 2004]. However, Demouchy et al. [2005] show that although H_2O contents can exceed 2.2 % by weight in wadsleyite at 900°C, water contents at 1400°C on a more realistic geotherm would be less than one percent, similar to what is observed in forsterite at 1250°C at 12GPa [*Mosenfelder et al.*, 2006, *Smyth et al.*, 2006]. At water contents of 2000 to 5000 ppmw, a small amount of partial melt would not occur near 410 km, but might occur near the low velocity zone [*Smyth et al.*, 2006; *Mierdel et al.*, 2006]. Water contents of mid-ocean ridge basalts at 1000 to 2500 ppmw would appear to be inconsistent with such a high water contents in the upper mantle. However, mid-ocean ridge basalts might not be representative of ocean crustal magmatism, but rather

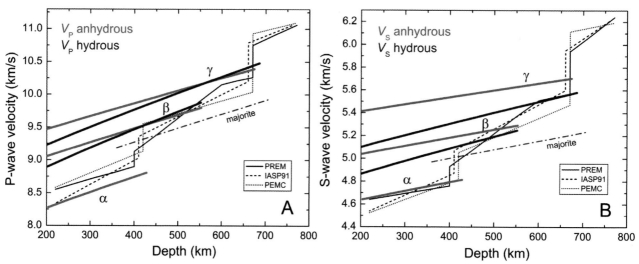

Plate 1. The effect of water on seismic velocities in Earth's Transition Zone (TZ), 410–660 km depth. Plot of monomineralic P-wave velocities (*A*) and shear velocities (*B*) versus depth for anhydrous (green) and hydrous (~1 wt% H_2O, blue) phases of wadsleyite (β), and ringwoodite (γ) using assumed thermoelastic parameters given in Table 1 and finite strain equations of state along an adiabat with 1673 K foot temperature. Velocities for anhydrous olivine (α) and majorite are also shown (see text for details). For comparison, several seismic velocity models are also shown. The figure illustrates that effect of water, which is to reduce elastic moduli, but elevate their pressure derivatives. The difference between anhydrous and hydrous P-wave velocities diminishes in the transition zone, whereas hydrous S-wave velocities remain several percent lower than anhydrous velocities throughout the TZ. Therefore, low S-wave velocities along with elevated V_P/V_S ratios would be indicative of a hydrous transition zone.

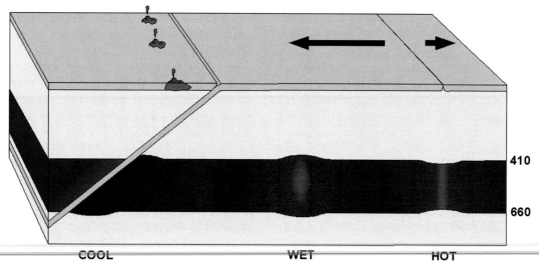

Plate 2. A schematic drawing of the contrasting effects of hot upwelling (right), hydrous (center), and a cold, subducting (left) regions of the Transition Zone, where blue indicates seismically fast and red indicates slow regions. A shallow 410 and deep 660 with intervening low S-wave velocities and elevated V_P/V_S ratios would indicate hydrous conditions in the Transition Zone.

result from nearly dry melting in the center of the up welling convection beneath the ridge [*Lee and DePaolo*, 2005].

A little water goes a long way in Earth's interior because of the enormous volumes of rock involved. Even if just 0.1 wt% H_2O resides in Earth's Transition Zone as residue of differentiation or has been later carried into the TZ in the basaltic crustal portion of the subducting lithosphere, the amount of H_2O between 410 and 660 km depth would be equivalent to almost one kilometer of liquid water on the surface. This amount seems to be a lower limit, and it is likely that fluxes are several times this amount. Geochemically, hydrogen at low pressure behaves as an incompatible element, but there is strong evidence that above 10 GPa the hydroxyl anion is a considerably more compatible species with two protons substituting for Mg in octahedral coordination. Because the proton does not occupy the position of the cation for which it substitutes in the crystal structure, H does not behave geochemically like other incompatible lithophile elements [*Smyth* et al., 2006]. In particular, H solubility in nominally anhydrous is much more strongly affected by pressure than that of other incompatible elements. Does the amount of liquid water on Earth's surface reflect a dynamic equilibrium with a deeper reservoir in Earth's interior? The question is now finally a major focus of interdisciplinary studies bridging mineral physics, seismology and geochemistry. We are only now beginning to gain a quantitative handle on the effects of hydration on the thermoelastic properties of the dense silicate minerals of Earth's interior.

Acknowledgments. The authors thank D. Rubie, F. Nestola, D. J. Frost, and M. M. Hirschmann for useful discussions and G. Bromiley for artistic help and discussions. This work was supported by U.S. National Science Foundation grants EAR and 03-37611 to JRS, the Alexander von Humboldt Foundation (JRS), and the Bavarian Geological Institute Visitors Program. SDJ acknowledges support form NSF EAR-0440112, a fellowship from the Carnegie Institution of Washington, and the Carneige/DOE Alliance Center (CDAC).

REFERENCES

Abe, Y., E. Ohtani, T. Okuchi, K. Righter, and Drake, M. *Origin of the Earth and Moon* (eds Canup, R. M. and Righter, K.) 413–433 (University of Arizona Press, Tucson, 2000.

Alt, J. C., Alteration of the upper oceanic crust; mineralogy, chemistry, and process. In Hydrogeology of the Oceanic Lithosphere (E. E. Davis and H. Elderfields, eds.) Cambridge University Press, 495–533, 2004.

Anderson, D. L. *Theory of the Earth,* Blackwell, Boston, 1989.

Anderson, D. L. and J. D. Bass, Transition region of the Earth's upper mantle. *Nature 320,* 321–328, 1986.

Asimow, P. D., and C. H. Langmuir, The importance of water to oceanic mantle melting regimes. *Nature 421,* 815–820, 2003.

Bell, D. R. and G. R. Rossman, Water in the Earth's mantle: Role of nominally anhydrous minerals. *Science, 255,* 1392–1396, 1992.

Bell, D., G. Rossman, J. Maldener, D. Endisch and F. Rauch Hydroxide in olivine: a quantitative determination of the absolute amount and calibration of the IR spectrum, *J. Geophys. Res., 108,* doi: 10.1029/2001JB000679, 2003.

Bercovici, D. and S. I. Karato, Whole mantle convection and the transition zone water filter. *Nature 425,* 39–44, 2003.

Chen, J., T. Inoue, H. Yurimoto, and D. J. Weidner, Effect of water on olivine-wadsleyite phase boundary in the $(Mg,Fe)_2SiO_4$ system. *Geophys. Res. Lett., 29,* 1875, 2002.

Crichton, W. A., N. L. Ross, D. J. Frost, and J. Kung, Comparative compressibilities of anhydrous and hydrous wadsleyites. *Journal of Conference Abstracts (EUG) 657,* 1999.

Davies, G. F., and A. M. Dziewonski, Homogeneity and constitution of the Earth's lower mantle and outer core. *Phys. Earth Planet. Int., 10,* 336–343, 1975.

Delsemme, A. H. The deuterium enrichment observed in recent comets is consistent with the cometary origin of seawater. *Planetary and Space Science 47,* 125–131, 1999.

Demouchy, S., E. Deloule, D. J. Frost, and H. Keppler, Pressure and temperature-dependence of water solubility in Fe-free wadsleyite. *Am. Mineral. 90,* 1084–1091, 2005

Dixon, J. E. and D. A. Clague, Volatiles in basaltic glasses from Loihi Seamount, Hawaii: Evidence for a relatively dry plume component. *J. Petrology, 42,* 627–654, 2001.

Dixon, J. E., L. Leist, C. Langmuir, and J. G. Schilling, Recycled dehydrated lithosphere observed in plume-influenced mid-ocean-ridge basalt. *Nature, 420,* 385–389, 2002.

Downs, R. T., C. S. Zha, T. S. Duffy, and L. W. Finger, The equation of state of forsterite to 17.2 GPa and effects of pressure media. *Am. Mineral., 81,* 51–55, 1996.

Drake M. J. and K. Righter Determining the composition of the Earth. *Nature, 416,* 39–44, 2002.

Duffy, T. S. and D. L. Anderson, Seismic velocities in mantle minerals and the mineralogy of the upper mantle. *J. Geophys. Res., 94,* 1895–1912, 1989.

Duffy, T. S., C. S. Zha, R. T. Downs, H. K. Mao, and R. J. Hemley, Elasticity of forsterite to 16 GPa and the composition of the upper mantle. *Nature, 378,* 170–173, 1995.

Dziewonski, A. M. and D. L. Anderson, Preliminary reference Earth model. *Phys. Earth Planet. Inter., 25,* 297–356, 1981.

Dziewonski, A. M., A. L. Hales, and E. R. Lapwood, Parametrically simple earth models consistent with geophysical data. *Phys. Earth Planet. Int., 10,* 12–48, 1975.

Gilbert, H. J., A. F. Sheehan, K. G. Dueker, and P. Molnar, Receiver functions in the western United States, with implications for upper mantle structure and dynamics, *J. Geophys. Res., 108* (B5), 2229, doi:10.1029/2001JB001194 2003.

Gordon, R. G. Present plate motions and plate boundaries. *Global Earth Physics: A Handbook of Physical Constants,* AGU Washington, 66–87, 1995.

Gwanmesia, G. D., G. Chen, and R. C. Liebermann, Sound velocities in $MgSiO_3$-garnet to 8 GPa. *Geophys. Res. Lett., 25,* 4553–4556.

Hazen R.M., Zhang, J. and Ko, J. Effects of Fe/Mg on the compressibility of synthetic wadsleyite β-$(Mg_{1-x}Fe_x)_2SiO_4$, with x = 0 to 0.25 *Phys. Chem. Minerals, 17*, 416–419, 1990.

Hazen, R. M., M. Weinberger, H. Yang, and C. T. Prewitt, Comparative high pressure crystal chemistry of wadsleyite, β-$(Mg_{1-x}Fe_x)_2SiO_4$, with x = 0 to 0.25. *Am. Mineral., 85*, 770–777, 2000.

Hirschmann, M. M., C. Abaud, and A. C. Withers, Storage capacity of H_2O in nominally anhydrous minerals in the upper mantle. *Earth Planet. Sci. Lett., 236*, 167–181, 2005.

Holl, C. M., J. R. Smyth, and S. D. Jacobsen, The effect of water on the compressibility of wadsleyite. *Eos Trans. Am. Geophys. Union Fall Meet. Suppl., 84*, T11C-0414, 2003.

Inoue, T. Effect of water on melting phase relations and melt composition in the system Mg_2SiO_4-$MgSiO_3$-H_2O up to 15 GPa. *Phys. Earth Planet. Int. 85*, 237–263, 1994.

Inoue, T., H. Yurimoto, and Y. Kudoh, Hydrous modified spinel, $Mg_{1.75}SiH_{0.5}O_4$: a new water reservoir in the mantle transition region. *Geophys. Res. Lett., 22*, 117–120, 1995.

Inoue, T., D. J. Weidner, P. A Northrup, and J. B.Parise, Elastic properties of hydrous ringwoodite (γ phase) in Mg_2SiO_4. *Earth Planet. Sci. Lett., 160*, 107–113, 1998.

Isaak, D. G. High-temperature elasticity of iron-bearing olivines. *J. Geophys. Res. 97*, 1871–1885, 1992.

Iwamori, H. Transportation of H_2O and melting in subduction zones, *Earth Planet. Sci. Lett. 160*, 65–80, 1998.

Jacobsen, S. D. and J. R. Smyth, Effect of water on sound velocities of ringwoodite in the transition zone. (This volume), 2006.

Jacobsen, S. D., S. Demouchy, D. J. Frost, T. Boffa-Ballaran, and J. Kung, A systematic study of OH in hydrous wadsleyite from polarized FTIR spectroscopy and single-crystal X-ray diffraction: oxygen sites for hydrogen storage in the Earth's interior. *Am. Mineral. 90*, 61–70, 2005.

Jacobsen, S. D., J. R. Smyth, H. A. Spetzler, C. M. Holl, and D. J. Frost, Sound velocities and elastic constants of iron-bearing hydrous ringwoodite. *Phys. Earth Planet. Int.*, 143–144, 77–91, 2004a.

Jacobsen, S. D., H. A. Spetzler, H. J. Reichmann, and J. R. Smyth Shear waves in the diamond-cell reveal pressure-induced instability in (Mg,Fe)O. *Proc. Natl. Acad. Sci. USA, 101*, 5867–5871, 2004b.

Karato, S., M. S. Patterson, and D. Fitzgerald, Rheology of synthetic olivine aggregates: Influence of grain size and water. *J. Geophys. Res., 91*, 8151–8176, 1986.

Karato, S. I. Plastic deformation of silicate spinel under transition-zone conditions of the Earth's mantle. *Nature, 395*, 266–269, 1998.

Katayama, I. and S. Nakashima, Hydroxyl in clinopyroxene from the deep subducted crust: evidence for H_2O transport into the mantle. *Am. Mineral., 88*, 229–234, 2003.

Katsura, T., N. Mayama, K. Shouno, M. Sakai, A. Yoneda, and I. Suzuki, Temperature derivatives of elastic moduli of $(Mg_{0.91}Fe_{0.09})_2SiO_4$ modified spinel. *Phys. Earth Planet. Int., 124*, 163–166, 2001.

Kavner, A. Elasticity and strength of hydrous ringwoodite at high pressure. *Earth Planet. Sci. Lett., 214*, 645–654, 2003.

Kawamoto, T., R. L. Hervig, and J. R. Holloway, Experimental evidence for a hydrous transition zone in the early Earth's mantle. *Earth Planet. Sci. Lett., 142*, 587–592, 1996.

Kennett, B. L. N. and E. R. Engdahl, Travel times for global earthquake location and phase identification. *Geophys. J. Int., 105*, 429–465, 1991.

Kohlstedt, D. L., H. Keppler, and D. C. Rubie, The solubility of water in α, β and γ phases of $(Mg,Fe)_2SiO_4$. *Contrib. Mineral. Petrol., 123*, 345–357, 1996.

Kudoh, Y., and T. Takéuchi The crystal structure of forsterite under high pressure to 149 kbar. *Z. Kristallogr., 171*, 291–302, 1985.

Kudoh, Y., T. Kuribayashi, H. Mizohata, and E. Ohtani, Structure and cation disorder of hydrous ringwoodite, γ-$Mg_{1.89}Si_{0.97}H_{0.34}O_4$. *Trans. Am Geophys Union*, F895, 1998.

Lee, V. E., and D.J. DePaolo, Isotopic modeling of dynamic geochemical layering in the mantle. , *Eos Trans. AGU* 86(52), Fall Meet. Suppl., Abstract DI 43A-08.

Li, B. Compressional and shear wave velocities of ringwoodite γ-Mg_2SiO_4 to 12 GPa. *Am. Mineral., 88*, 1312–1317, 2003.

Li, B., and R. C. Liebermann, Sound velocities of wadsleyite β-$(Mg_{0.88}Fe_{0.12})_2SiO_4$ to 10 GPa. *Am. Mineral., 85*, 292–295, 2000.

Li, B., R. C. Liebermann, and D. J. Weidner, Elastic moduli of wadsleyite (β-Mg_2SiO_4) to 7 gigapascals and 873 Kelvin. *Science 281*, 675–677, 1998.

Libowitzky, E. and G. R. Rossman, An IR absorption calibration for water in minerals. *Am. Mineral., 82*, 1111–1115, 1997.

Liu, W., J. Kung, and B. Li, Elasticity of San Carlos olivine to 8 GPa and 1073 K. *Geophys. Res. Lett., 32*, L16301, 2005.

Locke, D. R., J. R. Holloway, and R. Hervig, Oxidation independent solution mechanism for hydrogen in olivine: examples from simple and peridotite systems. *Eos Trans. AGU, 83(47)*, Fall Meet. Suppl., Abstract T21C-1116, 2002.

Manghnani, M. H., G. Amulele, J. R. Smyth, C. M. Holl, G. Chen, V. Prakapenka, and D. J. Frost, Equation of state of hydrous Fo_{90} ringwoodite to 45 GPa by synchrotron powder diffraction. *Min. Mag., 69*, 317–323, 2005.

Mayama, N., I. Suzuki, T. Saito, I. Ohno, T. Katsura, and A. Yoneda, Temperature dependence of the elastic moduli of ringwoodite. *Phys Earth Planet. Int., 148*, 353–359, 2005.

Mei, S. and D. L. Kohlstedt, Influence of water on plastic deformation of olivine aggregates, *J. Geophys. Res. 105*, 21,457-21,481, 2000.

Mojzsis, S. J., M. Harrison, and R. T. Pidgeon, Oxygen-isotope evidence from ancient zircons for liquid water at the Earth's surface 4,300 Myr ago. *Nature, 409*, 178–181, 2001.

Mosenfelder, J. L., N. I. Deligne, P. D. Asimow, and G. R. Rossman, Hydrogen incorporation in olivine from 2–12 GPa, Am. Mineral., 91, 285–294, 2006.

Parsons, B. Causes and consequences of the relation between area and age of the ocean floor. *Jour. Geophys. Res., 87*, 289–302, 1982.

Paterson, M. The determination of hydroxyl by infrared absorption in quartz, silicate glasses, and similar materials, *Bulletin de Minéralogie, 105*, 20–29, 1982.

Richard, G., M. Monnereau, and J. Ingrin, Is the transition zone an empty water reservoir? Inferences from numerical model of mantle dynamics. *Earth Planet. Sci. Lett. 205*, 37–51, 2002.

Ringwood, A. E. *Composition and Petrology of the Earth's Mantle.* McGraw Hill, New York, 619 pp., 1975.

Robert, F. The origin of water on Earth. *Science, 293*, 1056–1058, 2001.

Rossman, G. R., Smyth, J. R. Hydroxyl contents of accessory minerals in mantle eclogites and related rocks. *Am Mineral 75*, 775–780, 1990.

Schmidt, M. W. Experimental constraints on recycling of potassium from subducted ocean crust, *Science, 272*, 1927–1930, 1996.

Schmidt, M. W. and S. Poli Experimentally based water budgets for dehydrating slabs and consequences for arc magma generation. *Earth Planet. Sci. Lett. 163*, 361–179, 1998.

Sinogeikin, S. V., J. D. Bass, and T. Katsura, Single-crystal elasticity of ringwoodite to high pressures and high temperatures: implications for 520 km seismic discontinuity, *Phys. Earth Planet. Int., 136*, 41–66, 2003.

Sinogeikin, S. V., J. D. Bass, B. O'Neill, and T. Gasparik, Elasticity of tetragonal end-member majorite and solid solutions in the system $Mg_4Si_4O_{12}$-$Mg_3Al_2Si_3O_{12}$. *Phys. Chem. Minerals, 24*, 115–121, 1997.

Sinogeikin, S. V., T. Katsura, and J. D. Bass, Sound velocities and elastic properties of Fe-bearing wadsleyite and ringwoodite. *J. Geophys. Res., 103*, 20819–20825, 1998.

Song, A. T. R., D. V. Helmberger, and S. P. Grand, Low-velocity zone atop the 410-km seismic discontinuity in the northwestern United States, *Nature 427*, 530–533, 2004.

Smyth, J. R. Beta-Mg_2SiO_4; a potential host for water in the mantle? *Am. Mineral. 72*, 1051–1055, 1987.

Smyth, J. R. and D. J. Frost, The effect of water on the 410-km discontinuity: An experimental study. *Geophys. Res. Lett. 29*, 123, doi: 10.1029/2001GL014418, 2002.

Smyth, J. R., G. R. Rossman, and D. R. Bell, Incorporation of hydroxyl in upper mantle clinopyroxenes, *Nature 351*, 732–735, 1991.

Smyth, J. R., C. M. Holl, D. J. Frost, S. D. Jacobsen, F. Langenhorst, and C. A. McCammon, Structural systematics of hydrous ringwoodite and water in the Earth's interior. *Am. Mineral. 88*, 1402–1407, 2003.

Smyth, J. R. C. M. Holl, D. J. Frost, and S. D. Jacobsen, High pressure crystal chemistry of hydrous ringwoodite and water in the Earth's interior. *Phys. Earth Planet. Int.* 143–144, 2004.

Smyth, J. R., D. J. Frost, and F. Nestola, Hydration of olivine and the Earth's deep water cycle. *Geochim. Cosmochim. Acta, 69*, A746, 2005a.

Smyth, J. R., C. M. Holl, F. Langenhorst, H. M. S. Laustsen, G. R. Rossman, A. Kleppe, C. A. McCammon, T. Kawamoto, and P. A. van Aken Crystal chemistry of wadsleyite II and water in the Earth's interior. *Phys. Chem. Minerals, 31*, 691–705, 2005b.

Smyth, J. R., D. J. Frost, F. Nestola and G. Bromiley Olivine hydration in the deep upper mantle: effects of temperature and silica activity. *Geophys Res Lett,* 2006 (in press).

Suzuki, I., and O. L. Anderson, Elasticity and thermal expansion of a natural garnet up to 1000 K. *J Phys. Earth, 31*, 125–138, 1983.

Wang, J., S. V. Sinogeikin, T. Inoue, and J. D. Bass, Elastic properties of hydrous ringwoodite. *Am. Mineral., 88*, 1608–1611, 2003.

Wen L. and D. L. Anderson, The fate of slabs inferred from seismic tomography and 130 million years of subduction. *Earth Planet Sci Lett, 133*, 185–198, 1995.

Wilde S. A., Valley, J. W., Peck, W. H., and Graham, C. M. Evidence from detrital zircons for the existence of continental crust and oceans on the Earth 4.4 Gyr ago. *Nature 409*, 175–178, 2001.

Will, G., W. Hoffbauer, E. Hinze, and J. Lauterjung, The compressibility of forsterite up to 300 kbar measured with synchrotron radiation. *Physica 139 and 140B*, 193–197, 1986.

Williams, Q. and R. J. Hemley, Hydrogen in the deep Earth, *Ann. Rev. Earth Planet. Sci. 29*, 365–418, 2001.

Wood, B. J. The effect of H_2O on the 410-kilometer seismic discontinuity, *Science 268*, 74–76, 1995.

Yusa, H. and T. Inoue, Compressibility of hydrous wadsleyite (β-phase) in Mg_2SiO_4 by high-pressure X-ray diffraction. *J. Geophys. Res., 24*, 1831–1834, 1997.

Zha, C. S., T. S. Duffy, R. T. Downs, H. K. Mao, and R. J. Hemley, Brillouin scattering and X-ray diffraction of San Carlos olivine: direct pressure determination to 32 GPa. *Earth Planet. Sci. Lett., 159*, 25–33, 1998.

Joseph Smyth, University of Colorado, Dept. Geological Sciences, 2200 Colorado Ave., Boulder, CO 80309-0399 (smyth@colorado.edu)

Seismological Constraints on Earth's Deep Water Cycle

Suzan van der Lee

Department of Geological Sciences, Northwestern University, Evanston, Illinois, USA

Douglas A. Wiens

Department of Earth and Planetary Sciences, Washington University, St. Louis, Missouri, USA

Water can be present in the mantle in the form of hydrous melts, hydrous phases, or incorporated into the crystal structure of nominally anhydrous minerals of the major mantle mineralogy. The first two forms are likely in the uppermost mantle, where water solubility in major mantle minerals is low, whereas the latter form may be more important deeper in the upper mantle and transition zone. Seismological data contain unique and valuable information on the amount and distribution of water in the Earth's mantle. It is, however, challenging to extract this information because of limitations in the amount and density of available seismic data, multiple interpretations of similar observations, and limited quantification of the effects of water and other parameters on the seismic properties. While increased water content and elevated temperatures both lower seismic velocities, they have opposing effects on the depths of the discontinuities that bound the transition zone. And, while they both increase attenuation, they have opposing effects on the sharpness of these discontinuities. Independent geophysical observations, such a gravity, electrical conductivity, and surface heat flow, can further help to discriminate between temperature, water, and other compositional anomalies as the cause of observed seismic heterogeneity. Various types of observations have been combined to infer water content in the mantle, ranging from a few hundredths of weight percent to several weight percent. Altogether, the seismological literature suggests that the mantle is heterogeneously hydrated. However, with the limited studies available, there does not appear to be an obvious correlation between present tectonic environment and water content, though the literature shows a tendency to interpret inferred anomalously hydrous regions in the mid mantle as being related in one way or another to past subduction of oceanic lithosphere.

1. INTRODUCTION

The Earth's bulk composition of lithophile, siderophile, and chalcophile elements and its state of differentiation are fairly well understood from cosmo- and geochemistry, as well as geophysics. The Earth's mantle consists primarily of magnesium and iron silicates, with the olivine-wadsleyite-ringwoodite mineralogy comprising about half of the mantle assemblage in the top 660 km.

In contrast, the bulk water content of and its distribution within Earth is virtually unknown because of the high volatility of water. It is possible that the equivalent of several

Earth's Deep Water Cycle
Geophysical Monograph Series 168
Copyright 2006 by the American Geophysical Union.
10.1029/168GM03

ocean masses of water resides within mantle minerals, most prominently in the transition zone [e.g. *Smyth and Jacobsen*, this volume], though some argue that the solid Earth has effectively differentiated its water to the crust, hydro- and atmosphere [e.g. *Rüpke et al.*, this volume].

If an amount of water equal to that in the Pacific Ocean were homogeneously distributed over lattice defects in transition zone minerals wadsleyite and ringwoodite, the transition zone's water content would be around 0.2 wt % (i.e. 2000 ppm H_2O by weight). As shown by numerous papers in this volume, an amount of water as small as a few tenths of wt % in the Earth's mantle can significantly affect the bulk and shear moduli, density, electrical and thermal conductivity, anelasticity, anisotropy, and phase/state of mantle minerals. With the exception of thermal and electrical conductivity, all of these properties are most directly and accurately assessed through studies of seismic wave propagation.

Seismic wave propagation, however, is also affected by temperature, pressure, composition, structure, and mineralogy. In the upper third of Earth's mantle, first-order changes in seismic velocities are a gradual increase of seismic velocity with increasing pressure (depth), and sudden increases at the top and bottom of the transition zone near depths of about 410 and 660 km, which result from phase transitions in the olivine component of mantle mineralogy. Near 410 km, olivine transforms into wadsleyite, which transitions to ringwoodite over a broader depth range in the mid transition zone. Near 660 km ringwoodite transforms into the lower-mantle assemblage of perovskite and ferropericlase. Second order changes in seismic velocity and attenuation also occur laterally, under the predominant influence of lateral variations in mantle temperature, phase transitions in secondary mantle minerals such as garnet and pyroxene, variations in grain size, and compositional heterogeneity. Hydrogen from water is thus far the most important compositional factor governing the seismic and rheological properties of its host material [*Karato*, this volume; *Shito et al.*, this volume; *Jacobsen and Smyth*, this volume].

Water has a profound influence on the rheology of the upper mantle [*Hirth and Kohlstedt*, 1996; *Karato and Jung*, 1998, *Mei and Kohlstedt*, 2000a, 2000b, *Karato*, this volume]. Thus, the amount of water that resides in or cycles through the Earth's interior is an important consideration for understanding Earth's early evolution, the initiation and longevity of plate tectonics, low-velocity zones beneath oceans, and the lithospheric roots of Precambrian cratons, for example. In this chapter we review the ways in which seismology can constrain the water contents of the Earth's mantle. Two of the largest challenges that seismologists face are 1) extracting robust inferences from limited amounts of data, and 2) to discriminate the effects of water from the

effects that temperature, composition, and mineralogy have on these properties.

Lateral temperature variations are widely thought to provide the largest contribution to upper mantle seismic velocity and attenuation variations, as the temperature derivatives of these quantities are non-linear and become large at high temperatures and low (uppermost mantle) pressures [*Jackson et al.*, 2002; *Cammarano et al.*, 2003]. Observational studies typically confirm this thermal dominance on seismic heterogeneity [e.g. *Goes and Van der Lee*, 2002; *Priestley and McKenzie*, 2006; *Wiens et al.* 2006b]. Water may provide the next-largest contribution. In this paper we discuss the effects of water on deep seismic structure.

2. THE EFFECTS OF WATER AND OTHER VARIABLES ON SEISMIC PROPERTIES

2.1 The Effects of Composition Other Than Water

Early research focused on the influence of iron because of its relative abundance in the mantle's magnesium silicates. In a solid-molten mantle mixture iron readily partitions into the melt [*Jordan*, 1988] and lateral variations in mantle iron content should be a fairly straightforward phenomenon if the mantle underwent varying degrees of partial melting. Initially, the effects of iron on the seismic velocity of shear waves (S velocity) were thought to be considerable [*Jordan*, 1988]. A survey of the seismic properties of upper mantle xenoliths shows a small dependence of the S velocity on iron content (0.3% per Mg number in olivine) [*Lee*, 2003]. Recent calculations suggest that the effects of melt depletion on *S* velocity are even smaller and negligible for realistic degrees of iron depletion compared to the effects of realistic variations in temperature [*Cammarano et al.*, 2003; *Schutt and Lesher*, 2006; *Matsukage et al.*, 2005], especially in the garnet lherzolite stability field. This is mostly because melting-related iron depletion in mantle olivine is accompanied by a loss of relatively Si-poor garnet, which is seismically fast [*Schutt and Lesher*, 2006]. Even the S-velocity dependence found by *Lee* [2003] suggests that realistic changes in iron for upper mantle rocks can produce only a 1–2% change in S-velocity, much smaller than expected from temperature variations.

For deeper levels in the upper mantle, data on the effects of Fe content exist for only the olivine component. These effects are considerable at transition zone depths [*Sinogeikin et al.*, 2001], but not quite as dramatic as the effects of hydrogen content [*Jacobsen and Smyth*, this volume; *Shito et al.*, this volume]. However, if ferric iron (Fe^{3+}) is abundant at such depths, Fe might have more dramatic effects on seismic properties, comparable to those of H [*Frost*, 2003].

Little is known about the influence of other minor elements in the mantle such as Al, Ca, and Na. Recent experimental work shows that the presence of ~5 wt % Al_2O_3 in perovskite reduces the shear modulus by about 6 %, but has no discernable effect on the bulk modulus [*Jackson et al.,* 2004]. The presence of Al might also affect the solubility of water in mantle phases [e.g. *Rauch and Keppler,* 2002; *Bolfan-Casanova et al.,* this volume]. Furthermore, calcium-silicate perovskite ($CaSiO_3$) probably exists in the lower mantle as a minor but separate phase, with a shear modulus 25 to 40 % below that of $(Mg,Fe)SiO_3$ perovskite [e.g. *Li et al.,* 2006; *Sinelnikov et al.,* 1998]. Carbon, like H, is a volatile element whose presence in the upper mantle is evidenced by CO_2 degassing at mid-ocean ridges, CO_2-rich magmas such as carbonatites and kimberlites, and diamond-bearing xenoliths. Recent experimental work supports carbon recycling to as deep as 300 km [*Hammouda,* 2003] or deeper [*Dasgupta et al.,* 2004], though the solubility of C into major mantle minerals such as olivine is exceptionally low (about 1 ppm) [*Keppler et al.,* 2003]. Thus, H-accommodating minerals appear more stable in the upper mantle than C-accommodating minerals, allowing a water cycle in the mantle to extend much deeper than a carbon cycle.

2.2 The Effects of Water

Hydrogen, or water, has emerged as one of the most influential mantle elements in terms of seismic properties. While the major-element (Si, Mg, Fe, O) composition of the mantle is relatively well constrained, its hydrogen content is much more speculative. The presence of hydrogen in mantle minerals directly affects seismic properties, as discussed, but also affects these properties indirectly by catalyzing mantle melting under particular pressure-temperature and kinematic conditions [*Eiler,* 2004; *Karato et al.,* this volume; *Revenaugh and Sipkin,* 1994]. In some regions of the mantle, such as the wedges between subducting oceanic plates and the lithosphere of the overriding plate, the presence of water has been corroborated through the analysis of rock samples erupted from arc volcanoes [*Morris et al.,* 1990; *Gaetani and Grove,* 1998; *Dixon et al.,* 2004].

Small amounts of water in mid-ocean ridge magmas suggest a modest water content of about 0.005–0.02 wt % for the MORB source [*Saal et al.,* 2002]; levels as high as 0.17 wt % are inferred for magmas erupted at back-arc spreading centers near slabs [*Kelley et al.,* 2006]. For deeper levels in the mantle we do not have such corroborating material and claims of water in the mantle from seismic data are mostly arrived at by excluding other known factors that influence seismic properties, such as temperature or Fe heterogeneity [e.g. *Nolet and Zielhuis,* 1994].

The effects of water content and possibly associated fluid-rich melts on seismic observables in the upper mantle are incompletely understood. Free water may be present in a few extremely hydrous regions of the upper mantle such as the mantle wedge immediately above a slab [*Iwamori,* 1998; *Cagniocle et al.,* 2006]. In this case the effect of free water can be calculated using a similar poroelastic calculation, which depends on the geometry of the fluid pores [*Takei,* 2002]. However, in most cases the large water contents in excess of the water solubility in upper mantle minerals will instead produce a fluid-rich melt [*Hirschmann et al.,* 2005]. The effect of melt can result from both a poroelastic effect [*Mavko,* 1980; *Hammond and Humphreys,* 2000; *Takei,* 2002] and a grain-boundary sliding mechanism [*Faul et al.,* 2004]. The effects of melt are difficult to constrain since the poroelastic effect depends on the melt geometry and the effect of the grain-boundary sliding mechanism on the shear modulus is incompletely characterized at the present time.

For most of the upper mantle, water concentrations are low enough or pressure high enough that the water is accommodated in the mineral structure of nominally anhydrous major mantle minerals, primarily olivine and pyroxene [*Hirschmann et al.,* 2005] or, in particularly cool regions such as subducting lithospheric plates, in dense hydrous magnesium silicates [*Komabayashi,* this volume]. Although little known, the effects of water incorporated in the olivine crystal structure at uppermost mantle depths are likely reduction of elastic moduli. Recent compressibility studies show that the bulk modulus of forsterite is reduced by about 6% in the presence of 0.8 wt% H_2O [*Smyth et al.,* 2005; *Smyth and Jacobsen,* this volume]. *Karato* [2003] estimated the indirect velocity reduction induced by water using the assumption that water lowers the seismic velocity through the dispersion effect of attenuation, and that the micro-creep associated with anelasticity is proportional to the macroscopic creep inferred from rheological experiments. This result suggests that water should lower the S velocity (Vs), and quality factor (Q), more than the P velocity (Vp). Using the relationships in *Karato* [2003], increasing the upper mantle water content from normal MORB (0.005 wt %) to 0.15 wt % will reduce the shear velocity by about 1.5 % in the absence of melting. However, typical mantle temperatures at shallow depths are above the wet solidus and the melt triggered by 0.15 wt% of water could further lower the S velocity significantly. This suggests that variations in the water dissolved in nominally anhydrous minerals probably produce less heterogeneity than variations in temperature for the uppermost mantle, but may be more significant than other types of compositional variations. However, the relationships proposed by *Karato* [2003] rely on a number of assumptions and extrapolations, and additional relaxation mechanisms associated with water may be possible.

For the transition zone, ultrasonic experiments [*Jacobsen and Smyth*, this volume] on ringwoodite show that the unrelaxed seismic velocities can also be lowered by the incorporation of H from water into the mineral structure. *Jacobsen and Smyth* [this volume] report that 1 wt % of water in ringwoodite would reduce its Vs by over 100 m/s (~ 2 %) while having a negligible effect on Vp at transition zone pressures. Water thus has an elevating effect on the ratio of *P* velocity over *S* velocity, and thus on Poisson's ratio, of ringwoodite.

Hydrogen also has important effects on the properties of the upper mantle discontinuities, which may allow recognition of the presence of water. Thermodynamic calculations [*Wood*, 1995; *Helffrich and Wood*, 1996] as well as laboratory experiments [*Smyth and Frost*, 2002] show that under hydrous conditions, the thickness of the 410-km discontinuity can expand from the expected ~10 km for dry conditions to as much as 40 km, if water from olivine partitions tenfold into wadsleyite. However, 40 km may be an overestimate of the maximum thickness of the 410-km discontinuity because the hydrogen storage capacity of olivine relative to that of wadsleyite has probably been underestimated [*Hirschmann et al.*, 2005]. If the water content near 410 km exceeds the storage capacities of olivine and wadsleyite, the 410-km discontinuity would sharpen and carry a fluid or melt atop it [*Hirschmann et al.*, 2005; *Chen et al.*, 2002]. The thickness of the 660-km discontinuity would also increase under hydrous conditions [*Higo et al.*, 2001], but the thickening is only about a fraction of that for the 410-km discontinuity. In contrast, the 520-km discontinuity is expected to sharpen under hydrous conditions [*Inoue et al.*, 1998].

2.3 Discriminating Between the Effects of Water and Temperature

Since the first-order effects of water and temperature on seismic velocity and attenuation are of the same sign, it is necessary to discuss possible ways to discriminate between these possibilities. In principle, such discrimination is possible if the ratios of partial derivatives of the observables to their causes are not alike. For example, if the observables are Vs and Vp, we must have $(\partial Vp/\partial T)/(\partial Vs/\partial T) \neq (\partial Vp/\partial w)/(\partial Vs/\partial w)$ to allow discrimination between water content w, and temperature T. Here we consider the effects of temperature and water content anomalies on the mid and upper mantle. The dependence of seismic velocities on water and temperature is non-linear because at high water content or high temperature, attenuation is stronger, which amplifies the effects of small perturbations in temperature and water content on seismic velocities relative to their effects at lower temperature and lower water content [*Karato*, 2003, 2006; *Jackson et al.*, 1992]. Thus while the values of the partial derivatives, and thus the

contrast between them, depends on the actual temperature and water content, the following examples are nonetheless representative, in terms of their potential for discrimination, for a fair range of temperatures and water content.

If the mantle temperature at a depth of 600 km is elevated by 400 °C, the S- and P-velocities would be lowered by approximately 2.8 and 1.8 %, respectively [*Cammarano et al.*, 2003], which includes the effects of thermal expansion (anharmonic term) and increased attenuation (anelastic term). Water content elevated by 1 wt % in ringwoodite lowers ultrasonic S- and P-velocities by about 2.4 % and a negligible amount, respectively [*Jacobsen et al.*, 2004; *Jacobsen and Smyth*, this volume]. Actual S- and P-velocities (at seismic frequencies) would be further lowered by an anelastic effect of roughly 3 and 1.5 %, respectively [*Karato*, 2006]. Thus, the ratio of the P to S-velocity of a wet transition zone mantle would increase by several percent more than that for a hot transition-zone mantle. However, a temperature anomaly of 400 °C would depress the 410-km discontinuity by 30 to 50 km [*Litasov et al.*, this volume], elevate the 660-km discontinuity by an amount between 7 and 40 km [*Litasov et al.*, this volume], and thus reduce the transition thickness. In contrast, the wet mantle would thicken the transition-zone through elevation of the 410 km discontinuity up to anywhere from 10 to 30 km [*Smyth and Frost*, 2002; *Hirschmann et al.*, 2005] and depression of the 660-km discontinuity by up to 4 km [*Higo et al.*, 2001]. Furthermore, in a wet mantle these discontinuities would broaden, with widths of up to 40 km for the 410 km [*Smyth and Frost*, 2002; *Hirschmann et al.*, 2005] and up to 8 km for the 660 km discontinuity [*Higo et al.*, 2001], while in a hot mantle these discontinuities, in particular the 410-km discontinuity, would sharpen by around 5 km [*Helffrich and Bina*, 1994]. Thus it appears that for the transition zone, the effects of temperature and water can be distinguished by the different Vp/Vs ratios of the anomalies and by their differing effects on the mantle discontinuities.

At a depth of 100 km and 1250 °C, a 400 °C temperature increase would lower the S- and P-velocity by about 9 and 6 %, respectively [*Jackson et al.*, 2002; *Faul and Jackson*, 2005]. The effect of water on upper mantle velocities has not been fully established. Because water significantly lowers the bulk modulus of forsterite [*Smyth et al.*, 2005; *Smyth and Jacobsen*, this volume], we expect its shear modulus to be lowered as well. Temporarily assuming the effect of water on ringwoodite velocities at room pressure [Jacobsen et al., 2004] for olivine yields S and P velocities that are 1.8 and 1% lower, respectively, in the presence of 0.2 wt % of water. *Karato* [2003] estimated the additional effects of water on seismic velocities through shear modulus anelasticity, which suggests changes of 1.7% and 0.8% for Vs and Vp respectively for a water content of 0.2 wt % in the mantle, which is about the maximum water content that could be incorporated without producing hydrous melting [*Hirschmann et al.*, 2005]. Altogether 0.2 wt % of water might lower Vs and Vp by 3.5 and 1.8 %, respectively. Thus, for the uppermost mantle, limitations on the amount of water that can be taken up in mantle minerals suggest that the effects of water heterogeneity on seismic velocities and attenuation will be less than the effects of temperature variations, unless hydrous minerals or a hydrous melt is produced.

It may be possible to use the relationship between Vp, Vs, and anelasticity (1/Q) to distinguish temperature and water effects in the upper mantle, since attenuation depends linearly on water content but exponentially on the temperature anomaly [*Karato*, 2006]. The effect of water on olivine will produce a larger percentage change in Vs than in Vp when compared to the temperature effect, particularly at lower temperatures where the anelastic effect of temperature is minor. This suggests that the relationship between Vp, Vs, and Q can be used to distinguish water from temperature in the upper mantle [*Karato*, 2003; *Shito et al.,* this volume]. However, the differences between $(\partial Vp/\partial T)/(\partial Vs/\partial T)$ and $(\partial Vp/\partial w)/(\partial Vs/\partial w)$ are probably reduced at high temperatures in the upper mantle, where the temperature effects also reflect a large anelastic effect.

To discriminate seismologically between water content and temperature it thus appears most instructive in general to study S velocities along with P velocities and Q, as well as properties of mantle discontinuities. For the uppermost mantle, additional data such as surface heat flow, gravity, and the water content of erupted mantle basalts and xenoliths may be helpful in discriminating between different effects. Surface heat flow would be elevated for a thermal anomaly but not for a hydration anomaly. Likewise, gravity may provide constraints since a 400 °C thermal anomaly would lower the density by 1.8%, whereas a few tenths of wt % water taken up in the crystal structure of mantle minerals will have a negligible effect on gravity.

A definitive interpretation of tomographic images alone in terms of temperature, water content, and, in the shallow mantle, melt, are limited by incomplete data and understanding of the effects of melt and water, and to some extent temperature on seismic velocities and attenuation. Another limiting factor is the low resolution of and non-uniqueness in the seismic images or differences in resolution between Vs, Vp, and Q images. However, spatial coherence, depths and sharpness of seismic discontinuities, surface heat flow, density, and estimates of electrical conductivity can help favor one interpretation over another. Interpreting the seismic results in the context of known and publicly available surface heat flow and gravity data, as well as published

xenolith analyses, would greatly enhance the potential of the seismological data for assessing the hydration state of the mantle.

3. SEISMOLOGICAL CONSTRAINTS ON MANTLE WATER CONTENT

Seismic clues that could point to the presence of water in the mantle include lowered seismic velocities, elevated or depressed discontinuities, broadened or sharpened discontinuities, enhanced attenuation, and unconventional anisotropy patterns. On their own, each of these clues can be explained with alternatives to hydrogen, such as heightened or lowered temperatures, but in combination with each other and/or in a context provided by non-seismic data the clues can make a strong case for the presence of hydrogen/water in the mantle.

3.1 The Mantle Wedge

The mantle wedge, lying above the partially-hydrated oceanic crust of the downgoing slab, undoubtably receives the largest water input of any mantle region, and probably represents the most hydrated part of the upper mantle. This is perhaps the only part of the mantle in which substantial water may be present in three fundamental forms: as a free fluid [*Iwamori*, 1998], incorporated in hydrous minerals [*Davies and Stevenson*, 1992], and dissolved in nominally anhydrous minerals like olivine and pyroxene [*Kohlstedt et al.*, 1996]. Dehydration of the slab and fluxing of the mantle wedge is responsible for most island arc volcanism, as demonstrated by the predominance of fluid mobile elements in island arc volcanics [*Morris et al., 1990; Plank and Langmuir*, 1993; *Tatsumi and Eggins*, 1995].

The distribution of water within the mantle wedge is highly uncertain. It is well established that the downgoing slab releases water into the wedge beneath the island arc, which generally occurs where the slab depth is between 90–130 km [*England et al.*, 2004], but several different mineralogical reactions are involved in the dehydration [*Schmidt and Poli*, 1998]. Several subduction zones show distinct low velocity crustal layers that decrease in amplitude near 120–150 km depth, probably due to dehydration and eclogitization reactions in the subducting crust [*Yuan et al.*, 2000; *Ferris et al.*, 2003]. It is still unclear how much the slab releases of its water in the island arc region, and how much it retains and releases deeper in the mantle. This question has major implications for the amount and distribution of water in the deeper mantle.

The occurrence of water and fluid-mobile elements in various arc and backarc volcanics provide some clue about the distribution of water within the mantle wedge. Backarc spreading centers, which sample the mantle wedge, show decreasing levels of water and fluid mobile elements with increasing distance into the backarc, reaching levels typical of nominally dry mantle (MORB) at several hundred kilometers distance from the arc [*Pearce et al.*, 1995; *Taylor and Martinez*, 2003; *Kelley et al.*, 2005]. This evidence suggests that water is concentrated near the subducted slab and does not permeate the entire mantle wedge into the far backarc, at least at shallow (< 100 km) depths. In some cases, however, the mantle can be found hydrated many hundreds of km away from the slab if the slab's dip angle recently steepened, such as in the central Andes [*Kay et al., 1999; James and Sacks, 1999*].

Seismology has the potential of resolving many of the questions about water in the mantle wedge. Most seismic tomographic models of arcs show an inclined region of low Vp and Vs about 50 km above the slab, extending from the backarc to beneath the volcanic front [*Zhao et al.*, 1992; *Zhao et al.*, 1995; *Nakajima et al.*, 2001]. Larger-scale tomographic models show low seismic velocities and high attenuation throughout a large region of the mantle wedge, occasionally extending many hundreds of kilometers into the backarc [*Zhao et al.,* 1997; *Van der Lee et al.,* 2001, 2002; *Conder and Wiens,* 2006] (Plates 1 and 2). The prominent low-velocity anomaly in Plate 1 characterizes the mantle wedge beneath the central Andes and above the Nazca slab. The strength and spatial isolation of the anomaly have lead to it being interpreted as a relatively water- and melt-rich rather than a hot zone [*Van der Lee et al.*, 2001, 2002]. On the other hand, *Wiens et al.* [2006b] find that for various backarc spreading centers the lowest seismic velocities correlate well with petrological indicators of mantle temperature and fails to correlate with petrological estimates of mantle water content [*Kelley et al.*, 2006], suggesting that the seismic anomalies in backarcs may be largely controlled by temperature variations and not by water content.

In the upper 100–150 km beneath the arc and backarc, large seismic velocity and attenuation anomalies (Plate 2) have variously been interpreted as the effects of temperature, melt, and water. The effect of temperature on seismic velocities and attenuation are non-linear in the upper mantle, such that the effects become large at higher temperatures (1200–1400 °C) [*Jackson, et al.*, 1992; *Karato*, 1993; *Jackson, et al.*, 2002]. Furthermore, the seismic anomalies are also a function of grain size [*Jackson, et al.*, 2002; *Faul and Jackson*, 2005]. However, the large magnitude of the velocity and attenuation anomalies in the mantle wedge suggests that temperature effects alone cannot be entirely responsible, and that water and/or melt provide an important effect [*Conder and Wiens*, 2006; *Wiens et al.*, 2006a; *Van der Lee et al.*, 2000; 2001].

3.2 Upper Mantle Below the Mantle Wedge, Including the Transition Zone

Much less is known about the presence, cycling, and role of water in the mantle below the mantle wedge. This is in part because these mantle depths are not nearly as well sampled by magmatism and for another part because typical earthquake-seismometer distributions allow for higher resolution in imaging mantle wedge structure than in imaging deeper parts of the mantle.

Nevertheless, seismologists have imaged deep seismic anomalies that are unlikely results of temperature, iron, or grain size anomalies and thus more likely the result of water. As the solubility of water in olivine increases strongly with depth [*Kohlstedt, et al.*, 1996; *Chen et al.*, 2002; *Mosenfelder et al.*, this volume; *Hirschmann et al.*, this volume], water at depths below the mantle wedge would likely be absorbed into the olivine crystal structure, and would cause increased attenuation, lower seismic velocities, thickened transition zone, as well as broadened 410- and 660-km phase transitions. For example, *Nolet and Zielhuis* [1994] argued that a low S-velocity anomaly near depths of 300 km was caused by water enrichment of formerly depleted tectosphere by ruling out other known explanations for S-velocity anomalies. *Revenaugh and Sipkin* [1994] interpreted a low-velocity layer atop the 410-km discontinuity as an accumulation of melt from water-triggered melting of the deep upper mantle. *Zhao et al.* [1997] suggested that low P velocities extending to depths of 400 km in Tonga-Fiji likely result from the presence of water. The nearby subducting slab was implicated as the source of the water, though it is unclear whether the water-containing minerals in the slab would dehydrate at these depths or the water or water-rich mantle might have migrated up to these depths after having formed from slab dehydration near or in the lower mantle. A more recent analysis shows that these large low-velocity anomalies extend to only 250-300 km depth in Tonga-Fiji [*Conder and Wiens*, 2006] (Plate 2). However, *Shito and Shibutani* [2003] found high attenuation at depths between 200–400 km beneath the Philippine Sea and also suggested this may result from water from the subducting Pacific slab [*Shito et al.*, this volume].

Assuming that subducting slabs remain somewhat hydrous after passage through the mantle wedge above 150 km, *Komabayashi* [this volume] shows that dehydration reactions between 150 and 600 km are atypical and occur only in special circumstances. For example, if a slab flattens in the transition zone, it would eventually be warm enough for the dense hydrous magnesium silicates it might contain to break down and release water to the overlying upper mantle. This intermediate-depth dehydration could explain the low-velocities imaged by *Zhao* [2004] in the upper 400 km beneath

China, right above the Pacific slab, which appears to lie flat in the transition zone here. *Zhao* [2004] proposes that intraplate volcanism at the Chinese Wudalianchi and Changbai volcanoes is a result of this hydrated mantle. Cold slabs that do not heat up sufficiently in the transition zone would not dehydrate until they reached depths near the bottom of the transition zone or top of the lower mantle [*Komabayashi*, this volume]. *Van der Lee et al.* [2006] propose that the overlying mantle hydrated by the dehydrating slab might form a slow upwelling and eventually interact with lithosphere at the surface. Such a deep hydrous upwelling would have low S velocities, as imaged above the Farallon Plate beneath the eastern US (Plate 3). Subduction-induced deep hydrous upwellings could be related to intraplate volcanism and at a continent-ocean transition it could initiate a new subduction zone [*Van der Lee et al.*, 2006], thus implying that an advanced subduction process can eventually trigger a new subduction zone because of this deep water cycle.

Most of these inferences for hydrous material in the upper mantle are based on low seismic-velocity or low-Q anomalies. However, low velocities and low Q can have alternative explanations, such as elevated temperatures. Reasons for the mentioned studies to preferentially explain the observations with the presence of water are based upon the exclusion of other potential explanations and could thus change if this potential were expanded. These reasons range from the anomalies being close to a subducting slab and its associated volcanism [*Zhao*, 2004; *Shito and Shibutani*, 2003; *Conder and Wiens*, 2006], incomplete consistency of temperatures inferred from co-located S, P and Q anomalies [*Conder and Wiens*, 2006], the anomalies being so slow that only unrealistic temperature elevations or iron enrichment can explain them [*Nolet and Zielhuis*, 1994; *Revenaugh and Sipkin*, 1994; *Van der Lee and Nolet*, 1997], the absence of a hot source and of elevated heat flow as well as the properties of transition zone seismic discontinuities [*Van der Lee et al.*, 2006].

In fact, the properties of transition zone discontinuities themselves have been cited as primary evidence for a wet, as well as for a dry mantle. For example, *Courtier and Revenaugh* [this volume] infer water deep beneath the eastern US from their observing an elevated 410-km discontinuity and anomalously strong 520-km discontinuity. This is consistent with low S velocities imaged here by *Van der Lee and Nolet* [1997]. *Van der Meijde et al.*'s [2003] observations of a broadened 410-km discontinuity imply about 0.05–0.09 wt % or more [*Hirschmann et al.*, 2005] of water in olivine around depths of 410 km beneath parts of the Mediterranean region (Figure 1). The transition zone beneath this region is densely populated by recently subducted slab fragments [*Marone et al.*, 2004], which could be a likely source for the water but are not sufficiently cold to explain the observed broadened discontinuity. *Blum and Shen* [2004] also argue that an elevated 410-km discontinuity beneath southern Africa cannot be solely explained by lowered temperatures inferred from seismic tomography [*James et al.*, 2001] and infer that part of the uplift is caused by 0.3 to 0.7 wt % of water in the transition zone. *Suetsugu et al.* [this volume] infer about 1.2 ± 0.2 wt % of water in the bottom of the transition zone from tomographically imaged relatively normal P velocities and a depressed 660-km discontinuity. *Song and Helmberger* [this volume] deduced a layer of low velocities atop the 410-km discontinuity beneath the western US. This low-velocity layer might be caused by water released by the subducted Farallon Plate, although the expected melt layer [*Hirschmann et al.*, this volume] would be significantly thinner than that inferred by *Song and Helmberger* [this volume].

In contrast, *Gilbert et al.* [2003] imply that a lack of correlation between discontinuity width and uplift beneath the western US suggests that water does not play a dominant role there. A review of discontinuity widths and depths worldwide by *Helffrich* [2000] suggests that the seismologically inferred sharpness of the 410-km discontinuity precludes the transition zone from being extensively hydrated. *Tibi and Wiens* [2005] and *Braunmiller et al.* [this volume] investigated discontinuities in active subduction zones (Tonga and Andes, respectively) and found that their characteristics are consistent with a dry transition zone. Independently from the seismological evidence, *Dixon et al.* [2002] use geochemical analyses of volcanic rocks to argue that little surface water is subducted into the deep upper mantle.

4. DISCUSSION

Seismic observations have inferred [*Suetsugu et al.*, this volume; *Song and Helmberger*, this volume; *Zhao et al.*, 1997; *Shito et al.*, this volume; *Van der Meijde et al.*, 2003] and excluded [*Tibi and Wiens*, 2005; *Gilbert et al.*, 2003] water in the deep upper mantle next to or near subducting slabs. Away from slabs, inferences on water content of the mantle also vary [*Blum and Shen*, 2004; *Gao et al.*, 2002; *Chambers et al.*, 2005; *Shearer and Flanagan*, 1999; *Helffrich*, 2000]. Differences between these studies are in 1) the region studied, 2) the data analyzed, and 3) the way the data were processed. It is quite possible that there are regional differences between subducting slabs, in terms of their level of hydration, their temperatures at depth, and associated dehydration reactions. Some slabs may not carry much water downwards from the mantle wedge, others could dehydrate upon flattening and associated warming in the transition zone, while yet others would not dehydrate until reaching the lower mantle [*Komabayashi*, this volume]. It is also possible that different data highlight different aspects

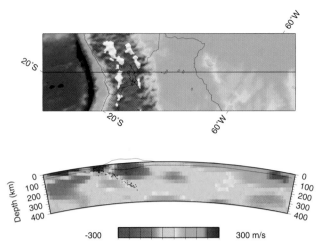

Plate 1. Cross section through South America showing large-scale anomalous *S*-velocity structure and Moho depth, jointly inferred from regional *S* and surface wave forms, receiver functions, and surface wave group velocities. The tomographic model is from *Feng et al.* [submitted manuscript]. The grey dots are earthquake hypoconeters.

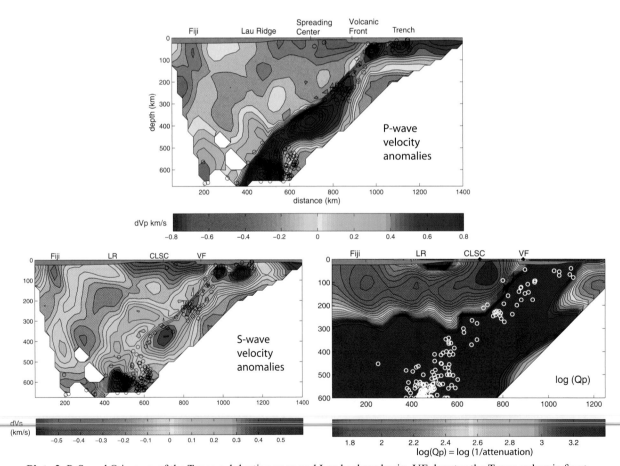

Plate 2. P, S, and Q images of the Tonga subduction zone and Lau backarc basin. VF denotes the Tonga volcanic front, CLSC denotes the Central Lau Spreading Center, and LR denotes the Lau Ridge (relict arc). Note that P and S images show anomalies relative to an average model, whereas the attenuation image shows log(Q) without a reference model. Low seismic velocities and low Q extending to depths of 250–300 km may result from water in the mantle wedge. P and S wave images are from *Conder and Wiens* [2006]; Q structure is re-inverted from data in *Roth et al.*, [1999].

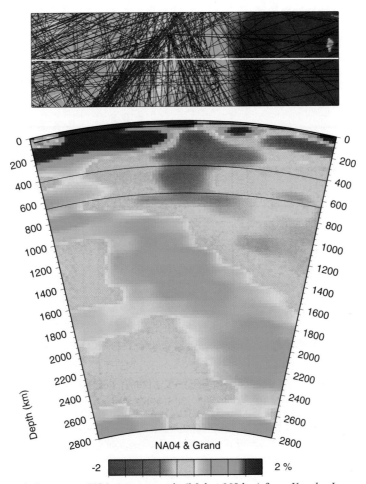

Plate 3. *S* velocity beneath the eastern USA: upper mantle (Moho-660 km) from *Van der Lee and Frederiksen* [2005], lower mantle (660–2800 km) from *Grand* [2002]. The green, east-dipping anomaly in the lower mantle represents cold lithosphere from the subducted Farallon Plate. Above it, between 100 and 750 km, a red, low-velocity anomaly could represent a water-rich (1.3 wt % at depth to 0.3 wt % near its top) upwelling related to dehydration of the subducted Farallon Plate in the top of the lower mantle, where hydrous phases potentially in it break down [*Komabayashi et al.,* this volume]. Uppermost mantle heterogeneity is stronger than 2% and saturates the color scale.

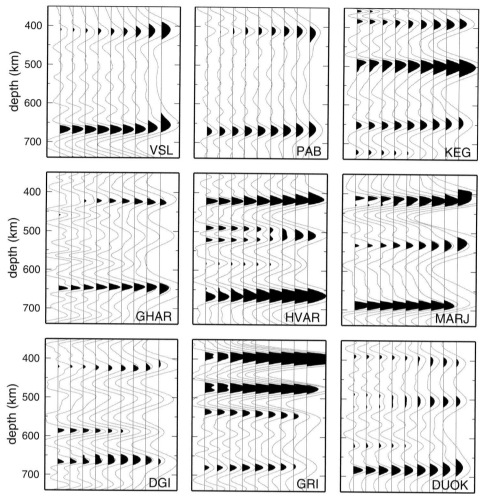

Figure 1. Receiver functions (grey lines) from *Van der Meijde et al.* [2003]. Positive signals are filled with black wherever they are significantly larger than zero with more than 95 % confidence. Black fill reaches to 2 standard deviations below the signal. Seismic stations are from the MIDSEA network [*Van der Lee et al.*, 2001]. Multiple receiver functions for each station represent a decreasing low-pass frequency from left to right. The positive signal (indicating a *P* to *S* conversion) generated by the 410-km discontinuity gets stronger with decreasing frequency, which is not observed to the same extent for discontinuities at other depths. This frequency-dependence indicates that the 410-km phase transition occurs over larger than standard depth intervals, which in turn could point to the presence of water in this depth region, which is known to significantly broaden the olivine to wadsleyite transition interval in water-undersaturated conditions [*Wood*, 1995; *Smyth and Frost*, 2002; *Hirschmann et al.*, this volume; *Karato*, 2006b]. The inferred widths of over 20 km could imply between 0.05 [*Wood*, 1995] and 0.2 [*Hirschmann et al.*, this volume] wt % of water in olivine near 410 km.

of the mantle. For example, receiver functions [*Gilbert et al.*, 2003] would record a water-triggered low-velocity layer on the 410 only if its top was sharply defined, while triplicated S waves would sense the layer no matter how sharp its top was [*Song and Helmberger*, this volume]. More disconcerting is that different inferences have been drawn because of different data processing methods for the same regions and very similar data sets. For example *Blum and Shen* [2004] find a transition zone beneath southern Africa that is 25 km thicker than a normal thickness of 245 km found there by

Gao et al. [2002] from virtually the same data. The different result must be due to differences in data processing, yet the superiority of one processing method over another is not established.

Rather than inferring one type of seismic property at a time, it seems advisable to simultaneously infer several properties from a diverse data set as such would yield more mutually consistent results and subsequently more robust models of mantle water content. For example, ignored, underestimated, or spatially unresolved velocity heterogeneity in a mantle-discontinuity study can bias the inferred depths of the discontinuity. Likewise, unaccounted discontinuity relief can bias seismic-velocity models.

Water affects many physical properties of the mantle, including the thickness of phase transitions. Studies of long-wavelength seismic waves, such as *SS* precursors, would record lower impedance contrasts for thicker phase transitions. On a global scale, *Chambers et al.* [2005] used such waves to map the impedance contrast at the 410-km discontinuity to be about 30 % lower than that found by *Shearer and Flanagan* [1999] from a somewhat sparser but very similar data set. Shorter-wavelength seismic waves, such as *P* to *S* converted waves or *PP* precursors, can place bounds on the thickness and sharpness of the 410-km discontinuity, as reviewed by *Helffrich* [2000]. *Helffrich* [2000] quotes maximum sharpness estimates from the literature of 4 to 5 km to support the 410-km discontinuity representing a phase change. Such a sharp discontinuity would also imply a dry mantle, as even a slightly wet mantle would thicken the phase transition [*Wood*, 1995; *Smyth and Frost*, 2002; *Karato*, 2006b]. However, the quoted 4–5 km are typically presented in the literature as the minimum thickness that would be consistent with the data analyzed. For example, *Benz and Vidale* [1993] analyzed data from two earthquakes, of which *P410P* precursors of one lead to a thickness inference of 4 km, while the other implies a much thicker 410-km discontinuity through the absence of *P410P* precursors. The data of *Neele and Snieder* [1992] as well as of *Vidale et al.* [1995] imply a thickness of 4 to 10 km, if not more, while *Yamazaki and Hirahara* [1994] show that the 410-km discontinuity must be significantly wider than 5 km. Globally, thicknesses of 4–5 km for the 410-km discontinuity thus seem more of an exception than a rule, in essence leaving it somewhat open whether the mid upper mantle is typically wet or dry.

5. OUTLOOK

Seismic data analysis provides without doubt unique and valuable ways to constrain the water content of the solid Earth. However, many challenges remain for this assessment. Some can be conquered only through collaborative cross-disciplinary research between, for example, seismologists and mineral physicists. Other challenges must be addressed by seismologists benchmarking their data processing methods and reducing the non-uniqueness of their models. Based on our review of seismologists' attempts to infer the water content of the upper mantle, we propose two different approaches that would contribute to the success of these attempts.

The first way is combining a variety of geophysical data that together can distinguish between hydrogen and other explanations. For example, a wet transition zone would be thicker and have low S velocities, while a cold transition zone would be thicker and have high S velocities. Or, a low-velocity anomaly below a region with high surface heat flow could be hot while a low-velocity anomaly under a region with low surface heat flow is more likely to be wet. Electrical conductivity provides independent information on the hydration state of at least the uppermost mantle. A steady-state situation requires smooth spatial changes due to temperature, which dissipates by nature, while abrupt spatial change is more likely compositional in origin. *Rychert et al.* [2005] used this method of discrimination to argue that the asthenosphere beneath the northeastern US is wet. A powerful approach is to simultaneously map heterogeneity in P velocity, S velocity, and anelasticity with comparable resolving power for the same region [*Conder and Wiens*, 2006; *Shito et al.*, this volume]. Water, temperature and melt have different relative effects on these properties.

A second way is for seismologists to use their data as a verification/falsification and correction tool for mineral physics models rather than independently invert their data for a regularized seismic model. For example, seismologists will use the petrologically derived seismic velocities of *Gerya et al.* [2006] as a starting model to explain their regional travel-time data and use discrepancies to adjust the model's petrological parameters rather than velocities on a regular, but non-physical grid. For another example, models for the entire upper mantle will start from velocities based on heat-flow derived laterally-varying geotherms, then adjust the geotherm parameters where indicated by the data and explain the remaining discrepancies with compositional anomalies.

6. CONCLUSIONS

For assessing the water content of the deep Earth seismologists have some of the most powerful tools, but also face considerable challenge in 1) extracting robust inferences from limited amounts of data and 2) discriminating the effects of water from those of other mantle properties. The first challenge can be lessened by more extensive benchmarking of data processing techniques, and simultaneous rather than independent inference of various mutually consistent seismic properties. Perhaps even more important is to expand seismic data sets, in particular by acquiring data in unexplored terrain, including aseismic regions, seas, and oceans, and by widening and densifying existing seismic arrays. The second challenge can be lessened by simultaneously interpreting different types of seismic data and combining them with cross-disciplinary inferences from mineral physics and other branches of geophysics. This task is greatly facilitated when seismologists and mineral physicists communicate, as this book intends to encourage. Seismological inferences on the Earth's deep water content and cycle depend completely on mineral physicists characterizing the seismic properties of hydrated mantle minerals and assessing the impact of compositional anomalies other than hydration on those seismic properties. Lastly, we encourage mineral physicists to construct physically plausible hypothetical seismic models for the Earth's mantle that seismologists can test against seismic data.

Acknowledgements. We thank Steven Jacobsen, Guust Nolet, and Heather Watson for the promptest ever reviews. Their thoughtful, critical, and helpful comments significantly improved this chapter.

REFERENCES

Benz, H. M., and J. E. Vidale, Sharpness of upper-mantle discontinuities determined from high-frequency reflections, nature, 365, 147–150, 1993.

Blum, J., and Y. Shen, Thermal, hydrous, and mechanical states of the transition zone beneath southern Africa, Earth Planet. Sci. Lett., 217, 367–378, 2004.

Bolfan-Casanova, N., C. McCammon, and S. Mackwell, Water in transition zone and lower mantle minerals, this volume, 2006.

Braunmiller, J., S. van der Lee, and L. Doermann, Transition-zone thickness in the central South American Subduction Zone from receiver functions, this volume.

Cagnioncle, A.-M., E. M. Parmentier, and L. T. Elkins-Tanton, The effect of solid flow above a subducting slab on water distribution and melting at convergent plate boundaries, Earth Planet. Sci. Lett., in press, 2006.

Cammarano F., S. Goes, P. Vacher, and D. Giardini, Inferring upper mantle temperatures from seismic velocities, Phys. Earth Planet. Inter., 138, 197–222, 2003.

Conder, J. A., and D. A. Wiens, Seismic structure beneath the Tonga arc and Lau backarc basin determined from joint Vp, Vp/Vs tomography, Geochem. Geophys. Geosystems, 7, Q03018, doi: 10.1029/2005GC001113, 2006.

Chambers, K., A. Deuss, and J.H. Woodhouse, Reflectivity of the 410-km discontinuity from PP and SS precursors, J. Geophys. Res., 110, doi:10.1029/2004JB003345, 2005.

Chen, J., T. Inoue, H. Yurimoto, and D. J. Weidner, Effect of water on olivine-wadsleyite phase boundary in the (Mg,Fe)2SiO4 system, Geophys. Res. Lett., 29, doi:1810.1029/2001GL014429, 2002.

Dasgupta, R., M. M. Hirschmann, and A. C. Withers, Deep global cycling of carbon constrained by the solidus of anhydrous, carbonated eclogite under upper mantle conditions, Earth Planet. Sc. Lett., 227, 73–85, 2004.

Davies, J. H., and D. J. Stevenson, Physical model of source region of subduction zone volcanics, *Journal* of Geophysical Research, 97, 2037–2070, 1992.

Dixon, J. E., Leist, L., Langmuir, C., and Schilling, J. G., Recycled dehydrated lithosphere observed in plume-influenced mid-ocean ridge basalt. Nature, 420, 385–389, 2002.

Dixon, J. E., T. H. Dixon, D. Bell, R. Malservisi, Lateral variation in upper mantle viscosity: role of water, Earth and Planetary Science Letters, 222, 451–467, 2004.

Eiler, J., Inside the subduction factory, AGU Geophysical Monograph Series, 138, 324 pp., 2004.

England, P., R. Engdahl, and W. Thatcher, Systematic variation in the depths of slabs beneath arc volcanoes, *Geophysical Journal International*, *156*, 377–408, 2004.

Evans, R., G. Hirth, K. Baba, D. Forsyth, A. Chave, R. Mackie, Geophysical evidence from the MELT area for compositional controls on oceanic plates*, Nature, 437*, 249–253, 2005.

Faul, U. H., J. D. Fitz Gerald, and I. Jackson, Shear wave attenuation and dispersion in melt-bearing olivine crystals: 2. Microstructural interpretation and seismological implications, *J. Geophys. Res.*, *109*, doi: 10.1029/2003JB002407, 2004.

Faul, U. H., and I. Jackson, The seismological signature of temperature and grain size variations in the upper mantle, *Earth Planet. Sci. Lett.*, *234*, 119–134, 2005.

Ferris, A., G. A. Abers, D. H. Christensen, and E. Veenstra, High resolution image of the subducted Pacific(?) plate beneath central Alaska, 214, Earth Planet. Sci. Lett., 575–588, 2003.

Frost, D. J., The structure and sharpness of (Mg,Fe)$_2$SiO$_4$ phase transformations in the transition zone, Earth Planet. Sci. Lett., 216, 313–328, 2003.

Gaetani, G. A., and T. L. Grove, The influence of water on melting of mantle peridotite. Contrib. Mineral. Petrol., 131, 323–346, 1998.

Gao, S. S., P. G. Silver, K. H. Liu, and the Kaapvaal Seismic Group, Mantle Discontinuities Beneath Southern Africa, Geophys. Res. Lett., 29, 10.1029/2001GL013834, 2002.

Gerya, T. V., Connolly, J. D., Yuen, D. A. , Gorczyk, W. and A. M. Capel, Seismic implications of mantle wedge plumes, Phys. Earth Planet. Inter., 156, 50–74, 2006.

Gilbert, H. J., A. F. Sheehan, K. G. Dueker, and P. Molnar, Receiver functions in the western United States, with implications for upper mantle structure and dynamics, J. Geophys. Res., 108, doi:10.1029/2001JB001194, 2003.

Goes, S., and S. van der Lee, Thermal structure of the North American uppermost mantle inferred from seismic tomography, J. Geophys. Res., 107, DOI:10.1029/2000JB000049, 2002.

Grand, S.P., Mantle shear-wave tomography and the fate of subducted slabs, Philosophical Transactions Royal Society of London, Series A, 360, 2475–2491, 2002

Hammond, W. C., and E. D. Humphreys, Upper mantle seismic wave velocity: Effects of realistic partial melt geometries, *J. Geophys. Res.*, *105*, 10,975–910,986, 2000.

Hammouda, T., High-pressure melting of carbonated eclogite and experimental constraints on carbon recycling and storage in the mantle, Earth Planet. Sc. Lett., 214, 357–368, 2003.

Helffrich, G. Topography of the transition zone seismic discontinuities Rev. Geophys., 38, 141–158, 2000.

Helffrich, G., and C. R. Bina, Frequency dependence of the visibility and depths of mantle seismic discontinuities, Geophys. Res. Lett., 21, 2613–2616, 1994.

Helffrich, G., and B. Wood, 410-km discontinuity sharpopness and the form of the olivine phase diagram: Resolution of apparent seismic contradictions, Geophys. J. Int., 126, F7–F12, 1996.

Higo, Y., T. Inoue, T. Irifune, and H. Yurimoto, Effect of water on the spinel-postspinel transformation in Mg_2SiO_4, Geophys. Res. Lett., 28, 3505–3508, 2001.

Hirschmann, M. M., C. Aubaud, and A. C. Withers, Storage capacity of H_2O in nominally anhydrous minerals in the upper mantle, Earth Planet. Sci. Lett., 236, 167–181, 2005.

Hirschmann, M. M., A. C. Withers, and C. Aubaud, Petrologic Structure of a Hydrous 410 km Discontinuity, this volume, 2006.

Hirth, G., and D. Kohlstedt, Water in the oceanic upper mantle: implications for rheology, melt extraction and the evolution of the lithosphere, Earth Planet Sci Lett, 144, 93–108, 1996.

Huang, X., Y. Xu, and S. Karato, Water content in the transition zone from electrical conductivity of wadsleyite and ringwoodite, Nature, 434, 746–749, 2005.

Inoue, T., D. J. Weidner, P.A. Northrup, J. B. Parise, Elastic properties of hydrous ringwoodite (γ-phase) in Mg_2SiO_4, Earth Planet. Sci. Lett., 160, 107–113, 1998.

Inoue, T., Y. Tanimoto, T. Irifune, T. Suzuki, H. Fukui, and O. Ohtaka, Thermal expansion of wadsleyite, ringwoodite, hydrous wadsleyite and hydrous ringwoodite, Phys. Earth Planet. Inter., 143–144, 279–290, 2004.

Iwamori, H., Transportation of H2O and melting in subduction zones, Earth Planet. Sci. Lett., 160, 65–80, 1998.

Jackson, I., J. D. Fitz Gerald, U. H. Faul, and B. H. Tan, Grain-size-sensitive seismic wave attenuation in polycrystaline olivine, J Geophys. Res., 107, doi:10.1029/2001JB001225, 2002.

Jackson, I., M. S. Paterson, and J. D. FitzGerald, Seismic wave dispersion and attenuation in Aheim Dunite: an experimental study, Geophys. J. Int., 108, 517–534, 1992.

Jackson, J. M., J. Zhang, and J. D. Bass, Sound velocities and elasticity of aluminous $MgSiO_3$ perovskite: Implications for aluminum heterogeneity in Earth's lower mantle, Geophys. Res. Lett., 31, L10614, doi: 10.1029/2004GL019918, 2004.

Jacobsen, S. D., J. R. Smyth, H. Spetzler, C.M. Holl, and D. J. Frost, Sound velocities and elastic constants of iron-bearing hydrous ringwoodite, Phys. Earth Planet. Inter., 143–144, 47–56, 2004.

Jacobsen, S. D., and J. R. Smyth, Effect of water on the sound velocities of ringwoodite in the transition zone, this volume, 2006.

James, D. E. and I. S. Sacks, Cenozoic formation of the central Andes: A geophysical perspective, Spec. Pub. 7 of the SEG, 1–25, 1999.

James, D. E., M. J. Fouch, J. C. VanDecar, S. van der Lee, and Kaapvaal Seismic Group, Tectospheric structure beneath southern Africa, Geophys. Res. Lett. 28, 2485–2488, 2001.

Jordan, T. H., Structure and formation of the continental tectosphere, J. Petrol. Special Lithospheric Issue, 11–37, 1988.

Karato, S.-I., The role of hydrogen in the electrical conductivity of the upper mantle, Nature, 347, 272–273, 1990.

Karato, S., Importance of anelasticity in the interpretation of seismic tomography, Geophysical Research Letters, 20, 1623–1626, 1993.

Karato, S.-I., Mapping water content in the upper mantle, in Inside the Subduction Factory, edited by J. M. Eiler, pp. 135–152, American Geophysical Union, Washington D.C., 2003.

Karato, S., and H. Jung, Water, partial melting, and the origin of the seismic low velocity zone in the upper mantle, Tectonophysics, 157, 193–207, 1998.

Karato, S., Hydrogen-related defects and their influence on the electrical conductivity and plastic deformation of mantle minerals: a critical review, this volume, 2006a.

Karato, S., Microscopic models for the influence of hydrogen on physical and chemical properties of minerals, in Superplume: Beyond Plate Tectonics, edited by D. A. Yuen, et al., Springer, in press, 2006b.

Kay, S. M., C. Mpodozis, and B. Coira, Magmatism, tectonism, and mineral deposits of the Central Andes (22°–33°S latitude. In Skinner, B. (ed.), Geology and Ore Deposits of the Central Andes, Society of Economic Geology Special Publication (SEG) No. 7, 27–59, 1999.

Kelley, K. A., T. Plank, T. L. Grove, E. M. Stolper, S. Newman, and E. Hauri, Mantle melting as a function of water content at subduction zones I: Back-arc basins, J. Geophys. Res., in press, 2006.

Keppler, H., M. Wiedenbeck, and S.S. Shcheka, Carbon solubility in olivine and the mode of carbon storage in the Earth's mantle, Nature 424, 414–416, 2003.

Kohlstedt, D. L., H. Keppler, and D. C. Rubie, Solubility of water in the α, β, and γ phases of $(Mg, Fe)_2 SiO_4$, Contrib. Mineral. Petrol., 123, 345–357, 1996.

Komabayashi, T., Phase Relations of Hydrous Peridotite: Implications for Water Circulation in the Earth's Mantle, this volume, 2006.

Lee, C. –T. A., Compositional variation of density and seismic velocities in natural peridotites at STP conditions: implications for seismic imaging of compositional anomalies in the upper mantle, J. Geophys. Res., 108, B9, 2441, doi: 10.1029/2003JB002413, 2003.

Li, L., D. J. Weidner, J. Brodholt, D. Alfe, G.D. Price, R. Caracas, and R. Wentzcovitch, Elasticity of $CaSiO_3$ perovskite at high pressure and high temperature. Phys. Earth Planet. Int., 155, 249–259, 2006.

Lizarralde, D., A. Chave, G. Hirth, and A. Schultz, Northeastern Pacific mantle conductivity profile from long-period magnetotelluric sounding using Hawaii-to-California submarine cable data, J. Geophys. Res., 100 (B9), 17,837–17,854, 1995.

Mei, S., and D. L. Kohlstedt, Influence of water on plastic deformation of olivine aggregates 1. Diffusion creep regime, J. Geophys. Res., 105, 21,457–21,469, 2000a.

Mei, S., and D. L. Kohlstedt, Influence of water on plastic deformation of olivine aggregates 2. Dislocation creep regime, J. Geophys. Res., 105, 21,471–21,481, 2000b.

Marone, F., S. van der Lee, and D. Giardini, Three-dimensional upper-mantle S-velocity model for the Eurasia-Africa plate boundary region, Geophys. J. Int., 158, 109–130, 2004.

Matsukage, K. N., Y. Nishihara, and S. Karato, Seismological signature of chemical differentiation of Earth's upper mantle, J. Geophys. Res., 110, doi:/10.1029/2004/JB003504, 2005.

Mavko, G. M., Velocity and attenuation in partially molten rocks, *J. Geophys. Res.*, *85*, 5412–5426, 1980.

Morris, J. D., W. P. Leeman and F. Tera, The subducted component in island arc lavas: constraints from Be isotopes and B-Be systematics, Nature, 344, 31–36, 1990.

Nakajima, J., T. Matsuzawa, A. Hasegawa, and D. Zhao, Three-dimensional structure of Vp, Vs, and Vp/Vs beneath the northeastern Japan arc: implications for arc magmatism and fluids, Journal of Geophysical Research, 106, 21843–21857, 2001.

Neele, F., and R. Snieder, Topography of the 400 km discontinuity from the observations of long period P400P phases, Geophys. J. Int., 109, 670 – 682, 1992.

Nolet, G., and A. Zielhuis, Low S velocities under the Tornquist-Teisseyre zone: evidence for water injection into the transition zone by subduction, J. Geophys. Res., 99, 15813–15820, 1994.

Pearce, J. A., M. Ernewein, S. H. Bloomer, L. M. Parson, B. J. Murton, and L. E. Johnson, Geochemistry of Lau Basin volcanic rocks: influence of ridge segmentation and arc proximity, in Volcanism Associated with Extension at Consuming Plate Margins, *Geol. Soc. Spec. Publ.*, *81*, 53–75, 1995.

Plank, T., and C. H. Langmuir, Tracing trace elements from sediment input to volcanic output at subduction zones, *Nature, 362*, 739–742, 1993.

Priestley, K., and D. McKenzie, The thermal structure of the lithosphere from shear wave velocities, Earth Planet. Sci. Lett., 244, 285–301, 2006.

Rauch, M. and H. Keppler, Water solubility in orthopyroxene. Contrib. Mineral. Petrol., 143, 525–536, 2002.

Rychert, C.A., K. M. Fischer, and S. Rondenay, A sharp lithosphere-asthenosphere boundary imaged beneath eastern North America, Nature, doi:10.1038/nature03904, 2005.

Revenaugh, J., and S. A. Sipkin, Seismic evidence for silicate melt atop the 410-km mantle discontinuity, Nature, 369, 474–476, 1994.

Rüpke, L., J. Phipps Morgan, and J. Dixon, Implications of Subduction Rehydration for Earth's Deep Water Cycle, this volume, 2006.

Saal, A. E., E. H. Hauri, C. H. Langmuir, M. R. Perfit, Vapour undersaturation in primitive mid-ocean-ridge basalt and the volatile content of Earth's upper mantle, *Nature, 419*, 451–455, 2002.

Schmidt, M. W., and S. Poli, Experimentally based water budgets for dehydrating slabs and consequences for arc magma generation, *Earth Planet Sci Lett, 163*, 361–379, 1998.

Schutt, D. L., and C. E. Lesher, Effects of melt depletion on the density and seismic velocity of garnet and spinel lherzolite, J. Geophys. Res., 111, 10.1029/2003JB002950, 2006.

Shearer, P., and M. Flanagan, Seismic velocity and density jumps across the 410- and 660-kilometer discontinuities, Science, 285, 1545–548, 1999.

Shito, A., and T. Shibutani, Nature of heterogeneity of the upper mantle beneath the northern Philippine Sea as inferred from attenuation and velocity tomography, *Phys. Earth and Planet. Int., 140*, 331–341, 2003.

Shito, A., S. Karato, K. Matsukage, and Y. Nishihara, Toward mapping three-dimensional distribution of water in the upper mantle from combined velocity attenuation tomography: Applications to subduction zone upper mantle, this volume, 2006.

Sinelnikov, Y. D., G. Chen, and R.C. Liebermann, Elasticity of $CaTiO_3$-$CaSiO_3$ perovksites, Phys. Chem. Minerals 25, 515–521, 1998.

Sinogeikin, S.V., J. D. Bass, and T. Katsura, Single-crystal elasticity of gamma -$(Mg_{0.91}Fe_{0.09})_2SiO_4$ to high pressures and to high temperatures, Geophysical Research Letters, 28, 22, 4335–4338, 2001.

Smyth, J. R., and D. J. Frost, The effect of water on the 410-km discontinuity: An experimental study, Geophys. Res. Lett., doi:10.1029/2002GL014418, 2002.

Smyth, J. R., D. J. Frost, and F. Nestola, Hydration of olivine and the Earth's deep water cycle, Geochim. Cosmochim. Acta, 69, A746, 2005.

Smyth, J. R. and S. Jacobsen, Nominally Anhydrous Minerals and Earth's Deep Water Cycle, this volume, 2006.

Song, T. A., and D. V. Helmberger, Low velocity zone atop the transition zone in the western US from S waveform triplication, this volume, 2006.

Suetsugu, D., T. Inoue, A. Yamada, D. Zhao, and M. Obayashi, A study of temperature anomalies and water content in the mantle transition zone beneath subduction zones as inferred from P-wave velocities and mantle discontinuity depths, this volume, 2006.

Taylor, B., and F. Martinez, Back-arc basin basalt systematics, Earth *Planet. Sci. Lett., 210*, 481–497, 2003.

Tibi, R. and D. A. Wiens, Detailed structure and sharpness of upper mantle discontinuities in the Tonga subduction zone from regional broadband arrays, J. Geophys. Res., 110, doi:10.1029/ 2004JB003433, 2005.

Takei, Y., Effect of pore geometry on Vp/Vs: From equilibrium geometry to crack, *J. Geophys. Res., 107*, 10.0129/2001JB000522, 2002.

Tatsumi, Y., and S. Eggins, *Subduction Zone Magmatism*, 211 pp., Blackwell, Cambridge, MA, 1995.

Trampert, J., F. Deschamps, J. Resovsky, and D. Yuen, Probabilistic Tomography Maps Chemical Heterogeneities Throughout the Lower Mantle, Science, 306, 853–856, 2004.

Vidale, J. E., X. Y. Ding, and S.P. Grand, The 410-km-depth discontinuity: A sharpness estimate from near-critical reflections, Geophys. Res. Lett., 22, 2557–2560, 1995.

Van der Lee, S. and G. Nolet, Upper-mantle S-velocity structure of North America, J. Geophys. Res., 102, 22,815–22,838, 1997.

Van der Lee, S., D. James and P. Silver, Upper-mantle S-velocity structure of western and central South America, *J. Geophys. Res.*, 106, 30821–30834, 2001.

Van der Lee, S., D. James and P. Silver, Correction to "Upper-mantle S-velocity structure of western and central South America", by Suzan van der Lee, David James, and Paul Silver, J. Geophys. Res., 107, DOI 10.1029/2002JB001891, 2002.

Van der Lee, S., F. Marone, M. van der Meijde, D. Giardini, A. Deschamps, L. Margheriti, P. Burkett, S. C. Soloman, P. M.

Alves, M. Chouliaras, A. Eshwehdi, A. Suleiman, H. Gashut, M. Herak, R. Ortiz, J. M. Davila, A. Ugalde, J. Vila, and K. Yelles, Eurasia-Africa plate boundary region yields new seismographic data, Eos Trans. AGU, 82, 637–646, 2001.

Van der Lee, S. and A. Frederiksen, Surface wave tomography applied to the North America upper mantle, in Seismic Earth: Array *Analysis of Broadband Seismograms*, edited by A. Levander and G. Nolet, American Geophysical Union, GM 157, 67–80, 2005.

Van der Lee, S., K. Regenauer-Lieb, and D. Yuen, The role of water in connecting past and future episodes of subduction, submitted manuscript, 2006.

Van der Meijde, M., F. Marone, D. Giardini, and S. van der Lee, Seismic evidence for water deep in Earth's upper mantle, Science, 300, 1556–1558, 2003.

Wiens, D. A., N. Seama, and J. A. Conder, Mantle structure and flow patterns beneath active backarc basins inferred from passive seismic and electromagnetic methods, in *Interactions among physical, chemical, biological, and geological processes in back-arc spreading systems*, edited by D. Christie, p. in press, American Geophysical Union, Washington D.C., 2006a.

Wiens, D. A., K. Kelley, and T. Plank, Mantle Temperature Variations Beneath Back-arc Spreading Centers Inferred from Seis-mology, Petrology, and Bathymetry, Earth and Planetary Science Letters, in press, 2006b.

Wood, B. J. The effect of H_2O on the 410-kilometer seismic discontinuity, Science, 268, 74–76, 1995.

Yamazaki, A., and K. Hirahara, The thickness of upper-mantle discontinuities, as inferred from short-period J-Array data, Geophys. Res. Lett., 21, 1811–1814, 1994.

Yuan, X., S. V. Sobolev, R. kind, O. oncken, G. bock, G. Asch, B. Schurr, F. Graeber, A. Rietbrock, P. Giese, P. Wigger, P. Rower, G. Zandt, S. Beck, T. Wallace, M. Pardo, and D. Comte, Subduction and collision processes in the Central Andes constrained by converted seismic phases, *Nature, 408,* 958–561, 2000.

Zhao, D., Global tomographic images of mantle plumes and subducting slabs: insight into deep Earth dynamics, Phys. Earth Planet. Inter., 146, 3–34, 2004.

Zhao, D., D. Christensen, and H. Pulpan, Tomographic imaging of the Alaska subduction zone, *J. Geophys. Res,* 100, 6487–6504, 1995.

Zhao, D., A. Hasegawa, and S. Horiuchi, Tomographic imaging of P and S wave velocity structure beneath Northeastern Japan, *J. Geophys. Res.,* 97, 19,909–19928, 1992.

Zhao, D., Y. Xu, D. A. Wiens, L. Dorman, J. Hildebrand, and S. Webb, Depth extent of the Lau back-arc spreading center and its relationship to the subduction process, *Science, 278,* 254–257, 1997.

Phase Relations of Hydrous Peridotite: Implications for Water Circulation in the Earth's Mantle

Tetsuya Komabayashi

Department of Earth and Planetary Sciences, Tokyo Institute of Technology, 2-12-1 Ookayama, Meguro, Tokyo 152-8551, Japan

Using recent results from high-pressure experiments and thermodynamic calculations, phase diagrams of simplified hydrous peridotite in the system MgO-SiO_2-H_2O have been constructed to lower mantle conditions. On the basis of constructed phase relations, I discuss possible water subduction and circulation processes in the deep mantle. For water transportation by subduction of peridotites, important phase relations are (1) of antigorite (serpentine) at lower pressures (< 10 GPa) and (2) of up to seven different high-pressure hydrous phases at higher pressures. In cold subducting slabs, water in antigorite will be partially transferred to hydrous phase A at depths greater than 160-km. With further subduction, water in phase A may be transported to the bottom of the transition zone via solid-solid reactions among seven high-pressure hydrous phases. If the slab temperature at 30 GPa is lower than 1000°C, hydrous phase D will carry water into the deep lower mantle. Along the cold slab geotherm, large fluid fluxes are predicted at shallower (~300-km depth) and deeper (~700-km depth) levels, depending on the slab temperature. The depth distributions of dehydration reactions suggest that observed subduction zone seismicity could be related to dehydration reactions in the slab. The lower water activities in the fluid phases at deep mantle conditions imply that such fluid phases could dissolve significant amounts of silicate components. Therefore, fluid phases released by dehydration reactions at deeper levels should have different physical properties from those at shallower levels.

1. INTRODUCTION

In order to evaluate possible water circulation in the deep mantle, accurate phase relations in hydrous peridotite systems are required. The system MgO-SiO_2-H_2O (MSH) has been studied now for 40 years [*Ringwood and Major*, 1967; *Yamamoto and Akimoto*, 1977; *Akaogi and Akimoto*, 1980; *Liu*, 1987; *Kanzaki*, 1991; *Luth*, 1995; *Shieh et al.*, 1998; *Wunder*, 1998; *Bose and Navrotsky*, 1998; *Frost and Fei*, 1998; *Gasparik*, 1993; *Inoue*, 1994; *Irifune et al.*, 1998;

Earth's Deep Water Cycle
Geophysical Monograph Series 168
Copyright 2006 by the American Geophysical Union.
10.1029/168GM04

Ohtani et al., 2000; *Angel et al.*, 2001; *Komabayashi et al.*, 2004; *Komabayashi et al.*, 2005a; *Komabayashi et al.*, 2005b]. Basically, water circulation may be divided into three processes: (1) water transport into the mantle by subduction, (2) storage in the mantle, and (3) return to the exosphere from the mantle. Since the second and third issues are discussed in other chapters of this volume [*e.g.* Hirschmann et al.; Bolfan-Casanova and McCammon; Karato and Bercovici; Regenauer-Lieb et al.], in this chapter I will focus on the first issue, water transportation into the deep mantle in subduction zones.

For water subduction, stability relations of hydrous phases in subducting slabs are essential because they carry water into the deep mantle. The MSH system may be divided into two pressure ranges: below 10 GPa for a serpentine mineral (antigorite) and above 10 GPa for high-pressure hydrous phases which are dense hydrous magnesium silicates (DHMS), and also hydrous wadsleyite, and hydrous ringwoodite. While antigorite phase relations have been investigated by both experimental and theoretical studies [*Ulmer and Trommsdorff*, 1995; *Wunder and Schreyer*, 1997; *Bose and Navrotsky*, 1998; *Bromiley and Pawley*, 2003; *Komabayashi et al.*, 2005a], phase relations of high-P hydrous phases have previously been studied by high-pressure experiments only. Thermodynamic calculations are required to fully describe stability relations of high-P hydrous phases. Theoretical investigation of phase stability in the MSH system at high P-T is complicated by the fact that the composition of the fluid phase is no longer pure H_2O. Fluid phases at high P-T conditions dissolve significant amounts of magnesia and silica components from coexisting minerals. Therefore, efforts to construct consistent thermodynamic calculations of dehydration reactions of high-P hydrous phases require a thermodynamic model of the magma-like fluid phase. Despite extensive previous work on

the MSH system over the past 40 years, no study has made a comprehensive phase diagram for hydrous peridotite to deep mantle conditions.

In this chapter, I will construct phase diagrams of hydrous peridotite in the system MSH to mid-lower mantle conditions on the basis of recent in-situ X-ray diffraction measurements and thermodynamic calculations of phase relations for high-P hydrous phases. In the thermodynamic calculations, the activity of water in the fluid phase at deep mantle conditions is considered. In addition, the effect of the difference in the equation of state (EOS) of fluid H_2O on the estimation of the water activity is discussed. On the basis of the phase diagram, water circulations in the mantle will be discussed. Finally, the relationship between dehydration reactions in subducting slabs and subduction zone seismicity will also be discussed. This was investigated previously with semi-quantitative phase diagram [*Komabayashi et al.*, 2004]. In this chapter, I revisit it with the newly constructed phase diagram. The phases encountered in the system MSH are shown in Figure 1 with the following abbreviations: antigorite (Atg), phase A (phA), clinohumite (cHu), phase E (phE), phase D (phD), superhy-

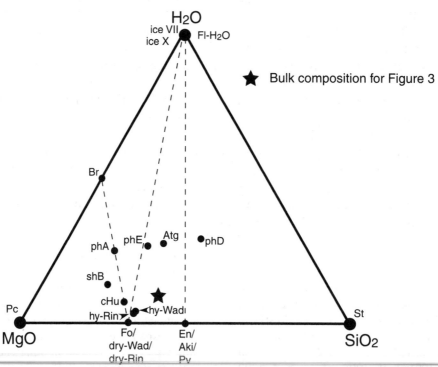

Figure 1. Compositions of the phases in the system MgO-SiO₂-H₂O modified after *Komabayashi and Omori* [2006]. Filled circles denote the composition of various phases. The bulk composition of the phase relations in Figure 3 is shown by the star. Phases are Atg, antigorite; phA, phase A; phE, phase E; phD, phase D; shB; superhydrous phase B; cHu, clinohumite; Br, brucite; hy-Wad, hydrous wadsleyite; hy-Rin, hydrous ringwoodite; Pc, periclase; Fo, forsterite; dry-Wad, dry wadsleyite; dry-Rin, dry ringwoodite; En, enstatite; Aki, akimotoite; Pv, magnesium-perovskite; St, stishovite; Fl-H₂O, fluid-H₂O; ice VII; ice X. Note that the composition of the fluid phase at higher P-T conditions is not pure H₂O. See text for details.

drous phase B (shB), hydrous wadsleyite (hy-Wad), hydrous ringwoodite (hy-Rin), forsterite (Fo), dry wadsleyite (dry-Wad), dry ringwoodite (dry-Rin), enstatite (En), akimotoite (Aki), magnesium-perovskite (Pv), periclase (Pc), Brucite (Br), stishovite (St), fluid (Fl), and high-pressure ices (VII and X). For simplicity, the coefficients of chemical reactions are omitted throughout this chapter. See *Komabayashi and Omori* [in press] for the coefficients of reactions.

2. ANTIGORITE PHASE RELATIONS

Near the surface, serpentine minerals dominate hydrous peridotite (i.e., serpentinites). In subduction zones, subducting hydrous peridotites become antigorite serpentinites at depths around 30-km [*Evans et al.*, 1976]. Dehydration reactions of antigorite may be linked to island arc magmatism [*Ulmer and Trommsdorff*, 1995]. Therefore, many studies have focused on antigorite stability [*Ulmer and Trommsdorff*, 1995; *Wunder and Schreyer*, 1997; *Bose and Navrotsky*, 1998; *Bromiley and Pawley*, 2003]. Dehydration reactions of antigorite have also been invoked to explain the double seismic zone [*Seno and Yamanaka*, 1996; *Peacock*, 2001; *Omori et al.*, 2002].

For water subduction into the deep mantle, the most important phase relation of antigorite occurs at about 5 GPa and 550°C. *Bose and Ganguly* [1995] showed on the basis of thermodynamic calculations that the antigorite dehydration reaction intersects a reaction $Fo + Fl-H_2O = phA + En$ to form an invariant point where Atg, Fo, En, phA, and $Fl-H_2O$ are stable (shown by the large star in Figure 2). Subducting antigorite thus transfers water to phase A only when the subduction geotherms are lower than this invariant point. If the subduction geotherms are higher, antigorite dehydrates to $Fo + En + Fl-H_2O$ and the solid phase assemblage becomes nominally dry. Therefore, the P-T location of this invariant point determines the fate of subducting water. Note that the reaction $Fo + Fl-H_2O = phA + En$ was termed the "water-line" [*Liu*, 1987] because it bounds the P-T region in the upper mantle where all the H_2O components in the system behave as a free fluid (Figure 2). Beyond the "water-line", water storage capacities of solid phase assemblages are significantly larger because DHMS are stabilized, so that up to several wt.% H_2O can be stored in the solid phases, and no free fluid exists. In addition, *Komabayashi et al.* [2005b] showed that a reaction $Fo + Fl-H_2O = cHu + En$ is another "water-line" (Figure 2).

In spite of their importance to water transportation, there are large discrepancies among previous experimental results on the antigorite dehydration reactions and the "water-line" [*Ulmer and Trommsdorff*, 1995; *Luth*, 1995; *Wunder and Schreyer*, 1997; *Wunder*, 1998; *Bose and Navrotsky*, 1998; *Bromiley and Pawley*, 2003]. The reported

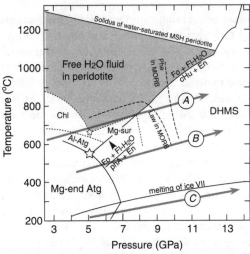

Figure 2. Phase relations for hydrous peridotite and MORB modified after *Komabayashi et al.* [2005a]. Stability of Mg-end member antigorite above melting temperature of ice VII and the "water-line" reaction $Fo + Fl-H_2O = phA + En$ are from *Komabayashi et al.* [2005a]. Stability of antigorite below melting temperature of ice VII is tentatively assumed. The melting curve of ice VII is taken from *Fei et al.* [1993]. Solidus curve for water-saturated MSH peridotite is from *Inoue* [1994]. The stability limits of lawsonite (Law) and phengite (Phe) in the hydrous MORB are shown by broken lines [*Schmidt and Poli*, 1998]. The stability of chlorite, Mg-sursassite, and Al-antigorite in the system MASH (3 wt.% Al_2O_3) which were modified after *Bromiley and Pawley* [2002], *Pawley* [2003], and *Bromiley and Pawley* [2003] are shown by dotted lines. Shaded region denotes where all the H_2O in the system is a free fluid. Relatively hot, cold, and very cold temperature profiles in subducting slabs are shown as profiles A, B, and C, respectively. Large and small stars indicate the critical conditions for the hydrous phase stability in the peridotites in the subduction zone in Mg-end member and Al-bearing systems, respectively. See text for details. Atg, antigorite; Fo, forsterite; En, enstatite; phA, phase A; DHMS, dense hydrous magnesium silicates; Chl, chlorite; Mg-sur, Mg-sursassite; $Fl-H_2O$, fluid water.

invariant point in these studies ranges from 4.5-6.2 GPa and 550-620°C. Possible sources for these discrepancies have been considered to be due to differences in composition of the starting natural antigorite and other experimental factors [*Mysen et al.*, 1998; *Wunder and Schreyer*, 1997; *Bromiley and Pawley*, 2003]. *Komabayashi et al.* [2005a] made an in-situ X-ray diffraction measurements of the "water-line" reaction $Fo + Fl-H_2O = phA + En$. In addition, they partly solved the discrepancies in the antigorite stability among previous studies by considering the pressure effect on the electromotive force of the thermocouple used in each experiment. The precisely determined P-T location of the invariant point is 5.1 GPa and 550°C in the Mg-end member system [*Komabayashi et al.*, 2005a] (Figure 2).

Addition of aluminum to the MSH system expands the stability of antigorite [*Bromiley and Pawley*, 2003]. In the system MgO-Al$_2$O$_3$-SiO$_2$-H$_2$O (MASH), aluminous hydrous phases, chlorite and Mg-sursassite are stabilized as post-antigorite hydrous phases [*Bromiley and Pawley*, 2002; *Bromiley and Pawley*, 2003; *Pawley*, 2003]. At pressures above the "water-line", Mg-sursassite transfers water to phase A by a solid-solid reaction Mg-sursassite + forsterite = phase A + pyrope + enstatite [*Bromiley and Pawley*, 2002]. Figure 2 summarizes these phase relations in both the Mg-end member and the 3 wt. % Al$_2$O$_3$-bearing systems. These phase relations indicate that the hydrous phases in the MASH system have higher thermal stability than in the MSH system. The critical conditions for water transportation in the subducting slabs will shift toward higher temperatures by the addition of aluminum to the MSH system. It is located at 5.1 GPa and 660°C in the system MASH.

Another important phase relation related to antigorite stability was proposed by *Bina and Navrotsky* [2000]. At very low temperature (below 300°C at 7 GPa), the dehydration reaction of antigorite intersects the melting curve of ice VII [e.g., *Fei et al.*, 1993], leading to formation of ice VII at the breakdown of antigorite by a reaction Atg = phA + En + ice VII (Figure 2). This reaction is not a fluid-forming reaction but a solid-solid reaction. Therefore if the slab temperature at the breakdown of antigorite is lower than the melting curve of ice VII, whole water in antigorite would be transferred into the deeper mantle by phase A and ice VII [*Bina and Navrotsky*, 2000]. The chapter by Lin et al. of this volume gives more information about the properties and melting curve of ice H$_2$O at high P-T.

3. PHASE RELATIONS OF HIGH-PRESSURE HYDROUS PHASES INCLUDING DHMS

At pressures higher than 10 GPa, hydrous phases encountered in the system MSH are DHMS, hydrous wadsleyite, and hydrous ringwoodite. Phase A is formed as the post-antigorite or post-Mg-sursassite hydrous phase in the slab depending on temperature as discussed above. Here I discuss the stability relations of high-P hydrous phases beyond antigorite and Mg-sursassite stability. They have been investigated by empirical high-pressure experimental studies only, due to their complicated phase relations [*Liu*, 1987; *Kanzaki*, 1991; *Ohtani et al.*, 1995; *Frost and Fei*, 1998; *Irifune et al.*, 1998; *Ohtani et al.*, 2000].

Komabayashi et al. [2004] constructed a net of chemical reactions between high-P hydrous phases with chemographic consistency by doing Schreinemakers analysis on the existing experimental data. According to *Komabayashi et al.* [2004; 2005b], seven high-pressure hydrous phases are encountered in peridotitic compositions (i.e., Mg/Si ~1.4) beyond antigorite stability to 30 GPa, 1600°C. Those are phase A, clinohumite, phase E, superhydrous phase B, phase D, hydrous wadsleyite, and hydrous ringwoodite.

Recently, an internally consistent thermodynamic model for these seven high-pressure hydrous phases was constructed up to 35 GPa and 1600°C [*Komabayashi*, 2005; *Komabayashi et al.*, 2005b; *Komabayashi and Omori*, in press]. Using this new thermodynamic data set, phase relations for these high-pressure hydrous phases are calculated.

Figure 3 shows the resulting phase relations for a peridotitic bulk composition (Mg/Si = 1.4) with a water content of 3.66 wt.%, corresponding to a mixture of phase A (Mg$_7$Si$_2$O$_8$(OH)$_6$) + enstatite (MgSiO$_3$) (phA : En = 7 : 71, in mole) as the post-antigorite phase assemblage. Reactions in Figure 3 are listed in Table 1. For the calculations of water-bearing reactions, the EOS of fluid-H$_2$O is necessary. We proposed that an EOS of fluid-H$_2$O by *Brodholt and Wood* [1993] might give a more accurate volume of water among the existing EOSs [*Komabayashi et al.*, 2005b], since this EOS can reproduce all the phase relations ever reported by the high-pressure experiments as shown below. As discussed later, the fluid phases at high P-T conditions are no longer pure H$_2$O. The calculated dehydration reactions in Figure 3 are with correspondingly lower water activities.

Phase D, phase E, hydrous wadsleyite, and hydrous ringwoodite are known to have compositional variations depending on pressure, temperature, and bulk composition [e.g., *Frost*, 1999]. However, *Komabayashi and Omori* [in press] argued that the compositional variations of these phases have a small effect on the Gibbs free energy of the phases because their thermodynamic data set with fixed compositions for these phases can reproduce almost all the experimentally determined phase relations in the wide P-T region ever reported [*Kanzaki*, 1991; *Gasparik*, 1993; *Ohtani et al.*, 1995; *Luth*, 1995; *Irifune et al.*, 1998; *Ohtani et al.*, 2001; *Higo et al.*, 2001; *Ohtani et al.*, 2003].

At pressures higher than the "water-line" reaction (3) (Figure 3), an assemblage of phase A + enstatite is stable down to the transition zone (~410 km). Above 12~17 GPa (the reactions 15, 17, and 23), a tie-line of shB + St is stable over a wide pressure range to 25.5~29 GPa (the reaction 38) below 1200°C. In the transition zone, the high-temperature stability of DHMS is about 1200°C, almost independent of pressure. While in the upper part of the transition zone (< 17 GPa), it is defined by dehydration reactions of phase E (19 and 22), it is defined by dehydrations of superhydrous phase B or phase D (24, 28, 31, and 34) in the lower part. Above 1200°C, hydrous wadsleyite or hydrous ringwoodite is the stable hydrous phase depending on pressure. In the lower mantle, the stability of DHMS is defined by reac-

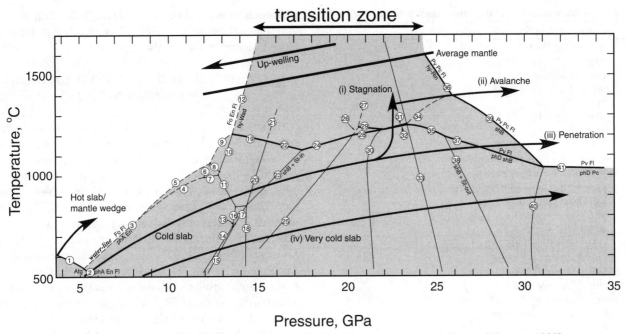

Figure 3. Phase relations for the subducting hydrous peridotite with an average mantle [*Ito and Katsura*, 1989], representative cold subducting, and hot up-welling P-T paths. The bulk composition after antigorite breakdown (beyond the reaction 2) along the cold slab path is shown in Figure 1. Three cold slab geotherms at the MBL correspond to, (i) the slab stagnation at the MBL, (ii) the avalanche of the stagnant slab into the lower mantle, and (iii) the slab penetration into the lower mantle. The shaded region illustrates where hydrous minerals are stable. The reactions are listed in Table 1. Thick lines are dehydration reactions, while thin solid lines are fluid-absent solid-solid reactions. Thin broken lines denote the water-bearing reaction which does not occur in the subduction process. Note that free fluids generated by the dehydration reactions are assumed to escape from the system such that no re-hydration reaction occurs in the slabs.

tions (39) and (41). The reaction (41) defines the high-pressure stability of DHMS (phase D) at the deep lower mantle condition. This reaction was experimentally investigated by *Shieh et al.* [1998], using a laser-heated diamond anvil cell. They showed that phase D + periclase assemblage is stable up to 44 GPa and 1200°C and that in the peridotitic composition no post-phase D hydrous phases occurred at the reaction (41). The existence of undiscovered further high-pressure hydrous phases should be investigated. However, if the breakdown of phase D occurs in the stability field of ice X [*Lin et al.*, 2005], it is not the dehydration reaction but a solid-solid reaction expressed as phD + Pc = Pv + ice X. See the chapter of this volume by Lin et al. for information about the melting of the high-pressure ice.

The low-pressure stability of hydrous wadsleyite above 1200°C is defined by a reaction (12). In the upper mantle above 410-km, there is no stable hydrous phase above 1200°C, therefore, this reaction is a dehydration reaction of hydrous wadsleyite. Note, however, that forsterite which is a nominally anhydrous mineral has been reported to accommodate water up to close to 0.8 wt% at the conditions of about 410-km depth [*Chen et al.*, 2002; *Smyth et al.*,

2005; Mosenfelder et al., 2006]. In addition, the maximum water content of hydrous wadsleyite at high temperatures (~1400°C) is ca. 1 wt. % [*Demouchy et al.*, 2005]. Therefore, the dehydration of hydrous wadsleyite might be concealed if the water storage capacity of olivine is similar to wadsleyite [*Hirschmann et al.*, 2005]. See the chapter of this volume by Hirschmann et al. for details. However, in our thermodynamic model, the nominally anhydrous phases were treated as completely anhydrous, as a first-order approximation. Further thermodynamic models should include the effect of the solid-solutions between hydrous forsterite and hydrous wadsleyite. On the other hand, as discussed later, the dehydration of hydrous wadsleyite is suggested from seismological observations [*Revenaugh and Sipkin*, 1994; *Song et al.*, 2004; *Chambers et al.*, 2005].

In Figure 3, the stability of hydrous ringwoodite is defined by a reaction (36). Similar to hydrous wadsleyite, this reaction is a dehydration reaction. In the system MSH, Mg-perovskite and periclase do not accommodate significant amounts of water [*Bolfan-Casanova et al.*, 2000]. However, in multi-component systems, both nominally anhydrous phases were found to accommodate water up to

Table 1. Chemical reactions in the petrogenetic grid

No.	Chemical Reaction[a]
1	Atg = Fo + En + Fl
2	Atg = phA + En + Fl
3	phA + En = Fo + Fl
4	phA + En = cHu + Fl
5	cHu + En = Fo + Fl
6	phE = cHu + En + Fl
7	phA + En = cHu + phE
8	cHu + En = phE + Fo
9	phE = Fo + En + Fl
10	hy-Wad = Fo + En + phE
11	phA + En = hy-Wad + phE
12	hy-Wad = Fo + En + Fl
13	phD + shB = phE + En
14	shB + phD = phA + En
15	shB + St = En + phD
16[b]	phA + En = shB + phE
17	shB + St = phE + En
18[b]	shB + En = hy-Wad + St
19	phE = hy-Wad + En + Fl
20	hy-Wad + St = phE + En
21	hy-Wad + St = En + Fl
22	phE = hy-Wad + St + Fl
23	shB + St = phE + hy-Wad
24	shB + St = hy-Wad + Fl
25	hy-Rin + shB + St = hy-Wad
26	hy-Rin + St + Fl = hy-Wad
27[b]	hy-Wad + St = Aki + Fl
28	shB + St = hy-Rin + Fl
29	hy-Rin + St = Aki + Fl
30[b]	hy-Rin + St = Aki + shB
31	shB + Aki = hy-Rin + Fl
32	shB + St = Aki + Fl
33	Aki = Pv
34	shB + Pv = hy-Rin + Fl
35	shB + St = Pv + Fl
36	hy-Rin = Pv + Pc + Fl
37	shB + phD = Pv + Fl
38	shB + St = phD + Pv
39	shB = Pv + Pc + Fl
40[b]	shB = Pv + Pc + phD
41	phD + Pc = Pv + Fl
42[b]	phA + En + Fl = phE
43	phA + phD + Fl = phE
44	shB + Fl = phA + phD
45	shB + phD + Fl = phE
46	shB + St + Fl = phE
47	phD = shB + St + Fl
48	phD + Br = phA + Fl
49	phD + Br = shB + Fl

[a] Left term is low-temperature side of equilibrium. Atg, antigorite; Fo, forsterite; En, enstatite; phA, phase A; phE, phase E; cHu, clinohumite; phD, phase D; shB, superhydrous phase B; hy-Wad, hydrous wadsleyite; hy-Rin, hydrous ringwoodite; Br, brucite; St, stishovite; Aki, akimotoite; Pv, Mg-perovskite; Pc, periclase; Fl, fluid.

[b] Left term is low-pressure side of equilibrium.

thousands of p.p.m. [*Murakami et al.*, 2002; *Litasov et al.*, 2003]. See the chapter of this volume by Bolfan-Casanova and McCammon for details.

4. WATER ACTIVITY IN THE FLUID PHASE AT MANTLE CONDITIONS

High-pressure experiments show that free fluid phases coexisting with mantle minerals at high P-T conditions dissolve significant amounts of silicate component, as evidenced by the presences of quench crystals [*Irifune et al.*, 1998; *Ohtani et al.*, 2000; *Stalder and Ulmer*, 2001; *Schmidt and Ulmer*, 2004; *Komabayashi et al.*, 2005b]. Therefore, in the calculations of water-bearing reactions, we must consider such an effect on the Gibbs energy of the fluid phase. *Komabayashi et al.* [2005b] showed, dehydration reactions calculated with pure H_2O do not match the fluid-present experimental results at high P-T conditions, above 10 GPa and 1000°C. As the fluid phase in the experiments at these P-T conditions contain significant amounts of silica and magnesia, the water activity in the fluid phase must be lower. Water activity in the fluid phase is defined as

$$a(H_2O) = \gamma X(H_2O)$$

where $a(H_2O)$ is a water activity in the fluid and $X(H_2O)$ is a mole fraction of water in the fluid, and γ is an activity coefficient. Figure 4 shows a schematic relationship between water activity and dehydration reaction. At lower P-T conditions, the dehydration reaction calculated with pure H_2O ($a(H_2O) = 1$) matches the experimental results. With increasing pressure and temperature, the calculated boundary with pure H_2O deviates from the experiments (Figure 4). The deviation of experiments from the calculations with pure H_2O is the effect of decrease in $a(H_2O)$ [*Komabayashi et al.*, 2005b]. Thus, comparing the P-T conditions of fluid-excess experiments with those of the calculated dehydration reactions, the water activity in the fluid phase can be estimated. With this procedure, namely by calculating dehydration reactions of the high-pressure hydrous phases to be consistent with the fluid-present experiments, the water activity is mapped across P-T space in Figure 5 [*Komabayashi et al.*, 2005b; *Komabayashi and Omori*, in press]. In Figure 5, chemical reactions other than in Figure 3 were calculated (42~49) in order to estimate the water activity over a wider P-T range. The reaction (3) determined by in-situ x-ray diffraction measurements [*Komabayashi et al.*, 2005a] is also shown in Figure 5. The EOS of fluid H_2O by *Brodholt and Wood* [1993] with unity $a(H_2O)$ reproduces the in-situ experiments [*Komabayashi et al.*, 2005b] (Figure 5a).

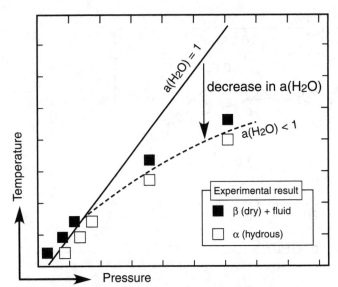

Figure 4. Schematic relationship between the water activity in the fluid phase and the dehydration reaction. Solid and broken lines denote the dehydration reactions calculated with $a(H_2O) = 1$ and < 1 assumed, respectively. The dehydration reaction calculated with pure H_2O, i.e., $a(H_2O) = 1$, does not match the fluid-present experiments at higher P-T conditions. Dissolution of silicate components into the fluid phase at higher P-T conditions decreases the water activity in the fluid phase. See text for details.

In Figure 5a, the water activity was unity up to 25 GPa below 700°C. At higher T conditions, the water activities were estimated as follows, at 13.5 GPa and 1200°C, $a(H_2O) = 0.60$; at 17 GPa and 1100°C, $a(H_2O) = 0.54$; at 20 GPa and 1200°C, $a(H_2O) = 0.13$; at 25 GPa and 1200°C, $a(H_2O) = 0.1$; at 24 GPa and 1600°C, $a(H_2O) = 0.012$ [*Komabayashi et al.*, 2005b; *Komabayashi and Omori*, in press]. At higher pressures, the water activity may further decrease. Thus, dehydration reactions in Figures 3 and 5a are with appropriate lower water activities, consistent with the fluid-excess experiments. In this procedure, an important point is that the water activity is mapped in the P-T space, independently from the experimental results on the composition of the fluid phase [e.g., *Stalder et al.*, 2001; *Mibe et al.*, 2002].

Komabayashi et al. [2005b] showed that *Brodholt and Wood* [1993]'s EOS of fluid H_2O reproduces the in-situ experiments on the reaction (3). However, I later noted that another EOS by *Pitzer and Sterner* [1994] also reproduces the in-situ experiments (Figure 5b). Here I made another water activity map with the EOS by *Pitzer and Sterner* [1994] in order to investigate the effect of the difference in the EOS of fluid H_2O on the water activity estimation (Figure 5b). Figure 5 shows the comparison of two sets of water activities with the EOS of fluid H_2O by

(a) *Brodholt and Wood* [1993] and (b) *Pitzer and Sterner* [1994]. The topologies of the fluid-excess phase relations calculated with two different EOSs are consistent. The water activity estimated with the EOS by *Pitzer and Sterner* [1994] are higher than by *Brodholt and Wood* [1993], because the former gives lower water fugacity at high P-T conditions. A drastic decrease in water activity with *Pitzer and Sterner* [1994]'s EOS occurs between 17 and 20 GPa at 1150°C from 0.81 to 0.21. More importantly, *Pitzer and Sterner* [1994]'s equation does not reproduce the reaction (49) phD + Br = shB + Fl consistently with the experiments by *Irifune et al.* [1998]. It should be located above 800°C at 26 GPa [*Irifune et al.*, 1998; *Komabayashi et al.*, 2004]. However, the reaction calculated with the equation by *Pitzer and Sterner* [1994] with unity $a(H_2O)$ is located below 300°C at 20 GPa, much lower temperature. In contrast, the EOS by *Brodholt and Wood* [1993] reproduces this reaction within the experimental uncertainty [*Komabayashi and Omori*, in press]. In this chapter, I adopted the EOS of H_2O by *Brodholt and Wood* [1993] because it reproduces all the fluid-saturated phase relations based on the high-pressure experiments (Figure 5a).

The water activities based on both equations show the same trend with increasing pressure and temperature. In addition, both equations give very low water activity at higher P-T conditions. For example, at 25 GPa and 1400°C, it is 0.05 and 0.1 by *Brodholt and Wood* [1993] and *Pitzer and Sterner* [1994], respectively. The data sources for these two EOSs are different. The EOS by *Brodholt and Wood* [1993] is based on their MD calculation, while that by *Pitzer and Sterner* [1994] is based on the existing experiments which measured the density of H_2O at both low pressure and extremely high pressure. Therefore, water activities based on the two EOSs with different data sources are similar at high P-T conditions, implying that the pure water fugacity (i.e., water activity) at such P-T condition may be reliable.

Figure 5a illustrates that the water activity decreases with increasing both pressure and temperature, except below 700°C. Below 700°C, unity $a(H_2O)$ is maintained up to 25 GPa. The composition of the free fluid may be very close to pure H_2O at such low T conditions. Above 700°C, with increasing pressure or temperature, the water activity systematically decreases. This strongly suggests that the fluid composition becomes silica- and magnesia-rich, namely magmas, as pressure or temperature increases. Activity-composition relationship for such a fluid phase should be investigated, although *Komabayashi and Omori* [in press] argue that the fluid phase at 25 GPa and 1200°C is not an ideal solution.

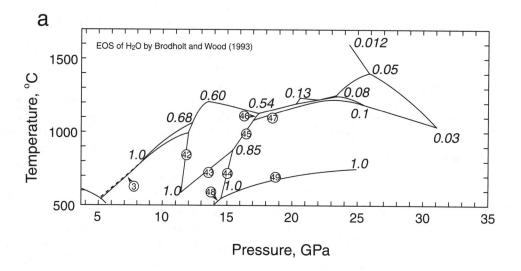

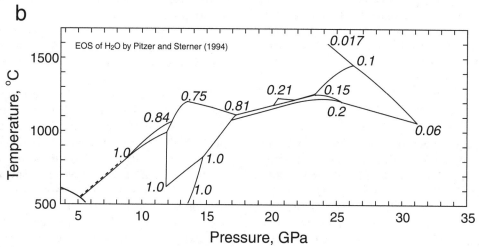

Figure 5. Maps of the water activity in the P-T space. Selected water-bearing reactions with water activities are shown. The EOSs of fluid H_2O used for the calculations are (a) by *Brodholt and Wood* [1993] and (b) by *Pitzer and Sterner* [1994]. The map of (a) is after *Komabayashi and Omori* [in press]. Note that the EOS by *Pitzer and Sterner* [1994] locates the reaction (49) below 300°C at 20 GPa, which is too low from the high-pressure experiments by *Irifune et al.* [1998]. See text for details. The reaction (3) which was determined by in-situ X-ray diffraction measurements [*Komabayashi et al.*, 2005a] is shown as broken line.

5. WATER CIRCULATION THROUGH THE MANTLE

5-1. Phase Relations in a Multi-Component System

Natural peridotite is a multi-component system including iron, aluminum, and calcium. As mentioned above, *Bromiley and Pawley* [2003] reported that aluminum has a significant effect on the stability of antigorite and it will also stabilize other MASH hydrous phases as post-antigorite phases (Figure 2). Here, the effect of other components on the phase relations for DHMS is evaluated. *Komabayashi and*

Omori [in press] investigated this effect by comparing fluid-saturated phase relations in the system MSH (i.e., Figure 5a) with those in a multi-component peridotite (KLB-1) + H_2O system [*Kawamoto et al.*, 1995; *Kawamoto*, 2004]. In the KLB-1 + H_2O system, chondrodite was found together with clinohumite or phase E below 11 GPa, 1000°C. In the system MSH, chondrodite was reported only in MgO-rich bulk compositions (Mg/Si > 2) [*Wunder*, 1998]. This indicates that chondrodite is stabilized in the multi-component system due to its preferential accommodation of titanium [*Kawamoto et al.*, 1995]. At higher P-T conditions, the P-T stability of

DHMS in the KLB-1 + H_2O system is identical to those in the system MSH [*Komabayashi and Omori*, in press]. This suggests that the simultaneous additions of iron, aluminum, and calcium to the MSH system do not have a remarkable effect on P-T locations of the water-bearing reactions above 14 GPa although other dry phases such as garnet are encountered in the multi-component system. To conclude, the P-T conditions of stability of DHMS in the natural peridotite can be approximated by those in the system MSH (Figure 3).

5-2. Hydration of Peridotites in Subduction Zones

The mantle wedge peridotite is believed to be hydrated by water released from the subducting hydrated mid-oceanic ridge basalt (MORB) layer [e.g., *Poli and Schmidt*, 1995; *Schmidt and Poli*, 1998; *Okamoto and Maruyama*, 1999]. This has been discussed in relations to island arc magmatism [e.g., *Schmidt and Poli*, 1998] and hydrous low-T plume model [*Gerya and Yuen*, 2003]. In contrast, whether or not the sub-crustal slab peridotite is hydrated upon entering the subduction zone is still a matter of debate [*Kerrick*, 2002]. However, previous studies implicitly included hydrated slab peridotite in their water transport models [*Shieh et al.*, 1998; *Irifune et al.*, 1998; *Frost and Fei*, 1998; *Frost*, 1999; *Ohtani et al.*, 2000; *Ohtani et al.*, 2001; *Angel et al.*, 2001]. Substantially hydrated slab peridotite down to 30-50-km depth from the plate surface was suggested from the consequence of the origin of the double seismic zone observed in the slab, on the basis of the dehydration embrittlement hypothesis [*Seno and Yamanaka*, 1996; *Peacock*, 2001; *Omori et al.*, 2002]. *Omori et al.* [2002] presented a tomographic image of hydrated Pacific plate beneath Tokyo, using the Poisson's ratio as an indicator of serpentinization. *Ranero et al.* [2003] have discussed the role of bending-related faulting along a trench on the water penetration into the subducting plate to substantial depth. Although the mechanism of hydration is still enigmatic, in the following sections, it is assumed that the slab peridotite is hydrated to substantial depths (30-50-km) and that hydration is heterogeneous within the slab. In addition, we assume that the free fluid released by the dehydration reactions escapes upward due to its low density, such that no hydration reaction occurs in the subducting slab.

5-3. Water Circulation in the Mantle

Water circulation in the Earth's mantle is discussed on the basis of the stability relations of hydrous phases. Given the phase relations of hydrous peridotite, the water transport path and the amount of water carried into the deep mantle by the subducting slab depend on the pressure-temperature trajectory of subduction. For water subduction, the first critical P-T condition is the antigorite choke point at 5.1 GPa and 550°C or 660°C in the systems MSH or MASH, respectively, which determines if the water can be transferred into the transition zone or not (Figure 2).

The mantle wedge and hot slab serpentinites could not clear the antigorite choke point [e.g., *Peacock and Wang*, 1999; *Iwamori*, 2000]. Antigorite in the hot P-T path will dehydrate to form a dry solid assemblage of Fo + En + H_2O (profile A in Figure 2). However, if the water from the subducting MORB layer comes up to the mantle wedge at pressure higher than the "water-line", phase A is likely to be formed in the mantle wedge peridotite by the hydration reaction of forsterite (profile A in Figure 2). Then, the water would be transported into deep mantle by the down-dragged thin layer (ca. < 10 km) of the mantle wedge peridotite. Figure 2 shows that lawsonite and phengite in subducting MORB dehydrate at pressures higher than the "water-line", if the temperature is below 800°C [*Schmidt and Poli*, 1998; *Komabayashi et al.*, 2005b].

In contrast, the coldest portion of the cold slab has been suggested to clear the critical condition of the antigorite choke point (profile B in Figure 2) [*Bina and Navrotsky*, 2000; *Iwamori*, 2000; *Peacock*, 2001; *Omori et al.*, 2002; *Komabayashi et al.*, 2005a]. Water in antigorite will be partially transferred to phase A by a reaction (2) Atg = phA + En + Fl. Moreover, in the case of much colder slab (profile C in Figure 2), ice VII is formed at the breakdown of antigorite. The numerical simulations by *Bina and Navrotsky* [2000] showed that the temperature of the slab beneath Tonga subduction zone may be about 200°C at 5 GPa, which is cold enough for the formation of ice VII. Then, whole water in antigorite will be carried into the deeper mantle by phase A and ice VII as suggested by *Bina and Navrotsky* [2000].

Here, I discuss water transportation by the cold slab peridotite after antigorite breakdown. An average mantle geotherm [*Ito and Katsura*, 1989] along assumed P-T paths of a cold subducting slab and an up-welling mantle is shown in Figure 3. For simplicity, the melting relation of ice VII is not included in Figure 3 but it will be discussed later. Figure 3 shows that only solid-solid reactions including DHMS occur along the P-T path of the cold slab down to the mantle boundary layer (MBL: 660 ± 40-km depth) after antigorite breakdown. With increasing pressure, they are given by reactions (11), (20), (23), (25), and (30). The solid phase assemblages change from antigorite decomposition to phA + En, phE + hy-Wad + En, phE + hy-Wad + St, shB + hy-Wad + St, shB + hy-Rin + St, and finally shB + hy-Rin + Aki at about 21 GPa, 1100°C. The water content of 3.66 wt.% in the slab peridotite does not change during subduction when passing through the upper mantle and the transition zone.

Three possible P-T paths of the cold subducting slab at the MBL are considered in Figure 3: P-T path (i), stagnation; P-T path (ii), avalanche; P-T path (iii), penetration. In the P-T path (i), a series of dehydrations occur at the MBL: the reactions (32) shB + St = Aki + Fl and (31) shB + Aki = hy-Rin + Fl. Even though the temperature of slab reaches the average mantle temperature, hydrous ringwoodite is still the stable hydrous phase. In the P-T path (ii), another dehydration reaction occurs following the reactions (32) and (31). It is the reaction (36) hy-Rin = Pv + Pc + Fl or (39) shB = Pv + Pc + Fl depending on temperature. This results in the complete dehydration of the slab at the top of the lower mantle. In the P-T path (iii), the dehydration reactions (37) shB + phD = Pv + Fl and (39) occur. This P-T path crosses the reaction (39) at around 800-km depth. Therefore, the free fluid would be released at the uppermost of the lower mantle.

In addition, consider a very cold slab P-T path (P-T path (iv)) that gives 800°C at 660-km depth (Figure 3). In this P-T path, the very cold subduction results in no dehydration when entering the lower mantle and thus carries water into the deep lower mantle. The first dehydration after antigorite decomposition is given by the reaction (41) phD + Pc = Pv + Fl at around 50 GPa, corresponding to 1250-km [Shieh et al., 1998]. However, at lower temperatures, this reaction is possibly the solid-solid reaction including ice X expressed as phD + Pc = Pv + ice X as discussed above. In this case, whole water in the slab will be transported down to the core-mantle boundary without fluid-forming reaction during subduction after antigorite breakdown.

The above case studies of the P-T paths show that possibility for transporting water into the lower mantle strongly depends on the dehydration reactions (39) and (41). The P-T paths (iii) and (iv) can carry water into the lower mantle. In the P-T paths (i), (ii), and (iii), two large fluxes of free fluids from the slab peridotite occur during subduction down to the MBL. These fluxes occur at both the initiation of subduction (antigorite breakdown) and the MBL (DHMS breakdowns). The free fluid originating from dehydration of antigorite will circulate to the surface, sometimes with island arc magmatism [Ulmer and Trommsdorff, 1995; Wunder and Schreyer, 1997], whereas that from DHMS will be bound in wadsleyite or ringwoodite in the mantle transition zone.

On the other hand, in an up-welling mantle, if wadsleyite or ringwoodite in the transition zone contains water, free fluids will be generated by the dehydration reaction of (12) hy-Wad = Fo + En + Fl around 410-km depth (Figure 3). This fluid will facilitate the up-welling movement through the upper mantle because water decreases the viscosity of rocks [e.g., Karato et al., 1986]. When the plume reaches at the bottom of the crust, partial melts may segregate to volcanoes and the water is released to the Earth's surface. Komatiitic magmatism has been suggested to be a consequence of the dehydration reaction of hydrous wadsleyite [Kawamoto et al., 1996; Shimizu et al., 2001]. Recently, seismological observations showed that

low velocity zones were observed in the upper mantle or just above 410-km depth in subduction zones [Nolet and Zielhuis, 1994; Revenaugh and Sipkin, 1994]. These authors ascribed these observations to the presence of the free fluid which was once subducted from the surface. See the chapters of this volume by Song and Helmberger; Zhao. The phase diagram in Figure 3 can be useful for such a discussion. Consider the scenario where water is carried to the transition zone by the cold slab. In the transition zone, the dehydration of phase E (19 or 22) or of superhydrous phase B (24, 28, or 31) should occur first in the slab by heating from the surrounding mantle and finally the dehydration of hy-Wad (12) occur at 410-km depth to produce the free fluid. This water cycle is a kind of island arc magmatism, but induced by much deeper dehydrations. Another scenario of the fate of the water in the up-welling mantle was proposed by Bercovici and Karato [2003]. They suggested that the fluid phase generated by the dehydration of hydrous wadsleyite might be denser than the surrounding mantle, therefore, it was possible that the fluid was gravitationally stable at 410-km depth. See the chapter of this volume by Karato and Bercovici. The fluid phase generated in the deep mantle dissolves significant amounts of silicate component as inferred from the lower water activities (Figure 5). Therefore, such a fluid phase should have different physical properties from that at shallower levels.

6. ORIGIN OF THE DEEP FOCUS SEISMICITY

The origin of subduction zone seismicity is one of the central issues of Earth's dynamics. As mentioned above, the double seismic zone may be related to the antigorite dehydration. The dehydration of serpentine causes acoustic emission, which may induce the seismicity [Meade and Jeanloz, 1991; Dobson et al., 2002].

In addition dehydration reactions of any other hydrous phase such as chlorite may also be related to seismicity in the slab [Omori et al., 2004; Komabayashi et al., 2005a]. Therefore, if phase A is formed in the slab as the post-antigorite phase, the origin of the deeper seismicities should first be considered in relation to dehydration of high-P hydrous phases. The relationship between P-T locations of dehydration reactions and those of subduction zone seismicities was investigated with previously constructed phase diagrams [Omori et al., 2004; Komabayashi et al., 2004]. However, these phase diagrams were predicted from the limited experimental results by Irifune et al. [1998] and Ohtani et al. [2001]. Therefore, they were semi-quantitative and a possibility that some of the predicted reactions were metastable still remains so far. Here I have newly constructed phase diagram by the thermodynamic calculation (Figure 6). All the reactions appearing in Figure 6 are quantitative and

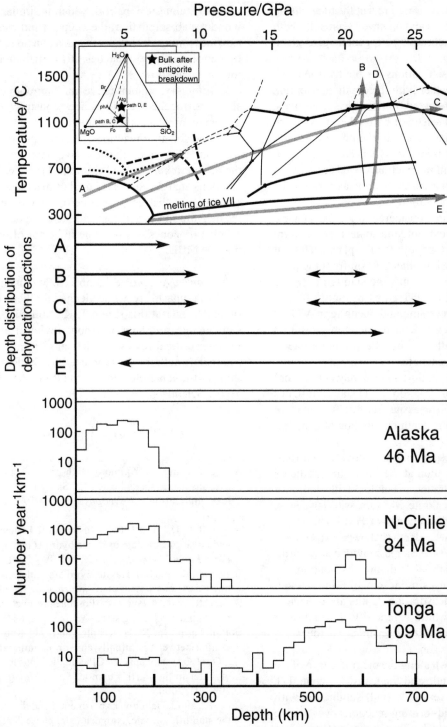

Figure 6. Phase relations for the peridotitic bulk compositions represented in the ternary pictured inset at upper left. Middle panel shows predicted distribution of the dehydration reactions in the slab peridotite along each P-T path labeled A through E. Lower panel shows the frequency mode of subduction zone seismicity beneath Alaska, northern Chile, and Tonga after *Omori et al.* [2004]. Thick solid and thin lines in the phase diagram are the same as in Figure 3. Thick dotted and broken lines are dehydration reactions in the MASH (peridotite) and MORB systems, respectively, after Figure 2. Each P-T path denotes the lowest T portion in the slab. See text for details.

thermodynamically consistent. Here I revisit the relationship between dehydration reactions and observed seismicity with the newly constructed quantitative petrogenetic grid given in Figure 6.

Figure 6 shows the phase relations for the hydrous slab peridotite and MORB, together with the depth distribution of the dehydration events predicted from the phase diagrams, and the depth frequency of seismicities in Alaska, northern Chile, and Tonga subduction zones. In the phase diagram, the melting curve of ice VII is included [*Fei et al.*, 1993]. The melting of ice is a fluid-forming reaction as well as dehydration reactions of hydrous phases, so that it is assumed to be another possible source of seismicity.

Five possible P-T paths are considered in Figure 6 (A-E). Note the paths in the figure denote the coldest temperature conditions in the slab, corresponding to the plate surface at the trench and to the 30-50-km inner portion from the plate surface at the transition zone. Thus the slab temperature should have ranges along the P-T paths in Figure 6. The dehydration reactions of lawsonite and phengite in MORB does not occur along path A where the temperature of MORB (~300°C higher than the path) is above the dehydration temperatures of lawsonite and phengite.

Along P-T path A, corresponding to relatively hot (young) subduction, the dehydration reaction in the slab terminates with the dehydration of Mg-sursassite. Seismic frequency in Alaska is consistent with this depth distribution of dehydration.

Along P-T path B, corresponding to relatively cold (old) subduction with slab stagnation at the MBL, the predicted depth distribution of dehydration is bimodal with shallower dehydrations by antigorite, chlorite, Mg-sursassite, lawsonite, and phengite, and deeper dehydrations by DHMS. The deeper dehydrations would terminate at depth shallower than 660-km because of slab stagnation. The seismic activity below northern Chile is consistent with this dehydration distribution.

Along P-T path C, which is similar to path B but without slab stagnation, bimodal dehydration depths are expected. The deeper dehydration would extend 700-km depth or deeper. No reported subduction zone seismicity in the world matches this distribution.

P-T paths D and E go down along the melting curve of ice VII through the upper mantle and the transition zone. P-T path D, corresponding to very cold (old) subduction with slab stagnation, shows fluid-forming reactions over a wide pressure range, from shallow depth to around 660-km. The Tonga subduction zone seismicity is consistent with this fluid-forming reactions distribution. On the basis of the P-wave velocity structure, *Zhao et al.* [1997] reported that the subducted slab beneath Tonga subduction zone is stagnant at 660-km depth.

Finally, along P-T path E, which is similar to path D but without slab stagnation, the deeper fluid-forming reaction extends to much deeper depth, but again no reported subduction zone seismicity matches this distribution of the fluid-forming reactions.

In summary, a quantitative and thermodynamically consistent petrogenetic grid has been constructed for the MSH system to lower mantle conditions involving both solid-solid and fluid-forming reactions. Various P-T paths along possible subduction geotherms have been evaluated against the new petrogenetic grid shown in Figure 6. If dehydration reactions are related to earthquakes in subduction zones, the results show that deep seismicities could be explained by the dehydration hypothesis and therefore deep seismicity in subduction zones is a possible indicator of water transport into the MBL.

Acknowledgements. S. Omori, S. Maruyama, K. Hirose, and E. Takahashi are acknowledged for their fruitful discussions throughout this study. T. Kawamoto, S. Jacobsen, S. van der Lee, and an anonymous reviewer are appreciated for their critical reviews which significantly improved the paper. The author was supported by the Research Fellowships of the Japan Society for the Promotion of Science for Young Scientists.

REFERENCES

Akaogi, M., and S.-i. Akimoto, High-pressure stability of a dense hydrous magnesian silicate $Mg_{23}Si_8O_{42}H_6$ and some geophysical implications, *J. Geophys. Res.*, *85* (B12), 6944–6948, 1980.

Angel, R.J., D.J. Frost, N.L. Ross, and R.J. Hemley, Stabilities and equations of state of dense hydrous magnesium silicates, *Phys. Earth Planet. Inter.*, *127*, 181–196, 2001.

Bercovici, D., and S.-i. Karato, Whole-mantle convection and the transition-zone water filter, *Nature*, *425*, 39–44, 2003.

Bina, C.R., and A. Navrotsky, Possible presence of high-pressure ice in cold subducting slabs, *Nature*, *408*, 844–847, 2000.

Bolfan-Casanova, N., H. Keppler, and D.C. Rubie, Water partitioning between nominally anhydrous minerals in the MgO-SiO_2-H_2O system up to 24 GPa: implications for the distribution of water in the Earth's mantle, *Earth Planet. Sci. Lett.*, *182*, 209–221, 2000.

Bose, K., and J. Ganguly, Experimental and theoretical studies of the stabilities of talc, antigorite and phase A at high pressures with applications to subduction processes, *Earth Planet. Sci. Lett.*, *136*, 109–121, 1995.

Bose, K., and A. Navrotsky, Thermochemistry and phase equilibria of hydrous phases in the system MgO-SiO_2-H_2O: Implications for volatile transport to the mantle, *J. Geophys. Res.*, *103* (B5), 9713–9719, 1998.

Brodholt, J., and B. Wood, Simulations of the structure and thermodynamic properties of water at high pressures and temperatures, *J. Geophys. Res.*, *98* (B1), 519–536, 1993.

Bromiley, G.D., and A.R. Pawley, The high-pressure stability of Mg-sursassite in a model hydrous peridotite: a possible mechanism for the deep subduction of significant volumes of H_2O, *Contrib. Mineral. Petrol.*, *142*, 714–723, 2002.

Bromiley, G.D., and A.R. Pawley, The stability of antigorite in the systems MgO-SiO_2-H_2O (MSH) and MgO-Al_2O_3-SiO_2-H_2O (MASH): The effects of Al^{3+} substitution on high-pressure stability, *Am. Mineral.*, *88*, 99–108, 2003.

Chambers, K., A. Deuss, and J.H. Woodhouse, Reflectivity of the 410-km discontinuity from PP and SS precursors, *J. Geophys. Res.*, *110*, B02301, doi:10.1029/2004JB003345, 2005.

Chen, J., T. Inoue, H. Yurimoto, and D.J. Weidner, Effect of water on olivine-wadsleyite phase boundary in the $(Mg, Fe)_2SiO_4$ system, *Geophys. Res. Lett.*, *29* (18), 1875, doi:10.1029/2001GL014429, 2002.

Demouchy, S., E. Deloule, D.J. Frost, and H. Keppler, Pressure and temperature-dependence of water solubility in Fe-free wadsleyite, *Am. Mineral.*, *90*, 1084–1091, 2005.

Dobson, D.P., P.G. Meredith, and S.A. Boon, Simulation of subduction zone seismicity by dehydration of serpentine, *Science*, 298, 1407–1410, 2002.

Evans, B.W., W. Johannes, H. Oterdoom, and V. Trommsdorff, Stability of chrysotile and antigorite in the serpentinite multisystem, *Schweiz. Mineral. Petrogr. Mitt.*, *56*, 79–93, 1976.

Fei, Y., H.-k. Mao, and R.J. Hemley, Thermal expansivity, bulk modulus, and melting curve of H_2O-ice VII to 20 GPa, *J. Chem. Phys.*, *99* (7), 5369–5373, 1993.

Frost, D.J., The stability of dense hydrous magnesium silicates in Earth's transition zone and lower mantle, in *Mantle Petrology: Field Observations and High Pressure Experimentation: A Tribute to Francis R. (Joe) Boyd*, edited by Y. Fei, C.M. Bertka, and B.O. Mysen, pp. 283–296, The Geochemical Society, 1999.

Frost, D.J., and Y. Fei, Stability of phase D at high pressure and high temperature, *J. Geophys. Res.*, *103* (B4), 7463–7474, 1998.

Gasparik, T., The role of volatiles in the transition zone, *J. Geophys. Res.*, *98* (B3), 4287–4299, 1993.

Gerya, T.V., and D.A. Yuen, Rayleigh-Taylor instabilities from hydration and melting propel 'cold plumes' at subduction zones, *Earth Planet. Sci. Lett.*, *212*, 47–62, 2003.

Higo, Y., T. Inoue, and T. Irifune, Effect of water on the spinel-postspinel transformation in Mg_2SiO_4, *Geophys. Res. Lett.*, *28* (18), 3505–3508, 2001.

Hirschmann, M.M., C. Aubaud, and A.C. Withers, Storage capacity of H_2O in nominally anhydrous minerals in the upper mantle, *Earth Planet. Sci. Lett.*, *236*, 167–181, 2005.

Inoue, T., Effect of water on melting phase relations and melt composition in the system Mg_2SiO_4-$MgSiO_3$-H_2O up to 15 GPa, *Phys. Earth Planet. Inter.*, *85*, 237–263, 1994.

Irifune, T., N. Kubo, M. Isshiki, and Y. Yamasaki, Phase transformations in serpentine and transportation of water into the lower mantle, *Geophys. Res. Lett.*, *25* (2), 203–206, 1998.

Ito, E., and T. Katsura, A temperature profile of the mantle transition zone, *Geophys. Res. Lett.*, *16* (5), 425–428, 1989.

Iwamori, H., Deep subduction of H_2O and deflection of volcanic chain towards backarc near triple junction due to lower temperature, *Earth Planet. Sci. Lett.*, *181*, 41–46, 2000.

Kanzaki, M., Stability of hydrous magnesium silicates in the mantle transition zone, *Phys. Earth Planet. Inter.*, *66*, 307–312, 1991.

Karato, S., M.S. Paterson, and J.D. Fitz Gerald, Rheology of synthetic olivine aggregates-influence of grain-size and water, *J. Geophys. Res.*, *91*, 8151–8176, 1986.

Kawamoto, T., Hydrous phase stability and partial melt chemistry in H_2O-saturated KLB-1 peridotite up to the uppermost lower mantle conditions, *Phys. Earth Planet. Inter.*, *143–144*, 387–395, 2004.

Kawamoto, T., R.L. Hervig, and J.R. Holloway, Experimental evidence for a hydrous transition zone in the early Earth's mantle, *Earth Planet. Sci. Lett.*, *142*, 587–592, 1996.

Kawamoto, T., K. Leinenweber, R.L. Hervig, and J.R. Holloway, Stability of hydrous minerals in H_2O-saturated KLB-1 peridotite up to 15 GPa, in *Volatiles in the Earth and Solar system*, edited by K.A. Farley, pp. 229–239, American Institute of Physics, New York, 1995.

Kerrick, D., Serpentinite seduction, *Science*, 298, 1344–1345, 2002.

Komabayashi, T., Phase relations in the system MgO-SiO_2-H_2O up to 100 GPa: implications for water circulation in the Earth's mantle, Ph.D. thesis, Tokyo Institute of Technology, 2005.

Komabayashi, T., K. Hirose, K.-i. Funakoshi, and N. Takafuji, Stability of phase A in antigorite (serpentine) composition determined by in-situ X-ray pressure observations, *Phys. Earth Planet. Inter.*, *151*, 276–289, 2005a.

Komabayashi, T., and S. Omori, Internally consistent thermodynamic data set for dense hydrous magnesium silicates up to 35 GPa, 1600°C: implications for water circulation in the Earth's deep mantle, *Phys. Earth Planet. Inter.*, *156*, 89–107, 2006.

Komabayashi, T., S. Omori, and S. Maruyama, Petrogenetic grid in the system MgO-SiO_2-H_2O up to 30 GPa, 1600°C: Applications to hydrous peridotite subducting into the Earth's deep interior, *J. Geophys. Res.*, *109*, B03206, doi:10.1029/2003JB002651, 2004.

Komabayashi, T., S. Omori, and S. Maruyama, Experimental and theoretical study of stability of dense hydrous magnesium silicates in the deep upper mantle, *Phys. Earth Planet. Inter.*, *153*, 191–209, 2005b.

Lin, J.-F., E. Gregoryanz, V.V. Struzhkin, M. Somayazulu, H.-k. Mao, and R.J. Hemley, Melting behavior of H_2O at high pressures and temperatures, *Geophys. Res. Lett.*, *32* L11306, doi:10.1029/2005GL022499, 2005.

Litasov, K., E. Ohtani, F. Langenhorst, H. Yurimoto, T. Kubo, and T. Kondo, Water solubility in Mg-perovskites and water storage capacity in the lower mantle, *Earth Planet. Sci. Lett.*, *211*, 189–203, 2003.

Liu, L.-G., Effects of H_2O on the phase behaviour of the forsterite-enstatite system at high pressures and temperatures and implications for the Earth, *Phys. Earth Planet. Inter.*, *49*, 142–167, 1987.

Luth, R.W., Is phase A relevant to the Earth's mantle? *Geochim. Cosmochim. Acta*, *59* (4), 679–682, 1995.

Meade, C. and R. Jeanloz, Deep-focus earthquakes and recycling of water into Earth's mantle, *Science*, *252*, 68–72, 1991.

Mibe, K., T. Fujii, and A. Yasuda, Composition of aqueous fluid coexisting with mantle minerals at high pressure and its bearing on the differentiation of the Earth's mantle, *Geochim. Cosmochim. Acta*, *66* (12), 2273–2285, 2002.

Mosenfelder, J.L., N.I. Deligne, P.D. Asimow, and G.R. Rossman, Hydrogen incorporation in olivine from 2–12 GPa, *Am. Mineral.*, *91*, 285–294, 2006.

Murakami, M., K. Hirose, H. Yurimoto, S. Nakashima, and N. Takafuji, Water in Earth's lower mantle, *Science*, *295*, 1885–1887, 2002.

Mysen, B.O., P. Ulmer, J. Konzett, and M.W. Schmidt, The upper mantle near convergent plate boundaries, in *Ultrahigh-pressure mineralogy*, edited by R.J. Hemley, pp. 97–138, 1998.

Nolet, G., and A. Zielhuis, Low S velocities under the Tornquist-Teisseyre zone: evidence for water injection into the transition zone by subduction, *J. Geophys. Res.*, *99* (B8), 15813–15820, 1994.

Ohtani, E., H. Mizobata, and H. Yurimoto, Stability of dense hydrous magnesium silicate phases in the systems Mg_2SiO_4-H_2O and $MgSiO_3$-H_2O at pressures up to 27 GPa, *Phys. Chem. Minerals*, *27*, 533–544, 2000.

Ohtani, E., T. Shibata, T. Kubo, and T. Kato, Stability of hydrous phases in the transition zone and the uppermost part of the lower mantle, *Geophys. Res. Lett.*, *22* (19), 2553–2556, 1995.

Ohtani, E., M. Toma, T. Kubo, T. Kondo, and T. Kikegawa, In situ X-ray observation of decomposition of superhydrous phase B at high pressure and temperature, *Geophys. Res. Lett.*, *30* (2), 1029, doi:10.1029/2002GL015549, 2003.

Ohtani, E., M. Toma, K. Litasov, T. Kubo, and A. Suzuki, Stability of dense hydrous magnesium silicate phases and water storage capacity in the transition zone and lower mantle, *Phys. Earth Planet. Inter.*, *124*, 105–117, 2001.

Okamoto, K., and S. Maruyama, The high-pressure synthesis of lawsonite in the MORB + H_2O system, *Am. Mineral.*, *84*, 362–373, 1999.

Omori, S., S.-i. Kamiya, S. Maruyama, and D. Zhao, Morphology of the Intraslab Seismic zone and Devolatilization Phase Equilibria of the Subducting Slab Peridotite, *Bull. Earthq. Res. Inst. Univ. Tokyo*, *76*, 455–478, 2002.

Omori, S., T. Komabayashi, and S. Maruyama, Dehydration and earthquakes in the subducting slab: empirical link in intermediate and deep seismic zones, *Phys. Earth Planet. Inter.*, *146* (1–2), 297–311, 2004.

Pawley, A., Chlorite stability in mantle peridotite: the reaction clinochlore + enstatite = forsterite + pyrope + H_2O, *Contrib. Mineral. Petrol.*, *144*, 449–456, 2003.

Peacock, S.M., Are the lower planes of double seismic zones caused by serpentine dehydration in subducting oceanic mantle? *Geology*, *29* (4), 299–302, 2001.

Peacock, S.M., and K. Wang, Seismic consequences of warm versus cool subduction metamorphism: Examples from southwest and northeast Japan, *Science*, *286*, 937–939, 1999.

Pitzer, K.S., and S.M. Sterner, Equations of state valid continuously from zero to extreme pressures for H_2O and CO_2, *J. Chem. Phys.*, *101*, 3111–3116, 1994.

Poli, S., and M.W. Schmidt, H_2O transport and release in subduction zones: Experimental constraints on basaltic and andesitic systems, *J. Geophys. Res.*, *100* (B11), 22299–22314, 1995.

Ranero, C.R., J. Phipps Morgan, K. McIntosh, and C. Reichert, Bending-related faulting and mantle serpentinization at the Middle America trench, *Nature*, *425*, 367–373, 2003.

Revenaugh, J., and S.A. Sipkin, Seismic evidence for silicate melt atop the 410-km mantle discontinuity, *Nature*, *369*, 474–476, 1994.

Ringwood, A.E., and A. Major, High-pressure reconnaissance investigations in the system Mg_2SiO_4-MgO-H_2O, *Earth Planet. Sci. Lett.*, *2*, 130–133, 1967.

Schmidt, M.W., and S. Poli, Experimentally based water budgets for dehydrating slabs and consequences for arc magma generation, *Earth Planet. Sci. Lett.*, *163*, 361–379, 1998.

Schmidt, M.W., and P. Ulmer, A rocking multianvil: elimination of chemical segregation in fluid-saturated high-pressure experiments, *Geochim. Cosmochim. Acta*, *68* (8), 1889–1899, 2004.

Seno, T., and Y. Yamanaka, Double seismic zones, compressional deep trench-outer rise events, and superplumes, in *Subduction: Top to Bottom*, edited by G.E. Bebout, D.W. Scholl, S.H. Kirby, and J.P. Platt, pp. 347–355, AGU, Washington, D. C., 1996.

Shieh, S.R., H.-k. Mao, R.J. Hemley, and L.C. Ming, Decomposition of phase D in the lower mantle and the fate of dense hydrous silicates in subducting slabs, *Earth Planet. Sci. Lett.*, *159*, 13–23, 1998.

Shimizu, K., T. Komiya, K. Hirose, N. Shimizu, and S. Maruyama, Cr-spinel, an excellent micro-container for retaining primitive melts-implications for a hydrous plume origin for komatiites, *Earth Planet. Sci. Lett.*, *189*, 177–188, 2001.

Smyth J.R., D.J. Frost, and F. Nestola, Hydration of olivine and Earth's deep water cycle, *Geochim. Cosmochim. Acta, 69 (10), Suppl.*, A746–A746, 2005.

Song, T.-R.A., D.V. Helmberger, and S.P. Grand, Low-velocity zone atop the 410-km seismic discontinuity in the northwestern United States, *Nature*, *427*, 530–533, 2004.

Stalder, R., and P. Ulmer, Phase relations of a serpentine composition between 5 and 14 GPa: significance of clinohumite and phase E as water carriers into the transition zone, *Contrib. Mineral. Petrol.*, *140*, 670–679, 2001.

Stalder, R., P. Ulmer, A.B. Thompson, and D. Guenther, High pressure fluids in the system MgO-SiO_2-H_2O under upper mantle conditions, *Contrib. Mineral. Petrol.*, *140*, 607–618, 2001.

Ulmer, P., and V. Trommsdorff, Serpentine stability to mantle depths and subduction-related magmatism, *Science*, *268*, 858–861, 1995.

Wunder, B., Equilibrium experiments in the system MgO-SiO$_2$-H$_2$O (MSH): stability fields of clinohumite-OH [Mg$_9$Si$_4$O$_{16}$(OH)$_2$], chondrodite-OH [Mg$_5$Si$_2$O$_8$(OH)$_2$] and phase A (Mg$_7$Si$_2$O$_8$(OH)$_6$), *Contrib. Mineral. Petrol.*, *132*, 111–120, 1998.

Wunder, B., and W. Schreyer, Antigorite: High-pressure stability in the system MgO-SiO$_2$-H$_2$O (MSH), *Lithos*, *41*, 213-227, 1997.

Yamamoto, K., and S.-i. Akimoto, The system MgO-SiO$_2$-H$_2$O at high pressures and temperatures-stability field for hydroxyl-chondrodite, hydroxyl-clinohumite and 10 Å-phase, *Am. J. Sci.*, *277*, 288–312, 1977.

Zhao, D., Y. Xu, D.A. Wiens, L. Dorman, J. Hildebrand, and S. Webb, Depth extent of the Lau back-arc spreading center and its relation to subduction processes, *Science*, *278*, 254–257, 1997.

Tetsuya Komabayashi, Department of Earth and Planetary Sciences, Tokyo Institute of Technology, 2-12-1 Ookayama, Meguro, Tokyo 152-8551, Japan, (tkomabay@geo.titech.ac.jp)

Hydrogen Incorporation in Natural Mantle Olivines

Jed L. Mosenfelder[1], Thomas G. Sharp[2],
Paul D. Asimow[1], and George R. Rossman[1]

Constraints on water storage capacity and actual content in the mantle must be derived not only from experimental studies, but also from investigation of natural samples. Olivine is one of the best-studied, OH-bearing "nominally anhydrous" minerals, yet there remain multiple hypotheses for the incorporation mechanism of hydrogen in this phase. Moreover, there is still debate as to whether the mechanism is the same in natural samples vs. experimental studies, where concentrations can reach very high values (up to ~0.6 wt% H_2O) at high pressures and temperatures. We present new observations and review IR and TEM data from the literature that bear on this question. Hydrogen incorporation in natural olivine clearly occurs by multiple mechanisms, but in contrast to some previous assertions we find that there are strong similarities between the IR signatures of experimentally annealed olivines and most natural samples. At low pressures (lower than ~2 GPa) in both experiments and natural olivines, hydrogen incorporation might be dominated by a humite-type defect, but the nature of the defect may vary even within a single sample; possibilities include point defects, planar defects and optically detectable inclusions. IR bands between 3300 and 3400 cm^{-1}, ascribed previously to the influence of silica activity, are apparently related instead to increased oxygen fugacity. At higher pressures in experiments, the IR band structure changes and hydrogen is probably associated with disordered point defects. Similar IR spectra are seen in olivines from xenoliths derived from deeper parts of the mantle (below South Africa and the Colorado Plateau) as well as in olivines from the ultra-high pressure metamorphic province of the Western Gneiss Region in Norway.

1. INTRODUCTION

The existence of structurally bound hydroxide (OH) groups in natural olivine crystals was established by spectroscopic investigation over 35 years ago [*Beran*, 1969]. The importance of nominally anhydrous minerals such as olivine as storage sites for water (in the form of trace amounts

of hydrogen) in the mantle was suggested soon thereafter [*Martin and Donnay*, 1972]. This hypothesis is now widely accepted, largely as a result of work on natural samples [e.g., *Beran and Putnis*, 1983; *Miller et al.*, 1987; *Bell and Rossman*, 1992] as well as experimental studies at high pressures and temperatures [e.g., *Mackwell et al.*, 1985; *Bai and Kohlstedt*, 1992; 1993; *Kohlstedt et al.*, 1996]. Several review papers are now available on this topic [*Rossman*, 1996; *Ingrin and Skogby*, 2000; *Bolfan-Casanova*, 2005]. As testified to by the papers in this volume, it is also now recognized by workers in multiple disciplines across Earth science that trace amounts of hydrogen can profoundly affect a variety of geophysical processes, including deformation, melting, electrical conductivity, and the propagation of seismic waves.

[1]California Institute of Technology, Division of Geological and Planetary Sciences, Pasadena, California, USA

[2]Department of Geology, Arizona State University, Tempe, Arizona, USA

Earth's Deep Water Cycle
Geophysical Monograph Series 168
Copyright 2006 by the American Geophysical Union.
10.1029/168GM05

As a result of the scarcity of very high-pressure natural samples as well as the possibility of sample alteration during ascent to the surface [*Demouchy et al., 2006; Peslier and Luhr, 2006*], our view of water storage capacity throughout the upper mantle [e.g., *Hirschmann et al.,* this volume] is informed primarily by experimental studies. Much of the attention in this field has been focused on olivine, the predominant mineral of the upper mantle. However, recent work has highlighted potential discrepancies between earlier experimental work and nature with regard to mechanisms of hydrogen incorporation in olivine [*Matveev et al., 2001; Lemaire et al., 2004; Berry et al., 2005; Matveev et al., 2005;* for a contrasting view see *Mosenfelder et al., 2006*]. Differences in opinion among various research groups may lead to confusion for the wider audience interested in water storage in the mantle. Our intent is not to provide a complete review of this topic but to highlight the most important

issues raised by recent studies, as well as present some new observations on natural OH-bearing mantle olivines.

2. METHODS

As part of this study, we reinvestigated some of the samples originally surveyed by *Kitamura et al.* [1987] and *Miller et al.* [1987]. Results on some of the samples prepared for the latter study (particularly GRR1629 and infrared (IR) spectra of CIT15089a and CIT15089b) have not been previously published. In addition, new samples from the Buell Park and Green Knobs localities (Arizona, USA) and from Alpe Arami (Central Alps, Switzerland) were prepared for examination using Fourier-transform infrared spectroscopy, optical microscopy, electron microprobe analysis (EMPA), and transmission electron microscopy (TEM). Table 1 lists the sample localities,

TABLE1. Samples studied

Sample #	Locality	Occurrence (host)	IR bands[a]	ppm H_2O[b]	Features[c]	Reference[d]
GRR949	Almklovdalen, Norway	UHP peridotite body	(I), Ia, serp	20[e]	Lily pads, Mt inclusions	1
CIT15089a	Almklovdalen, Norway	UHP peridotite body	I, Ia, serp	26	Lily pads, Mt inclusions	1
CIT15089b	Almklovdalen, Norway	UHP peridotite body	I, Ia, serp	23[e]	Lily pads, Mt inclusions	1
GRR999a	Zabargad Island, Egypt	hydrothermal vein in peridotite	Ia, humite?	14	Clear	2
GRR1001	Morales, Mexico	basalt	Ia, II	14	Clear	3
GRR1006	Monastery Farm, South Africa	kimberlite	I, Ia	61[e]	Ti-Chu inclusions	3
GRR1007	Vesuvius, Italy	1776 eruption leucitite basalt	(I), Ia, humite?	55	Clear	3
GRR1390	Buell Park, AZ	minette diatreme	I, Ia, Ti-Chu, serp	471	Clear, orange-brown,Ti-Chu planar defects	4
GRR1629	Buell Park, AZ	minette diatreme	I	39	Di, Chr inclusions	4
GRR1784a	Buell Park, AZ	minette diatreme	I, Ia, Ti-Chu	80	Di, Chr inclusions, Ti-Chu inclusions and planar defects	4
GRR1784b	Buell Park, AZ	minette diatreme	I, Ia, Ti-Chu	188	clear, orange-brown, Ti-Chu planar defects (?)	4
GRR1784e	Buell Park, AZ	minette diatreme	I	54	Chr, Di inclusions	4
GK01	Green Knobs, AZ	minette diatreme	I, Ia	54	Chr, Di inclusions	5
Alpe Arami	Alpe Arami, Switzerland	UHP Alpine peridotite	Ia	25	Ilm, Chr inclusions	6

[a]See text and Figure 1 for more information; serp = serpentine, Ti-Chu = Ti-clinohumite. (I) = minor component in spectra

[b]expressed as ppm by weight, calculated using the *Bell et al.* [2003] calibration as discussed in text

[c]Mt = magnetite, Di = diopside, Cr = chromite, Ilm = ilmenite. "Lily pads" are disk-shaped stress fractures around small inclusions such as chromite or spinel.

[d]References: 1 = *Medaris* [1999]; 2 = *Beran and Putnis [1983]*; 3 = *Miller et al.* [1987]; 4 = *Kitamura et al.* [1987]; 5 = *Smith and Levy* [1976]; 6 = *Dobrzhinetskaya et al.* [1996]

[e]value from *Kent and Rossman* [2002]

spectral features, calculated OH concentrations, inclusion features and references to petrological studies of the localities. Orientation of the crystals was achieved using Raman spectroscopy [see *Mosenfelder et al.*, 2006] and by optical techniques, particularly by identification of growth faces or the (010) cleavage. The uncertainty in orientations is estimated to be 5° or less. Crystal faces were cut with a wire saw and polished using alumina grinding papers and 0.25 μm diamond powder. Olivine in garnet lherzolite from Alpe Arami, collected by the first author during the Fifth International Eclogite Conference, was studied *in situ* in a 150-μm thin section prepared using Crystalbond™, which was subsequently dissolved from the underlying glass slide using acetone.

FTIR spectroscopy was conducted primarily using the main compartment of a Nicolet Magna 860 spectrometer. We collected polarized, mid-IR spectra from 4000 to 2100 cm^{-1} at 2 cm^{-1} resolution by averaging 512 scans, using a GLOBAR infrared light source, a CaF_2 beamsplitter, a $LiIO_3$ Glan-Foucault prism polarizer, and an MCT-A detector. Circular apertures with diameters of 200–1000 μm were used to select analysis areas. For each single-crystal sample, polarized spectra were collected with the E-vector (**E**) parallel to the [100], [010], and [001] directions (using the same convention for these directions as *Bell et al.* [2003]). For the Alpe Arami thin section, spectra were collected using an IR microscope on randomly oriented grains and corrected using the procedures outlined in *Mosenfelder et al.* [2006]. OH concentrations were calculated using the calibration of *Bell et al.* [2003] applied to three orthogonal spectra, baseline corrected using the procedure outlined in *Asimow et al.* [2006] and integrated from 3100 to 3700 cm^{-1}. The comparison between this calibration and the more commonly used calibration of *Paterson* [1982] has been discussed by *Bell et al.* [2003] and *Mosenfelder et al.* [2006]. The precision of the absorption coefficient determined by *Bell et al.* [2003] is 6.5%, but the accuracy of the values in Table 1 also depends on other factors that are difficult to evaluate, such as choice of baseline correction, choice of integration interval and a possible (uncalibrated) dependence of absorption coefficient on wavenumber.

Wavelength-dispersive EMP analyses were obtained using a JEOL 733 microprobe operating at 15 kV (Table 2). Major element analyses were collected using a spot size of 10-μm and a beam current of 25 nA. For minor and trace elements (Ti, Al, Mn, Ca, Ni, Na, and Cr) in olivine, we used a beam current of 300 nA and increased counting times to four minutes on peak and two minutes on background. Detection limits under these conditions are in the range of 10 to 30 ppm (by weight, for the corresponding oxide). Data processing followed the CITZAF method [*Armstrong*, 1988]. We used well-characterized natural and synthetic standards: Ni_2SiO_4, Mn_2SiO_4, Mg_2SiO_4, TiO_2, Fe_2SiO_4, albite, microcline and

Cr_2O_3. San Carlos olivine with a composition of Fo_{90} was used as a secondary standard. Note that chemical analyses of some of the olivines in Table 1 have also been published by *Kent and Rossman* [2002], who additionally analyzed for P, Li and B but not for Ti, Cr or Al.

TEM was performed at the Center for High Resolution Microscopy at Arizona State University, using a Philips CM-200 FEG microscope operating at 200 kV. Qualitative energy-dispersive chemical analyses were collected using a Kevex EDS detector and an EmiSpec analytical system. The samples for TEM were oriented thin sections mounted to copper grids and thinned to electron transparency by dimpling and Ar-ion bombardment using a Gatan Precision Ion Polishing System with an acceleration potential of 5 KeV and an incidence angle of 5°.

3. INFRARED SIGNATURES OF NATURAL OLIVINE CRYSTALS

Infrared spectroscopic studies of olivines have revealed a large number of variably occurring bands in the OH-stretching vibrational region, in the range from ~3750 to 3100 cm^{-1}. For instance, *Matsyuk and Langer* [2004] designated 70 different bands present in varying combinations in a suite of olivines from kimberlitic xenoliths from the Siberian shield. Most of the IR bands are sharp (small full-width at half maximum, FWHM) and exhibit strong polarization, most typically showing strongest absorption with the E-vector (**E**) parallel to [100]. The greater variety of band positions in comparison to other nominally anhydrous minerals complicates attempts to uniquely assign bands to specific OH defects within the olivine crystal structure [*Libowitzky and Beran*, 1995]. Consequently, disagreement remains about band assignments and modes of hydrogen incorporation in olivine. A further complication is that some natural olivines contain inclusions of water or hydrous solid phases such as serpentine, talc, and humite-series minerals (for a recent discussion of the nomenclature of humite-series minerals, see *Matsyuk and Langer* [2004]). As these inclusions often can only be observed at the nm-scale using electron microscopy, some of the OH in olivine may be falsely interpreted as being integral to its structure.

Extensive tables of band positions and FWHM in olivine can be found in several papers [*Miller et al.*, 1987; *Libowitzky and Beran*, 1995; *Kurosawa et al.*, 1997; *Khisina et al.*, 2001; *Matsyuk and Langer*, 2004]. In an attempt to simplify the situation, we adopt and modify the simple scheme of *Bai and Kohlstedt* [1993], who divided the bands in their experimentally annealed olivines into two groups: Group I, including bands from 3450 to 3650 cm^{-1}, and

TABLE 2. Average electron microprobe analyses of selected olivine crystals and Ti-clinohumite in GRR1784a

Sample no.	GRR999a	GRR1007	GRR1006	GRR1629	GRR1784e	GRR1390	GRR1784a	GRR1784A - Ti-Chu inclusion
Locality	Zabargad	Vesuvius	Monastery	Buell Park	Buell Park	Buell Park	Buell Park	Buell Park
No. analyses[a]	3/3	3/3	2/3	3/3	3/3	3/6	3/4	2/0
SiO_2	40.63	40.09	40.54	40.74	40.89	40.40	40.84	36.17
TiO_2	0.004	0.009	0.004	n.d.	0.005	0.058	0.012	5.20
Al_2O_3	n.d.	0.013	n.d.	n.d.	n.d.	n.d.	n.d.	n.d.
FeO	9.38	12.41	7.72	8.27	7.63	9.08	7.86	7.88
MnO	0.125	0.219	0.101	0.113	0.109	0.143	0.094	0.10
MgO	50.01	46.55	50.85	50.83	50.42	49.91	50.81	48.56
CaO	0.009	0.310	n.d.	n.d.	n.d.	n.d.	n.d.	n.d.
NiO	0.315	0.179	0.354	0.404	0.389	0.367	0.305	0.30
Na_2O	n.d.	n.d.	n.d.	n.d.	n.d.	n.d.	n.d.	n.d.
Cr_2O_3	n.d.	0.012	n.d.	n.d.	n.d.	0.009	n.d.	0.05
Total	100.48	99.81	99.59	100.37	99.46	99.99	99.93	98.28
Mg#[b]	90.5	87.0	92.2	91.6	92.2	90.7	92.0	91.7

Notes:
all concentrations expressed in weight percent
n.d. = below detection limit (typically 10-30 ppm by weight)
[a]Number of major/trace analyses averaged for table.
[b]Mg# = 100 x molar Mg/(Mg+Fe)

Group II, over the range from 3200 to 3450 cm^{-1}. In addition, we distinguish a subset of Group I bands, hereafter referred to as "Group Ia", which comprises two strong bands near 3573 and 3525 cm^{-1} with shoulders at 3563 and 3541 cm^{-1}. Other bands that have been conclusively linked to the presence of extrinsic hydrous phases [*Kitamura et al., 1987; Miller et al., 1987; Matsyuk and Langer, 2004*] are also noted below and in Table 1.

The delineation of the band groups is illustrated in Figure 1, where we show that the classification can be applied not only to experimental samples [*Zhao et al., 2004; Mosenfelder et al., 2006*] but also to natural olivines. Further examples of the distinction between I and Ia bands can be seen in Figures 2 and 3. Figure 2 illustrates spectra from a suite of olivines from the Buell Park and Green Knobs diatremes in Arizona. Figure 3 shows spectra from olivines that come from the Western Gneiss Region in Norway. Group I bands have been suggested to be a feature indicative of high pressures [*Mosenfelder et al., 2006*]. Group Ia bands, which are strongly correlated in peak height [*Matsyuk and Langer, 2004*], have been associated with humite-like defects [*Kitamura et al., 1987; Miller et al., 1987*], but the exact nature of the defects may vary between samples or even within a single sample. The significance of Group II bands is controversial [*Matveev et al., 2001; Lemaire et al., 2004*] and may be related either to varying silica activity or oxygen fugacity. These issues are discussed in the next section.

4. MECHANISMS OF HYDROGEN INCORPORATION IN OLIVINE

Incorporation of hydrogen in olivine can occur via the following classes of mechanisms: 1) as point defects; 2) in planar defects, comprised of either a separate hydrous phase or an ordered array of point defects; 3) via inclusions of hydrous minerals; and 4) as inclusions of fluid H_2O. Evidence to support these mechanisms comes primarily from IR spectra and TEM studies. Below we review each of these mechanisms, using new observations from the present study as well as literature data.

4.1 Point Defects

Point defects have long been linked to hydrogen incorporation in olivine [*Beran, 1969; Beran and Putnis, 1983; Bai and Kohlstedt, 1992; 1993*]. Possible defects that can balance the excess charge attending incorporation of hydrogen bonded to oxygen include coupled substitution with non-divalent impurity cations (e.g., Al^{3+} and Cr^{3+} in tetrahedral sites, or Na^+ and Li^+ in octahedral sites), metal (Fe and Mg) vacancies, Si vacancies, ferric iron (Fe^{3+}) in tetrahedral sites, and oxygen interstitials. OH associated with line defects has also been suggested [*Beran and Putnis, 1983*], although there is no direct evidence for this mechanism.

The question of which—if any—types of defects dominate under different conditions is still under debate and it is not straightforward to assign IR bands uniquely to specific point defects based on dipole directions and wavenumber positions. For instance, *Libowitzky and Beran* [1995] thoroughly analyzed the IR spectra of a nearly pure Mg-end member forsterite crystal and could not assign all bands uniquely to either Mg or Si vacancies. Attempts to correlate trace element concentrations in olivine with hydrogen content have met with mixed success. Coupled substitution of hydrogen and boron has been clearly demonstrated in some olivines [*Sykes et al.*, 1994; *Kent and Rossman*, 2002], but this element is unlikely to be present at high levels throughout the mantle. *Kurosawa et al.* [1997] showed a positive correlation

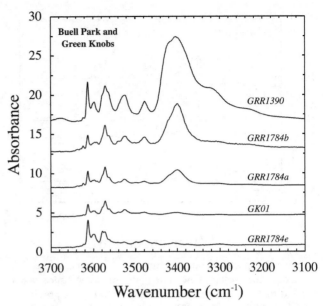

Figure 2. IR spectra of olivines from Buell Park and Green Knobs, polarized with **E** ‖ [100]. The sample numbers are listed directly above each spectrum. Note gradation from spectrum dominated by Group I bands (GRR1784e) to spectra showing Group Ia bands and bands attributed to Ti-clinohumite.

of trivalent cations (Al^{3+}, Cr^{3+}) with monovalent cations (H^+, Li^+, Na^+) in a suite of olivines from garnet peridotites, but a more recent study [*Bell et al.*, 2004] failed to reveal any such correlation. Even more recently, *Hauri et al.* [in press] obtained ion microprobe analyses on experimental samples showing both a rough equality between molar H and Al contents as well as a correlation between Al content in olivine and partitioning of hydrogen between olivine and melt. They used this evidence to argue for a prominent role of Al in hydrogen incorporation in olivine, but the correlation shows considerable scatter and application to natural samples is uncertain.

A prominent role for either oxygen interstitials or metal vacancies has been argued on the basis of thermodynamics, point defect studies under anhydrous conditions, and IR data on experimental samples [*Bai and Kohlstedt*, 1993; *Kohlstedt et al.*, 1996; *Kohlstedt and Mackwell*, 1998; *Zhao et al.*, 2004]. On the other hand, the *ab initio* calculations of *Brodholt and Refson* [2000] suggest that the concentration of silicon vacancies is enhanced under hydrous conditions. In this case, silica activity would be expected to play an important role in hydrogen incorporation, as argued by *Matveev et al.* [2001], who ascribed Group I bands to Si vacancies formed under conditions of low silica activity (in equilibrium with MgO) and Group II bands to metal vacancies formed under higher silica activity conditions (in equilibrium with orthopyroxene). This conflicts with the previous experi-

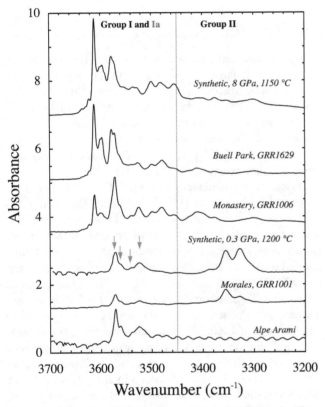

Figure 1. Comparison between Group I, Ia, and II IR bands, illustrated in both natural and experimental samples. The line divides Group II bands from Groups I and Ia, and the grey arrows point to the four Group Ia bands. All spectra normalized to 1 cm with the exception of the experimental sample [*Mosenfelder et al.*, 2006] from 8 GPa and 1150 °C, normalized to 0.5 mm. Spectra in this and other figures are offset in absorbance arbitrarily for the sake of clarity and comparison. All spectra polarized with **E** ‖ [100] except for the spectrum taken from *Zhao et al.* [2004] of an experimental sample annealed at 0.3 GPa and 1200 °C, which is nominally unpolarized with the incident beam parallel to [010]. Polarized spectra with **E** ‖ [010] and [001] for GRR1629 and the 8 GPa synthetic sample published in *Mosenfelder et al.* [2006]. Natural samples from localities discussed in Table 1 and in the text.

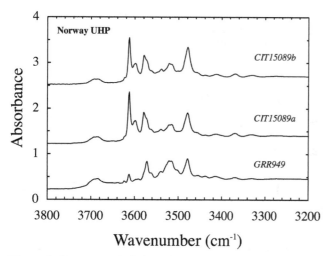

Figure 3. IR spectra of olivines from Almklovdalen, in the UHP province of the Western Gneiss Region, Norway. All spectra normalized to 1 cm and polarized with **E** ‖ [100]. The previously published [*Miller et al.*, 1987] spectrum for sample GRR949 predominantly shows Group Ia bands, but the other two crystals predominantly show Group I bands. The broad band at 3685 cm⁻¹ is assigned to serpentine.

mental results of *Bai and Kohlstedt* [1993], who showed no evidence for a significant effect of silica activity on incorporation mechanism. Moreover, it implies that the vast majority of natural olivines, which are dominated by Group I or Ia bands (Figures 1–3; [*Miller et al.*, 1987]), were equilibrated at low silica activity. This conclusion is problematic as most of these olivines come from orthopyroxene-bearing peridotite sources.

A possible resolution to this problem, as argued by *Mosenfelder et al.* [2006] [see also *Berry et al.*, 2005], is that the Group II bands are related to elevated oxygen fugacity rather than differences in silica activity. This conclusion is consistent with the faster increase in peak heights of Group II bands (compared to Group I) with increasing oxygen fugacity documented in the original study of *Bai and Kohlstedt* [1993], and with the oxygen buffer used in the experiments of *Matveev et al.* [2001]. However, this does not explain another set of low-wavenumber bands seen in two more recent experimental studies. These bands, at 3160 and 3230 cm⁻¹, are present in both Fe-free [*Demouchy and Mackwell*, 2003; *Lemaire et al.*, 2004] and Fe-bearing [*Berry et al.*, 2005] olivines synthesized under high-silica activity conditions at pressures below 2 GPa. Although *Berry et al.* [2005] only presented unpolarized spectra, the bands observed by *Demouchy and Mackwell* [2003] and *Lemaire et al.* [2004] are broad and strongly polarized with the greatest absorption ‖ [001], unlike most other bands in olivine. The band at 3160 cm⁻¹ has not been seen in any natural olivines, to our

knowledge, but the band at 3230 cm⁻¹ may correspond to a band with the same polarization seen in olivines from Buell Park (Figure 2), Zabargad Island, Egypt, and Vesuvius, Italy Figure 4). This band is correlated with the presence of a band near 3400 cm⁻¹ in natural samples (Figures 2 and 4) and in some experimental samples [*Lemaire et al.*, 2004, Figure 6]. Bands at nearly identical wavenumbers are also present in the spectra of Ti-clinohumite [*Matsyuk and Langer*, 2004]. Based on this correspondence, these bands may represent the presence of defects or inclusions related to Ti-clinohumite, as discussed further below, even though the 3230 cm⁻¹ band has also been attributed to OH defects [*Kitamura et al.*, 1987] or even molecular water [*Matsyuk and Langer*, 2004].

4.2. Planar Defects

4.2.1. Clinohumite Lamellae. OH-bearing, planar defects (stacking faults) in olivine were described by *Kitamura et al.* [1987], who studied sample GRR1390 (from Buell Park, Arizona) using TEM. The defects are aligned parallel to (001), (021), and (0–21) (Figure 5). Similar defects along (001) have been documented by *Drury* [1991], *Sykes et al.* [1994], and *Risold et al.* [2001]. They have also been reproduced in high-pressure experiments [*Wirth et al.*, 2001].

Based on the displacement vector (**R** = ¼<011>) of the defects, *Kitamura et al.* [1987] determined that their structure was consistent with layers of Ti-clinohumite. This conclusion is also consistent with IR spectra of the sample, which show strong Group Ia bands at 3571 and 3524 cm⁻¹, as well as a very strong band at 3402 cm⁻¹ with shoulders at 3422, 3319, and 3230 cm⁻¹. All of these bands can be ascribed to Ti-clinohumite [*Miller et al.*, 1987; *Matsyuk and Langer*, 2004]. Furthermore, the orange to orange-yellow pleochroism of GRR1390, unusual for olivine, is similar to that of Ti-clinohumite. We re-examined this sample with TEM and confirmed the structure and displacement vector of the defects using high-resolution imaging (Figure 5a–c). Moreover, qualitative EDS analyses of the stacking faults reveal high Ti concentrations, and EMP analyses of the bulk sample (Table 2) show unusually high Ti content for olivine (580 ppm TiO₂ by weight), consistent with the original attribution of the defects as Ti-clinohumite layers.

We have also detected planar defects in GRR1784a, another olivine crystal from Buell Park that shows IR bands corresponding to the presence of Ti-clinohumite (Figure 2). The density of planar defects in GRR1784a is lower, the defect segments are generally shorter, and there are no (021) and (0–21) segments; the sample also contains a significantly higher density of dislocations and subgrain boundaries. GRR1784a lacks the unusual pleochroism of GRR1390 (it is a "normal" green color) and EMP analyses reveal a lower Ti content (117 ppm wt TiO_2), although significantly above the detection limit (30 ppm). Furthermore, this sample contains optically detectable inclusions of Ti-clinohumite (Figure 5e–f), discussed in more detail in section 4.3. Figure 2 also shows another Buell Park olivine (GRR1784b) exhibiting the band at 3402 cm^{-1}, with intensity intermediate between GRR1784a and GRR1390; this olivine has similar pleochroism to GRR1390 and presumably contains planar defects, but we have not examined it using TEM or EMPA.

Risold et al. [2001] showed that the planar defects in olivine from the Alpe Arami garnet peridotite are linked directly to ilmenite ((Mg,Fe)TiO$_3$) inclusions, which have nucleated on the defects. We have collected IR spectra from several grains of olivine containing ilmenite inclusions. Fig. 1 shows the **E** ‖ [100] spectrum of Alpe Arami olivine, showing only type Ia bands without the band at 3402 cm^{-1}. Predominance of the latter band appears to be diagnostic of Ti-bearing humite-series minerals [*Miller et al.*, 1987]; although *Matsyuk and Langer* [2004] presented IR data from humite-series minerals nominally lacking Ti that also show this band, they did not present microprobe data for these samples. Moreover, synthetic, Ti-free clinohumite exhibits no band at this wavenumber [*Liu et al.*, 2003]. Therefore, the IR spectrum of the Alpe Arami olivine is consistent with the hypothesis of *Risold et al.* [2001] that the planar defects were originally Ti-clinohumite but have lost Ti via diffusion during precipitation of the ilmenite rods in the sample. This explanation obviates the need for very high pressures (7 GPa or higher) advocated by *Dobrzhinetskaya et al.* [1996]

[see also *Bozhilov et al.*, 2003] to explain the existence of the ilmenite rods, because Ti solubility in olivine has been shown to be a strong function not of pressure but of temperature [*Hermann et al.*, 2005].

4.2.2. Hydrous-modified Olivine. *Khisina et al.* [2001] and *Khisina and Wirth* [2002] documented a different type of OH-bearing planar defect in an olivine crystal in a kimberlitic xenolith from the Siberian Platform. The planar defects in this sample are grouped together in lamellae, wider than the clinohumite defects discussed above. EDS analyses of the lamellae show a deficiency of Mg and Fe relative to the surrounding olivine. In addition to the lamellae, discrete, nm-sized inclusions of the same phase are present. *Khisina and Wirth* [2002] interpreted the lamellae to be ordered arrays of M-site (Mg and Fe) vacancies in olivine and called this new phase "hydrous olivine". A similar structural model was devised by *Kudoh* [2002], who called the theoretical phase "hydrous-modified olivine." Although the structure of this phase is similar to clinohumite, it is chemically distinct in that Mg/Si < 2, whereas the humite-series minerals have Mg/Si > 2. Intriguingly, the IR spectra of the sample studied by *Khisina et al.* [2001] show a predominance of the same Group Ia bands seen in many other olivines, as well as in samples GRR1390 and GRR1784a, which contain distinct clinohumite layers. Thus, it is apparently not possible to distinguish between defects that are Mg/Fe-deficient and those that are Si-deficient, using IR spectroscopy alone.

4.3. Hydrous Mineral Inclusions

Effective "incorporation" of hydrogen in olivine via inclusion of, or intimate intergrowth with, hydrous minerals has also been long recognized. For instance, *McGetchin et al.* [1970] described intergrowths of Ti-clinohumite and olivine (as well as Ti-clinohumite inclusions in garnet) and suggested the importance of Ti-clinohumite as a carrier for water in the mantle, long before the importance of nominally anhydrous minerals was broadly recognized. Based on IR spectra, *Miller et al.* [1987] established the presence of other hydrous inclusions in some olivines that were not distinguishable using optical methods. These include serpentine, exhibiting a complex multiplet of bands around 3680 cm^{-1}, and talc, distinguished by a sharper band at 3678 cm^{-1} together with a much smaller band at 3662 cm^{-1}. The possible presence of humite-series minerals was also discussed at length by *Miller et al.* [1987], but the correspondence of band positions was not found to be as precise as for serpentine and talc. More recently, *Khisina et al.* [2001] documented nm-scale inclusions of talc, serpentine and 10-Å phase in olivines using TEM, and *Matsyuk and Langer* [2004] presented evidence from IR spectra for sub-microscopic inclu-

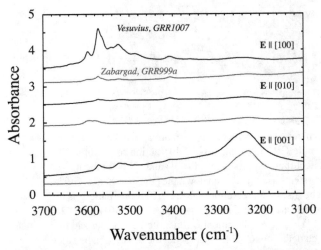

Figure 4. IR spectra of olivines from Vesuvius, Italy (black) and Zabargad Island, Egypt (grey). All spectra normalized to 1 cm. The top two spectra are polarized with **E** ‖ [100], the middle two with **E** ‖ [010], and the bottom two with **E** ‖ [001]. The band at 3404 cm^{-1} is weakly polarized, but the band near 3230 cm^{-1} is strongly polarized, showing greatest absorption with **E** ‖ [001].

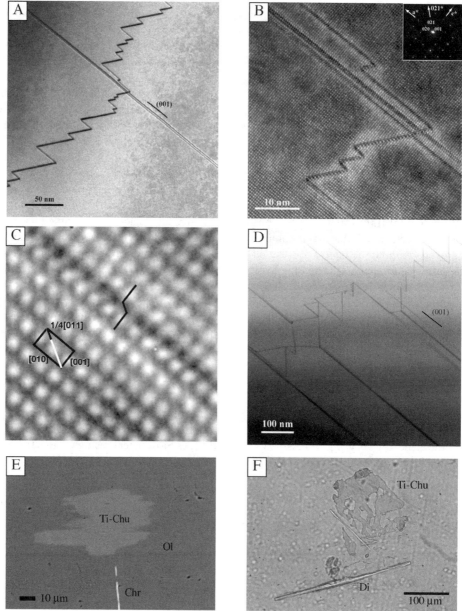

Figure 5. Inclusions of Ti-clinohumite (Ti-Chu) from the nanometer- to micrometer-scale in olivines from Buell Park, Arizona. **(a)** Bright-field TEM image of Ti-Chu defects in sample GRR1390 taken with diffraction to g = 004. The image shows two long straight defects on (001) and a zigzag defect that consists of short (001) and (021) segments. **(b)** High-resolution TEM (HRTEM) image of the same set of planar defects as in **(a)**, taken down the [100] zone axis. Ti-Chu layers parallel to (001) are present as are defects along (021). The inset selected-area electron diffraction pattern is the [100] zone axis. **(c)** Enlargement of HRTEM image in **(b)** demonstrating that the displacement of the (001) and (020) lattice fringes is consistent with a displacement vector R = ¼<011>. The displacement is illustrated by the zig-zag line in the middle of the image. **(d)** Bright-field TEM image of Ti-Chu defects in GRR1390 showing the interaction between planar (001) defects and complex zig-zag defects consisting of (001), (021) and (0–21) segments. Note that double (001) defects terminate into (021) + (0–21) pairs. Image taken with diffraction to g = 004. **(e)** BSE image of a Ti-Chu inclusion in sample GRR1784A. The bright, needle-like phase is chromite (Chr). An EMPA analysis of the Ti-Chu inclusion is shown in Table 2. **(f)** Optical photo-micrograph in plane-polarized light of a Ti-Chu inclusion in GRR1784A associated with diopside (Di) needles. The inclusions form as flattened plates in the (100) plane of the crystal. Similar Ti-Chu inclusions were also identified in GRR1006.

sions of serpentine, talc, amphibole and hydrous wadsleyite (one of the high-pressure forms of olivine). We consider the attribution of the latter two phases to be highly speculative; the IR bands of amphiboles overlap those of serpentine, while the bands attributed to hydrous wadsleyite (at 3327 cm^{-1} and 3357 cm^{-1}) are identical to two of the Group II bands, which are present in low-pressure olivines, show strong anisotropy, and probably are related to elevated oxygen fugacity as discussed in section 4.1 above.

Inclusions of talc and serpentine most likely represent secondary alteration of olivines, but some humite-series minerals may be stable at the high pressure and temperature conditions from which mantle xenoliths are extracted [*Ulmer and Trommsdorff*, 1999]. In addition to the planar defects described above, we have identified micrometer-scale inclusions of Ti-clinohumite (Figure 5e–f) in samples from Buell Park (GRR1784a and other crystals not yet prepared for IR or EMP analysis) and in a crystal from the Monastery kimberlite in South Africa (GRR1006). These orange-colored inclusions comprise very thin platelets confined to the (100) planes of the green-colored olivines. They have been positively identified by Raman spectroscopy as well as electron microprobe analysis (Table 2). In sample GRR1784a, they are associated with other oriented crystals identified by EDS as diopside and chromite (Figure 5e–f). This association suggests that they form by exsolution during cooling from higher temperatures to the conditions from which the xenoliths were extracted; the solubility of trace elements such as Ti, Al, Ca and Cr in olivine is expected to increase with temperature [e.g., *Hermann et al.*, 2005].

Although inclusions of Ti-clinohumite were predicted [*Miller et al.*, 1987; *Libowitzky and Beran*, 1995] to be present in olivines from Zabargad Island, Egypt and Vesuvius, Italy (Figure 4), we have not found any optically detectable inclusions in these samples (more than 20 crystals from Zabargad were examined for this purpose). We also studied Zabargad olivine using TEM [cf. *Beran and Putnis*, 1983] and did not find planar defects of the type discussed above. Furthermore, we did not observe any nanometer-scale precipitates of Ti-clinohumite such as those identified by *Wirth et al.* [2001] in an experimental sample. We tentatively suggest that the bands in these samples near 3400 cm^{-1} and 3230 cm^{-1} represent point defects related to humite, but more work is needed to model the nature of these defects; another possibility is that lamellae or inclusions of Ti-clinohumite are present but at such low density that we were unable to detect them.

4.4 Fluid-Inclusion H$_2$O

Many minerals exhibit a broad band of absorption that is centered at 3420 cm^{-1} and lacks polarization. This band

may underlie sharp bands that are polarized [*Miller et al.*, 1987; *Bell et al.*, 2004; *Rossman et al.*, in preparation]. Although this band has been associated with the presence of fluid inclusions, the inclusions may not be optically visible. *Miller et al.* [1987] documented such a feature in a sample from Kimberley, South Africa and suggested that it reflected the presence of submicroscopic fluid inclusions so small that ice could not form during cooling (to 77 K) of the olivine. Indeed, abundant nanometer-scale pores in this specimen have recently been imaged using high-resolution scanning electron microscopy [*Rossman et al.*, in preparation]. Although other broad bands at lower energies (between 3175 and 3260 cm^{-1}) have also been tentatively assigned to molecular water with "relatively strong bonds to the matrix" [*Matsyuk and Langer*, 2004], there is no direct evidence to support this suggestion. Moreover, the band near 3230 cm^{-1} with strong polarization ∥ [001] may be associated with a humite-type defect as discussed above.

The presence of a large broadband component that is not associated with visible fluid inclusion arrays or fractures appears to be a phenomenon restricted to olivines from very high-pressure xenoliths [*Miller et al.*, 1987; *Bell et al.*, 2004]. Nanometer-scale features such as those documented by *Rossman et al.* [in preparation] may represent water exsolved from the structure of the olivine during rapid decompression and cooling. Thus, while providing an intriguing signature, it is not clear whether these inclusions represent a valid mechanism for incorporating hydrogen in olivine *in situ* in the mantle.

5. DISCUSSION

5.1. Effect of Pressure on Hydrogen Incorporation in Olivine

Bai and Kohlstedt [1993] and *Kohlstedt et al.* [1996] established the systematic, dramatic increase of water storage capacity in olivine with increasing pressure, and the recent experiments of *Mosenfelder et al.* [2006], using an updated IR calibration [*Bell et al.*, 2003], confirm these results and indicate that the storage capacity of olivine is even higher than previously thought [cf. *Smyth et al.*, 2005]. Although the applicability of some of these experiments to the mantle has been questioned [*Matveev et al.*, 2001; *Lemaire et al.*, 2004; *Berry et al.*, 2005; *Matveev et al.*, 2005], our comparison of IR spectra from synthetic and natural samples (Figure 1) shows that whatever mechanism is responsible for hydrogen incorporation in experiments from about 2–13 GPa is also operative in some natural samples. *Mosenfelder et al.* [2006] suggested that there may be a transition in mechanisms between low pressures, where a humite-type defect represented by Group Ia bands dominates, and high pressures, where a different defect associated with Group I bands (particularly the bands at 3613, 3598, and 3579 cm^{-1}, but also, in some samples, a triplet at 3502, 3483 and 3456 cm^{-1}) dominates. *Berry et al.* [2005] also speculated on a change in hydrogen incorporation mechanism at high pressures, suggesting that the Group I bands may reflect the presence of Ti-free, "OH-clinohumite-like defects". This assignment is problematic based on comparison between the relatively simple IR spectra of OH-clinohumite, containing only three bands *[Liu et al., 2003]*, and the more complicated pattern of Group I bands noted above. Nevertheless, TEM studies of experimental samples are warranted in order to investigate the possibility that layers or inclusions of OH-clinohumite are present.

The natural olivines represented by the upper three spectra in Figure 1 and the spectra in Figures 2 and 3 all originally come from high-pressure environments. The exact pressure and temperature conditions of formation for the xenolithic olivines are difficult to constrain because they were collected as discrete grains without accompanying phases. However, for the Buell Park olivines, derivation from the garnet peridotite stability field is suggested based on association with pyropic garnet and Cr-diopside and the high Mg-number (~Fo$_{92}$, Table 2) of the olivines compared to the Mg-number of olivines in spinel-peridotite xenoliths from the same locality [*Smith and Levy*, 1976]. Even higher pressures and temperatures have been suggested for xenoliths from the Monastery kimberlite [*Bell et al.*, 2004], although GRR1006 presumably comes from a different xenolithic suite than the olivines studied by *Bell et al.* [2004]. We note that the olivines in that study, which are among the most OH-rich natural olivines ever measured, are dominated by Group Ia IR bands, which contradicts the hypothesis of *Mosenfelder et al.* [2006]. However, these olivines are also substantially more Fe-rich (Fo$_{80}$-Fo$_{82}$, Table 5 in *Bell et al.* [2004]) than the olivines (Fo$_{92}$, Table 2) that are dominated by Group I bands, and the presence of Fe may stabilize whatever defect is represented by the Group Ia bands, as suggested by the experimental data of *Zhao et al.* [2004]. The OH-rich sample from Vesuvius, Italy (Figure 4) also exhibits strong Group Ia bands and is relatively Fe-rich (Fo$_{87}$, Table 2).

The olivine crystals from Norway come from a garnet-peridotite body (Almklovdalen) in the ultrahigh-pressure province of the Western Gneiss Region. The IR spectrum of sample GRR949, originally published by *Miller et al.* [1987], shows very weak Group I bands and stronger Group Ia bands, but the other two olivines exhibit distinctly different IR spectra (Figure 3), with strong Group I bands at 3613, 3598 and 3579 cm^{-1}. These crystals may have experienced

peak pressures of 3.8 GPa [*Medaris,* 1999] or even higher [*van Roermund et al.*, 2000]. However, they obviously had a more protracted history of exhumation to the surface than the xenoliths discussed above and have likely been affected by fluid infiltration events at lower pressures [*Kostenko et al.*, 2002], reflected by the ubiquitous presence of serpentine in these samples. Because the diffusion of hydrogen in olivine is known to be rapid [*Kohlstedt and Mackwell*, 1998; *Demouchy and Mackwell, 2003*], at least on the several million-year timescale it must have taken for exhumation of these rocks, the proposed survival of a "high-pressure OH signature" is therefore surprising. Unfortunately the samples, received as donations from gem dealers, do not have a properly documented geologic context. A more thorough study of OH in olivines from ultra-high pressure garnet-peridotites could yield further information with respect to hydrogen incorporation mechanisms at high pressure.

5.2. Ti in Olivine and the Importance of Humite-Type Defects

The existence of planar defects and inclusions of Ti-clinohumite in olivine described above suggest a potentially important role for Ti in incorporation of hydrogen in olivine. We note that although these defects and inclusions have been identified in olivines from two different, far-apart localities (Buell Park, Arizona and Monastery Mine, South Africa), the ubiquity of such a phenomenon in the mantle is far from certain. For instance, *Matsyuk and Langer* [2004] studied a much larger suite of olivines in kimberlitic xenoliths from the Siberian shield and none of their samples show the band at 3402 cm^{-1} characteristic of Ti-clinohumite. The Ti-clinohumite defects and inclusions appear to reflect exsolution from olivines containing structural Ti and H; exsolution of other trace elements (Cr, Al, and Ca) in these samples has also clearly taken place as reflected by the presence of associated chromite and diopside inclusions (Figure 5). For the Buell Park samples such a process is consistent with petrological studies that document protracted cooling of the xenoliths prior to extraction from the mantle source [*Smith and Levy*, 1976]. Ti, which is highly soluble in olivine at high temperatures according to the experimental data of *Hermann et al.* [2005], may have been introduced into the olivines by localized metasomatism at high temperatures; this would be consistent with the highly variable concentration of Ti and Ti-clinohumite defects in these samples (Figure 2 and Table 2). Note that the history of the Alpe Arami garnet lherzolite, containing similar planar defects in olivine, is quite different and probably reflects a multi-stage process such as that proposed by *Hermann et al.* [2005]. In this case the Ti-clinohumite, perhaps formed by an exsolution process analogous to that in the Buell Park samples, must have become unstable during a later metamorphic event as evidenced by breakdown of the defects to form ilmenite (plus olivine). This secondary event may reflect either prograde metamorphism [*Hermann et al.*, 2005] or retrograde metamorphism at relatively high temperatures during rapid exhumation.

Berry et al. [2005] have suggested that a point defect associated with Ti is the most important defect site in the shallow upper mantle for incorporating hydrogen. The extremely high concentration of OH in olivines with presumably low Ti contents annealed at pressures up to 13 GPa indicates that this is not the only important mechanism for incorporating hydrogen in olivine in deeper parts of the mantle. Moreover, the exact nature of the proposed "humite-type" point defects, as we have discussed throughout this paper, is uncertain. There may be more than one variety of "humite-type" defect, as suggested by differences in IR spectra of low-pressure olivines (compare Figures 1 and 4). A "hydrous-modified olivine" defect [*Khisina and Wirth*, 2002; *Kudoh* 2002] is one strong possibility that should be investigated further with combined IR and TEM studies of natural olivines.

Acknowledgments. Financial support for this work was provided by NSF grants OCE-0095294 and OCE-0241716 to PDA, EAR-0337816 to GRR, and EAR-0208419 to TGS. Several of the Buell Park samples came from the thesis collection repository of David Bell at Caltech. Doug Smith graciously provided samples, enlightening discussion and encouragement. Erik Hauri provided a preprint and went beyond the call of duty discussing it. We also thank Maarten Broekmans, Bradley Hacker, Stephen Mackwell, Gordon Medaris, Michael Roden, and Michael Terry for helpful discussions and information about localities. Chi Ma assisted with electron microprobe analyses. Thorough reviews by Andrew Berry, Steven Jacobsen and an anonymous reviewer helped us refine the manuscript. Finally, we once again thank all of the original donors who provided samples for the study of *Miller et al.* [1987], and especially Masao Kitamura, for continuing use of his samples from Buell Park.

REFERENCES

Armstrong, J. T., Quantitative analysis of silicate and oxide minerals, comparison of Monte Carlo, ZAF and φ(ρz) procedures, in *Microbeam analysis*, edited by D. E. Newbury, pp. 239–246, San Francisco Press, San Francisco, CA, 1988.

Asimow, P. D., L. C. Stein, J. L. Mosenfelder, and G. R. Rossman, Quantitative polarized FTIR analysis of trace OH in populations of randomly oriented mineral grains, *Am. Mineral.*, *91*, 278–284, 2006.

Bai, Q., and D. L. Kohlstedt, Substantial hydrogen solubility in olivine and implications for water storage in the mantle, *Nature*, *357*, 672–674, 1992.

Bai, Q., and D. L. Kohlstedt, Effects of chemical environment on the solubility and incorporation mechanism for hydrogen in olivine, *Phys. Chem. Miner.*, *19*, 460–471, 1993.

Bell, D. R., and G. R. Rossman, Water in the Earth's mantle: the role of nominally anhydrous minerals, *Science*, *255*, 1391–1397, 1992.

Bell, D. R., G. R. Rossman, J. Maldener, D. Endisch, and F. Rauch, Hydroxide in olivine: a quantitative determination of the absolute amount and calibration of the IR spectrum, *J. Geophys. Res.*, *108*, doi: 10.1029/2001JB000679, 2003.

Bell, D. R., G. R. Rossman, and R. O. Moore, Abundance and partitioning of OH in a high-pressure magmatic system: megacrysts from the Monastery Kimberlite, South Africa, *J. Petrol.*, *45*, 1539–1564, 2004.

Beran, A., Über (OH)-Gruppen in Olivin, *Anzeiger der Osterreichischen Akademie der Wissenschaften, Mathematisch Naturwissenschaftliche Klasse*, 73–74, 1969.

Beran, A., and A. Putnis, A model of the OH positions in olivine, derived from infrared-spectroscopic investigations, *Phys. Chem. Miner.*, *9*, 57–60, 1983.

Berry, A.J., J. Hermann, H. O'Neill, and G. J. Foran, Fingerprinting the water site in mantle olivine, *Geology*, *33*(11), 869–872, 2005.

Bolfan-Casanova, N., Water in the Earth's mantle, *Min. Mag.*, *69*, 229–257, 2005.

Bozhilov, K. N., H. W. Green, and L. F. Dobrzhinetskaya, Quantitative 3D measurement of ilmenite abundance in Alpe Arami olivine by confocal microscopy: confirmation of high-pressure origin, *Am. Mineral.*, *88*, 596–603, 2003.

Brodholt, J. P., and K. Refson, An ab initio study of hydrogen in forsterite and a possible mechanism for hydrolytic weakening, *J. Geophys. Res.*, *105*, 18,977–918,982, 2000.

Demouchy, S.D., and S. Mackwell, Water diffusion in synthetic iron-free forsterite. *Phys. Chem. Miner.,* *30*, 486–494, 2003.

Demouchy, S.D., S.D. Jacobsen, F. Gaillard, and C.R. Stern, Rapid magma ascent recorded by water diffusion profiles in mantle olivine. *Geology,* *34*, 429–432, 2006.

Dobrzhinetskaya, L., H. W. Green II, and S. Wang, Alpe Arami: a peridotite massif from depths of more than 300 kilometers, *Science*, *271*, 1841–1844, 1996.

Drury, M. R., Hydration-induced climb dissociation of dislocations in naturally deformed mantle olivine, *Phys. Chem. Miner.*, *18*, 106–116, 1991.

Hauri, E.H., G.A. Gaetani, and T.H. Green, Partitioning of water during melting of the Earth's upper mantle at H_2O-undersaturated conditions, *Earth Planet. Sci. Lett.*, in press, 2006.

Hermann, J., H. S. C. O'Neill, and A. Berry, Titanium solubility in olivine in the system TiO_2-MgO-SiO_2: no evidence for an ultradeep origin of Ti-bearing olivine, *Contrib. Mineral. Petrol.*, *148*, 746–760, 2005.

Hirschmann, M.M., A.C. Withers, and C. Aubaud (this volume), Petrologic structure of a hydrous 410 km discontinuity.

Ingrin, J., and H. Skogby, Hydrogen in nominally anhydrous upper-mantle minerals: concentration levels and implications, *Eur. J. Mineral.*, *12*, 543–570, 2000.

Kent, A. J. R., and G. R. Rossman, Hydrogen, lithium, and boron in mantle-derived olivine: the role of couple substitutions, *Am. Mineral.*, *87*, 1432–1436, 2002.

Khisina, N. R., and R. Wirth, Hydrous olivine $(Mg_{1-y}Fe^{2+}_y)_{2-x}v_x$-$SiO_4H_{2x}$—a new DHMS phase of variable composition observed as nanometer-sized precipitations in mantle olivine, *Phys. Chem. Miner.*, *29*, 98–111, 2002.

Khisina, N. R., R. Wirth, M. Andrut, and A. V. Ukhanov, Extrinsic and intrinsic mode of hydrogen occurrence in natural olivines: FTIR and TEM investigation, *Phys. Chem. Miner.*, *28*, 291–301, 2001.

Kitamura, M., S. Kondoh, N. Morimoto, G. H. Miller, G. R. Rossman, and A. Putnis, Planar OH-bearing defects in mantle olivine, *Nature*, *328*, 143–145, 1987.

Kohlstedt, D. L., H. Keppler, and D. C. Rubie, Solubility of water in the α, β, and γ phases of $(Mg, Fe)_2SiO_4$, *Contrib. Mineral. Petrol.*, *123*, 345–357, 1996.

Kohlstedt, D. L., and S. J. Mackwell, Diffusion of hydrogen and point defects in olivine, *Zeitschrift für Physikalische Chemie*, *207*, 147–162, 1998.

Kostenko, O., B. Jamtveit, H. Austrheim, K. Pollok, and C. Putnis, The mechanism of fluid infiltration in peridotites at Almklovdalen, western Norway, *Geofluids*, *2*, 203–215, 2002.

Kudoh, Y., Predicted model for hydrous modified olivine (HyM-α), *Phys. Chem. Miner.*, *29*, 387–395, 2002.

Kurosawa, M., H. Yurimoto, and S. Sueno, Patterns in the hydrogen and trace element compositions of mantle olivines, *Phys. Chem. Miner.*, *24*, 385–395, 1997.

Lemaire, C., S. C. Kohn, and R. A. Brooker, The effect of silica activity on the incorporation mechanisms of water in synthetic forsterite: a polarised infrared spectroscopic study, *Contrib. Mineral. Petrol.*, *147*, 48–57, 2004.

Libowitzky, E., and A. Beran, OH defects in forsterite, *Phys. Chem. Miner.*, *22*, 387–392, 1995.

Liu, Z., G. A. Lager, R. J. Hemley, and N. L. Ross, Synchrotron infrared spectroscopy of OH-chondrite and OH-clinohumite at high pressure, *Am. Mineral.*, *88*, 1412–1415, 2003.

Mackwell, S. J., D. L. Kohlstedt, and M. S. Paterson, The role of water in the deformation of olivine single crystals, *J. Geophys. Res.*, *90*, 11,319–11,333, 1985.

Martin, R. F., and G. Donnay, Hydroxyl in the mantle, *Am. Mineral.*, *57*, 554–570, 1972.

Matsyuk, S. S., and K. Langer, Hydroxyl in olivines from mantle xenoliths in kimberlites from the Siberian platform, *Contrib. Mineral. Petrol.*, *147*, 413–437, 2004.

Matveev, S., H. S. C. O'Neill, C. Ballhaus, W. R. Taylor, and D. H. Green, Effect of silica activity on OH- IR spectra of olivine: implications for low-$aSiO_2$ mantle metasomatism, *J. Petrol.*, *42*, 721–729, 2001.

Matveev, S., M. Portnyagin, C. Ballhaus, R. Brooker, and C. A. Geiger, FTIR spectrum of phenocryst olivine as an indicator of silica saturation in magmas, *J. Petrol.*, doi:10.1093/petrology/egh090, 2005.

McGetchin, T. R., L. T. Silver, and A. A. Chodos, Titanoclinohumite: a possible mineralogical site for water in the upper mantle, *J. Geophys. Res.*, *75*, 255–259, 1970.

Medaris, L. G., Garnet peridotites in Eurasian ultrahigh-pressure terranes: a diversity of origins and thermal histories, *Int. Geol. Rev.*, *41*, 799–815, 1999.

Miller, G. H., G. R. Rossman, and G. E. Harlow, The natural occurrence of hydroxide in olivine, *Phys. Chem. Miner.*, *14*, 461–472, 1987.

Mosenfelder, J. L., N. I. Deligne, P. D. Asimow, and G. R. Rossman, Hydrogen incorporation in olivine from 2–12 GPa, *Am. Mineral.*, *91*, 285–294, 2006.

Paterson, M., The determination of hydroxyl by infrared absorption in quartz, silicate glasses and similar materials, *Bull. Mineral.*, *105*, 20–29, 1982.

Peslier, A.H. and J.F. Luhr, Hydrogen loss from olivines in mantle xenoliths from Simcoe (USA) and Mexico: mafic alkalic magma ascent rates and water budget of the sub-continental lithosphere, *Earth Planet. Sci. Lett.*, *242*, 302–319, 2006.

Risold, A.-C., V. Trommsdorf, and B. Grobety, Genesis of ilmenite rods and palisades along humite-type defects in olivine from Alpe Arami, *Contrib. Mineral. Petrol.*, *140*, 619–628, 2001.

Rossman, G. R., Studies of OH in nominally anhydrous minerals, *Phys. Chem. Miner.*, *23*, 299–304, 1996.

Rossman, G. R., C. Verdel, and E. A. Johnson (in preparation), Nanopores: an important reservoir of water in some nominally anhydrous minerals.

Smith, D., and S. Levy, Petrology of the Green Knobs diatreme and implications for the upper mantle below the Colorado Plateau, *Earth Planet. Sci. Lett.*, *29*, 107–125, 1976.

Smyth, J.R., D.J. Frost, and F. Nestola, Hydration of olivine at 12 GPa, *EOS Trans. AGU*, *85*(47), Fall Meet. Suppl., Abstract T32B-04, 2005.

Sykes, D., G. R. Rossman, D. R. Veblen, and E. S. Grew, Enhanced hydrogen and fluorine incorporation in borian olivine, *Am. Mineral.*, *79*, 904–908, 1994.

Ulmer, P., and V. Trommsdorff, Phase relations of hydrous mantle subducting to 300 km, in *Mantle petrology: field observations and high pressure experimentation: a tribute to Francis R. (Joe) Boyd*, edited by Y. Fei, et al., pp. 259–282, Geochemical Society, Houston, Texas, 1999.

van Roermund, H. L. M., M. R. Drury, A. Barnhoorn, and A. A. de Ronde, Super-silicic garnet microstructures from an orogenic garnet peridotite, evidence for an ultra-deep (>6 GPa) origin, *J. Metamorph. Geol.*, *18*, 135–147, 2000.

Wirth, R., L. F. Dobrzhinetskaya, and H. W. Green, Electron microscope study of the reaction olivine + H_2O + TiO_2 -> titanian clinohumite + titanian chondrodite synthesized at 8 GPa, 1300 K, *Am. Mineral.*, *86*, 601–610, 2001.

Zhao, Y.-H., S. B. Ginsberg, and D. L. Kohlstedt, Solubility of hydrogen in olivine: dependence on temperature and iron content, *Contrib. Mineral. Petrol.*, *147*, 155–161, 2004.

Jed L. Mosenfelder, California Institute of Technology, Division of Geological and Planetary Sciences, Pasadena, CA 91125-2500, USA (jed@gps.caltech.edu)

Water in Transition Zone and Lower Mantle Minerals

Nathalie Bolfan-Casanova

Laboratoire Magmas et Volcans, CNRS, Clermont-Ferrand, France

Catherine A. McCammon

Bayerisches Geoinstitut, Bayreuth, Germany

Stephen J. Mackwell

Lunar and Planetary Institute, Houston, Texas, USA

Wadsleyite and ringwoodite have the potential to contain up to 3.2 weight percent of water as structurally bound hydroxyl (OH). However, at transition zone pressures the solubility of water in these phases decreases with increasing temperature to ~2 wt% H_2O at 1460°C for wadsleyite and to ~1 wt% H_2O at 1600°C for ringwoodite, the approximate temperatures of the 410 and 660 km discontinuities, respectively. Majorite garnet can only contain up to ~0.1 wt% water or less, and is therefore not likely to control water storage in the transition zone. In the lower mantle, water solubility in magnesium silicate perovskite and in ferropericlase are much lower. OH solubility in ferropericlase increases with pressure and water fugacity, with up to 20 ppm wt H_2O in $(Mg_{0.93}Fe_{0.07})O$ at 25 GPa and 1200°C, and new results show that it does not seem to depend on iron content. The solubility of water in mantle perovskite is still controversial, and depends on interpretation of the infrared data. Maximum water contents reported are ~0.2 wt% H_2O, but could be overestimated. Contrary to previous expectations, the solubility of water in silicate perovskite probably does not increase with Al content; Al-rich perovskites synthesized in the MORB system appear to contain less water than peridotitic perovskite. Similar to transition zone minerals, water solubility in perovskite and (Mg,Fe)O decreases with increasing temperature. A partition coefficient of 710(±180) between coexisting ringwoodite and perovskite+ferropericlase in the Al-free system implies that ringwoodite will be stabilized to greater depths in the presence of water.

1. INTRODUCTION

This chapter summarizes what is known about the maximum water content, or water storage capacity, of the nominally anhydrous minerals composing the transition zone and lower mantle. Such data are obtained by measuring experimentally the solubility of water in minerals as a function of pressure, temperature and composition. The definition of water solubility, or storage capacity, is the maximum H_2O concentration that a phase can incorporate and implies water saturation; that is, the presence of a co-existing hydrous fluid phase during the experiment. The previous chapter

Earth's Deep Water Cycle
Geophysical Monograph Series 168
Copyright 2006 by the American Geophysical Union.
10.1029/168GM06

by *Mosenfelder et al.* [this volume] concentrates on hydration mechanisms and water storage capacity of olivine in the upper mantle. Here we review data on the solubility of water in the major transition zone and lower mantle phases: wadsleyite, ringwoodite, majorite, silicate perovskite, and ferropericlase. We will focus on ferropericlase-(Mg,Fe)O and present new results on the effect of iron content and temperature on OH incorporation into this lower mantle phase.

Infrared (IR) spectroscopy is highly suitable for studying water in minerals because (1) IR absorption in the 3000-4000 cm^{-1} range is extremely sensitive to OH in minerals and water contents at the ppm level can be measured; (2) the technique is site specific, so various OH species or point defects can be identified, thus constraining the mechanism of OH incorporation and allowing the detection of hydrous inclusions; (3) the spatial resolution is routinely 30 μm, but 10 μm is achievable. A major challenge of this technique is that absorption is anisotropic in non-cubic minerals (i.e. depends on the orientation of the absorber in the structure and the polarization of the light) and it does not provide absolute hydrogen concentration without prior calibration. The calibration of *Paterson (1982)* has been extensively used, and its analytical form allows it to be applied to any OH stretching band in many silicates and oxides. A similar calibration by *Libowitzky and Rossmann (1996)* uses polarized-IR spectra. *Rauch and Keppler (2002)* measured the water solubility in pyroxenes up to 10 GPa and the calculated water contents using Paterson's (1982) calibration were on average 1.4 times lower than those calculated using the extinction coefficient of *Bell et al. (1995),* determined by H gas extraction and manometric measurements on pyroxenes. In the case of olivine, the choice of calibration is even more important considering the previous results of *Kohsltedt et al. (1996)* on olivine using Paterson's (1982) calibration need to be multiplied by a factor of ~3.5 according to the recent calibration *of Bell et al. (2003)* using Nuclear Reaction Analysis. IR calibration for olivine is a particularly important issue because the storage capacity of the upper mantle is largely determined by the high proportion of olivine. Since mineral-specific calibrations are not yet available for high-pressure phases considered in this study, we shall continue using the method of Paterson (1982) because it at least provides a basis for comparison and consistent extinction coefficients for all minerals (see *Bolfan-Casanova, 2005,* for a review).

2. WATER IN TRANSITION ZONE MINERALS

Wadsleyite

Using simple electrostatic considerations, *Smyth (1987)* observed that wadsleyite possesses an unusual oxygen site

(O1) that is highly underbonded because it is not coordinated to Si, unlike most oxygen sites in silicates. This non-silicate oxygen is coordinated to 5 Mg atoms and has a Pauling bond strength sum of only 1.67 instead of 2, which enhances its potential for hydroxylation. *Smyth (1987; 1994)* proposed that replacement of the O1 oxygen in wadsleyite by an OH group could lead to significant storage of water in this phase since full protonation of O1 (1/8 of the oxygen sites) yields 3.3 wt % H$_2$O represented by Mg$_{2-x}$SiH$_{2x}$O$_4$, with x up to 0.25. The presence of H in wadsleyite was confirmed experimentally by infrared studies (*McMillan et al.,* 1991; *Young et al.,* 1993), and *Inoue et al. (1995)* reported the synthesis of hydrous wadsleyite containing up to 3.3 wt% H$_2$O (~500,000 H/10^{6}Si), as measured by secondary ion mass spectrometry (SIMS). Hydration of wadsleyite occurs mainly via creation of Mg^{2+} vacancies for charge balance, as shown by low Mg/Si ratios measured with an electron microprobe (*Inoue et al.,* 1995; *Demouchy et al.,* 2005) and single-crystal X-ray diffraction (*Kudoh et al., 1996; Smyth and Kawamoto, 1997*). Hydration of wadsleyite also leads to minor structural changes, namely a lowering from orthorhombic to monoclinic symmetry (*Smyth et al., 1997; Kudoh and Inoue 1999*). Deviation from orthorhombic symmetry appears to result from ordering of vacancies, divalent cations and possibly Si into two non-equivalent M3 sites (*Smyth et al., 1997*). *Jacobsen et al. (2005)* later reported that monoclinic symmetry was detected only in the case of wadsleyite containing more than ~0.5 wt% H$_2$O.

Structural hydrogen in wadsleyite is characterized by two groups of IR absorption bands at 3270–3330 and 3580–3615 cm^{-1} (*McMillan et al.,* 1991; *Kohlstedt et al.,* 1996; *Bolfan-Casanova et al.,* 2000; *Kohn et al.,* 2002; *Jacobsen et al.,* 2005), as shown in Figure 1. Increasing water content in wadsleyite leads to proton disorder (i.e., the number of different H positions increases as shown by an increase in the number of IR bands) as observed by *Kohn et al.* (2002) who used FTIR and 1H MAS NMR. Indeed, *Downs* (1989) early on suggested that other oxygen sites than O1 would preferentially be hydroxylated. Also, *Jacobsen et al.* (2005) observed that the relative intensities of the main OH stretching bands in the IR spectra change from low water contents to high water contents, which indicates a difference in the substitution mechanism upon hydration. Such disorder could affect the entropy of formation of hydrous wadsleyite, and thus the olivine to wadsleyite transition, but the effect has not been investigated in detail.

Kawamoto et al. (1996) and recently Demouchy et al. (2005) measured the effect of temperature on the solubility of water in wadsleyite using SIMS and found that solubility decreases rapidly with increasing temperature (see Figure 2); whereas the effect of pressure is insignificant given the

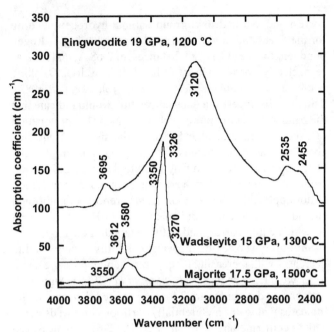

Figure 1. Unpolarized infrared spectra of wadsleyite, ringwoodite and majorite single crystals synthesized in the system MgO-SiO₂-H₂O from the study of *Bolfan-Casanova et al. (2000)*.

uncertainties in water contents. At 1400°C the solubility of water in Fe-bearing wadsleyite is 2.5 wt% H_2O; whereas at 1500°C it decreases to 1.5 wt% H_2O. Such behavior is in contrast to olivine, where the solubility of water is reported to increase with temperature from 1100 to 1300°C at 0.3 GPa (*Zhao et al.*, 2004). One explanation for this difference is that at the low pressures of the *Zhao et al.* (2004) experiments, excess water is present as an almost pure fluid; whereas at high pressures the fluid phase dissolves a significant component of silicate leading to reduced water fugacity (see e.g. *Mibe et al.*, 2002; *Komabayashi*, this volume). Moreover, the high-pressure study of Demouchy et al. (2005) indicates that water solubility increases in the range of 900 to 1000°C, before decreasing between 1000 and 1400°C. Such behavior is expected in binary systems, where at temperatures below the eutectic the solubility of water in the solid increases with temperature, and at temperatures above the eutectic solubility decreases with increasing temperature (Figure 3). Of course, above the critical point where fluid and melt form only one phase and cannot be distinguished, the definition of the eutectic point is different (see *Hirschmann et al.*, 2005). Comparing various SIMS data found in the literature (see Fig. 2) shows that samples with iron display higher water contents, indicating that incorporation of iron increases the solubility of water in wadsleyite.

It is difficult to quantify the water content in wadsleyite above ~0.5 wt% H_2O with IR because the high concentra-

tions of OH can absorb all the IR radiation, even when samples are only 15 μm thick. Also, unlike for olivine (*Bell et al., 2003*), an absolute calibration for water in wadsleyite has not been established. In such cases, the study of *Jacobsen et al. (2005)* provides a correlation between the crystallographic *b/a* axial ratio and (IR) water content ($b/a = 2.008 + 1.25 \times 10^{-6} C_{H2O}$), yielding an alternative to SIMS measurements for very hydrous wadsleyite.

Ringwoodite

Like wadsleyite, ringwoodite could also be a major repository for water in the mantle since it has been observed to contain up to 2.7 wt % H_2O, i.e. ~450,000 H/10⁶Si (*Kohlstedt et al.*, 1996; *Bolfan-Casanova et al.*, 2000; *Ohtani et al.*, 2000). The infrared spectra from *Bolfan-Casanova et al.* (2000) display typically three types of bands: one at 3695 cm⁻¹, a very broad (major) band at 3120 cm⁻¹, and two smaller bands at 2455 and 2535 cm⁻¹ (see Figure 1). From the electrostatic point of view ringwoodite was not expected to

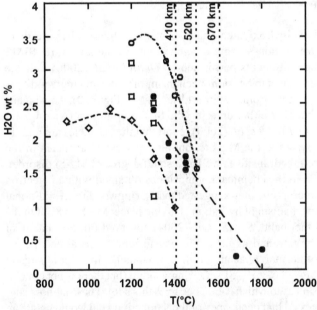

Figure 2. SIMS derived water contents as a function of temperature for pure Mg_2SiO_4 wadsleyite (*Inoue et al., 1995:* open squares; *Demouchy et al., 2005:* open diamonds); $(Mg,Fe)_2SiO_4$ wadsleyite in the KLB-1 system (*Kawamoto et al., 1996:* open circles); and pure Mg_2SiO_4 ringwoodite (*Ohtani et al., 2000:* filled circles). The dashed curves are guidelines to show the change of water solubility with temperature. The dashed vertical lines indicate the approximate temperatures at which the phase transitions from olivine to wadsleyite (410 km), wadsleyite to ringwoodite (520 km) and ringwoodite to perovskite plus ferropericlase (670 km) take place in the mantle.

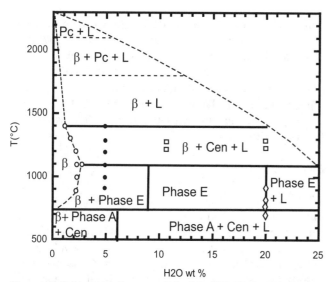

Figure 3. Schematic diagram of the Mg$_2$SiO$_4$-H$_2$O system at 15 GPa. This plot is only approximate and is derived from the available data shown as open circles (*Demouchy et al., 2005*), filled circles (*Inoue et al., 1995*) and open diamonds and squares (*Kanzaki, 1994*, and *Ohtani et al. 1995*). Cen = clinoenstatite, L= liquid, Pc= periclase, β=wadsleyite

be a hydroxylated phase, because unlike wadsleyite there are no non-silicate oxygen atoms in the structure. SIMS measurements performed by *Ohtani et al.* (2000) indicate that the substitution mechanism of water in ringwoodite is different from that in wadsleyite (see Figure 2). This means that H is either entering the Si site or an otherwise vacant site. *Kudoh et al.* (2000) report that the vacancies are mainly located at the Mg site, but also observed vacancies in the tetrahedral site as well as a small degree of Mg-Si disorder. From the chemical compositions obtained with an electron microprobe they suggested that in ringwoodite an additional mechanism of hydration is taking place: Mg$_2$Si = 8H. On the other hand, *Smyth et al.* (2004) observed full occupancy of the tetrahedral sites, but instead report partial occupancy of normally vacant tetrahedral voids i.e. interstitial sites in Mg$_{1.63}$Fe$_{0.22}$H$_{0.4}$Si$_{0.95}$O$_4$ ringwoodite (but there is not conclusion as to which cation occupies this site). These authors suggested that occupancy of the tetrahedral void would create an Si$_2$O$_7$ group, subsequently requiring a non-silicate oxygen, indicating that ringwoodite may hydrate by creation of a wadsleyite-like defect. The broadness of the OH stretching band centered at 3120 cm^{-1} as observed in the IR spectrum of hydrous ringwoodite (*Bolfan-Casanova et al.*, 2000, about 400 cm^{-1} full width at half height) is unusual and indicates substantial disorder of the proton positions. According to the latest correlation between OH stretching frequency and O-H...O hydrogen bond distance of *Libowitzky* (1999), the IR stretching bands at 3695 cm^{-1}, 3120 cm^{-1} and at ~2500 cm^{-1}

correspond to a hydroxyl group without hydrogen bonding for the first band, and to hydroxyl groups having hydrogen bond lengths, i.e., O-H...O distances, of 2.68 Å, and 2.57 Å, respectively, for the two latter bands. Since d(O...O) along octahedral edges are typically 2.8–3.0 Å (*Sazaki et al.*, 1982), using the distance-frequency correlation would indicate that the band at 3120 cm^{-1} more likely corresponds to protonation of a short tetrahedral edge. However, the distance frequency correlation assumes there is linear hydrogen bonding; if the O-H...O environments in ringwoodite deviate significantly from 180°, it may be that the distance-frequency correlation is not applicable. *Smyth et al.* (2003) report an additional shoulder at 3500 cm-1 equivalent to an O-H distance of ~3 Å, which could correspond to the octahedral edge.

Similar to wadsleyite, the water solubility in ringwoodite decreases with increasing temperature at high pressure down to ~1 wt% H$_2$O at 1600°C (*Ohtani et al.*, 2000) (Figure 2). Comparing data in Figure 2 for wadsleyite and ringwoodite indicates that water preferentially partitions into wadsleyite. This is confirmed by *Kawamoto et al.* (1996) who measured a partition coefficient of water of 2.5 between wadsleyite and ringwoodite synthesized at 15.5 GPa and 1300°C. The data in Figure 2 also indicate that the water distribution between both phases may become equal with increasing temperature if the solubilities do not vary much with pressure. Using the solubility data in Figure 2, one can derive an approximate temperature dependence of water solubility in wadsleyite and ringwoodite. However the unspecified change in water fugacity upon dissolution of the silicate phase as temperature increases prohibits direct calculation of the enthalpy of solution (ΔH).

Majorite

Bolfan-Casanova et al. (2000) reported one broad absorption band in pure magnesium (MgSiO$_3$) tetragonal majorite garnet, synthesized at 17.5 GPa and 1500 °C. This band is located near 3550 cm^{-1} with a full width at half height of ~120 cm^{-1} (see Figure 1). When integrated this band yields a water content of ~7500 H/10^{6}Si (~677 ppm wt H$_2$O). The same feature was observed for majoritic garnet grown in a peridotitic composition assemblage (*Bolfan-Casanova et al., 2000*). The frequency of the OH absorption band in majoritic garnet is 100 cm^{-1} lower than that of the hydrogarnet component in low-pressure synthetic cubic garnets, which is located around 3650 cm^{-1} (for comparison, a negligible frequency shift was observed between the OH stretching bands of the two polymorphs ortho- and clinoenstatite, *Bolfan-Casanova et al., 2000*). Also, this band is much broader than the hydrogarnet band in synthetic hydrous garnets, for which the full widths at half height are 7-60 cm^{-1} (*Withers et al., 1998*). A previous study has shown that the hydrogarnet component is not stable in mantle garnets such as pyrope, which dehydrates at pressures above 6 GPa (*Withers et al., 1998*). Thus, high-pressure garnet, majorite, is likely not an important water storage phase within the mantle transition zone.

3. WATER IN LOWER MANTLE MINERALS

A peridotitic lower mantle is composed of around 80% by volume orthorhombic magnesium silicate perovskite $(Mg,Fe)(Si,Al)O_3$, 15% $(Mg,Fe)O$ and 5% cubic calcium silicate perovskite $CaSiO_3$ (*Wood, 2000*). In contrast to minerals of the upper mantle and transition zone, it is difficult at present to study the solubility of water in perovskite and $(Mg,Fe)O$ at lower-mantle conditions experimentally (from 670 km, *i.e.* 24 GPa and 1600°C, to 2900 km, *i.e.* 140 GPa and 3000°C). Hence our knowledge about the water storage capacity of the lower mantle is very limited, especially for the solubility of water in Mg-silicate perovskite, which requires synthesis conditions of at least 24 GPa. Studying $CaSiO_3$ perovskite is even more difficult because it is unquenchable. On the other hand, ferropericlase is stable above ~600°C even at ambient pressure. The wide stability range of $(Mg,Fe)O$ enables systematic studies of its physical and chemical properties, up to 25 GPa using the multi-anvil apparatus, and thus allows more confident extrapolations

to deeper mantle pressures. Compared with upper mantle and transition zone minerals, the lower-mantle assemblage appears to be much drier.

Ferropericlase

The solubility of water in $(Mg_{0.93}Fe_{0.07})O$ ferropericlase has been studied by Bolfan-Casanova et al. (2002); the results are summarized below together with recent findings. The experiments consisted of annealing a single crystal of ferropericlase embedded in powder of the same composition to which water was added. The annealing was performed as a function of pressure from 5 to 25 GPa at 1200°C in a double capsule of Re (lined foil) and Pt (welded tube) inside which rhenium oxide was added to buffer the oxygen fugacity. The recent experiments were performed in a multi-anvil press at the Laboratoire Magmas and Volcans, except for the experiments at 25 GPa, using the 10/4 assembly (10 mm edge octahedral pressure medium together with WC cubes of 4 mm truncation), which were performed at the Bayerisches Geoinstitut. Bolfan-Casanova et al. (2002) observed that OH vibrations occur at 3320 and 3480 cm^{-1}, together with a broad band at 3400 cm^{-1} and one peak at ~3700 cm^{-1} associated with brucite $(Mg,Fe)(OH)_2$ inclusions (see Figure 4a). Broad bands are more difficult to interpret in terms of structural OH, but because the intensity of the 3400 cm^{-1}

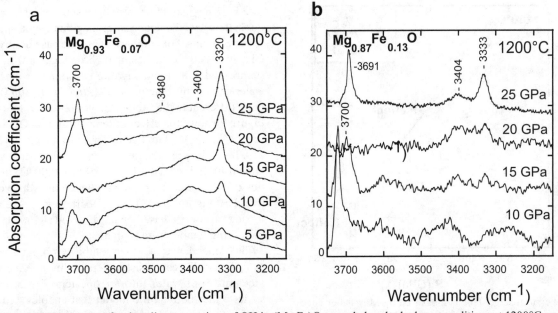

Figure 4. IR spectra, after baseline correction, of OH in $(Mg,Fe)O$ annealed under hydrous conditions at 1200°C and Re-ReO$_2$ buffered conditions up to 25 GPa (a) $X_{Fe} = 0.07\%$ from the study of *Bolfan-Casanova et al. (2002)* and (b) $X_{Fe} = 0.13\%$. The degradation of spectroscopic quality from samples with higher iron content is due to the fact that they are less transparent inducing a lower signal to noise ratio. Typical thickness is 100 to 50 μm for 7% FeO and 100 to 25 μm for 13% FeO samples. The minimum spot size is 30 μm.

band decreases with increasing pressure, it was decided not to include it in the calculation of water at ~25 GPa. The water content associated with the bands at 3320 and 3480 cm^{-1} increases with increasing pressure, and hence water fugacity under saturated conditions, reaching only 20 ppm wt H$_2$O (100 H/10^{6}Me, where Me= Mg and Fe). Also, the oxidation state of iron is important for the point defect chemistry of iron-bearing phases in the mantle. We observed that whereas the OH content in (Mg,Fe)O increases with pressure, the ferric iron (Fe^{3+}) content decreases (see Figure 5), as shown by Mössbauer spectroscopy, even if oxidizing conditions are maintained in the capsule by using the Re-ReO$_2$ buffer (f_{O_2} between Ni-NiO and Fe$_2$O$_3$-Fe$_3$O$_4$, Pownceby and O'Neill, 1994). This observation suggests that H replaces Fe^{3+} defects. The decrease in Fe^{3+}/Fe$_{tot}$ with pressure of hydrothermal annealing at 1200°C is observed for all the three compositions studied (X_{Fe} = 0.01, 0.07, 0.13). At low pressures, samples with greater total iron concentration contain more ferric iron (with similar proportions of Fe^{3+}/Fe$_{tot}$), but under lower mantle pressures, i.e., at 25 GPa, all samples tend to have the same oxidation state. A somewhat unusual behavior is that with increasing temperature, the ferric iron content slightly increases. *McCammon et al.* (1998) explain the reduction of ferric iron solubility in Mg$_{0.8}$Fe$_{0.2}$O at high pressures to be due to the enhanced stability of magnesiofer-

rite, MgFe$_2$O$_4$, relative to (Mg,Fe)O. However, no magnesioferrite was observed in the present samples.

The effect of varying iron content from 0 to 13% has little effect on the solubility of water in periclase (MgO). The incorporation mechanism appears to be the same, judging from the similarity in the IR spectra, except for pure MgO where the relative intensities of the two main bands are reversed compared to those in ferropericlase (see Figure 6). This indicates that in pure MgO the substitution mechanism of H is different than in (Mg,Fe)O. The OH bands are located at 3372, 3309 and 3295 cm^{-1} in MgO. The two latter bands are clearly observed in iron-bearing periclase, but at higher wavenumbers. This shift is consistent with an expansion of the unit cell of periclase upon incorporation of iron. Overall, as iron content increases, the OH stretching bands shift to higher wavenumbers and become broader. The bands at 3295 and 3372 cm^{-1} in MgO correspond to OH...O bond distances of 2.74 and 2.78 Å, according to the distance-frequency correlation of *Libowitzky (1999)*. These are much shorter than the O-O edges of the MgO octahedron, e.g., 2.982 Å. The reason may be that the hydrogen bond is not linear because bent bonds constantly yield higher wavenumbers (*Libowitzky, 1999*). Comparison of OH spectra between samples containing 7 and 13% Fe substituting for Mg shows that the OH content decreases either very slightly with increasing iron content, or else remains constant within the uncertainties (see Fig. 6). These samples have a ferric iron content around 0±2 Fe^{3+}/Fe$_{tot}$, and therefore have similar defect concentrations; thus it would not be surprising that they display similar water contents. For comparison, in the case of olivine at 0.3 GPa and 1200°C, increasing the iron content from 0 to 12% (expressed as percent fayalite component) increases the water content by a factor of 4.5 (*Zhao et al., 2005*), which is quite different to the subtle changes observed for (Mg,Fe)O. Figure 7 shows the effect of temperature on OH incorporation in (Mg,Fe)O with two different iron contents of 1% and 7%. It can be seen that at 25 GPa when the temperature of annealing is increased from 1200 to 1600°C, the intensities of OH bands decrease and hence the water content of (Mg,Fe)O is lowered. This appears to be correlated with ferric iron content based on its observed slight increase with increasing temperature (Fig. 6).

The dependence of water solubility on parameters such as P, T, and X_{Fe} can assist understanding the incorporation mechanism of water into ferropericlase in order to construct a thermodynamic model that enables extrapolation of the data to conditions beyond the present experimental capabilities. The point defect equations describing the thermodynamics of the incorporation of OH in (Mg,Fe)O have been described by *Bolfan-Casanova (2005)* and are briefly summarized in the following. The dependencies of

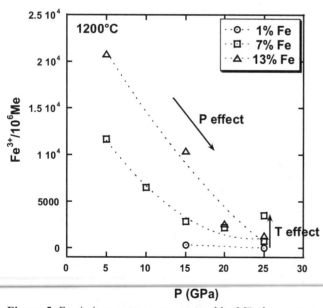

Figure 5. Ferric iron content as measured by Mössbauer spectroscopy, in (Mg,Fe)O samples with three different iron contents annealed at high pressures up to 25 GPa and 1200°C under hydrous conditions and oxygen fugacity (f_{O_2}) buffered at Re-ReO$_2$. The effect of increasing temperature in oxidizing (Mg$_{0.93}$Fe$_{0.07}$)O is also shown.

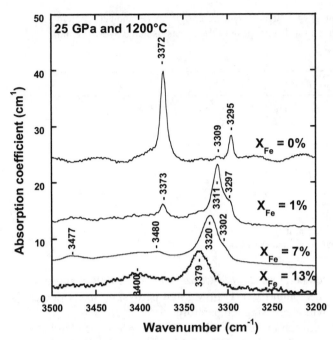

Figure 6. IR spectra of periclase with different iron contents. Samples were annealed at 25 GPa and 1200°C, except for MgO which was annealed at 1500°C (the intensity ratios are the same at 1200°C and lower pressures).

H solubility on oxygen fugacity, water fugacity and iron content as a function of the charge neutrality condition are given in Table 1. Three observations can be made: (1) the dependence on oxygen fugacity is very weak, to nil; (2) For the three plausible charge neutrality conditions examined, the dependence on iron content is negative or nil, so that OH solubility either decreases with increasing Fe content or does not depend on Fe content; (3) The water fugacity exponent varies from 1/3 to 1.

At a given temperature, the dependence of hydrogen solubility on pressure is governed by the activation volume term $P\Delta V_{solid}$ and a water fugacity term, yielding:

$$C_{OH}(P,T,f_{H_2O}) = A(T) f_{H_2O}^n \exp(-P\Delta V_{solid}/RT) \quad (1)$$

where $A(T)$ is a function of temperature, and n is dependent on the substitution mechanism. By fixing n according to Table 1, we found, for $Mg_{0.93}Fe_{0.07}O$, $A(T) = 0.2$ H/10^6 Me, $V_{solid} = 4.0$ cm^3/mol at 1200°C for $n = 1/2$ (*Bolfan-Casanova et al., 2002*). In dry (Mg,Fe)O, ferric iron in octahedral site is the major defect, $Fe_{Me}^{\cdot}$ (*Gourdin and Kingery, 1979*), charge balanced by metal vacancies, $V_{Me}^{''}$, and this defines the electroneutrality of the material $[Fe_{Me}^{\cdot}] = 2 [V_{Me}^{''}]$ (*i.e.* two ferric iron, positively charged, are charge balanced by one metal vacancy, twice negatively charged). Here, a water exponent of 1/2 means that OH is incorporated as an isolated defect. This is in agreement with the

low OH contents observed. For example, in olivine (*Zhao et al., 2004*) and enstatite (*Rauch and Keppler, 2002*) the water exponent equals 1, meaning that H is incorporated as pairs, i.e., 2 H$^+$ = Mg^{2+}, and the water solubilities are 1-2 orders of magnitude higher. However, there are other possibilities, and an exponent of 1/3 still fits the solubility data, so it is possible that a change in the charge neutrality condition from $[Fe_{Me}^{\cdot}] = 2 [V_{Me}^{''}]$ to $[(OH)_O^{\cdot}] = 2 [V_{Me}^{''}]$ occurs, where $(OH)_O^{\cdot}$ describes an hydroxyl group, positively charged. Such a change would have implications for the transport properties of ferropericlase. Water fugacity exponents of 2/3 and 1, involving defect associates clearly do not fit the data; however, we note that the available equations of state of water are not necessarily representative of hydrous fluids at high pressure and temperature.

Thus, so far, the available dataset on (Mg,Fe)O indicates that the water exponent is 1/2 or 1/3 and that there is no dependence of H solubility on iron content, *i.e.* $m = 0$. As shown in Table 1, there are two solutions for the charge neutrality condition which fulfill the above mentioned observations. Thus, for $n = 1/2$ and $m = 0$ the charge neutrality condition is $[H_i^{\cdot}] = [H_{Me}^{'}]$; and for $n = 1/3$ and $m = 0$ the charge neutrality condition is $[H_i^{\cdot}] = 2 [V_{Me}^{''}]$. Both solutions imply that H is the major defect species together with metal vacancies.

Magnesium Silicate Perovskite

The first study to propose structurally bound H in MgSiO$_3$ perovskite was *Meade et al. (1994)* who examined single

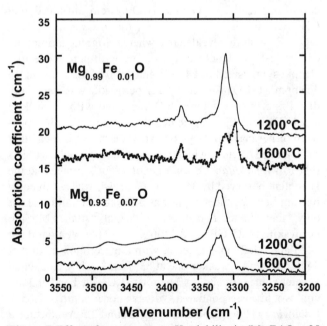

Figure 7. Effect of temperature on H solubility in (Mg,Fe)O at 25 GPa. Spectra are shown for two different iron contents.

Table 1. Dependence of hydroxyl point defect concentrations on oxygen fugacity (p), water fugacity (n) and FeO content (m) in (Mg,Fe)O for different charge neutrality conditions, given for three types of OH defects, isolated hydroxyl and defect associates (dimers and trimers).

$$C_{OH}(P,T,f_{H_2O}) = A(T) f_{O2}{}^p f_{H_2O}^n x_{Fe}{}^m \exp(-P\Delta V_{solid}/RT)$$

Charge neutrality condition	$[H_i^{\cdot}]$			$[(H_{Mg})']$			$[(2H_{Mg})^X]$		
	p	n	m	p	n	m	p	n	m
$[Fe_{Me}^{\cdot}] = 2\,[V_{Me}'']$	-1/12	1/2	-1/3	1/12	1/2	1/3	0	1	0
$[H_i^{\cdot}] = 2\,[V_{Me}'']$	0	1/3	0	0	2/3	0	0	1	0
$[H_i^{\cdot}] = [H_{Me}']$	0	1/2	0	0	1/2	0	0	1	0

where $H_i^{\cdot} = (OH)_O^{\cdot}$ is an interstitial hydrogen, *i.e.* simply an hydroxyl group
$(H_{Mg})' = \{(OH)_O^{\cdot} - V_{Me}''\}'$ dimers *i.e.* one hydroxyl coupled to one metal vacancy
$(2H_{Mg})^X = \{(OH)_O^{\cdot} - V_{Me}'' - (OH)_O^{\cdot}\}^X$ trimers *i.e.* two hydroxyls coupled to one metal vacancy

crystals of perovskite that were quenched from an H$_2$O-rich melt. They observed two pleochroic bands located at 3483 and 3423 cm^{-1} consistent with 60 ppm wt H$_2$O (700 ± 170 H/10^6 Si). The OH stretching frequencies are consistent with O...H distances of 2.82 and 2.87 Å using the distance-frequency correlation of *Libowitzky (1999)*, and the bands have different pleochroic behavior. *Ross et al. (2003)* analyzed the Laplacian of the electron density of MgSiO$_3$ perovskite in order to locate the most likely position of nonbonding electron pairs. They found that this structure is the first high-pressure silicate to have no critical point (i.e potential docking sites for H) at either of the two oxygen atoms.

Subsequent experiment performed under various *P-T* conditions in the MgO-SiO$_2$-H$_2$O system did not detect OH peaks in MgSiO$_3$ perovskite (*Bolfan-Casanova et al., 2000*). This study revealed that whereas MgSiO$_3$ akimotoite (ilmenite-type structure) coexisting with perovskite in a sample synthesized at 24 GPa and 1600°C dissolves up to 425 ppm wt H$_2$O (4740 H/10^{6}Mg), perovskite was essentially dry. Experiments where MgO coexisted with perovskite at 24 GPa and 1500°C yielded the same result. The spectrum of periclase displays two very sharp bands located at 3295 and 3372 cm^{-1}, which correspond to ~2 ppm wt H$_2$O (9H/10^{6}Mg). *Bolfan-Casanova et al. (2000)* argued that the H content observed by *Meade et al. (1994)* was not in equilibrium because the sample was allowed to equilibrate for only a few minutes. *Litasov et al. (2003)* also studied MgSiO$_3$ perovskite at 25 GPa and 1300°C and observed bands at 3423 and 3482 cm^{-1}, together with a more intense band at 3448 cm^{-1} and a weaker one at 3397 cm^{-1}. They calculated a water content of about 40 ppm wt H$_2$O. The band at 3448 cm^{-1} would correspond to a hydrogen bond length of 2.85 Å (*Libowitzky, 1999*), and clustering around 2.8 Å indicates a disorder of the proton positions within the dodecahedron. In comparison to the water contents in MgSiO$_3$ grown at higher

temperature (*Bolfan-Casanova et al., 2000*), it appears that the solubility of H in MgSiO$_3$ perovskite likely decreases with increasing temperature.

In the presence of iron, the IR spectra of (Mg,Fe)SiO$_3$ perovskite synthesized at 24 GPa and 1400°C display only one very weak peak at 3388 cm^{-1} (*Bolfan-Casanova et al., 2003*), yielding 2 ppm wt water. This study also determined the partition coefficient of water between (Mg,Fe)$_2$SiO$_4$ ringwoodite and (Mg,Fe)SiO$_3$ perovskite to be ~1050, or 1400 depending on the presence of melt or not. Given that water was measured to partition 60 times more in (Mg,Fe)O ferropericlase than in (Mg,Fe)SiO$_3$ perovskite (*Bolfan-Casanova et al., 2003*), the bulk partition coefficient of water between ringwoodite and its decomposition products perovskite + ferropericlase is estimated to be between 530 and 710.

Litasov et al. (2003) analyzed Al-perovskites containing 2, 4.5 and 7.2 wt% Al$_2$O$_3$, synthesized at temperatures of 1200–1400°C and pressures of 25–26 GPa. They reported OH peaks located at 3397, 3404 and 3448 cm^{-1}, superimposed on a very broad band centered around 3400–3450 cm^{-1}, whose intensity increases with increasing Al$_2$O$_3$ content in the perovskite. The integrated water content yields 100, 1100 and 1400 ppm wt H$_2$O as a function of increasing Al content. However, it seems that the intensity of the sharp peaks decreases with increasing Al content. If these peaks are considered as the only structural water in perovskite, then it follows that H solubility in Al-perovskite decreases with increasing Al content.

Litasov et al. (2003) also studied Al-Fe bearing perovskites for MORB and peridotite compositions. In the MORB system, perovskites synthesized at pressures of 25–26 GPa and temperatures of 1000, 1200 and 1300°C display IR peaks at 3397, 3423 and 3448 cm^{-1}. The water content associated with the most intense band at 3397 cm^{-1} is 40–110 ppm wt H$_2$O, decreasing in strength with increasing temperature of synthesis. In perovskite synthesized at 25 GPa and temperatures of 1400 and 1600°C in the peridotite system, the IR bands are more intense than in MORB-related perovskite and yield 1400-1800 ppm wt H_2O, with water contents decreasing slightly with increasing temperature. These perovskites also display very broad bands in the IR spectra. *Murakami et al. (2002)* reported about 0.2 wt % H_2O in $MgSiO_3$ grown in a natural peridotitic composition at ~25 GPa and 1600–1650°C. The IR absorption features are very broad with a major broad band centered on 3400 cm^{-1} and a sharp peak at 3690 cm^{-1}. The latter band resembles the band of brucite $Mg(OH)_2$ at 3698 cm^{-1} (*Farmer, 1974;* brucite microinclusions have also been identified in (Mg,Fe)O hydrothermally annealed at high pressures and temperatures, see above). The infrared spectra measured on different crystals display different intensities, and the SIMS measurements also show a large scatter in water contents (from 0.1 to 0.36 wt% H_2O). These heterogeneities either indicate that the samples are not homogeneous in water content, which means that they are not in equilibrium, or that there are some impurity phases. The feature around 3400 cm^{-1} is not well understood, but is also observed in (Mg,Fe)O coexisting with perovskite, and shows significant pleochroism despite the fact that ferropericlase is cubic. Therefore, whether this band is due to structurally bound hydroxyl in perovskite or not still remains to be proved.

It has been suggested that the incorporation of trivalent ions, especially Al^{3+}, will enhance the solubility of hydrogen in magnesium silicate perovskite (*Navrotsky, 1999*) thus promoting H incorporation by the following mechanism :

$$Mg^{2+} + Si^{4+} = Al^{3+} + H^+ \qquad (2)$$

This idea is based on the observation that in ceramic perovskites the trivalent cation occupies the B site in the ABO_3 structure, resulting in the creation of oxygen vacancies (see *Smyth, 1989*). However, the picture may not be as simple in magnesium silicate perovskite because Al^{3+} can also enter the A site. Indeed most of the experimental data indicate that OH content decreases with increasing Al content. The increase in OH solubility with incorporation of Al in $MgSiO_3$ perovskite is verified by experiment only for low Al contents and low temperatures, as seen above. The H contents reported by *Murakami et al. (2002)* decrease with increasing Al content in perovskite synthesized in the peridotitic system at ~25 GPa and ~1650°C. The same behavior is observed in the data of *Litasov et al. (2003)*. Such effects can be due to coupling of Al^{3+} defects ($Al_2O_3=MgSiO_3$) at high Al contents or to the decrease of water fugacity upon silicate dissolution at high pressures and high temperatures. It is therefore difficult to relate the water content in alumi-

nous perovskite to a simple model of protonation of oxygen vacancies if it is not known which site Al occupies. Also, the interplay between Al^{3+} and Fe^{3+} plays an important role in the point defect chemistry of perovskite (*Lauterbach et al., 2000*). *Litasov et al. (2003)* find that in (Al,Fe)-perovskite, coupling of Al^{3+} with Fe^{3+} (up to 60% of the total iron being ferric) controls the number of oxygen vacancies. In MORB-related perovskite, the high Al and Fe contents are sufficient to charge compensate the Al^{3+} defects on the B site by Fe^{3+} defects on the A site, and the number of oxygen vacancies is lower than in peridotitic perovskite.

4. SUMMARY AND CONCLUSIONS

The upper part of Earth's transition zone (410-520 km depth) is thought to be composed of ~60% by volume of β-$(Mg,Fe)_2SiO_4$ wadsleyite, which has the potential to contain up to 3.2 wt% H_2O, incorporated as hydroxyl. With increasing temperature, the solubility of water in $(Mg,Fe)_2SiO_4$ wadsleyite decreases to ~2 wt% H_2O at temperatures corresponding to a depth of 410 km, *i.e.*, ~1460°C. At ~520 km depth, the stable polymorph ringwoodite also dissolves up to 1.5 wt% water. With increasing temperature the solubility of water in ringwoodite decreases to 1 wt% H_2O at 1600°C, approximately the temperature at the 660 km discontinuity. Roughly 40% of the transition zone consists of the high-pressure form of garnet, majorite, which has been shown to contain little water (~0.07 wt% H_2O) compared to the high-pressure polymorphs of olivine. Thus, if water is present in the mantle, wadsleyite and ringwoodite are the major hosts for water in the transition zone and define the water budget in that region. If the transition zone were saturated with water, then melting (formation of a silicate+fluid phase) should occur not only during the transition from wadsleyite to olivine in the passively upwelling mantle as proposed by *Bercovici and Karato (2003)*, but also during the transition from ringwoodite to perovskite+ferropericlase during downward convection, in subducting slabs.

In the lower mantle, water solubility in magnesium silicate perovskite and ferropericlase is much lower than in transition zone minerals. Concerning ferropericlase, the new results presented here show that (1) the incorporation mechanism of water in pure MgO is different from (Mg,Fe)O, based on the difference in IR absorption bands in the OH stretching region. (2) OH solubility increases with pressure and water fugacity under water saturated conditions, up to 20 ppm wt H_2O at 25 GPa and 1200°C; but (3) decreases with increasing temperature to 1600°C, and (4) remains constant with increasing iron content from 1 to 13 mol% FeO at 25 GPa and 1200°C; (5) Ferric iron content increases with increasing iron content and increasing temperature, but decreases with increasing pressure. Also, while it may be expected that Al^{3+} in ferropericlase will enhance water solubility, this question needs to be further investigated.

The solubility of water in lower mantle perovskite is still controversial due to the difficulty of synthesizing samples free of impurities, as shown by the large number of different infrared spectra published. Furthermore, it may be necessary to study OH incorporation *in situ*, rather than on quenched samples where perovskite is far from its stability field and crystals are typically highly strained. SIMS analyses always yield higher water contents than FTIR, mostly because of the impossibility of eliminating the water associated with impurities. The infrared spectra display broad bands of varying shape whose significance is not yet understood. Integrating over such broad bands likely over-estimates the structurally bound OH determined by FTIR studies. Therefore, further investigation of the IR spectra of perovskite synthesized in hydrous systems is required. Contrary to previous expectations, the solubility does not appear to increase with Al content, and Al-rich perovskites such as those synthesized in the MORB system contain less water than peridotitic perovskite. Such effects are likely related to the site occupied by Al, which is difficult to assess, and to the coupling between Fe^{3+} and Al^{3+}. Similar to transition zone minerals, H solubility in perovskite decreases with increasing temperature, which is contrary to the common belief that high temperatures favor higher defect concentrations and thus higher water solubility. Thus, it is likely that the decrease in water fugacity upon dissolution of the oxide into the fluid is the dominant factor controlling the water solubility. A partition coefficient of 710 (±180) between co-existing ringwoodite and perovskite+ferropericlase in the Al-free system implies that ringwoodite will be stabilized to greater depths than in the dry system. The low water solubility in lower mantle phases as well as the negative effect of increasing temperature on water solubility indicates that the lower mantle is fairly dry.

The solubility data presented in this chapter are summarized in Figure 8, where a water saturation profile is constructed through the transition zone and lower mantle. The data compiled for transition zone minerals was obtained from SIMS whereas the water solubility in lower mantle minerals is based on IR data, thus there is obviously a lack of internal consistency in this water profile, however the order of magnitude is probably correct. Since there is no measurable effect of pressure on H solubility in transition zone phases, the profile of water storage capacity versus depth depends mostly on the choice of geotherm. This is not straightforward, since the geotherm is anchored by phase transitions, which, in turn depend on water content (*see e.g. Litasov et al., this volume*). We choose, here, a recent

geotherm model from *Katsura et al. (2004)* to calculate the $[H_2O]$ versus $T(z)$, *i.e.* 1460°C at 400 km, 1546°C at 500 km and 1586°C at 650 km depth. At the boundaries, deflections are expected which arise from two sources: the change in mineral proportion across the two-phase loop and the latent heat effects. Qualitatively, across the 410 km discontinuity, the proportion of the most hydrous phase, *i.e.*, wadsleyite, increases at the expense of the less hydrous phase, *i.e.*, olivine. *Chen et al. (2002)* reported a partition coefficient of water between wadsleyite and olivine of ~5 at 1200°C, but note that given the strong decrease of water solubility in wadsleyite with increasing temperature, it is expected that this partition coefficient decreases with increasing temperature. Also, the increase of storage capacity across the 410 km discontinuity should be minimized by the fact that, at the same time, the water solubility in wadsleyite should decrease during this exothermic reaction. A ΔT of 60 K such as for the dry olivine to wadsleyite transition (*Katsura et al., 2004*) should have a non negligible effect given that the dependence in temperature of the water solubility in wadsleyite is a very steep function (see Fig. 2). The net effect on the water profile at seismic discontinuities is thus probably more complex than the present calculation. Concerning the 660 km discontinuity, the reasoning is inverse as the low-pressure hydrous phase,

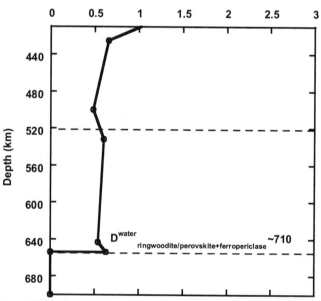

Figure 8. Saturation profile within the transition zone and lower mantle as compiled from the available literature. This profile was calculated using temperature-dependent water solubilities in wadsleyite and ringwoodite obtained from SIMS (shown in Figure 2) along the geotherm given by *Kastura et al. (2004)* for the transition zone. Water contents in the lower mantle were recalculated using the partition coefficient between ringwoodite and the perovskite + (Mg,Fe)O assemblage as obtained by IR.

i.e., ringwoodite, transforms to an almost anhydrous phase assemblage, *i.e.*, perovskite+ferropericlase, with a partition coefficient of ~710 and the reaction is endothermic.

Acknowledgments. The authors thank Tony Withers, Steve Jacobsen, Federica Marone and an anonymous reviewer for their helpful comments.

REFERENCES

Bell, D.R., Ihingher, P.D. and Rossman, G.R. (1995) Quantitative analysis of trace OH in garnet and pyroxenes, *Am. Mineral.* **80**, 465–474.

Bell, D. R., G. R. Rossman, A. Maldener, D. Endisch, and F. Rauch (2003) Hydroxide in olivine: A quantitative determination of the absolute amount and calibration of the IR spectrum, *J. Geophys. Res.* **108**, 2106.

Bercovici D. and Karato S-I. (2003) Whole-mantle convection and the transition-zone water filter, *Nature* **425**, 39–44.

Bolfan-Casanova, N., Keppler, H. and Rubie, D.C. (2000) Partitioning of water between mantle phases in the system $MgO-SiO_2-H_2O$ up to 24 GPa: Implications for the distribution of water in the Earth´s mantle, *Earth and Planetary Science Letters*, **182**, 209–221.

Bolfan-Casanova, N., Mackwell S., Keppler H., McCammon, C. and Rubie, D.C. (2002) Pressure dependence on H solubility in magnesiowüstite up to 25 GPa: Implications for the storage of water in the Earth's lower mantle, *Geophysical Research Letters*, **29**, No.10, 10.1029.

Bolfan-Casanova, N., Keppler, H., and Rubie, D.C. (2003) Water partitioning at 660 km depth and evidence for very low water solubility in magnesium silicate perovskite, *Geophysical Research Letters*, **30**, doi: 10.1029/2003GL017182.

Bolfan-Casanova N. (2005): Water in the Earth's mantle, *Min. Mag.* **69**, 227–255.

Demouchy, S., Deloule, E., Frost, D. J. and Keppler, H. (2005) Temperature and pressure dependence of the water solubility on iron free-wadsleyite. *American Mineralogist* **90**, 1084–1091.

Downs, J.M. (1989) Possible sites for protonation in $\beta-Mg_2SiO_4$ from an experimentally determined derived electrostatic potential, *American Mineralogist*, **74**, 1124–1129.

Chen J., Inoue T., Yurimoto H. and Weidner D.J. (2002) Effect of water on olivine-wadsleyite phase boundary in the $(Mg,Fe)_2SiO_4$ system, *Geophys. Res. Lett.* **29**, doi: 10.1029/GL014429.

Farmer, V.C. (1974) The infrared spectra of minerals, Mineralogical Society, London.

Gourdin, W.H. and Kingery, W.D. (1979) The defect structure of MgO containing trivalent defect: solutes shell model calculations, *Journal of Material Science*, **14**, 2053–2073.

Hirschmann, M.M., Aubaud, C., and Withers, A.C. (2005) Storage capacity of H_2O in nominally anhydrous minerals in the upper mantle *Earth and Planetary Science Letters*. **236**, 167–181.

Inoue, T., Yurimoto, H., Kudoh, Y. (1995) Hydrous modified spinel, $Mg_{1.75}SiH_{0.5}O_4$, a new water reservoir in the mantle transition zone, *Journal of Geophysical Research*, **22**, 117–120.

Jacobsen, S. D, Demouchy, S., Frost, D. J. Boffa Ballaran, T. and Kung, J. (2005). A systematic study of OH in hydrous wadsleyite from polarized infrared spectroscopy and single-crystal X-ray diffraction. *American Mineralogist*, **90**, 61–70.

Kanzaki M. (1991) Stability of hydrous magnesium silicates in the mantle transition zone, *Phys. Earth Planet. Int.* **66**, 307–312.

Katsura T. Yamada H., Nishikawa O., Song M., Kubo A., Shinmei T., Yokoshi S., Aizawa Y., Yoshino T., Walter M.J. and Ito E. (2004) Olivine-Wadsleyite transition in the system $(Mg,Fe_2)SiO_4$, *Jour. Geophys. Res.* **109**, doi: 10.1029/2003JB002438.

Kawamoto, T., Hervig, R.H. and Holloway, J.R. (1996) Experimental evidence for a hydrous transition zone in the early Earth´s mantle, *Earth Planet Science Letters*, **142**, 587–592.

Kohlstedt, D.L., Keppler, H. and Rubie, D.C. (1996) Solubility of water in the α, β and γ phases of $(Mg,Fe)_2SiO_4$, *Contributions to Mineral Petrology*, **123**, 345-357.

Kohn, S.C., Brooker, R.A., Frost, D.J., Slesinger, A.E. and Wood B.J. (2002) Ordering of hydroxyl defects in hydrous wadsleyite $(\beta-Mg_2SiO_4)$, *American Mineralogist*, **87**, 293–301.

Kudoh, Y., Inoue, T. and Arashi, H. (1996) Structure and crystal chemistry of hydrous wadsleyite, $Mg_{1.75}SiH_{0.5}O_4$ possible hydrous magnesium silicate in the mantle transition zone, *Physics and Chemistry of Minerals*, **23**, 461–469.

Kudoh, Y., and Inoue T. (1999) Mg-vacant structural modules and dilution of the symmetry of hydrous wadsleyite, $\beta-Mg_2SiO_4$ with $0.00<x<0.25$, *Phys. Chem. Min.*, **26**; 382–388.

Kudoh, Y., Kuribayashi, T., Mizobata, H. and Ohtani, E. (2000) Structure and cation disorder of hydrous ringwoodite, $\gamma-Mg_{1.89}Si_{0.98}H_{0.30}O_4$, *Physics and Chemistry of Minerals*, **27**, 474–479.

Lauterbach S., McCammon C., van Aken P., Langenhorst F. and Seifert F. (2000) Mössbauer and ELNES spectroscopy of $(Mg,Fe)(Si,Al)O_3$ perovskite: a highly oxidized component of the lower mantle, *Contrib. Mineral Petrol.* **138**, 17–26.

Libowitzky, E. (1999) Correlation of O-H stretching frequencies and O-H...O hydrogen bond lengths in minerals, *Monatshefte für. Chemie*, **130**, 1047–1059.

Libowitzky, E. and Rossman, G.R. (1996) Principles of quantitative absorbance measurements in anisotropic crystals, *Phys. Chem. Minerals*, **23**, 319–327.

Litasov, K., Ohtani, E., Langenhorst, F., Yurimoto, H., Kubo ,T. and Kondo, T. (2003) Water solubility in Mg-perovskites and water storage capacity in the lower mantle, *Earth and Planetary Science Letters*, **211**, 189–203.

Meade, C, Reffner, J.A. and Ito, E. (1994) Synchrotron infrared absorbance measurements of hydrogen in $MgSiO_3$ perovskite, *Science*, **264**, 1558–1560.

McCammon, C., Peyronneau, J. and Poirier, J-P. (1998) Low ferric iron content of $(Mg,Fe)O$ at high pressures and temperatures, *Geophys. Res. Letters* **25**, 1589–1592.

McMillan P.F., Akaogi M., Sato R.S., Poe B., Foley J. (1991) Hydroxyl groups in $\beta-Mg_2SiO_4$, *Am. Mineral.* **76**, 354–360.

Mibe, K., Fujii, T. and Yasuda, A. (2002) Composition of aqueous fluids coexisting with mantle minerals at high pressure and its bearing on the differentiation of the Earth's upper mantle, *Geochimica et Cosmochimica Acta*, **66**, 2273–2285.

Murakami, M., Hirose, K., Yurimoto, H., Nakashima, S. and Takafuji, N. (2002) Water in the Earth's lower mantle, *Science,* **295**, 1885–1887.

Navrotsky, A. (1999) A lesson from Ceramics, *Science,* **284**, 1788–1789.

Ohtani E., Shibata T., Kubo T. and Kato T. (1995) Stability of hydrous phases in the transition zone and the upper part of the lower mantle, *Geophys. Res. Lett.* **22**, 2553–2556.

Ohtani, E., Mizobata, H. and Yurimoto, H. (2000) Stability of dense hydrous magnesium silicate phases in the systems Mg_2SiO_4-H_2O and $MgSiO_3$-H_2O at pressures up to 27 GPa, *Physics and Chemistry of Minerals,* **27**, 533–544.

Paterson, M.S. (1982) The determination of hydroxyl by infrared absorption in quartz, silicate glasses and similar materials, *Bull. Mineral.* **105**, 20–29.

Pownceby, M.I., O'Neill, H.St.C. (1994) Thermodynamic data from redox reactions at high temperatures. IV. Calibration of the Re-ReO_2 oxygen buffer from EMF and NiO+Ni-Pd redox sensor measurements, *Contributions to Mineral Petrology,* **118**, 130–137.

Rauch, M., and Keppler, H. (2002) Water solubility in orthopyroxene, *Contrib Mineral Petrol* **143**, 525–536.

Ross N.L., Gibbs G.V., and Rosso K. (2003) Potential docking sites and positions of hydrogen in high-pressure silicates, *Am. Min* **88**, 1452–1459.

Sasaki S., Prewitt C., Sato Y. and Ito E. (1982) Single crystal study of γ-Mg_2SiO_4, *J. Geophys. Res.* **87**, 7829–7832.

Smyth, D.M. (1989) Defect equilibria in perovskite oxides, *in* Perovskite, AGU Monograph Series 45.

Smyth, J.R. (1987) β-Mg_2SiO_4 a potential host for water in the mantle ? *American Mineralogist,* **72**, 1051–1055.

Smyth, J.R. (1994) A crystallographic model for hydrous wadsleyite (β-Mg_2SiO_4): An ocean in the Earth's interior? *American Mineralogist,* **79**, 1021–1024.

Smyth, J.R. and Kawamoto, T. (1997) Wadsleyite II :A new high-pressure hydrous phase in the peridotite-H_2O system, *Earth and Planeteray Science Letters,* **146**, E9–E16.

Smyth, J.R., Kawamoto, T. , Jacobsen S.D., Swope J.R., Hervig R.L. and Holloway J.R. (1997) Crystal strcuture of monoclinic hydrous wadsleyite [β-$(Mg,Fe)_2SiO_4$], *Am. Min.* **82**, 270–275.

Smyth, J.R., Holl, C.M., Frost, D.J., Jacobsen, S.D., Langenhorst, F. and McCammon C.A. (2003) Structural systematics of hydrous ringwoodite and water in Earth's interior, *Am. Min* **88**, 1402–1407.

Smyth, J.R., Holl, C.M., Frost, D. and Jacobsen, S. (2004) High pressure crystal chemistry of hydrous ringwoodite and water in the Earth's interior, *Physics of the Earth and Planetary Interiors,* **143**,–**144**, 271–278.

Withers A.C., Wood B.J. and Carroll M.R. (1998) The OH content of pyrope at high pressure, *Chem. Geol.* **147**, 161–171.

Wood, B.J. (2000) Phase transformations and partitioning relations in peridotite under lower mantle conditions, *Earth and Planetary Science Letters,* **174**, 341–354.

Young, T.E., Green, H.W. II, Hofmeister, A.M., and Walker, D. (1993) Infrared spectroscopic investigation of hydroxyl in β-$(Mg,Fe)_2SiO_4$ and coexisiting olivine: implications for mantle evolution and dynamics. Physics and Chemistry of Minerals, **19**, 409–422.

Zhao, Y.H., Ginsberg, S.B. and Kohlstedt, D.L. (2004) Solubility of hydrogen in olivine dependence on temperature and iron content, *Contributions to Mineral Petrology,* **147**, 155–161.

Raman Spectroscopic Studies of Hydrous and Nominally Anhydrous Deep Mantle Phases

Annette K. Kleppe

Department of Earth Sciences, University of Oxford, Oxford, UK

Andrew P. Jephcoat*

Department of Earth Sciences, University of Oxford, Oxford, UK

This chapter deals with Raman spectroscopic measurements on both hydrous and nominally anhydrous silicates that are relevant to water cycling and storage in the Earth's mantle. Raman spectroscopy is a sensitive tool for probing structure and chemistry of silicates. It is particularly useful in the identification of OH and H_2O in minerals and in the discrimination between structurally similar phases. Here we present a collection of room-pressure and -temperature Raman spectra and structure models followed by low-temperature Raman spectra of lawsonite, the only known phase containing molecular water groups that is stable below ~150 km depth. We compare the behavior of the Raman active OH stretching frequencies in chlorite across the pressure-induced transition at 9-10 GPa with the results of first-principles calculations, and we also present high-pressure Raman spectroscopic studies of transition zone silicate spinelloids and spinel. The latter studies show that the incorporation of iron (Fe^{2+}) into hydrous wadsleyite and ringwoodite may allow resonance electronic Raman scattering processes.

1. INTRODUCTION

The study of high-pressure hydrous and OH-bearing nominally anhydrous phases is central to understanding the Earth's deep water cycle, and in particular to how water is transported, stored, and recycled in the mantle. Even minor amounts of water, in the form of structurally bound water molecules (H_2O) or hydroxyl (OH) in minerals, are known to influence their physical and chemical properties [e.g., *Karato*, this volume; *Smyth et al.*, 2003; *Hirth and Kohlstedt*,

1996; *Karato et al.*, 1986], and the presence of water, whether global or localized, may profoundly affect melt generation, solid-state convection, and seismic velocities in the Earth. Other chapters in this volume have focused on water storage capacity [*Bolfan-Casanova et al.*, and *Mosenfelder et al.*, this volume], phase relations [*Komabayashi*, this volume], phase transformations and thermodynamics [*Litasov et al.*, and *Hirschmann et al.*, this volume] and seismic velocities [*Jacobsen and Smyth*, this volume]. Here, we focus on vibrational studies of hydrous and nominally anhydrous high-pressure phases from Raman spectroscopic methods as they pertain to understanding hydrogen incorporation mechanisms and the nature of the hydrogen bond under pressure. A collection of room-pressure and -temperature Raman spectra and crystallographic models for the most important hydrous and nominally anhydrous silicates is presented. Also, we offer two examples of crystal chemically

*Currently at Diamond Light Source Ltd., Chilton, Didcot, Oxfordshire, UK

Earth's Deep Water Cycle
Geophysical Monograph Series 168
Copyright 2006 by the American Geophysical Union.
10.1029/168GM07

interesting low-temperature (room-pressure) and high-pressure (room-temperature) behaviour related to H_2O and OH in minerals: We discuss briefly low-temperature Raman spectroscopy of lawsonite and experimental as well as theoretical investigations of the effect of pressure on hydrogen bonding in chlorite. The rest of the chapter concentrates on high-pressure Raman spectroscopic investigations of hydrous transition zone spinelloids and spinel.

1.1. Recycling of Hydrogen in Cold Subduction Zones

In the lower crust and upper mantle water is stored in the form of H_2O and OH in hydrous minerals such as chlorite, lawsonite, talc, and serpentine. These minerals can persist along relatively low-temperature high-pressure paths within old subduction zones to depths of 150-300 km, but ultimately break down upon dehydration [e.g., *Komabayashi*, this volume]. Most of the water is released on subduction causing partial melting of the overlying mantle wedge and delivered back to the surface in melts from arc volcanism, while some water may be transferred to dense hydrous magnesium silicates within the cool interior of the slab and dragged deeper down into the Earth. Dense hydrous magnesium silicates (DHMS, e.g., phases A, B, shyB, D, and E) might be important links in a chain of hydrous phases that could reach the transition zone; Phase D could even be stable well into the lower mantle, to at least 1200 km depth within subducting lithosphere [e.g., *Shieh et al.*, 1998]. In the transition zone significant quantities of water could then become incorporated in the form of structurally bound OH in the nominally anhydrous high-pressure olivine polymorphs, wadsleyite (β-$(Mg,Fe)_2SiO_4$) and ringwoodite (γ-$(Mg,Fe)_2SiO_4$) [e.g., *Bolfan-Casanova et al.*, and *Hirschmann et al.*, this volume]. Wadsleyite can accommodate up to 3.3 wt% H_2O and ringwoodite up to 2.2 wt% H_2O [e.g., *Bolfan-Casanova et al.*, 2000; *Kudoh et al.*, 1996; *Kohlstedt et al.*, 1996; *Inoue et al.*, 1995; *Smyth*, 1994, 1987]. Though the actual amount of water present in wadsleyite and ringwoodite at transition zone conditions is likely less [e.g., *Bolfan-Casanova et al.*, this volume], the transition zone has the potential to contain a significant proportion of the global water budget.

1.2. Micro-Raman Spectroscopy of Water and Hydroxyls in Minerals

1.2.1. Raman effect and infrared (IR) absorption process.
The Raman effect [e.g., *Gardiner and Graves*, 1989; *Brüesch*, 1986] arises from inelastic scattering of photons with energies of the order of 2-4 eV (visible light) by optical phonons. IR spectroscopy is based on an absorption process involving photons in the infrared region of the spectrum that

have energies comparable to phonon energies [e.g., *Brüesch*, 1986]. Both, Raman scattering and IR absorption process are governed by quantum mechanical selection rules for vibrational transitions [e.g., *Gardiner and Graves*, 1989; *Brüesch*, 1986; *Rousseau et al.*, 1981]. A vibration will be Raman active if there is a change in the polarizability of the molecule during the vibrational motion, and it will be IR active if there is a change in dipole moment. Hence, Raman and IR spectroscopy provide complementary information. However, the two sets of vibrations are not always mutually exclusive: for non-centrosymmetric molecules some vibrations can be both, Raman and IR active.

1.2.2. Vibrational modes of hydroxyl and water molecule.
The OH molecule has only one stretching vibration along the O-H bond which is Raman and IR active. The stretching mode of the free (gaseous) hydroxide ion occurs at 3555,6 cm^{-1} [*Lutz*, 1995]. Bonding in solids changes this wavenumber. In minerals the hydroxyl group is often hydrogen bonded to an oxygen ion (O-H$\cdots$O) and the exact wavenumber of the OH stretching vibration depends on the strength of the hydrogen bond. Non-hydrogen bonded hydroxyls show stretching modes above 3555 cm^{-1} (up to ~3700 cm^{-1}). A shift to higher wavenumbers compared to the free OH ion is mainly caused by an increase in the strength of the internal OH bond due to the electrostatic field of neighbouring metal ions [*Lutz*, 1995].

The non-linear H_2O molecule has three normal vibrations which are all Raman and IR active. There are two fundamental stretching and one bending mode: in symmetric vibration (v_s) the two O-H bonds lengthen and shorten simultaneously; in antisymmetric stretching vibration (v_{as}) one O-H bonds lengthens while the other shortens; and in bending vibration (δ), the H-O-H angle oscillates. The vibrations of the free water molecule have been observed at v_s=3657 cm^{-1}, v_{as}=3756 cm^{-1} and v_δ=1595 cm^{-1} [e.g., *Lutz*, 1988]. In crystalline solids the vibrations of the free H_2O molecule are affected by its structural environment where hydrogen bonding and interactions with adjacent metal-ions occur. In Raman and IR spectra the stretching wavenumber of OH (H_2O and OH) structurally bound in silicates is often observed between 2900 and 3700 cm^{-1}. The bending mode of the water molecule (1595-1650 cm^{-1}) makes possible to distinguish spectroscopically structural OH from H_2O.

Isotopic substitution of hydrogen by deuterium leads to frequency (v) shifts due to the mass ratio (v^2 is proportional to force constant/reduced mass); e.g., the stretching mode of the free OD molecule occurs at v=2625,3 cm^{-1} [*Lutz*, 1995]. In crystalline solids the H/D isotopic shift for OH stretching and librational modes is about 1,3 and for lattice (translational) modes which include OH ion motions 1,03 [*Lutz*, 1995] making OH related vibrational modes easily distinguishable from other vibrational modes in Raman and IR spectra. Extended discussions of the vibrational spectra of the water molecule and OH/OD ions in crystalline solids are presented in e.g., Lutz (1988, 1995) and Nakamoto et al. (1955).

1.2.3. Hydrogen bonding in minerals. The nature of hydrogen bonding in minerals at mantle pressures can exhibit considerable complexity and influences properties such as compressibility [e.g., *Smyth et al.,* 2003], electrical conductivity [e.g., *Karato,* this volume; *Huang et al.,* 2005], rheology [e.g., *Hirth and Kohlstedt,* 1996; *Karato et al.,* 1986] and, in a wider context, mineral stability as a function of pressure, temperature, and composition.

Conventionally, the hydrogen bond (O-H···O) consists of a strong, covalent O-H bond and a longer, weaker (H···O) bond. The strength of hydrogen bond, and hence the OH stretching vibration, depends on the O···O distance, the O-H···O bond angle, and type of metal-ions bonded to the OH group (synergetic effect) [*Lutz,* 1995]. For straight hydrogen bonds an empirical correlation between OH stretching frequencies (ν_{OH}) and O···O distances has been established [e.g., *Libowitzky,* 1999; *Nakamoto et al.,* 1955]; bent and bifurcated bonds deviate in their behaviour. The correlation between OH stretching frequencies and O···O distances can be used to help assign observed OH stretching modes to distinct crystal structural sites. But, O-H···O bonds are not always (close to) linear, and hence, care has to be taken when using this correlation. In the case of crystallographically different hydroxide ions involved in equally strong hydrogen bonds, their different Raman bands can only be distinguished from the variation in the orientational behaviour of their intensities. Maximum intensity is observed when the electric vector of the incident light is orientated parallel to the OH-bonding axis.

In high-pressure experiments, the behaviour of the OH stretching modes in response to decreasing nearest neighbour O···O distances is of particular interest. The correlation between OH stretching frequencies and O···O distances suggests for linear O-H···O bonds a decrease in the OH stretching frequency with reduced O···O distance due to increased hydrogen bonding. For large O···O distances, beyond 2.9 Å, a reduction of the O···O distance leads to a small, negative frequency shift that becomes increasingly negative at distances below 2.9 Å. Negative pressure dependencies of OH stretching frequencies due to pressure-enhanced hydrogen-bonding are much more common and better understood than positive pressure shifts of OH stretching frequencies. An important factor in the positive pressure dependence of OH stretching modes might be cation-H repulsion and in particular H-H repulsion under compression [e.g., *Prewitt and Parise,* 2000].

1.2.4. Raman vs infrared (IR) spectroscopy. Raman spectroscopy is a useful, non-destructive, fingerprinting technique for characterising the molecular species present within a mineral; in particular it makes possible identification of structural H_2O and OH in silicates. In addition, Raman spectroscopy is, due to its sensitivity to subtle changes in local or long-range symmetry, well suited to probe changes in mineral structure and bonding as function of water content and in-situ as function of pressure and/or temperature. However, a quantitative measurement of the concentrations of molecular species with Raman spectroscopy is complicated by practical difficulties such as determination of the absolute Raman scattering cross-section of different species.

In contrast to Raman spectroscopy, IR spectroscopy allows determining more routinely quantitative abundances of molecular species and has been preferentially used to measure concentrations and speciations of volatiles in minerals. IR spectroscopy is particularly powerful in determining small concentrations of H_2O and OH due to their high IR activity. However, the high sensitivity of this technique for hydrous species places constraints on sample thickness and sample dilution to avoid saturation. Accurate determination of OH concentrations with IR spectroscopy requires independent calibration. A limitation of IR spectroscopy is the sample size: synchrotron IR radiation requires samples >5 μm; a restriction caused by the physical diffraction limits of IR radiation. Consequently samples with sizes of the order of the wavelength of infrared light (e.g., some fluid inclusions) are more suited for Raman spectroscopic investigations. Confocal Micro-Raman spectroscopy allows analyzing diffraction limited sample sizes and areas as small as ~1 μm in diameter using visible excitation light with its much shorter wavelength (400-700 nm) than IR radiation. Significant efforts have been made to establish Raman spectroscopy as a quantitative, non-destructive compositional probe [e.g., *Zajacz et al.,* 2005; *Pasteris et al.,* 1988].

Micro-Raman techniques have distinct advantages in the field of high-pressure diamond-anvil cell studies, which are easier to perform than high-pressure IR spectroscopic experiments. Difficulties of high-pressure IR spectroscopic experiments include coupling the IR radiation via optical components with a small sample in the diamond-anvil cell and a more complex sample preparation than for high-pressure Raman spectroscopic experiments. Nevertheless, high-pressure IR and high-pressure Raman spectroscopic studies have and continue to provide valuable information on the variation of hydrogen bonding in minerals at mantle pressures. The interpretation of spectroscopic measurements can be difficult without complementary, direct structural information on the pressure dependence of the H atom positions.

1.2.5. Linking Raman and IR spectroscopy with macroscopic properties. While Brillouin scattering provides information on the elastic properties of minerals from optoacoustic shifts directly related to compressional and shear wave velocities (at GHz frequencies), derivation of macroscopic properties from Raman and IR spectroscopy is less straightforward. Knowledge of the vibrational spectrum as a function of pressure and temperature allows derivation of some thermodynamic properties such as the thermal conductivity, heat capacity, and vibrational entropy [e.g., *Hofmeister,* 2001; *Chopelas,* 1991b; *Kieffer,* 1979], but the full vibrational spectrum is rarely obtained for complex structures and compositions. Raman and IR spectroscopy are sensitive techniques widely used in detecting the onset of structural phase transformations at high-pressure and temperature. Vibrational spectroscopic techniques are also particularly sensitive to local, structural defects, which are important in understanding rheological properties, and provide critical tests of theoretical calculations on defect structure and energetics [e.g., *Braithwaite et al.,* 2003; *Brodholt and Refson,* 2000]. Here, we focus on the ability of Raman and IR spectroscopy to identify and characterize hydrous species in minerals.

2. MOTIVATION

Our motivation to investigate a range of hydrous and nominally anhydrous deep mantle phases with Micro-Raman spectroscopy has been stimulated by their possibly significant role as water carriers and reservoirs in the deep Earth. Spectroscopic studies are also aimed at understanding the nature of hydrogen bonding in minerals at high-pressures, which remains poorly understood. In order to constrain possible protonation sites and charge-balancing vacancies, we emphasize the importance of changes in hydrogen bonding and proton environments under pressure and test for pressure-induced structural changes using a local structural probe. The work draws upon ab-initio techniques in some cases to help interpret the experimentally observed Raman shifts. In this chapter, we include Raman spectroscopic studies of lawsonite, chlorite, 10-Å phase, phase E, olivine and the nominally anhydrous transition zone phases, wadsleyite, wadsleyite II, and ringwoodite. This selection of hydrous phases contains hydrous species either as an integral part of their crystal structure or as trace hydroxyls incorporated into a nominally anhydrous structure. In addition this suite of phases spans a wide range of hydrogen contents, from ~0.8 wt% in olivine to 2.4 wt% in wadsleyite and up to ~13.5 wt% H_2O in chlorite. The speciation of hydrogen in these phases is also variable: whereas lawsonite contains both molecular water

and hydroxyls, the denser mineral phases contain only OH groups. In order to provide a better control on the protonation mechanism we also studied deuterated samples (e.g., partly deuterated 10-Å phase). The important advantages of deuterated samples are that OH related vibrations can be distinguished from other spectral modes by their H/D isotopic shift and any hydrogen-related surface contamination is excluded.

3. EXPERIMENTAL METHODS

Here we describe the method of the typical experiments performed at Oxford: Unpolarized, room-pressure, low-temperature, and high-pressure, room-temperature Raman spectra were measured in 135° scattering geometry with a SPEX Triplemate equipped with a back-illuminated, liquid-N_2-cooled CCD detector in the range 50-4000 cm^{-1} from single-crystal fragments and/or compressed powders of the samples. Raman spectra were excited by the 514.5 and 488 nm line of an argon-ion laser focused to a ~5 μm spot on the sample and collected through a confocal aperture giving high spatial resolution at the sample. The laser power was low enough to avoid heating the samples. The intrinsic resolution of the spectrometer is 1.5 cm^{-1} and calibrations are accurate to ±1 cm^{-1}. The frequency of each Raman mode was obtained by Voigtian curve fitting using a least-squares algorithm. The fitted positions of the Raman bands can vary outside the intrinsic limits of accuracy by 2-25 cm^{-1} for bands with large half width and/or strong overlap (e.g., phase E). The relative intensities of Raman modes do change with changing crystal to laser beam orientation, and therefore room-pressure and -temperature Raman spectra were often collected in various, arbitrary orientations of a crystal relative to the incident laser-beam in order to reveal as many Raman modes as possible.

In each high-pressure Raman spectroscopic experiment a single-crystal fragment or a compacted powder disk of a sample was mounted in a diamond-anvil cell together with a <10 μm diameter ruby sphere for pressure calibration. Fluid helium or argon was loaded at 0.2 GPa [*Jephcoat et al.,* 1987] as a pressure-transmitting medium to ensure hydrostatic conditions. Natural, low-fluorescence diamonds, synthetic type IIa diamonds (Sumitomo Electric), or moissanite anvils together with a preindented stainless steel gasket formed the sample chamber. Data were recorded for both, increasing and decreasing pressure. The low-temperature Raman spectroscopic experiments were performed using a vacuum loading, continuous flow microscope cryostat system (Oxford Instruments) equipped with water-free quartz windows and an ITC temperature controller for monitoring and stabilizing the temperature to within ±0.1 K.

4. ROOM-PRESSURE AND -TEMPERATURE RAMAN SPECTRA OF HYDROUS AND NOMINALLY ANDHYROUS SILICATES

In this section we present a collection of high quality, room-pressure and -temperature Raman spectra and crystallographic details of selected, hydrous and nominally anhydrous silicate phases that might play an important role in transporting and storing water in the deep Earth. The spectra are taken either from previously published studies in the literature or are newly obtained. The choice of silicates in this collection is intended to cover the phases most relevant to Earth's deep water cycle, but it is not comprehensive. In addition to illustrating and characterizing the presence of hydroxyl in these phases, this collection is intended to aid other researchers in quick phase identification using Raman spectra. However, a word of caution: Raman spectra can differ significantly from sample to sample for various reasons. Care must be taken when collecting Raman spectra from very small diamond-anvil cell and multi-anvil press run products that are generally composed of several phases. Considerable differences between published Raman spectra might be due to probing multi-phase assemblages. Caution is also necessary when collecting Raman spectra of dark colored (e.g., blue) minerals using visible laser-excitation because these minerals absorb the laser-beam strongly, heating up locally and sometimes even back transforming. The Raman spectra shown in Table 1 have been collected using laser powers low enough to avoid heating of the samples, and, we believe, that they are single-phase spectra. A full characterization of each sample can be found in the references listed in Table 1.

The reported Raman spectra range from 50 to 4000 cm^{-1} encompassing both silicate stretching and bending modes, more complicated MgO$_6$ vibrations, and OH stretching and bending frequencies. In the range 50 to 1200 cm^{-1} the most characteristic stretching and bending modes of SiO$_4$, Si$_2$O$_7$, and SiO$_6$ groups can be used to infer their presence in a mineral. In the range 3100 to 3700 cm^{-1} OH stretching modes of hydroxide groups and H$_2$O molecules can be observed, while H$_2$O bending modes occur around 1590-1660 cm^{-1}.

4.1. Hydrous, Layered Silicates

4.1.1. Clinochlore. Chlorite, one of the most OH-rich silicates, is a major constituent of hydrated oceanic crust and may have a significant role as a water carrier in cold, old subduction zones [*Peacock*, 1990]. The Mg end-member chlorite, clinochlore (Mg$_5$Al)(Si$_3$Al)O$_{10}$(OH)$_8$, contains about 13.5% water by weight and is expected to be the major aluminous phase in low-alkali hydrous peridotites below

800 °C [*Jenkins*, 1981]. The crystal structure of chlorite is characterized by an alternation of a talc-like 2:1 layer (two opposing tetrahedral sheets with an octahedral sheet between them) and a brucite-like layer (B) (Tab. 1). Bonding between these layers involves only hydrogen bonds. In the case of end-member clinochlore, the octahedra of the 2:1 layer are fully occupied by Mg, and those of the brucite-like layer have the following occupancies: M3 = Mg, M4 = Al (M3: M4 = 2:1). Increasing Al substitution in clinochlore via the Mg-Tschermak's exchange, Mg$^{[6]}$ + Si$^{[4]}$ ↔ Al$^{[6]}$ + Al$^{[4]}$, involves only compositional changes in the 2:1 layer; the cation composition of the brucite-like layer remains fixed at Mg$_2$Al.

The Raman spectrum of clinochlore (Tab. 1) consists of three OH-stretching bands between 3400 and 3650 cm^{-1}, which are attributed to the hydrogen-bonded interlayer OH: the 3477 cm^{-1} mode results from OH-stretching vibrations of (Mg$_2$Al)O-H···O(SiAl), and the 3605 and 3647 cm^{-1} bands are due to OH-stretching vibrations of the (Mg$_2$Al)O-H···O(SiSi) hydroxyl groups. The highest- frequency and narrow OH band at 3679 cm^{-1} (full width at half maximum: ~20 cm^{-1}) is assigned to the non-hydrogen-bonded hydroxyls of the talc-like 2:1 layer. Its frequency is close to the frequency of the talc OH stretching mode. The Raman spectrum of aluminous clinochlore shows one significant difference in the OH stretching region: the presence of a (Mg$_2$Al)O-H contribution from the 2:1 layer at 3641 cm^{-1} and a corresponding reduced intensity for the Mg$_3$O-H peak at 3675 cm^{-1}. The increased intensity of the (Mg$_2$Al)OH···O(SiAl) Raman band at 3455 cm^{-1} in aluminous clinochlore agrees with the increased Al$^{[4]}$ content.

4.1.2. Talc. Talc is a common, hydrothermal alteration product of peridotite and may transport H$_2$O in subducting slabs to depths of ~150 km [e.g., *Pawley and Wood*, 1995]. The talc structure is characterised by alternating, charge-neutral 2:1 layers, which consist of two opposing tetrahedral sheets (T) with one octahedral sheet (O) between them. Adjacent 2:1 layers are held together by van der Waals bonds. All tetrahedral sites are occupied by Si and all octahedral sites by Mg ions (trioctahedral); the hydroxyl groups point perpendicularly towards the centre of each six-membered ring of tetrahedra.

The Raman spectrum of Mg end-member talc is characterised by one OH stretching mode due to non-hydrogen bonded Mg$_3$O-H at 3676 cm^{-1} and several modes between 100 and 1200 cm^{-1}. Lattice modes above 600 cm^{-1} have been assigned to SiO$_4$ tetrahedral vibrations [*Rosasco and Blaha*, 1980]; the most intense mode at 678 cm^{-1} is due to Si-O-Si symmetric stretching vibrations. If Mg^{2+} is partially replaced by Fe^{2+} in the talc structure the OH stretching mode can

Table 1. Overview of Raman spectra and structure models of some hydrous and nominally anhydrous silicates that might play an important role in recycling and storing water in the deep Earth. References containing information on synthesis (normal type), characterisation of the samples (normal type), and performed Raman spectroscopic studies (in bold) are given.

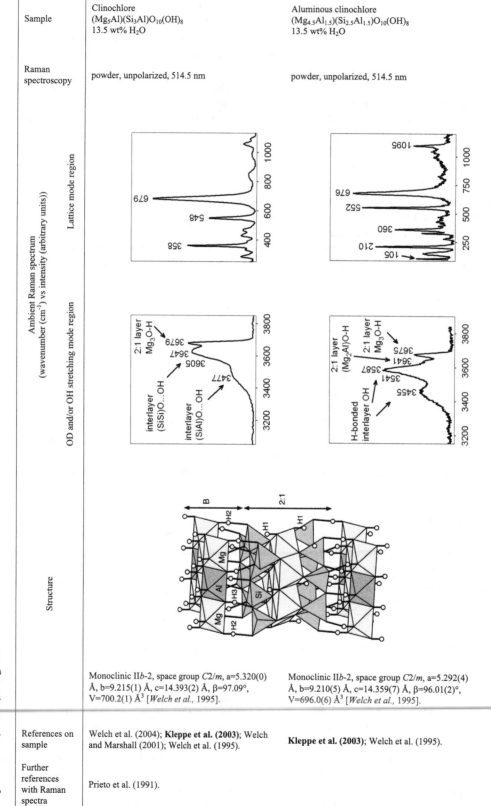

	Clinochlore	Aluminous clinochlore
Sample	$(Mg_5Al)(Si_3Al)O_{10}(OH)_8$ 13.5 wt% H_2O	$(Mg_{4.5}Al_{1.5})(Si_{2.5}Al_{1.5})O_{10}(OH)_8$ 13.5 wt% H_2O
Raman spectroscopy	powder, unpolarized, 514.5 nm	powder, unpolarized, 514.5 nm
Structure	Monoclinic IIb-2, space group $C2/m$, a=5.320(0) Å, b=9.215(1) Å, c=14.393(2) Å, β=97.09°, V=700.2(1) Å³ [Welch et al., 1995].	Monoclinic IIb-2, space group $C2/m$, a=5.292(4) Å, b=9.210(5) Å, c=14.359(7) Å, β=96.01(2)°, V=696.0(6) Å³ [Welch et al., 1995].
References on sample	Welch et al. (2004); **Kleppe et al. (2003)**; Welch and Marshall (2001); Welch et al. (1995).	**Kleppe et al. (2003)**; Welch et al. (1995).
Further references with Raman spectra	Prieto et al. (1991).	

Table 1. (Cont.)

	Talc $Mg_3Si_4O_{10}(OH)_2$ ~4.7 wt% H_2O	10-Å phase $Mg_3Si_{3.83}O_{9.33}(OD)_{2.67}\cdot1.13\,D_2O$ ~10 wt% D_2O
Sample		
Raman spectroscopy	single-crystal, unpolarized, 514.5 nm	single-crystal, unpolarized, 514.5 nm, collected at Oxford
Ambient Raman spectrum (wavenumber (cm^{-1}) vs intensity (arbitrary units))		
Structure		
	Triclinic, space group C1?, a~5.29 Å, b~9.173 Å, c~9.46 Å, α~90.46°, β~98.68°, γ~90.09° [*Perdikatsis and Burzlaff,* 1981]	Monoclinic, space group *C2/m* [after *Comodi et al.,* 2005], a=5.3052(8) Å, b=9.187(1) Å, c=10.172(2) Å, β=100.03(2)°, V=488.2(1) Å³. [*Welch et al.,* submitted 2006; *Pawley et al.,* 2004].
References on sample	**Fumagalli et al. (2001).**	Welch et al. (submitted 2006).
Further references with Raman spectra	Blaha and Rosasco (1978); Rosasco and Blaha (1980).	Fumagalli et al. (2001).

Table 1. (Cont.)

		Phase A $Mg_{6.85}Fe_{0.14}Si_2O_8(OH)_6$ >11 wt% H_2O	Phase E $Mg_{2.08}Si_{1.16}O_{2.8}(OH)_{3.2}$ up to 18 wt% H_2O
Sample			
Raman spectroscopy		single-crystal, unpolarized, 514.5 nm, collected at Oxford	single-crystal, unpolarized, 488 and 514.5 nm
Ambient Raman spectrum (wavenumber (cm^{-1}) vs intensity (arbitrary units))	Lattice mode region		
	OD and/or OH stretching mode region		
Structure		Hexagonal, space group $P6_3$, a=7.8678(4) Å, c=9.5771(5) Å, V=513.43(4) Å^3 [*Holl et al., 2006*].	Hexagonal, space group $R\bar{3}m$ [*Kudoh et al., 1993a*], a=2.975(1) Å, c=13.908(4) Å, V=106.58(7) Å^3 [*Kleppe et al., 2001b*].
References on sample		Holl et al., (2006).	**Kleppe et al. (2001b).**
Further references with Raman spectra		Liu et al. (1997a); Hofmeister et al. (1999).	Ohtani et al. (1995)[a]; Liu et al. (1997b)[b]; Mernagh and Liu (1998)[b]; Frost and Fei (1998); Shieh et al. (2000).

Table 1. (Cont.)

	Superhydrous phase B (= phase C)	Hydroxyl clinohumite	Hydroxyl chondrodite
Sample	$Mg_{10}Si_3O_{14}(OH)_4$ up to 5.8 wt% H_2O	$Mg_9Si_4O_{18}H_2$ 2.89 wt% H_2O	$Mg_5Si_2O_{10}H_2$ up to 5.3 wt% H_2O
Raman spectroscopy	single-crystal, unpolarized, 514.5 nm	single-crystal, unpolarized, 514.5 nm	single-crystal, unpolarized, 514.5 nm

Ambient Raman spectrum (wavenumber (cm^{-1}) vs intensity (arbitrary units))

Lattice mode region

OD and/or OH stretching mode region

Structure

	Orthorhombic, space group *Pnnm* [*Pacalo and Parise,* 1992] or *P2₁nm* [*Kudoh et al.,* 1993b].	Monoclinic, spacegroup *P2₁/c*, a=13.682(14) Å, b=4.755(6) Å, c=10.264(9) Å, β=100.56(86)° [*Lin et al.,* 2000].	Monoclinic, spacegroup *P2₁/c*, a=10.337(6) Å, b=4.750(5) Å, c=7.907(7) Å, β=108.68(6)° [*Lin et al.,* 1999].
References on sample	Jacobsen and Littlefield (personal communication).	**Lin et al. (2000).**	**Lin et al. (1999)**; Mernagh et al. (1999).
Further references with Raman spectra	Ohtani et al. (1995)[a]; Frost and Fei (1998); Hofmeister et al. (1999); Ohtani et al. (2001); Liu et al. (2002a).	Stalder and Ulmer (2001).	Cynn et al. (1996).

		Lawsonite CaAl$_2$Si$_2$O$_7$(OH)$_2$·H$_2$O 11.5 wt% H$_2$O	Olivine α-(Mg$_{0.97}$Fe$_{0.03}$)$_2$SiO$_4$ 0.8 wt% H$_2$O
Sample			
Raman spectroscopy		single-crystal, unpolarized, 514.5 nm, collected at Oxford	single-crystal, unpolarized, 514.5 nm, collected at Oxford

Table 1. (Cont.)

Ambient Raman spectrum (wavenumber (cm^{-1}) vs intensity (arbitrary units))

Lattice mode region

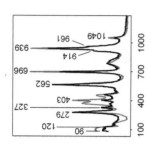
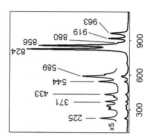

OD and/or OH stretching mode region

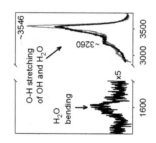
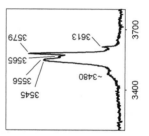

Structure

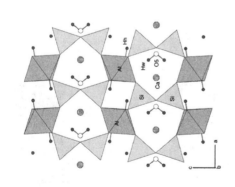
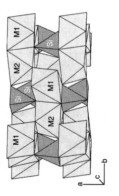

Orthorhombic, space group *Cmcm*, a=8.769(4) Å, b=5.860(2) Å, c=13.157(6) Å, V=676.0(4) Å^3 [*Chinnery et al.,* 2000].

Orthorhombic, space group *Pbnm*; a=4.75873(55) Å, b=10.22344(52) Å, c=5.99169(44) Å, V=291.50(4) Å^3 [*Smyth*, personal communication].

References on sample: Sample from a vein from Jenner (Northern California), same sample batch as in Chinnery et al. (2000) and Pawley and Allan (2001).

Smyth et al. (2005a).

Further references with Raman spectra: Le Cléach and Gillet (1990); Daniel et al. (2000).

			Hydrous wadsleyite	

Sample

Raman spectroscopy

β-Mg₂SiO₄, 1.65 wt% H₂O

β-Mg$_2$SiO$_4$, 1.65 wt% H$_2$O

single-crystal, unpolarized, 514.5 nm

β-(Mg$_{0.9}$Fe$_{0.1}$)$_2$SiO$_4$, 2.4 wt% H$_2$O

single-crystal, unpolarized, 488 and 514.5 nm

Ambient Raman spectrum (wavenumber (cm⁻¹) vs intensity (arbitrary units))

Lattice mode region

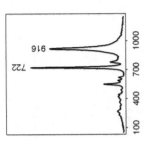

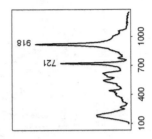

OD and/or OH stretching mode region

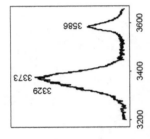

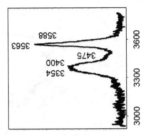

Structure

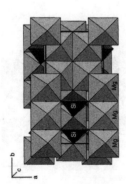

Orthorhombic, space group *Imma*, a=5.679(2) Å, b=11.548(2) Å, c=8.250(3) Å, V=541.1(4) Å³ [*Kleppe et al.*, 2001a].

Monoclinic, space group *I2/m*, a=5.7003(4) Å, b=11.5018(6) Å, c=8.2669(6) Å, β=90.153(4)°, V=542.007 Å³ [*Smyth*, personal communication].

References on sample

Kleppe et al. (2001a).

Kleppe et al. (2006a); Smyth et al. (2005b); McCammon et al. (2004; sample SZ0005).

Further references with Raman spectra

Mc Millan et al. (1991); Kudoh et al. (1996)[a]; Mernagh and Liu (1996); Liu et al. (1998b).

Table 1. (Cont.)

Sample		wadsleyite II $Mg_{1.71}Fe_{0.18}Al_{0.01}H_{0.33}Si_{0.96}O_4$, 2.1 wt% H_2O
Raman spectroscopy		single-crystal, unpolarized, 488 and 514.5 nm
Ambient Raman spectrum (wavenumber (cm^{-1}) vs intensity (arbitrary units))	Lattice mode region	
	OD and/or OH stretching mode region	
Structure		 Orthorhombic, space group *Imma*, a=5.6884(4), b=28.9238(15), c=8.2382(6) Å, V=1355.4 Å^3 [*Smyth and Kawamoto*, 1997].
References on sample		**Kleppe et al. (2006b)**; Smyth et al. (2005b); Smyth and Kawamoto, (1997).
Further references with Raman spectra		

Table 1. (Cont.)

Table 1. (Cont.)

		Hydrous ringwoodite	
Sample		γ-Mg$_2$SiO$_4$ 0.74 wt% H$_2$O	γ-(Mg$_{0.89}$Fe$_{0.11}$)$_2$SiO$_4$ 0.79 wt% H$_2$O
Raman spectroscopy		single-crystal, unpolarized, 514.5 nm	single-crystal, unpolarized, 488 and 514.5 nm
Ambient Raman spectrum (wavenumber (cm^{-1}) vs intensity (arbitrary units))	Lattice mode region		
	OD and/or OH stretching mode region		

Structure		Cubic, space group $Fd3m$, a=8.0687(4) Å, V=525.3 Å^3 [*Kleppe et al.,* 2002a].	Cubic, space group $Fd3m$, a=8.0944(6) Å, V=530.3 Å^3 [*Kleppe et al.,* 2002b].
References on sample		Smyth et al. (2003; sample ringby5); **Kleppe et al. (2002a)**.	Manghnani et al. (2005); McCammon et al. (2004; sample SZ0002); Smyth et al. (2003; sample SZ0002); **Kleppe et al. (2002b)**.
Further references with Raman spectra		Liu et al. (2002b); Kudoh et al. (2000)[a].	

[a] Only OH stretching bands.
[b] Poor agreement with the present spectrum.

split in as many as four peaks, depending on the degree of substitution. The four OH stretching modes correspond to the four possible combinations of the two different cations within the three adjacent octahedral sites [*Wilkins and Ito*, 1967]: $3Mg^{2+}$ (3676 cm^{-1}), $2Mg^{2+} + Fe^{2+}$ (3662 cm^{-1}), $Mg^{2+} + 2Fe^{2+}$ (3644 cm^{-1}), and $3Fe^{2+}$ (3623 cm^{-1}).

4.1.3. 10-Å phase. The 10-Å phase is thought to be a possible storage site of OH and H_2O in subducting slabs where it might play an important role in transferring water from serpentine to phase A bearing assemblages [*Ulmer and Trommsdorff*, 1995]. The 10-Å phase forms from talc plus water at 4-5 GPa and ~700 °C and is stable to at least 7 GPa [e.g., *Pawley and Wood*, 1995]. Recently, this phase has been reported to be stable in Al-bearing compositions [*Fumagalli and Poli*, 2005]. The amount of water that might be incorporated in the 10-Å phase, $Mg_3Si_4O_{10}(OH)_2 \cdot n\, H_2O$, appears to depend on experimental run times and values of $n=0.65$, 1 and 2 have been reported [*Fumagalli et al.*, 2001 and references therein]. The structure of the 10-Å phase ($n=1$) has been determined from single-crystal X-ray diffraction data [*Comodi et al.*, 2005]: It consists of 2:1 tetrahedral-octahedral layers parallel to (001) and water molecules at the 12-fold coordinated interlayer site. Proton positions have not been refined and are uncertain. In the structure plot in Table 1 one possible orientation of an interlayer water molecule is shown together with various, other O···O distances between the water oxygen and the basal oxygen acceptors (thin black lines). Recent NMR and XRD studies suggest a more complex structure of the 10-Å phase with Si vacancies in the tetrahedral sheet [*Welch et al.*, submitted 2006; *Kohn and Fumagalli*, 2002].

The Raman spectrum of the 10-Å phase closely resembles the Raman spectrum of talc at wavenumbers <800 cm^{-1} reflecting the structural similarity of the 10-Å phase to talc. The most intense mode at 677 cm^{-1} has been associated with Si-O-Si bending vibrations by analogy with talc [*Fumagalli et al.*, 2001]. A comparison of the lattice mode frequencies of hydrated [*Fumagalli et al.*, 2001] and highly deuterated 10-Å phase (present study) shows isotopic shifts for several modes: the modes at 116, 330, 359, 432, 453 and 467 cm^{-1} of the hydrated phase are shifted to lower frequencies in the deuterated phase indicating that these modes contain contributions from OH/OD vibrations.

In the OH and OD stretching frequency range the OH stretching mode at 3676 cm^{-1} and OD stretching mode at 2711 cm^{-1} occur at exactly the same frequency as the talc OH and OD stretching mode [e.g., *Ferrage et al.*, 2003] and therefore we suggest that these modes are due to Mg_3O-H and Mg_3O-D stretching vibrations in the 2:1 layer, respectively, with the local environment of the OH and OD ions

being identical to talc. The ratio $\nu(OH)/\nu(OD)$ for the modes 3676/2711 and 3666/2705 is 1.355 which is the ratio for the free molecules and is consistent with absent or weak hydrogen bonds. The OD/OH stretching frequencies at 2711/3676, 2705/3666, 2695/3658 and 2646 cm^{-1} are very sharp with full widths at half maximum (FWHM) between 3 and 6 cm^{-1}. These FWHM values reflect extremely regular environments of the hydroxyls and are consistent with a high degree of structural order. The intensity of the modes at 2711, 2646, and 3676 cm^{-1} is very sensitive to the crystal to laser-beam orientation (black and grey lines in Table 1 illustrate spectra measured at two different, arbitrary orientations of the 10-Å phase crystal with respect to the laser-beam). Detailed, unambiguous assignments of the OH/OD stretching modes await a full determination of the structure of the 10-Å phase (including proton positions) and polarized Raman and/or IR spectroscopic investigations on orientated single-crystals.

4.1.4. Phase A. Phase A is a potential breakdown product of serpentine [*Komabayashi*, this volume] and one of the most important DHMS phases stable in subduction zone regions in the deep upper mantle [e.g., *Ulmer and Trommsdorff*, 1995]. Phase A might transport water into the stability field of phase E or wadsleyite [e.g., *Ohtani et al.*, 2000]. The crystal structure of phase A, $(Mg,Fe)_7Si_2O_8(OH)_6$, consists of two types of layers, OT1 and OT2, perpendicular to the c direction: The OT1 layer is composed of Si1O$_4$ tetrahedra and Mg2O$_6$ as well as Mg3O$_6$ octahedra; the OT2 layer consists of Si2O$_4$ tetrahedra and Mg1O$_6$ octahedra. No ordering of Fe among the three distinct octahedra has been observed [*Holl et al.*, 2006]. From neutron diffraction studies two distinct hydrogen sites have been identified in the structure of phase A [e.g., *Kagi et al.*, 2000]: The O2···O3 edges of vacant octahedral sites and the O4···O3 edges of vacant tetrahedral sites within the OT2 layers are protonated. Both hydrogen sites are hydrogen bonded to the same oxygen O3 but the lengths of the hydrogen bonds differ being 1.96(2) Å for H1···O3 and 2.29(1) Å for H2···O3 [*Kagi et al.*, 2000], and hence the hydrogen bond for O2-H1···O3 is stronger than for O4-H2···O3.

The Raman spectra of phase A exhibit two OH stretching modes confirming the presence of two hydrogen sites with different degrees of hydrogen bonding in the structure. The OH stretching modes occur at 3400 and 3517 cm^{-1} agreeing with protonation of the underbonded oxygens O4 and O2, respectively. Unpolarized Raman spectra taken for different, arbitrary crystal to laser-beam orientations show strong changes in the relative intensities of the OH stretching modes (black and grey lines in Table 1) reflecting that the O-H vectors of the two hydroxyls are nearly perpendicular. In the lattice mode region Raman bands correspond to the various vibrations involving SiO_4 tetrahedra and MgO_6 octahedra. The Raman spectra are in agreement with previous Raman and IR spectroscopic studies of Phase A [e.g., *Liu et al.,* 1997a; *Cynn et al.,* 1996].

4.1.5. Phase E. Phase E is a potential carrier of water in a minor part of cold subducted oceanic lithosphere between approximately 350 and 500 km [e.g., *Shieh et al.,* 2000]. The importance of this hydrous silicate has been emphasized by its possible, natural occurrence as nanometer-sized inclusions in olivine from kimberlitic nodules [*Khisina and Wirth,* 1997]. Phase E possesses unique crystal chemical features: it is non-stoichiometric, of variable composition, and has a cation-disordered crystal structure with a type of long-range disorder not observed before in crystalline silicates [*Kudoh et al.,* 1993a; *Kanzaki et al.,* 1992]. The structure consists of layers with brucite-type units cross-linked by Si in tetrahedral coordination, Mg in octahedral coordination, and possibly hydrogen bonds.

The Raman spectrum of phase E comprises several broad bands (full width at half maximum: 55-100 cm^{-1}) between 50 and 1200 cm^{-1}. The broadness of the Raman modes reflects the proposed long-range cation disorder in the structure. In the OH stretching region we observe three bands between 2700 and 3650 cm^{-1}. The highest OH mode at 3617 cm^{-1} can be associated with non- or very weakly hydrogen bonded hydroxyls, and the mode at 3429 cm^{-1} with hydrogen-bonded hydroxyls; corresponding $O\cdots O$ distances would be greater than 3 Å and around 2.8 Å respectively using the correlation between OH stretching frequencies and $O\cdots O$ distances for linear hydrogen bonds [*Libowitzky,* 1999].

4.1.6. Superhydrous phase B (= phase C). This dense hydrous magnesium silicate might act as a water carrier between 550 to 800 km depth in Mg-rich ultramafic systems [e.g., *Ohtani et al.,* 2000; *Frost,* 1999]. The crystal structure of superhydrous phase B has been reported with two space groups (Tab. 1), and further polymorphs appear to exist [*Koch-Müller et al.,* 2004]. The structure consists of double, mixed octahedral-tetrahedral (O-T) layers composed of MgO_6 octahedra and SiO_4 tetrahedra. These layers alternate along the *b*-axis with O-layers containing both MgO_6 and SiO_6 octahedra. One third of the Si atoms in superhydrous phase B are octahedrally coordinated, unlike in the lower-pressure hydrous phases in which Si is only found in tetrahedral coordination. The protons are thought to be located between neighbouring O-T layers and are bonded to oxygens coordinated by three Mg ions.

The Raman spectrum of superhydrous phase B shows two OH stretching modes. Cynn et al. (1996) proposed that the 3410 cm^{-1} band arises from H being distributed symmetrically among the O3 and O5 site while the 3353 cm^{-1} band is due to an asymmetric distribution of H among these two O-sites. The asymmetric configuration agrees with the lower symmetry inferred by Kudoh et al. (1993b). In the lattice mode region peaks between 645-730 cm^{-1} have been assigned to bending of Si octahedra and modes between 800-1000 cm^{-1} to Si-O stretching vibrations of the SiO_4 tetrahedra [*Cynn et al.,* 1996].

4.1.7. Phase D (= phase F, phase G). Phase D has the highest pressure stability of the dense hydrous magnesium silicates and might exist within subducting lithosphere in the transition zone and upper part of the lower mantle [e.g., *Ohtani et al.,* 2001; 2000; *Shieh et al.,* 2000; 1998; *Frost,* 1999; *Frost and Fei,* 1998; *Irifune et al.,* 1998]. The ideal chemical composition of phase D is $MgSi_2H_2O_6$, but depending on synthesis conditions the composition varies: Mg/Si values range from 0.55 to 0.71 and the water content from 10 to 18 wt% H_2O [e.g., *Frost and Fei,* 1998]. Phase D is of trigonal symmetry, space group $P\bar{3}1m$. The structure is characterised by alternating SiO_6 and MgO_6 octahedral layers stacked along the *c*-axis [e.g., *Suzuki et al.,* 2001; *Frost and Fei,* 1998; *Yang et al.,* 1997]. It is the only DHMS where Si is found exclusively in octahedral coordination. One in three octahedra in the SiO_6 layer and two in three octahedra in the MgO_6 layer are vacant. The hydrogen atom position is disordered within the MgO_6 octahedral layers [*Suzuki et al.,* 2001; *Yang et al.,* 1997].

Frost and Fei (1998) reported a Raman spectrum of phase D between 200 and 3300 cm^{-1}; however, their sample contained more than one phase and superhydrous phase B modes are present in their spectrum. Liu et al. (1998a) reported Raman spectra of phase D in the lattice mode region with a different mode structure but did not observe any OH stretching mode. Though some uncertainty is associated with the Raman spectrum of this silicate, it is likely characterized by a very broad OH stretching mode at about 2850 cm^{-1} [*Frost and Fei,* 1998]; a wavenumber that is consistent with an $O\cdots O$ distance of ~2.68 Å for the proton site determined from single-crystal X-ray [*Yang et al.,* 1997] and powder neutron diffraction experiments [*Suzuki et al.,* 2001].

4.2. Humite Group Minerals

Within the humite group, clinohumite and chondrodite are the most important in terms of water storage and transport in subduction zones and the mantle wedge because of their high P-T stability compared to other water carriers such as serpentine [e.g., *Stalder and Ulmer,* 2001]. The characteristic structural unit of monoclinic clinohumite and chondrodite with space group $P2_1/c$ is a serrated chain of edge sharing

MgO_6 octahedra parallel to the *a*-axis cross-linked to chains in the same and adjacent layers by SiO_4 tetrahedra. The hydrogen atoms H1 and H2 in hydroxyl clinohumite and chondrodite are statistically disordered over two sites (50% occupancy), and hydrogen bond distances range from 2.01 to 2.58 Å in hydroxyl clinohumite [e.g., *Berry and James*, 2001] and from 1.97 to 2.49 Å in hydroxyl chondrodite [e.g., *Lager et al.*, 2001]. The slightly larger hydrogen bond distances in hydroxyl clinohumite are reflected by its slightly higher OH stretching mode frequencies compared to hydroxyl chondrodite. The Raman spectra in the lattice mode region of hydroxyl clinohumite and hydroxyl chondrodite are similar as is to be expected for those closely related structures.

4.3. Lawsonite

Lawsonite is unique among high-pressure hydrous phases because water molecules are incorporated in the structure, residing within large cages. In total, the mineral contains about 11.5 wt% water in the form of H_2O and OH. Lawsonite is stable to pressures of 12-13.5 GPa at temperatures of 800-960 °C [e.g., *Schmidt*, 1995; *Pawley*, 1994], and hence likely especially efficient at transporting water to depth in subduction zones. In addition to its role as a water carrier, lawsonite has attracted attention crystal-chemically because it exhibits two phase transitions at low-temperatures (<298 K) that are primarily related to proton ordering.

At room-pressure and -temperature the structure of lawsonite belongs to the orthorhombic crystal system, space group *Cmcm* [e.g., *Chinnery et al.*, 2000]. It consists of chains of edge-sharing AlO_6 octahedra linked by Si_2O_7 groups. Hydroxyl groups are bound to the edge-sharing AlO_6 octahedra (O4, H). Rings of two AlO_6 octahedra and two Si_2O_7 groups form cavities that each contain an isolated water molecule (O5, 2x Hw) and a calcium atom (Ca). The water molecules and hydroxyl groups are aligned parallel to the (100) symmetry plane [e.g., *Libowitzky and Rossman*, 1996; *Labotka and Rossman*, 1974].

The room-pressure and -temperature Raman spectrum of lawsonite is shown in Table 1 for two different, arbitrary orientations of the lawsonite crystal with respect to the laser-beam (black and grey lines). It is consistent with previous Raman spectra [*Daniel et al.*, 2000; *Le Cléach and Gillet*, 1990]. In the lattice mode region Le Cléach and Gillet (1990) assigned modes between 690 and 1050 cm^{-1} to vibrations of the Si_2O_7 unit with the exception of the broad and weak mode at 780-800 cm^{-1}, which is thought to originate from AlO_6 stretching motions. In the higher frequency range we observe an H_2O bending mode at about 1600 cm^{-1}, and resolve three OH stretching modes between 2700 and 3700 cm^{-1} using several spectra with a range of different, arbitrary

crystal to laser-beam orientations. In Table 2 the Raman modes between 2700 and 3700 cm^{-1} are compared with previously reported IR modes [*Libowitzky and Rossman*, 1996]. Libowitzky and Rossman (1996) assigned the H_2O and OH modes using low-temperature, polarized IR data and assuming single, uncoupled OH stretching frequencies and orientations for the asymmetric H_2O molecule in lawsonite. The Raman and IR modes in the OH stretching frequency region agree, but we do not resolve the two OH vibrations assigned to O4a-Hh and O5-Hwa at 298 K and room-pressure.

4.4. Olivine

Olivine ranks as the most abundant mineral phase in most models of the upper mantle. The solubility of H_2O in this nominally anhydrous mineral is pressure and temperature dependent [e.g., *Smyth et al.*, 2005a; *Zhao et al.*, 2004; *Kohlstedt et al.*, 1996] and might be as high as 0.7-0.8 wt% H_2O at 1250 °C and 12 GPa [*Smyth et al.*, 2005a; *Chen et al.*, 2002]. Hence, olivine may act as a significant reservoir for hydrogen in the upper mantle and as a global conduit for subduction of water into the deeper mantle. The olivine structure consists of a nearly hexagonal-close-packed arrangement of O^{2-} ions with octahedrally coordinated Mg^{2+} and Fe^{2+} and tetrahedrally coordinated Si^{4+}. The hydration mechanism of pure Mg end-member olivine has been found to depend on the silica-activity and appears to involve mainly Si vacancies at low silica-activity and Mg vacancies at high silica-activity [*Lemaire et al.*, 2004]. For iron-bearing olivines synthesized at 12 GPa and 1100-1600 °C the principal hydration mechanism appears to be protonation of M1 octahedral edges with substitution of $2H^+$ for Mg^{2+} [*Smyth et al.*, 2005a].

We studied Fo_{97} olivine with 0.8 wt% H_2O with Micro-Raman spectroscopy at room-pressure and -temperature. Table 1 shows the unpolarized Raman spectrum of the hydrous olivine sample for one chosen crystal-orientation relative to the incident laser beam. Modes between ~450 and 1000 cm^{-1} are generally associated with the internal stretching and bending vibrations of the SiO_4 tetrahedra. The observed Raman modes of our sample agree with those of the anhydrous Fo_{100} and Fo_{88} phase studied by Chopelas (1991a). Consequently the presence of OH seems to have only minor effect on the framework lattice dynamics at room-pressure and -temperature conditions. A similar observation has been made for the hydrous and anhydrous high-pressure polymorphs of olivine, wadsleyite and ringwoodite [*Kleppe et al.*, 2001a; 2002a]. We observe six OH stretching modes in the Raman spectra of hydrous Fo_{97} olivine and sometimes a weak band at ~3600 cm^{-1}. OH stretching bands at these frequencies are in agreement with FTIR spectroscopic studies by Lemaire et al. (2004) and Smyth et al. (2005a).

4.5. Nominally Anhydrous Transition Zone Spinelloids and Spinel

In the system MgO-FeO-SiO_2-H_2O the transformation from olivine (α-$(Mg,Fe)_2SiO_4$) to ringwoodite (γ-$(Mg,Fe)_2SiO_4$, spinel) can involve intermediated phases, the spinelloids (spinel derivatives) wadsleyite (β-$(Mg,Fe)_2SiO_4$) and wadsleyite II. Wadsleyite II is a hydrous magnesium-iron silicate with variable composition that occurs between the stability regions of wadsleyite and ringwoodite, if sufficient trivalent cations (Fe^{3+}, Al^{3+}) are available [Smyth and Kawamoto, 1997]. If present in the transition zone, wadsleyite II may account for the generally diffuse nature and splitting of the 520 km seismic discontinuity observed in some regions of the Earth's mantle [Smyth et al., 2005b]. The hydrogen as well as the iron content of the spinelloid phases and spinel is likely to influence mantle properties globally.

4.5.1 Spinelloids: wadsleyite and wadsleyite II. The crystal structures of the spinelloid phases are derived from the spinel structure. However, unlike the spinel structure that contains only isolated SiO_4 tetrahedra, the spinelloid structures contain polymerized tetrahedra: In wadsleyite only Si_2O_7 groups are present whereas in wadsleyite II both, single SiO_4 and coupled Si_2O_7 tetrahedra exist. The ratio of SiO_4 to Si_2O_7 is 1:5 in wadsleyite II so that the structure can be thought of as a mixture of one-sixth spinel and five-sixths wadsleyite. The presence of polymerized Si tetrahedra in the spinelloid structures is coupled with the occurrence of underbonded, non-silicate oxygens that can act as protonation sites [e.g., Smyth et al., 2005b; Smyth, 1987]. The principal charge-balancing mechanism for hydration of the spinelloid phases has been found to be octahedral site vacancy [e.g., Smyth et al., 2005b; Smyth et al., 1997; Kudoh et al., 1996]: In pure Mg wadsleyite the vacancies are strongly concentrated on the Mg3 site which is consistent with protonation of the non-silicate oxygen O1; in wadsleyite II charge-balancing vacancies occur in M5 and M6 agreeing with hydration of the underbonded oxygen O2. In addition, in wadsleyite II the oxygens surrounding the partially vacant tetrahedral site Si2 possibly act as protonation sites [Smyth et al., 2005b].

As expected from their closely related crystal structures, the Raman spectra of wadsleyite and wadsleyite II are very similar and dominated in the lattice mode region by the characteristic Si_2O_7 symmetric stretching mode (Si-O-Si stretching of the disilicate group) at 721 (709) cm^{-1} and the symmetric stretching mode of the SiO_3 terminal unit around ~915 cm^{-1}. However, the wadsleyite II spectrum is modified with bands (e.g., at 773 cm^{-1}) in frequency regions where the SiO_4 tetrahedral stretching vibrations of Fo_{89} hydrous ringwoodite occur reflecting the presence of isolated SiO_4 tetrahedra in its structure. Table 1 shows the Raman spectrum of a single-crystal (solids lines) and of a powder sample (dotted lines) of wadsleyite II. A comparison of the Raman spectrum of hydrous wadsleyite with anhydrous wadsleyite [e.g., Chopelas, 1991b] reveals that the presence of hydrogen and iron has only a minor effect on the lattice dynamics at room-pressure and -temperature: The lattice modes are identical except for an additional band at 487 cm^{-1} in hydrous wadsleyite that is thought to originate from a translational OH vibration [Kleppe et al., 2001a; Mernagh and Liu, 1996].

Raman and IR spectroscopic studies of hydrous wadsleyite (>0.2 wt% H_2O) show consistently two intense groups of OH stretching modes [Kleppe et al., 2006a; Jacobsen et al., 2005; Kohn et al., 2002; Kleppe et al., 2001a; Cynn and Hofmeister, 1994]: group 1 comprises a doublet around 3550-3650 cm^{-1} and group 2 consists of three components around 3300-3400 cm^{-1}. However, the number of resolved modes and individual mode frequencies as well as other spectral details vary depending on the water and iron (Fe^{2+}, Fe^{3+}) content of the wadsleyite sample. Jacobsen et al. (2005) could explain the main OH stretching bands by protonation of the O1 site using polarized FTIR spectroscopy and single-crystal X-ray diffraction on oriented single-crystals of hydrous Mg end-member wadsleyite.

The Raman spectrum of wadsleyite II matches in the OH stretching region a superposition of an iron-bearing hydrous wadsleyite and hydrous ringwoodite spectrum: between 3350 and 3600 cm^{-1} the wadsleyite II spectrum shows modes similar to those of Fo_{90} hydrous wadsleyite, but overlain with the broad OH stretching bands of ringwoodite that have been observed at 3165 and 3685 cm^{-1} (grey, dotted lines in Table 1; full width at half maximum around 350 and 150-200 cm^{-1} respectively [Kudoh et al., 2000]). The spectrum can be fitted with a minimum of six bands. Assuming linear O-H$\cdots$O topologies we can correlate the observed OH stretching frequencies with O$\cdots$O bond distances [Libowitzky, 1999]: The broad and weak OH band at 3243 cm^{-1} corresponds to an O$\cdots$O distance of about 2.7 Å, the triplet at 3351, 3408 and 3484 cm^{-1} fits with O$\cdots$O distances around 2.8 Å, and the doublet at 3564 and 3628 cm^{-1} only allows O$\cdots$O distances 3 Å. These distances are in overall agreement with protonation of the oxygens surrounding the partially vacant tetrahedral site Si2 (average edge length is about 2.73 Å) and the non-silicate oxygen O2, as suggested from single-crystal X-ray diffraction data [Smyth et al., 2005b]. However, hydrogen bonds might deviate significantly from being linear and the assumption of (close to) linear hydrogen bonds needs to be re-assessed when more detailed structural information on the proton positions becomes available.

4.5.2. Spinel: ringwoodite. The Raman spectra of hydrous ringwoodite are characterized by two intense bands at 796 and ~840 cm^{-1} that correspond to the asymmetric and symmetric stretching vibrations of the isolated SiO_4 tetrahedra of the spinel structure. The modes at 709, 881, and 939 cm^{-1} in the spectrum of the iron-bearing phase are not allowed for the ideal cubic spinel structure from symmetry analysis [e.g., *White and DeAngelis,* 1967]. They have been associated with the presence of Si_2O_7 groups and hence, the presence of non-silicate oxygens in the structure that could act as sites for protonation [*Smyth et al.,* 2003; *Kleppe et al.,* 2002b]. Non-silicate oxygens in the Fe-bearing ringwoodite as protonation sites differs from the structural model suggested for the incorporation of water into the Mg end-member phase by Kudoh (2001) and Kudoh et al. (2000). The existence of different hydration mechanisms for iron-bearing and iron-free ringwoodite is conceivable because the presence of Fe offers new possibilities for the incorporation of protons; for example, redox reactions in which the reduction of ferric iron accompanies the OH group creation could be important [e.g., *Blanchard et al.,* 2005]. The low-frequency Raman signal (100-250 cm^{-1}) in hydrous iron-bearing ringwoodite might be associated with localized modes generated by (Fe^{2+}, Fe^{3+}) substitution in the ideal, hydrous Mg end-member ringwoodite structure [*Kleppe et al.,* 2002b]. What these results do emphasize is that the actual dynamical properties of minerals are likely to depend strongly on composition, and it is no longer sufficient to study end-member phases alone.

Despite direct evidence from IR absorption spectroscopy on the same sample batch [*Smyth et al.,* 2003], we did not observe OH stretching vibrations expected to occur in the range 2500-3600 cm^{-1} and reported for the hydrous Mg end-member composition by Kudoh et al. (2000). The lack of observation in Raman is unusual but may not be surprising: The laser power required to avoid thermal instability may have been too low for the observation of weak and broad OH stretching bands. The observed IR absorption bands of hydrous ringwoodite samples from the same sample batch show a maximum at 3105-3140 cm^{-1} and correlate with an O···O distance of about 2.7 Å, consistent with protonation of the short O···O approach (2.71 Å) on the tetrahedral edge [*Smyth et al.,* 2003]. As charge-balancing vacancies occur dominantly on octahedral sites the presence of protons at a tetrahedral edge requires partial Mg^{2+}-Si^{4+} disorder in the structure [*Smyth et al.,* 2004, *Smyth et al.,* 2003]. A small degree of Mg^{2+}-Si^{4+} disorder in the structure of hydrous Mg end-member ringwoodite has been reported by Kudoh et al. (2000). Additional OH stretching bands at higher frequencies have been observed in samples with a water content >1.2 wt%: a Raman band at about 3685 cm^{-1} [*Kudoh*

et al., 2000] and an IR band at about 3645 cm^{-1} [*Kohlstedt et al.,* 1996]. These frequencies agree with hydration of the octahedral edges with O···O distances of 2.85-3.0 Å.

5. LOW TEMPERATURE RAMAN SPECTROSCOPY OF LAWSONITE

At low-temperatures (<298 K), two polymorphs of the lawsonite structure have been reported: at 273 K the structure changes from space group *Cmcm* to *Pmcn* and below 155 K it changes from *Pmcn* to *P2$_1$cn* [e.g., *Sondergeld et al.,* 2005; *Libowitzky and Rossman,* 1996; *Libowitzky and Armbruster,* 1995]. These phase transitions mainly involve changes in the proton positions and the hydrogen-bond system. Lawsonite also exhibits two high-pressure, room-temperature phase transitions around ~4 and 9.5 GPa [*Boffa-Ballaran and Angel,* 2003; *Pawley and Allan,* 2001; *Daniel et al.,* 2000; *Scott and Williams,* 1999]. While the pressure-induced transitions are related to discontinuities in the contraction of the lattice parameters, the low-temperature transitions result from freezing H_2O and OH units at the endpoints of their dynamic motions. Hence, for lawsonite, pressure-induced crystal-chemical effects differ significantly from temperature-induced effects. The three phases of lawsonite between 2 and 298 K have been investigated in great detail with different techniques including IR spectroscopy, and neutron and X-ray diffraction [e.g., *Sondergeld et al.,* 2005; *Meyer et al.,* 2001; *Libowitzky and Rossman,* 1996; *Libowitzky and Armbruster,* 1995]. In order to round off the existing studies we performed low-temperature, room-pressure Raman spectroscopic measurements of natural lawsonite down to 5 K.

Figure 1 shows representative Raman spectra for each of the three phases between 5 and 298 K; frequencies related to the H_2O and OH groups are reported in Table 2. The present low-temperature Raman data confirm previous, low-temperature IR studies: In the lattice-mode region the Raman spectra of the *Cmcm* phase and *Pmcn* phase are very similar while the spectrum of the *P2$_1$cn* phase appears more complex with additional modes evolving. Additional modes in the *P2$_1$cn* phase have also been observed in IR spectra [*Scott and Williams,* 1999]. With decreasing temperature the lattice modes shift to higher frequencies. The OH stretching mode region of the lawsonite spectra changes significantly with decreasing temperature as is expected for phase transitions that are primarily related to proton ordering. Modes assigned to O5-Hw···O4 and O4-Hh···O4a (Tab. 2) decrease with decreasing temperature reflecting the formation of strong hydrogen bonds. At low-temperatures (<200 K) the unresolved doublet associated with O4a-Hh and O5-Hwa vibrations becomes resolved

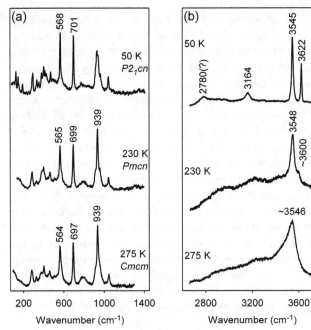

Figure 1. Selected, low-temperature Raman spectra of lawsonite (a) in the lattice mode region and (b) in the OH stretching mode region.

because the temperature-induced frequency shift of these two vibrations differs. Libowitzky and Rossman (1996) correlated the O···O and O···H distance in the structure with the observed vibrational frequencies and described the changing hydrogen bond system in detail. It would be useful to perform low-temperature high-pressure Raman studies to induce both volume- and dynamically-driven transitions.

6. HYDROGEN BONDING AT HIGH-PRESSURES IN CLINOCHLORE

High-pressure Raman spectroscopy and synchrotron X-ray diffraction studies on clinochlore to 26.5 GPa have shown that this mineral exhibits a reversible transformation between 9 and 10 GPa that is primarily associated

with a major reorganization of the interlayer hydrogen bonding and a change in the interlayer topology [*Welch et al.*, 2004; *Kleppe et al.*, 2003]. The transformation was first manifested in the Raman spectra by large, positive frequency shifts (100-140 cm⁻¹) of the interlayer OH vibrations [*Kleppe et al.*, 2003]. Positive pressure dependencies of OH-stretching frequencies are much less common and less well understood than negative dependencies. Recently, significant positive pressure-induced frequency shifts have been observed in humites [e.g., *Lin et al.*, 2000, 1999]. They were explained as being due to a combination of H-H repulsion and hydrogen-bond elongation toward an empty polyhedron. The positive shifts of the hydrogen-bonded hydroxyl groups observed for clinochlore are much higher than those in humites and the OH stretching modes reach frequencies of 3750-3800 cm⁻¹ above 16 GPa.

To further explore the nature and possible cause of the observed, pressure-induced changes in the OH-stretching region of clinochlore, and to provide constraints for a full structural refinement we used first-principles calculations based on density functional theory [*Refson et al.*, 2004]. Calculations were performed in the PBE GGA approximation using a plane-wave basis set and pseudopotentials with the CASTEP code [*Segall et al.*, 2002]. Variable cell geometry optimization under applied external pressure induced a change from type IIa to IIb stacking at around 10 GPa, close to the experimentally observed transition pressure. *Ab-initio* lattice dynamics calculations using density-functional perturbation theory were performed at a range of pressures through the transition to compute the frequencies of the OH stretching modes in the harmonic approximation [*Refson et al.*, in prep]. Figure 2 compares the pressure dependence of the calculated vibrational modes with the experimentally determined modes. The calculations reproduce the transition pressure well and also show the frequency increases for the hydrogen-bonded, interlayer OH groups at 9-10 GPa. The high vibrational frequencies above 9 GPa (>3725 cm⁻¹) together with their positive frequency shifts under further compression are characteristic of unbonded or multi-furcated hydroxyl groups.

Table 2. Comparison of the OH stretching modes of lawsonite observed in the present, low-temperature Raman spectroscopic study with a previous low-temperature IR spectroscopic study.

T (K)	Present Raman study					IR study by Libowitzky and Rossman (1996)		
	298	275	230	50	5	298	100	assignment
ν_i (cm⁻¹)	~2950	~2950	2920	2780(?)	2780	2960	2838	O5-Hw···O4
	~3260	~3260	3225	3164	3162	3245	3187	O4-Hh···O4a
	3546	~3546	3548	3545	3545	3540	~3554	O4a-Hh
			~3600	3621	3622	3575	~3612	O5-Hwa

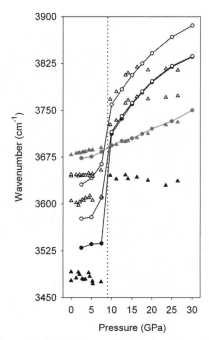

Figure 2. Theoretically (circles, solid lines) and experimentally (triangles) determined pressure-dependence of the OH stretching modes of clinochlore. The dotted line represents the transformation pressure.

7. HIGH-PRESSURE RAMAN SPECTROSCOPY OF TRANSITION ZONE SPINELLOIDS AND SPINEL

The characteristic Raman modes of the silicate spinelloid phases, wadsleyite and wadsleyite II (Si_2O_7 and SiO_3 symmetric stretching mode), and spinel, ringwoodite, (asymmetric and symmetric stretching vibrations of SiO_4 tetrahedra) shift continuously up to pressures of 50-60 GPa giving no indication for a crystal structural change despite compression well beyond the stability field of the transition zone phases in terms of pressure [*Kleppe et al.,* 2006a, b; 2002a, b; 2001a]. The pressure dependence of these modes for Fo_{90} hydrous and Fo_{100} hydrous wadsleyite agrees within the error of measurement. Also, the modes of hydrous and anhydrous Mg end-member ringwoodite are nearly identical up to 20 GPa, suggesting that protonation has only a minor effect on the lattice dynamics over the entire pressure stability range for wadsleyite and ringwoodite in the mantle at room temperature.

A striking feature in the high-pressure Raman spectra of Fo_{90} hydrous wadsleyite and wadsleyite II as well as Fo_{89} hydrous ringwoodite, is a significant growth in intensity of new bands in the mid-frequency range (about 300-650 cm⁻¹ at 1-bar and shifted to 400-750 cm⁻¹ at 51.4 GPa) relative to the Si_2O_7 and SiO_3 symmetric stretching modes and SiO_4 stretching vibrations, respectively, under compression (Fig.

3). The appearance of new Raman modes near 40 GPa is reversible on decompression without hysteresis. In contrast to the iron-bearing hydrous spinelloids and spinel, the high-pressure Raman spectra of the hydrous Mg end-member phases do not show enhanced intensity and emergence of pressure-induced Raman modes in the mid-frequency range under compression [*Kleppe et al.,* 2002a, 2001a]. Hence, our observations appear to be connected to iron (Fe^{2+}, Fe^{3+}) substitution in the Mg end-member structures. Previously the pressure-induced modes in Fo_{89} hydrous ringwoodite were interpreted as a possible increase in the coordination of Si related to the structural complexity generated by both Fe and proton substitution [*Kleppe et al.,* 2002b]. Here we argue that the observed, pressure-induced changes in the Raman spectra of iron-containing wadsleyite, wadsleyite II and ringwoodite are more likely a result of resonance electronic Raman scattering based on newly-reported optical and near infrared absorption spectra of wadsleyite II and ringwoodite [*Keppler and Smyth,* 2005; *Smyth et al.,* 2005b].

7.1. On Resonance Electronic Raman Scattering in Wadsleyite II and Ringwoodite at High-Pressures

Resonance electronic Raman scattering can originate from the presence of Fe^{2+} in octahedral sites and can cause striking intensity enhancement together with new modes at frequencies forbidden for vibrational Raman modes. Electronic transitions related to the substitution of Fe^{2+} for Mg^{2+} on octahedral sites have been observed in the optical and near infrared absorption spectra of wadsleyite II and ringwoodite [*Keppler and Smyth,* 2005; *Smyth et al.,* 2005b]: The absorption spectrum of ringwoodite is characterized by two bands, at 8678 and 12265 cm⁻¹, due to spin-allowed crystal-field transitions of octahedral Fe^{2+}, and by one band, at 17482 cm⁻¹, due to $Fe^{2+} \rightarrow Fe^{3+}$ intervalence charge transfer [*Keppler and Smyth,* 2005]. The pressure dependence of the crystal-field bands has been established to 21.5 GPa and is for the 12265 cm⁻¹ crystal-field band 77.5 cm⁻¹/GPa [*Keppler and Smyth,* 2005]. Extrapolation to 40 GPa shifts the 12265 cm⁻¹ crystal-field transition band to 15365 cm⁻¹. The full width at half maximum of this band does not change significantly under pressure and is about 5000 cm⁻¹ [*Keppler and Smyth,* 2005]. Hence near 40 GPa this crystal-field transition band starts to overlap significantly with the 514.5 nm (19436 cm⁻¹) laser excitation line. This overlap makes possible resonance excitation involving crystal-field levels of octahedral Fe^{2+}. Fluorescence as origin of the modes can be excluded as 488 nm (20492 cm⁻¹) laser excitation results in identical frequencies for all modes. The electronic Raman scattering is present in both, spectra excited with 514.5 and 488 nm, because the difference in the excitation energies is small with respect to

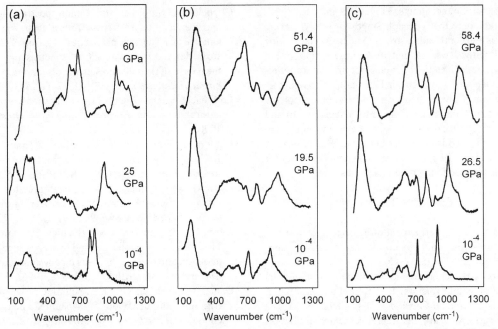

Figure 3. Selected Raman spectra of (a) Fo_{89} hydrous ringwoodite, (b) hydrous wadsleyite II, and (c) Fo_{90} hydrous wadsleyite as a function of pressure in the wavenumber range 50-1350 cm^{-1}.

the half width of the crystal-field transition band. The coincidence, and hence resonance, between the laser excitation line and the crystal-field levels becomes better under further compression causing a discrete, electronic Raman mode structure in the mid-frequency range at higher pressures, which is more intense than the vibrational Raman modes (Fig. 3). The fact that the phonon intensities remain about the same level while the electronic Raman intensities become strongly enhanced indicates that the coupling between the electronic states of Fe^{2+} and the lattice vibrations is weak.

Similar considerations hold true for wadsleyite II: The optical and near infrared absorption spectrum of wadsleyite II shows a band centred around 10638 cm^{-1} due to crystal field transitions of octahedral Fe^{2+} at room-pressure [*Smyth et al.,* 2005b]. High-pressure absorption spectra do not exist but a shift of the crystal-field band to higher frequencies under compression is expected. Assuming that the band has a pressure-dependence similar to that of the crystal-field transition band of ringwoodite, we can also explain the intense mode structure in the mid-frequency range above 40 GPa of wadsleyite II with resonant electronic Raman modes.

8. CONCLUSIONS

Raman spectroscopy is a sensitive tool for probing structure and chemistry of hydrous and nominally anhydrous silicates. It is particularly useful in the identification of OH and H_2O in minerals and in the discrimination between structurally closely related silicates (e.g., spinelloids) and small, multi-phase samples from high-pressure synthesis. High-pressure Raman spectroscopic studies of hydrous and nominally anhydrous deep-Earth silicates have shown that the nature of hydrogen bonding at mantle pressures can exhibit considerable complexity. A combined experimental-atomistic modelling approach can help to interpret modes and the role of hydrogen bonding in minerals. Calculations based on density functional perturbation theory can be applied to complex silicate systems reproducing well the experimental results. Further combined experimental-theoretical studies are necessary to assess the reliability of the recently developed quantum mechanical models regarding the prediction of phonon frequencies. Understanding the dynamics of H behaviour, particularly at high pressure, will require a multi-technique approach to experimental studies that is coupled with advanced computational methods. Investigations of nominally anhydrous transition zone spinelloids and spinels with more realistic mantle compositions (departing from the pure Mg end-members) containing iron and hydrogen have shown that complex scattering processes related to resonance electronic Raman scattering can occur in the hydrous, iron-bearing phases under compression.

Acknowledgement. This work was supported by Natural Environment Research Council fellowship NER/I/S/2001/00723, Natural Environment Research Council grant NER/B/S/2003/00258, and Royal Society research grant RSRG 24067 to A.K.K., and Natural

Environment Research Council grants GT59801ES and GR3/10912 to A.P.J. We would like to thank Joseph Smyth for freely sharing his samples with us and contributing structure plots, Alison Pawley for kindly providing the lawsonite and 10-Å phase sample, Steve Jacobsen and Elizabeth Littlefield for contributing the superhydrous phase B Raman spectrum, Patrizia Fumagalli and co-authors for the talc spectrum, Chung-Cheng Lin and co-authors for the chondrodite and clinohumite spectrum, and Joseph Smyth and co-authors for the IR spectra of ringwoodite. We also would like to thank Steve Jacobsen and Suzan van der Lee for helpful comments and discussions, and Hans Keppler and an anonymous reviewer for reviews which have helped to improve this manuscript.

REFERENCES

Berry, A.J., and James, M., Refinement of hydrogen positions in synthetic hydroxyl-clinohumite by powder neutron diffraction, *American Mineralogist*, 86, 181–184, 2001.

Blaha, J.J., and Rosasco, G.J., Raman microprobe spectra of individual microcrystals and fibers of talc, tremolite, and related silicate minerals, *Analytical Chemistry*, 50 (7), 892–896, 1978.

Blanchard, M., Wright, K., and Gale, J.D., A computer simulation study of OH defects in Mg_2SiO_4 and Mg_2GeO_4 spinels, *Physics and Chemistry of Minerals*, 32, 585–593, 2005.

Boffa-Ballaran, T., and Angel, R.J., Equation of state and high-pressure phase transitions in lawsonite, *European Journal of Mineralogy*, 15, 241–246, 2003.

Bolfan-Casanova, N., Keppler, H., Rubie, D.C., Water partitioning between nominally anhydrous minerals in the $MgO-SiO_2-H_2O$ system up to 24 GPa: implications for the distribution of water in the Earth's mantle, *Earth and Planetary Science Letters*, 182, 209–221, 2000.

Braithwaite, J.S., Wright, K., and Catlow, C.R.A., A theoretical study of the energetics and IR frequencies of hydroxyl defects in forsterite, *Journal of Geophysical Research*, 105 (B6), 2284, 2003.

Brodholt, J.P., and Refson, K., An ab initio study of hydrogen in forsterite and a possible mechanism for hydrolytic weakening, *Journal of Geophysical Research*, 105 (B8), 18977–18982, 2000.

Brüesch, P., Phonons: Theory and Experiments, *Springer Series in Solid-State Sciences 65*, Springer Verlag, Berlin, Heidelberg, 1986.

McCammon, C.A., Frost, D.J., Smyth, J.R., Laustsen, H.M.S., Kawamoto, T., Ross, N.L., and van Aken, P.A., Oxidation state of iron in hydrous mantle phases: implications for subduction and mantle oxygen fugacity, *Physics of the Earth and Planetary Interiors*, 143–144, 157–169, 2004.

Chen, J., Inoue, T., Yurimoto, H., and Weidner, D.J., Effect of water on olivine-wadsleyite phase boundary in the (Mg, Fe)$_2$SiO$_4$ system, *Geophysical Research Letters*, 29 (18), 1875, 2002.

Chinnery, N., Pawley, A.R., and Clark, S.M., The equation of state of lawsonite to 7 GPa and 873 K, and calculation of its high-pressure stability, *American Mineralogist*, 85, 1001–1008, 2000.

Chopelas, A., Single-crystal Raman spectra of forsterite, fayalite, and monticellite, *American Mineralogist*, 76, 1101-1109, 1991a.

Chopelas, A., Thermal properties of β-Mg_2SiO_4 at mantle pressures derived from vibrational spectroscopy: implications for the nature of the 400-km seismic discontinuity, *Journal of Geophysical Research*, 96 (B7), 11817–11829, 1991b.

Le Cléach, A., and Gillet, P., IR and Raman spectroscopic study of natural lawsonite, *European Journal of Mineralogy*, 2, 43–53, 1990.

Comodi, P., Fumagalli, S., Nazzareni, S., and Zanazzi, P.F., The 10-Å phase: Crystal structure form single-crystal X-ray data, *American Mineralogist*, 90, 1012–1016, 2005.

Cynn, H., and Hofmeister, A.M., High-pressure IR spectra of lattice modes and OH vibrations in Fe-bearing wadsleyite, *Journal of Geophysical Research*, 99 (B9), 17717–17727, 1994.

Cynn, H., Hofmeister, A.M., Burnley, P.C., and Navrotsky, A., Thermodynamic properties and hydrogen speciation from vibrational spectra of dense hydrous magnesium silicates, *Physics and Chemistry of Minerals*, 23, 361–376, 1996.

Daniel, I., Fiquet, G., Gillet, P., Schmidt, M.W., and Hanfland, M., High-pressure behaviour of lawsonite: a phase transition at 8.6 GPa, *European Journal of Mineralogy*, 12, 721–733, 2000.

Ferrage, E., Martin, F., Petit, S., Pejo-Soucaille, S., Micoud, P., Fourty, G., Ferret, J., Salvi, S., De Parseval, P., and Fortune, J.P., Evaluation of the talc morphology using FTIR and H/D substitution, *Clay Minerals*, 38, 141–150, 2003.

Frost, D.J., The stability of dense hydrous magnesium silicates in Earth's transition zone and lower mantle, *in Mantle Petrology: Field Observations and High Pressure Experimentation: A tribute to Francis R. Boyd*, edited by Fei, Y., Bertka, C.M., and Mysen, B.O., Special Publication No. 6, The Geochemical Society, 1999.

Frost, D.J., and Fei, Y., Stability of phase D at high pressure and high temperature, *Journal of Geophysical Research*, 103 (B4), 7463–7474, 1998.

Fumagalli, P., and Poli, S., Experimentally determined phase relations in hydrous peridotites to 6.5 GPa and their consequences on the dynamics of subduction zones, *Journal of Petrology*, 46 (3), 555–578, 2005.

Fumagalli, P., Stixrude, L., Poli, S., and Snyder, D., The 10-Å phase: a high-pressure expandable sheet silicate stable during subduction of hydrated lithosphere, *Earth and Planetary Science Letters*, 186, 125–141, 2001.

Gardiner, D.J., and Graves, P.R. (eds.), Practical Raman Spectroscopy, Springer Verlag, Berlin, 1989.

Hirth, G., and Kohlstedt, D.L., Water in the oceanic upper mantle: implications for rheology, melt extraction and the evolution of the lithosphere, *Earth and Planetary Science Letters*, 144 (1–2), 93–108, 1996.

Hofmeister, A.M., Thermal conductivity of spinels and olivines from vibrational spectroscopy: Ambient conditions, *American Mineralogist*, 86, 1188–1208, 2001.

Hofmeister, A.M., Cynn, H., Burnley, P.C., and Meade, C., Vibrational spectra of dense, hydrous magnesium silicates at high pressure: Importance of the hydrogen bond angle, *American Mineralogist*, 84, 454–464, 1999.

Holl, C.M., Smyth, J.R., Manghnani, M.H., Amulele, G.M., Sekar, M., Frost, D.J., Prakapenka, V.B., and Shen, G., Crystal structure and compression of an iron-bearing Phase A to 33 GPa, *Physics and Chemistry of Minerals*, 33, 2006.

Huang, X., Xu, Y., and Karato, S., Water content in the transition zone from electrical conductivity of wadsleyite and ringwoodite, *Nature*, 434, 746–749, 2005.

Inoue, T., Yurimoto, H., and Kudoh, Y., Hydrous modified spinel, $Mg_{1.75}SiH_{0.5}O_4$: a new water reservoir in the mantle transition region, *Geophysical Research Letters*, 22 (2), 117–120, 1995.

Irifune, T., Kubo, N., Isshiki, M., and Yamasaki, Y., Phase transformations in serpentine and transportation of water into the lower mantle, *Geophysical Research Letters*, 25 (2), 203–206, 1998.

Jacobsen, S.D., Demouchy, S., Frost, D.J., Boffa Ballaran, T., and Kung, J., A systematic study of OH in hydrous wadsleyite from polarized FTIR spectroscopy and single-crystal X-ray diffraction: Oxygen sites for hydrogen storage in Earth's interior, *American Mineralogist*, 90, 61–70, 2005.

Jenkins, D.M., Experimental phase relations of hydrous peridotites modelled in the system H_2O-CaO-MgO-Al_2O_3-SiO_2, *Contributions to Mineralogy and Petrology*, 77, 166–176, 1981.

Jephcoat, A.P., Mao, H.-K., and Bell, P.M., Operation of the Megabar Diamond-Anvil Cell, *in Hydrothermal Experimental Techniques*, edited by Ulmer, G.C., and Barnes, H.L., pp. 469–506, Wiley-Interscience, New York, 1987.

Kagi, H., Parise, J.B., Cho, H., Rossman, G.R., and Loveday, J.S., Hydrogen bonding interactions in phase A [$Mg_7Si_2O_8(OH)_6$] at ambient and high pressure, *Physics and Chemistry of Minerals*, 27, 225–233, 2000.

Kanzaki, M., Stebbins, J., and Xue, X., Characterization of crystalline and amorphous silicates quenched from high-pressure by ^{29}Si MAS NMR spectroscopy, *in High-Pressure Research: Application to Earth and Planetary Sciences*, edited by Syono, Y., and Manghnani, M.H., Geophysical Monograph 67, pp. 89–100, Terra Scientific Publishing Company and American Geophysical Union, Tokyo, Washington D.C., 1992.

Karato, S., Paterson, M.S., and Fitz Gerald, J.D., Rheology of synthetic olivine aggregates: influence of grain-size and water, *Journal of Geophysical Research*, 91 (B8), 8151–8176, 1986.

Keppler, H., and Smyth, J.R., Optical and near infrared spectra of ringwoodite to 21.5 GPa: implications for radiative heat transport in the mantle, *American Mineralogist*, 90, 1209–1212, 2005.

Khisina, N.R., and Wirth, R., Water-bearing inclusions in olivine from kimberlite: High-pressure hydrous silicates?, *EOS*, 78, 735, 1997.

Kieffer, S.W., Thermodynamics and lattice vibrations of minerals: 1. Mineral heat capacities and their relationships to simple lattice vibrational modes, *Reviews of Geophysics and Space Physics*, 17 (1), 1–19, 1979.

Kleppe, A.K., Jephcoat, A.P., Ross, N.L., Raman spectroscopic studies of phase E to 19 GPa, *American Mineralogist*, 86, 1275–1281, 2001b.

Kleppe, A.K., Jephcoat, A.P., and Smyth, J.R., Raman spectroscopic study of hydrous γ-Mg_2SiO_4 to 56.5 GPa, *Physics and Chemistry of Minerals*, 29, 473–376, 2002a.

Kleppe, A.K., Jephcoat, A.P., and Smyth, J.R., High-pressure Raman spectroscopic study of Fo_{90} hydrous wadsleyite, *Physics and Chemistry of Minerals*, 32, 700–709, 2006a.

Kleppe, A.K., Jephcoat, A.P., and Smyth, J.R., High-pressure Raman spectroscopic studies of hydrous wadsleyite II, *American Mineralogist, in press*, 2006b.

Kleppe, A.K., Jephcoat, A.P., and Welch, M.D., The effect of pressure upon hydrogen bonding in chlorite: A Raman spectroscopic study of clinochlore to 26.5 GPa, *American Mineralogist*, 88, 567–573, 2003.

Kleppe, A.K., Jephcoat, A.P., Smyth, J.R., and Frost, D.J., On protons, iron and the high-pressure behavior of ringwoodite, *Geophysical Research Letters*, 29 (21), 17,1–17,4, 2002b.

Kleppe, A.K., Jephcoat, A.P., Olijnyk, H., Slesinger, A.E., Kohn, S.C., and Wood, B.J., Raman spectroscopic study of hydrous wadsleyite (β-Mg_2SiO_4) to 50 GPa, *Physics and Chemistry of Minerals*, 28, 232–241, 2001a.

Koch-Müller, M., Dera, P., Fei, Y., Hellwig, H., Liu, Z., van Orman, J., and Wirth, R., Polymorphic phase transition in superhydrous phase B, *Lithos*, 73 (1–2), S59-S59, 2004.

Kohlstedt, D.L., Keppler, H., and Rubie, D.C., Solubility of water in the α, β and γ phases of $(Mg,Fe)_2SiO_4$, *Contributions to Mineralogy and Petrology*, 123, 345–357, 1996.

Kohn, S.C., and Fumagalli, P., New constraints on the structure of 10 Å phase from ^1H and ^{29}Si MAS NMR data, Abstract WP16, 18th EMPG meeting, Edinburgh, UK, 2002.

Kohn, S.C., Brooker, R.A., Frost, D.J., Slesinger, A.E., and Wood, B.J., Ordering of hydroxyl defects in hydrous wadsleyite (β-Mg_2SiO_4), *American Mineralogist*, 87, 293–301, 2002.

Kudoh, Y., Structural relation of hydrous ringwoodite to hydrous wadsleyite, *Physics and Chemistry of Minerals*, 28, 523–530, 2001.

Kudoh, Y., Finger, L.W., Hazen, R.M., Prewitt, C.T., Kanzaki, M., and Veblen, D.R., Phase E: A high-pressure hydrous silicate with unique crystal chemistry, *Physics and Chemistry of Minerals*, 19, 257–360, 1993a.

Kudoh, Y., Inoue, T., and Arashi, H., Structure and crystal chemistry of hydrous wadsleyite, $Mg_{1.75}SiH_{0.5}O_4$: possible hydrous magnesium silicate in the mantle transition zone, *Physics and Chemistry of Minerals*, 23, 461–469, 1996.

Kudoh, Y., Kuribayashi, T., Mizobata, H., and Ohtani, E., Structure and cation disorder of hydrous ringwoodite, γ-$Mg_{1.89}Si_{0.98}H_{0.30}O_4$, *Physics and Chemistry of Minerals*, 27, 474–479, 2000.

Kudoh, Y., Nagase, T., Ohta, S., Sasaki, S., Kanzaki, M., and Tanaka, M., Crystal structure and compressibility of superhydrous phase B, $Mg_{20}Si_6H_8O_{36}$, *in High-Pressure Science and Technology*, edited by Schmidt, S.C., Shaner, J.W., Samara, G.A., and Ross, M., American Institute of Physics, pp. 469–472, 1993b.

Labotka, T.C., and Rossman, G.R., The infrared pleochroism of lawsonite: the orientation of the water and hydroxide groups, *American Mineralogist*, 59, 799–806, 1974.

Lager, G.A., Ulmer, P., Miletich, R., and Marshall, W.G., O-D···O bond geometry in OD-chondrodite, *American Mineralogist*, 86, 176–180, 2001.

Lemaire, C., Kohn, S.C., and Brooker, R.A., The effect of silica activity on the incorporation mechanisms of water in synthetic forsterite: a polarized infrared spectroscopic study, *Contributions to Mineralogy and Petrology*, 147, 48–57, 2004.

Libowitzky, E., Correlation of O-H stretching frequencies and O-H···O hydrogen bond lengths in minerals, *Monatshefte für Chemie*, 130, 1047–1059, 1999.

Libowitzky, E., and Armbruster, T., Low-temperature phase transitions and the role of hydrogen bonds in lawsonite, *American Mineralogist*, 80, 1277–1285, 1995.

Libowitzky, E., and Rossman, G.R., FTIR spectroscopy of lawsonite between 82 and 325 K, *American Mineralogist*, 81, 1080–1091, 1996.

Lin, C.-C., Liu, L.-G., and Irifune, T., High-pressure Raman spectroscopic study of chondrodite, *Physics and Chemistry of Minerals*, 26, 226–233, 1999.

Lin, C.-C., Liu, L.-G., Mernagh, T.P., and Irifune, T., Raman spectroscopic study of hydroxyl-clinohumite at various pressures and temperatures, *Physics and Chemistry of Minerals*, 27, 320–331, 2000.

Liu, L.G., Lin, C.C., Mernagh, T.P., and Inoue, T., Raman spectra of phase C (superhydrous phase B) at various pressures and temperatures, *European Journal of Mineralogy*, 14, 15–23, 2002a.

Liu, L.G., Lin, C.C., Mernagh, T.P., and Inoue, T., Raman spectra of hydrous γ-Mg_2SiO_4 at various pressures and temperatures, *Physics and Chemistry of Minerals*, 29, 181–187, 2002b.

Liu, L.G., Lin, C.C., Mernagh, T.P., and Irifune, T., Raman spectra of Phase A at various pressures and temperatures, *Journal of Physics and Chemistry of Solids*, 58 (12), 2023–2030, 1997a.

Liu, L.G., Mernagh, T.P., Lin, C.C., and Irifune, T., Raman spectra of phase E at various pressures and temperatures with geophysical implications, *Earth and Planetary Science Letters*, 149, 57–65, 1997b.

Liu, L.G., Lin, C.C., Irifune, T., and Mernagh, T.P., Raman study of phase D at various pressures and temperatures, *Geophysical Research Letters*, 25 (18), 3453–3456, 1998a.

Liu, L.-G., Mernagh, T.P., Lin, C.-C., Xu, J., and Inoue, T., Raman spectra of hydrous β-Mg_2SiO_4 at various pressures and temperatures, *in Properties of Earth and Planetary Materials at High Pressure and Temperature*, edited by Manghnani, M.H., and Yagi, T., Geophysical Monograph 101, pp. 523–530, American Geophysical Union, 1998b.

Lutz, H.D., Bonding and structure of water molecules in solid hydrates: Correlation of Spectroscopic and structural data, *Structure and Bonding*, 69, 97–125, 1988.

Lutz, H.D., Hydroxide ions in condensed materials—Correlation of spectroscopic and structural data, *Structure and Bonding*, 82, 86–103, 1995.

Manghnani, M.H., Amulele, G., Smyth, J.R., Holl, C.M., Chen, G., Prakapenka, V., and Frost, D.J., Equation of state of hydrous Fo_{90} ringwoodite to 45 GPa by synchrotron powder diffraction, *Mineralogical Magazine*, 69 (3), 317–323, 2005.

Mernagh, T.P., and Liu, L.G., Raman and infrared spectra of hydrous β-Mg_2SiO_4, *Canadian Mineralogist*, 34, 1233–1240, 1996.

Mernagh, T.P., and Liu, L.G., Raman and infrared spectra of phase E, a plausible hydrous phase in the mantle, *Canadian Mineralogist*, 36, 1217–1223, 1998.

Mernagh, T.P., Liu, L.G., and Lin, C.-C., Raman spectra of chondrodite at various temperatures, *Journal of Raman Spectroscopy*, 30, 963–969, 1999.

Meyer, H.-W., Marion, S., Sondergeld, P., Carpenter, M.A., Knight, K.S., Redfern, S.A.T., and Dove, M.T., Displacive components of the low-temperature phase transitions in lawsonite, *American Mineralogist*, 86, 566–677, 2001.

McMillan, P.F., Masaki, A., Sato, R.K., Poe, B. and Foley, J., Hydroxyl groups in β-Mg_2SiO_4, *American Mineralogist*, 76, 354–360, 1991.

Nakamoto, K., Margoshes, M., and Rundle, R.E., Stretching frequencies as a function of distances in hydrogen bonds, *Journal of the American Chemical Society*, 77, 6480–6486, 1955.

Ohtani, E., Mizobata, H., and Yurimoto, H., Stability of dense hydrous magnesium silicate phases in the systems Mg_2SiO_4-H_2O and $MgSiO_3$-H_2O at pressures up to 27 GPa, *Physics and Chemistry of Minerals*, 27, 533–544, 2000.

Ohtani, E., Shibata, T., Kubo, T., and Kato, T., Stability of hydrous phases in the transition zone and the upper most part of the lower mantle, *Geophysical Research Letters*, 22, 2553–2556, 1995.

Ohtani, E., Toma, M., Litasov, K., Kubo, T., and Suzuki, A., Stability of dense hydrous magnesium silicate phases and water storage capacity in the transition zone and lower mantle, *Physics of the Earth and Planetary Interiors*, 124, 105–117, 2001.

Pacalo, R.E.G., and Parise, J.B., Crystal structure of superhydrous B, a hydrous magnesium silicate synthesized at 1400 °C and 20 GPa, *American Mineralogist*, 77, 681–684, 1992.

Pasteris, J.D., Wopenka, B., and Seitz, J.C., Practical aspects of quantitative laser Raman microprobe spectroscopy for the study of fluid inclusions, *Geochimica et Cosmochimica Acta*, 52 (5), 979–988, 1988.

Pawley, A.R., The pressure and temperature stability limits of lawsonite: implications for H_2O recycling in subduction zones, *Contributions to Mineralogy and Petrology*, 118, 99–108, 1994.

Pawley, A.R., and Allan, D.R., A high-pressure structural study of lawsonite using angle-dispersive powder-diffraction methods with synchrotron radiation, *Mineralogical Magazine*, 65 (1), 41–58, 2001.

Pawley, A.R., and Wood, B.J., The high-pressure stability of talc and 10-Å phase: potential storage sites for H_2O in subduction zones, *American Mineralogist*, 80, 998–1003, 1995.

Pawley, A.R., Welch, M.D., and Smith, R.I., The 10-Å phase: structural constraints from neutron powder diffraction, *Lithos*, 73, S86, 1–2, 2004.

Peacock, S.M., Fluid processes in subduction zones, *Science*, 248, 329–337, 1990.

Perdikatsis, B., and Burzlaff, H., Strukturverfeinerung am Talk $Mg_3[(OH)_2Si_4O_{10}]$, *Zeitschrift für Kristallographie*, 156, 177–186, 1981.

Prewitt, C.T., and Parise, J.B., Hydrous phases and hydrogen bonding at high-pressure, *in Reviews in Mineralogy and Geochemistry 41*, edited by Hazen, R.M., and Downs, R.T., pp. 309–334, 2000.

Prieto, A.C., Dubessy, J., and Cathelineau, M., Structure-composition relationships in trioctahedral chlorites: A vibrational spectroscopy study, *Clays and Clay Minerals*, 39 (5), 531–539, 1991.

Refson, K., O'Connor, M.V., Kleppe, A.K., and Jephcoat, A.P., Structure, Sorption and Spectroscopy of Phyllomanganates and Phyllosilicates, *in CECAM Workshop: First Principles Simulations: Perspectives and Challenges in Mineral Sciences*, Berichte aus Arbeitskreisen der DGK, pp. 13–19, 2004.

Rosasco, G.J., and Blaha, J.J., Raman microprobe spectra and vibrational mode assignments of talc, *Applied Spectroscopy*, 34 (2), 140–144, 1980.

Rousseau, D.L., Baumann, R.P., and Porto, S.P.S., Normal mode determination in crystals, *Journal of Raman Spectroscopy*, 10, 253–290, 1981.

Schmidt, M.W., Lawsonite: upper pressure stability and formation of higher density hydrous phases, *American Mineralogist*, 80, 1286–1292, 1995.

Scott, H.P., and Williams, Q., An infrared spectroscopic study of lawsonite to 20 GPa, *Physics and Chemistry of Minerals*, 26, 437–445, 1999.

Segall, M.D., Lindan, P.J.D., Probert, M.J., Pickard, C.J., Hasnip, P.J., Clark, S.J., and Payne, M.C., First-principles simulation: ideas, illustrations and the CASTEP code, *Journal of Physics: Condensed Matter*, 14, 2717–2744, 2002.

Shieh, S.R., Mao, H.K., Hemley, R.J., and Ming, L.C., Decomposition of phase D in the lower mantle and the fate of dense hydrous silicates in subducting slabs, *Earth and Planetary Science Letters*, 159 (1–2), 13–23, 1998.

Shieh, S.R., Mao, H.K., Konzett, J., and Hemley, R.J., In-situ high-pressure X-ray diffraction of phase E to 15 GPa, *American Mineralogist*, 85, 765–769, 2000.

Smyth, J.R., β-Mg_2SiO_4: a potential host for water in the mantle?, *American Mineralogist*, 72, 1051–1055, 1987.

Smyth, J.R., A crystallographic model for hydrous wadsleyite (β-Mg_2SiO_4): An ocean in the Earth's interior?, *American Mineralogist*, 79, 1021–1025, 1994.

Smyth, J.R., and Kawamoto, T., Wadsleyite II: A new high pressure hydrous phase in the peridotite-H_2O system, *Earth and Planetary Science Letters*, 146, E9–E16, 1997.

Smyth, J.R., Frost, D.J., and Nestola, F., Hydration of olivine and Earth's deep water cycle, *Geochimica et Cosmochimica Acta*, 69 (10), A746–A746, 2005a.

Smyth, J.R., Holl, C.M., Frost, D.J., Jacobsen, S.D., High-pressure crystal chemistry of hydrous ringwoodite and water in the Earth's interior, *Physics of the Earth and Planetary Interiors*, 143–144, 271–278, 2004.

Smyth, J.R., Holl, C.M., Frost, D.J., Jacobsen, S.D., Langenhorst, F., and McCammon, C.A., Structural systematics of hydrous ringwoodite and water in the Earth's interior, *American Mineralogist*, 88, 1402–1407, 2003.

Smyth, J.R., Holl, C.M., Langenhorst, F., Laustsen, H.M., Rossman, G.R., Kleppe, A.K., McCammon, C.A., Kawamoto, T., and van Aken, P.A., Crystal chemistry of wadsleyite II and water in the Earth's interior, *Physics and Chemistry of Minerals*, 31, 691–705, 2005b.

Sondergeld, P., Schranz, W., Tröster, A., Armbruster, T., Giester, G., Kityk, A., and Carpenter, M.A., Ordering and elasticity associated with low-temperature phase transitions in lawsonite, *American Mineralogist*, 90, 448–456, 2005.

Stalder, R., and Ulmer, P., Phase relations of a serpentine composition between 5 and 14 GPa: significance of clinohumite and phase E as water carriers into the transition zone, *Contributions to Mineralogy and Petrology*, 140, 670–679, 2001.

Suzuki, A., Ohtani, E., Kondo, T., Kuribayashi, T., Niimura, N., Kurihara, K., and Chatake, T., Neutron diffraction study of hydrous phase G: Hydrogen in the lower mantle hydrous silicate, phase G, *Geophysical Research Letters*, 28 (20), 3987–3990, 2001.

Ulmer, P., and Trommsdorff, V., Serpentine stability to mantel depths and subduction-related magmatism, *Science*, 268, 858–861, 1995.

Welch, M.D., and Marshall, W.G., High-pressure behavior of clinochlore, *American Mineralogist*, 86, 1380–1386, 2001.

Welch, M.D., Barras, J., and Klinowski, J., A multinuclear NMR study of clinochlore, *American Mineralogist*, 80, 441–447, 1995.

Welch, M.D., Kleppe, A.K., and Jephcoat, A.P., Novel high-pressure behavior in chlorite: a synchrotron XRD study of clinochlore to 27 GPa, *American Mineralogist*, 89, 1337–1340, 2004.

White, W.B., and DeAngelis, B.A., Interpretation of the vibrational spectra of spinels, *Spectrochimica Acta*, 23A, 985–995, 1967.

Wilkins, R.W.T., and Ito, J., Infrared spectra of some synthetic talcs, *American Mineralogist*, 52, 1649–1661, 1967.

Yang, H., Prewitt, C.T., and Frost, D.J., Crystal structure of the dense hydrous magnesium silicate, phase D, *American Mineralogist*, 82, 651–654, 1997.

Zajacz, Z., Halter, W., Malfait, W.J., Bachmann, O., Bodnar, R.J., Hirschmann, M.M., Mandeville, C.W., Morizet, Y., Muntener, O., Ulmer, P., and Webster, J.D., A composition-independent quantitative determination of water content in silicate glasses and silicate melt inclusions by confocal Raman spectroscopy, *Contributions to Mineralogy and Petrology*, 150 (6), 631–632, 2005.

Zhao, Y.H., Ginsberg, S.B., and Kohlstedt, D.L., Solubility of hydrogen in olivine: dependence on temperature and iron content, *Contributions to Mineralogy and Petrology*, 147 (2), 155–161, 2004.

Andrew P. Jephcoat, Diamond Light Source Ltd., Chilton, Didcot, Oxfordshire, UK

Annette K. Kleppe, University of Oxford, Department of Earth Sciences, Parks Road, Oxford, OX1 3PR, U.K.

Influence of Water on Major Phase Transitions in the Earth's Mantle

Konstantin D. Litasov, Eiji Ohtani, and Asami Sano

Institute of Mineralogy, Petrology and Economic Geology, Tohoku University, Sendai, Japan

In this chapter we summarize recent results on the influence of water on major phase transformations in the Earth's mantle with implications for seismic discontinuity structure and mantle dynamics. The experimental data are based on quench multianvil and *in situ* X-ray diffraction studies. Differences in water solubility between olivine and wadsleyite, and between ringwoodite and Mg-perovskite + ferropericlase may displace the phase transition boundaries, which are responsible for the 410- and 660-km discontinuities, respectively. The results show that water expands the stability field of wadsleyite to lower pressures, which is consistent with broadening of the 410-km discontinuity in some regions of the mantle. A significant shift of the wadsleyite-ringwoodite phase transition to higher pressure caused by water may also be responsible for depth variations or absence of the 520-km discontinuity. Study of the post-spinel transformation in hydrous pyrolite indicates that the phase boundary also shifts to higher pressures. Displacement of this boundary with ~2 wt.% H_2O corresponds to about 15 km at 1473 K. Thus, presence of water could account for half of the observed 30–40 km depressions at the 660-km discontinuity in subduction zones at this temperature. Study of the post-garnet transformation in anhydrous and hydrous MORB show that this phase boundary shifts to the lower pressures by ~2 GPa with the addition of 2–5 wt.% water. This observation demonstrates that the density crossover between peridotite and basaltic components near 660 km might be absent under hydrous conditions, inhibiting the separation of these components at the 660-km discontinuity.

1. INTRODUCTION

Water has played an important role in Earth's evolution due to its strong influence on the chemical and physical properties of crust and mantle constituents. Water has a remarkable ability to weaken rocks and minerals. Water-induced reduction of the viscosity and strength of mantle materials may be crucial to understanding the driving forces of plate tectonics [e.g. *Griggs and Blacic,* 1965; *Karato,* 1990; *Hirth and Kohlstedt,* 1996; *Chen et al.,* 1998; *Mei*

and Kohlstedt, 2000]. Water also enhances rates of phase transformations and dramatically depresses the melting temperature of silicate minerals [e.g. *Kushiro,* 1968; *Mysen,* 1973; *Mysen and Boettcher,* 1975; *Inoue,* 1994; *Kubo et al.,* 1998; *Litasov and Ohtani,* 2002].

Many recent studies provide information about the possible existence of water in the Earth's mantle, especially in the transition zone, where wadsleyite and ringwoodite can incorporate up to 3 wt% of H_2O in their structures [e.g. *Kohlstedt et al.,* 1996]. For example, low-velocity zones detected atop the 410-km discontinuity may indicate existence of trapped high-density melt [e.g. *Revenaugh and Sipkin,* 1994; *Song et al.,* 2004; see also *Song and Helmberger,* this volume], which can be hydrous according to differences in water solu-

Earth's Deep Water Cycle
Geophysical Monograph Series 168
Copyright 2006 by the American Geophysical Union.
10.1029/168GM08

bility between olivine and wadsleyite and a higher melting temperature of the anhydrous peridotite mantle compared to the present mantle geotherm. In addition, some electrical conductivity anomalies in the upper mantle and transition zone related to subduction zones [e.g. *Fukao et al.*, 2004; *Tarits et el.*, 2004] have been interpreted as water-rich regions. Kinetic studies of the hydrous olivine-wadsleyite transformation [*Ohtani et al.*, 2004] are consistent with a majority of seismological observations, which suggest the absence of a metastable olivine wedge (proposed from a sluggish olivine-wadsleyite transformation in the anhydrous system [*Rubie and Ross*, 1994]) in the descending subducting slabs. The thermoelastic properties of hydrous wadsleyite and ringwoodite indicate that P- and S- wave velocities are significantly reduced compared with anhydrous wadsleyite and ringwoodite [*e.g. Jacobsen et al.*, 2004; *Inoue et al.*, 2004]. Mass balance calculations indicate that the amount of water transported by subducting slabs into the deep mantle is greater than that returned to the surface by the magmatism in the island arcs and mid-ocean ridges [e.g. *Peacock*, 1990]. All these data show that a significant amount of water can be stored in the mantle even if the amount of primordial water, which was maintained in the deep Earth after major degassing events, is limited.

Recent analytical and experimental works demonstrate that many hydrous phases may be stabilized under the mantle conditions. In particular, several dense hydrous magnesium silicates like phase E, D, superhydrous B, are stable in the upper and lower mantle conditions [e.g. *Frost*, 1999; *Ohtani et al.*, 2001; 2004; see also *Komabayashi*, this volume]. A volume of water corresponding to at least several times the current hydrosphere could be stored in the upper mantle alone in nominally anhydrous minerals such as olivine, garnet, and stishovite [e.g. *Bell and Rossman*, 1992; *Pawley et al.*, 1993; *Ingrin and Skogby*, 2000; *Bolfan-Casanova et al.*, 2000].

The effect of water on phase transformations in the mantle transition zone and at the core-mantle boundary may lead to a significant revision of temperature distribution in the Earth and its spatial and temporal energy balance. The uncertainty on the abundance of hydrogen in the core material may give rise to variations of thousand degrees in the inferred temperature profile of the Earth's core [e.g. *Williams and Hemley*, 2001]. Thus, it is important to clarify the influence of water on major phase transition boundaries in the deep Earth interior.

Seismic discontinuities at 410- and 660-km depth (hereafter referred to as *d*410 and *d*660) divide the mantle into the upper mantle, transition zone and lower mantle. The presence of these two discontinuities on a global scale has long been established and is usually attributed to the phase transforma-

tions of olivine in a peridotite composition [e.g. *Helffrich*, 2000]. At approximately 410-km depth α-$(Mg,Fe)_2SiO_4$ (olivine) transforms to β-$(Mg,Fe)_2SiO_4$ (wadsleyite), hereafter referred to as the olivine-wadsleyite transformation (OWT), and at approximately 660-km depth γ-$(Mg,Fe)_2SiO_4$ (ringwoodite) decomposes to $(Mg,Fe)SiO_3$-perovskite and $(Mg,Fe)O$-ferropericlase (or magnesiowüstite). The latter transformation is referred to as the post-spinel transformation (hereafter PST) (Fig.1).

The depth, sharpness and amplitude of these discontinuities depend on the mantle potential temperature, chemical composition and mineral proportions. According to the most seismological studies *d*410 and *d*660 are sharp and the changes in physical properties associated with them occur over small depth interval. For *d*410, a width of 6-10 km is consistent with majority of seismic data and a width of <5 km has been suggested for *d*660 [e,g., *Agee*, 1998; *Helffrich*, 2000; *Weidner and Wang*, 2000; *Frost*, 2003]. Studies of discontinuity topography indicate relatively shallow *d*410 and depressed *d*660 beneath subduction zones, and the opposite structure within hot spots/mantle plumes [e.g. *Flanagan and Shearer*, 1998]. Until recently, these variations were con-

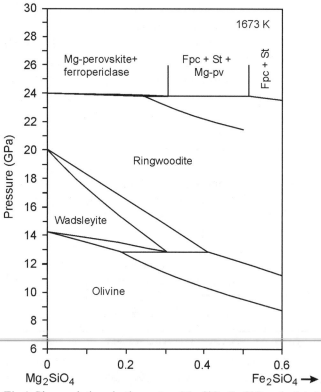

Fig.1. Phase relations in the system Mg_2SiO_4-Fe_2SiO_4 calculated from thermodynamic data at 1673 K after *Katsura and Ito* [1989], *Ito and Takahashi* [1989] and *Frost* [2003]. Fpc: ferropericlase; St: stishovite, Mg-pv: Mg-perovskite.

sidered to be consistent with the experimentlally observed Clapeyron slopes of OWT and PST. For example, *Ito and Takahashi* [1989] reported Clapeyron slope of -3.0 MPa/K, and *Bina and Helffrich* [1994] calculated Clapeyron slope of -2.0 MPa/K for PST. However, recent *in situ* X-ray diffraction studies of PST in Mg_2SiO_4 and pyrolite composition indicated that the negative slope of this boundary is much gentler (-0.4 to -1.3 MPa/K) [*Katsura et al.*, 2003; *Fei et al.*, 2004a; *Litasov et al.*, 2005a]. These studies suggest that it is difficult to explain the observed topography of *d*660 in terms of anhydrous Clapeyron slopes alone, and indicate that the PST may account for less than half of the observed variations in *d*660. Therefore, another explanation for the observed seismological variations at *d*660 is required, such as the influence of minor element or volatiles (e.g. water) or inaccuracy of pressure calibration scales used in high pressure experiments.

In addition, two other phase transformations important for mantle dynamics and velocity structure of the Earth are considered here: (a) the wadsleyite-ringwoodite transformation (hereafter WRT) in peridotite, which is believed to be responsible for seismic discontinuity at approximately 520-km (hereafter *d*520) [e.g. *Shearer*, 1996], and (b) post-garnet (garnet to perovskite) transformation (hereafter PGT) in eclogite (basaltic) composition [e.g. *Ringwood*, 1994; *Hirose et al.*, 1999; *Litasov et al.*, 2004]. Eclogite is formed in the top part of subducting slabs descending to the deep mantle and may accumulate above *d*660 due to a density crossover with peridotite mantle [e.g. *Ringwood*, 1994].

In this paper, we summarize recent advances in the study of mantle phase transformations under hydrous conditions with implications for seismic discontinuity structure and mantle dynamics. Recent and new experimental studies are based on the results of quench and *in situ* X-ray diffraction experiments using multianvil apparatus.

2. EXPERIMENTS AND PRESSURE CALIBRATION

Most of the *in situ* X-ray diffraction experiments were conducted at the synchrotron radiation facility 'SPring-8' in Hyogo prefecture, Japan. We used a Kawai-type multi-anvil apparatus, 'SPEED-1500', installed at a bending magnet beam line BL04B1. Details of the experiments in Spring-8 and description of phase assemblies are reported elsewhere [e.g. *Litasov et al.*, 2004; 2005a,b]. Starting material was a synthetic mixture of glass and olivine crystals representing $CaO-MgO-FeO-Al_2O_3-SiO_2$ pyrolite [*Litasov et al.*, 2005a,b] mixed with the Au-pressure marker at 15:1 or 20:1 by weight. For hydrous experiments 2-3 wt% of H_2O as $Mg(OH)_2$ (brucite) was added to the starting material adjusting the proportion of MgO. The starting material for anhydrous eclogite

was synthetic glass representing average MORB and for hydrous experiments 2, 5 and 10 wt.% H_2O was added to the capsule prior to welding [*Litasov et al.*, 2004; *Sano et al.*, 2006]. We used an AgPd capsule, which is semi-transparent to X-rays in hydrous experiments and graphite capsules for anhydrous experiments. For each experiment the cell assembly was first compressed to the desired press load at an ambient temperature and then heated gradually to the target temperature. Temperatures were maintained for typically 30 minutes before obtaining diffraction patterns of the stable mineral assemblages at high pressure and temperature. Thereafter, while continuously taking the diffraction patterns, the temperature and press load were changed following the various P-T paths.

In order to determine the phase boundaries accurately, a pressure scale with a reliable equation of state (EOS) must be employed. Au, Pt, and MgO are often used as pressure markers for *in situ* multianvil studies [e.g. *Fei et al.*, 2004b]. In these experiments, the pressure at high temperature was calculated from the unit cell volume of gold using an EOS by *Tsuchiya* [2003] and MgO using an EOS by *Speziale et al.* [2001] (hereafter AT-03 and MS-01 respectively). Uncertainty in the unit cell volumes of Au and MgO calculated by least squares typically results in less than 0.1 GPa uncertainty in pressure. Although the pressure scale is critical to determining phase transformation boundaries and Clapeyron slopes, there are notable discrepancies between different pressure scales, especially above 20 GPa at high temperature. For instance, the calculated pressure using different EOS of Au [e.g. *Jamieson et al.*, 1982; *Anderson et al.*, 1989; *Shim et al.*, 2002; *Tsuchiya et al.*, 2003; *Fei et al.*, 2004b] may vary by as much as 2.5 GPa at 25 GPa and 2000 K. Recently *Fei et al.* [2004b] made a comprehensive review of different pressure scales and re-calibrated Au and Pt scales against MgO using MS-01. We have found that the pressures calculated by MS-01 are generally consistent with those of AT-03. We estimated that P (MS-01) - P (AT-03) = 0.08 ± 0.36 GPa in the pressure range 20-25 GPa at temperatures between 1500 and 2200 K. However, at lower temperatures of 1200-1500 K, the difference P (MS-01) - P (AT-03) may exceed 0.4 GPa [*Litasov et al.*, 2005a]. In some studies (e.g. Katsura et al., 2004) the MgO EOS by *Matsui et al.* [2000] was used as the pressure scale. Pressures using the MgO scale of *Matsui et al.* [2000] are typically 0.1 GPa lower than those from MS-01.

All *in situ* experiments were consistent with the additional quench experiments made with the same starting compositions and under nearly identical conditions. The quench experiments were carried out using a 1000- and 3000-ton Kawai-type multianvil apparatus installed at Tohoku University. The pressure calibration was made using results of *in situ* experiments and is generally consistent with pressure versus load calibration using semiconductor to metal transitions at room temperature and phase transition of SiO_2, $MgSiO_3$, and Mg_2SiO_4 [e.g. *Litasov and Ohtani*, 2002, 2005].

3. EXPERIMENTAL RESULTS

3.1. Olivine-Wadsleyite Transformation

Water appears to have significant effect on the OWT due to differences in H_2O solubility between olivine and wadsleyite, as predicted earlier by thermodynamic calculations [*Wood*, 1995]. It was shown that at 1300-1500 K wadsleyite contain 5-40 times more water than olivine [e.g. *Kohlstedt et al.*, 1996; *Chen et al.*, 2002], therefore the phase transition boundary is expected to be shifted to lower pressures [*Smyth and Frost* 2002].

We performed several experiments on OWT in anhydrous and hydrous pyrolite compositions using *in situ* X-ray diffraction (Fig.2). In the anhydrous run at 1673–1773 K we observed nucleation of wadsleyite in the olivine-bearing assemblage at 14.2 GPa and 1673 K. Then, we heated the

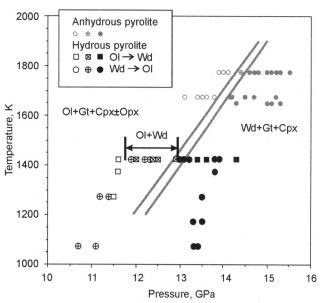

Fig.2. Olivine-wadsleyite transformation boundary in anhydrous and hydrous pyrolite determined by *in situ* X-ray diffraction experiments using pressures calculated from Au EOS by *Tsuchiya* [2003]. Open symbols: olivine; crossed symbols: olivine+wadsleyite; filled symbols: wadsleyite. Estimated olivine-wadsleyite transition loop is shown at ~1470 K by the thick black lines and arrow. Ol: olivine; Gt: garnet; Cpx: clinopyroxene; Opx: orthopyroxene; Wd: wadsleyite. Note that a fluid phase is present in hydrous assemblages at or above saturation.

sample to 1773 K to reverse the transformation. Growth of olivine diffraction peaks in the wadsleyite-bearing assemblage was detected at 14.4 GPa and 1773 K. After holding the temperature about 40 minutes, the relative intensity of olivine and wadsleyite diffraction peaks were stabilized, indicating conditions within the phase loop had been achieved. Following decompression, we observed a rapid increase in the intensity of the olivine diffraction peaks and decrease of intensity of wadsleyite diffraction peaks at 14.1 GPa and 1773 K (Fig.2). Although we could not accurately determine the Clapeyron slope of the OWT boundary with this experiment, it was consistent with the Clapeyron slope in $(Mg,Fe)_2SiO_4$ determined by *Katsura et al.* [2004]. Applying their Clapeyron slope we obtain a linear equation for the OWT boundary in pyrolite as P (GPa) = 0.0039 T (K) + 7.47 using the pressure scale of AT-03. The choice of the pressure scale does not have a significant influence on the slope of the phase transformation. The observed interval of olivine and wadsleyite coexistence is about 0.2-0.3 GPa in present experiment. The pressure of the OWT boundary in pyrolite was found to be slightly higher (by 0.3-0.5 GPa) than that obtained for Fo_{90} composition by *Katsura et al.* [2004] using the MgO scale by *Matsui et al.* [2000] (Fig.3).

In hydrous pyrolite (with about 3 wt% of H_2O) we performed two experiments approaching the phase transformation boundary from both directions at 1423 K. During the olivine to wadsleyite transformation we observed appearance of minor wadsleyite diffraction peaks at 12.0 GPa and 1423 K. It was difficult to clarify the pressure interval of coexistence of olivine and wadsleyite, while significant growth of wadsleyite peaks and disappearance of olivine peaks was observed by 13.4 GPa at the same temperature (Fig.3). In the experiment on the transformation from wadsleyite to olivine, we observed the appearance of olivine diffraction peaks at 12.4 GPa at 1423 K. Based on these experiments, we estimate a shift of the OWT to lower pressures by less than 0.5 GPa, but more significant broadening of the two phase loop by about 1.2 GPa in the hydrous system (Fig.3). Since we determined phase transition in hydrous pyrolite only at 1423 K, we could not determine the effect of water on the Clapeyron slope of OWT boundary.

Previous results on the OWT under anhydrous and hydrous conditions are summarized in Fig.3 along with new data presented here. The Clapeyron slope of the OWT boundary in Mg_2SiO_4 and $(Mg,Fe)_2SiO_4$ by *Katsura and Ito* [1989] indicated 2.5 MPa/K with an interval loop of 0.5 GPa at 1900 K for Fo_{90} composition. *Akaogi et al.* [1989] calculated the OWT boundary using calorimetric data and suggested a shallower Clapeyron slope of 1.5±0.5 MPa/K. *Bina and Helffrich* [1994] re-examined the data of *Akaogi et al.* [1989] obtaining a Clapeyron slope of 3.0 MPa/K. Recent *in situ*

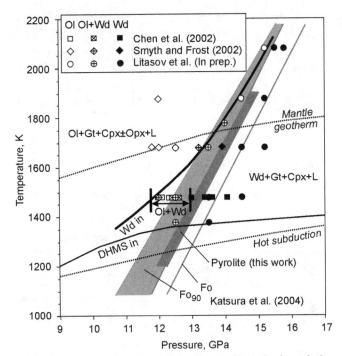

Fig.3. Comparison of olivine-wadsleyite transformation boundaries obtained in different studies. Bold line shows assumed wadsleyite stability based on in situ experiments (Fig.2). *Thin black line* shows stability limit of DHMS (dense hydrous magnesium silicate phases, such as phase A and E). Mantle geotherm is after *Akaogi et al.* [1989]. Hot subduction geotherm is after *Kirby et al.* [1996]. Note, that *Chen et al.* [2002] used olivine-wadsleyite phase boundary in Mg_2SiO_4 obtained by *Akaogi et al.* (1989) for pressure calibration, which is 0.7 GPa higher relative to phase boundary reported by *Morishima et al.* [1994] and *Katsura et al.* [2004]. Accordingly, we recalculated pressures of *Chen et al.* [2002] for compatibility. See Fig. 2 for other abbreviations.

X-ray diffraction studies showed that the Clapeyron slope may be as high as 3.6 MPa/K [*Morishima et al.*, 1994] to 4.0 MPa/K [*Katsura et al.*, 2004] in Mg_2SiO_4 and Fo_{90} systems. *Katsura et al.* [2004] estimated the width of the binary loops for OWT to be ~0.4 GPa at 1900 K and ~0.6 GPa at 1600 K for Fo_{90}, consistent with earlier studies.

Wood [1995] argued that hydrogen, being more soluble in wadsleyite than olivine, expands the stability field of wadsleyite to lower pressures through the effect of configuration entropy of disorder. *Wood* [1995] calculated that 500 ppm H_2O (0.05 wt%) in olivine would be necessary to expand the olivine-wadsleyite loop interval from 7 km (anhydrous) to about 22 km. *Smyth and Frost* [2002] observed a shift of the OWT boundary by ~1 GPa to lower pressures in a peridotite composition with Fo_{89} olivine and ~3 wt% H_2O, which is consistent with our *in situ* results (Fig.3), also containing about 3 wt% H_2O in the starting

material. *Chen et al.* [2002] determined the OWT boundary in pure forsterite and Fo_{90} compositions in both dry and hydrous (11 wt.% H_2O) experiments simultaneously using multiple charges at 1473 K and between 12.6–14.7 GPa (Fig.3). In the Fe-bearing system, they also observed a shift of the OWT boundary to lower pressure, but found the pressure width of the olivine-wadsleyite loop decreased to 0.3 GPa, which they attribute to the formation of fluid phase under saturated conditions [*Chen et al.*, 2002; also see *Hirschmann et al.*, this volume].

In our experiments influence of Fe^{3+} content in coexisting phases was ignored though they were performed under relatively oxidizing conditions provided by a $LaCrO_3$ heater. Fe/Mg partitioning between the phases in our experiments will be considered elsewhere. Here, we note that the partition coefficient K_{Ol-Wd}^{Fe-Mg} is consistent with those reported by *Katsura et al.* [2004], who maintained relatively reducing conditions in the experiments controlled by boron-nitride in the assembly, and K_{Ol-Wd}^{Fe-Mg} is in the range of 0.5–0.6 at 1873 K and 0.45-0.5 at 1373–1673 K. *Frost* [2003] calculated thermodynamic model for OWT using data on Fe-Mg-partitioning experiments and estimated the binary loop in anhydrous peridotite system of ~0.3 GPa at 1700 K, which is consistent with our *in situ* experiments.

Summarizing the data for OWT, we emphasize that at or below saturation water widens the pressure interval of the binary loop of the OWT interval and expands stability field of wadsleyite to lower pressures. OWT becomes increasingly non-linear with increasing H_2O concentration as suggested by *Wood* [1995]. The OWT interval is ~0.3 GPa at 1700 K (or ~6 km in the depth) in anhydrous pyrolite, whereas it is ~1.2 GPa at 1500 K (or 33 km) in pyrolite with 3 wt% of H_2O.

3.2. Wadsleyite – Ringwoodite Transformation

Data for the WRT is limited compared with those for OWT and PST for both olivine and peridotite systems. Original determination of the WRT phase boundary in $(Mg,Fe)_2SiO_4$ [*Katsura and Ito*, 1989] indicated a significant positive Clapeyron slope of +5 MPa/K. *Suzuki et al.* [2000] reported a linear equation for WRT boundary in Mg_2SiO_4 expressed as P (GPa) = 0.0069 T (K) + 8.43 using the NaCl pressure scale, which is in good agreement with estimations by MgO scales. *Litasov and Ohtani* [2003] observed a minor shift of the WRT boundary to higher pressure in the CaO-MgO-Al_2O_3-SiO_2-pyrolite system with 2 wt% H_2O, whereas *Kawamoto* [2004] observed wadsleyite at 20 GPa and 1573 K in pyrolite + 13 wt.% H_2O, which indicates 2.5 GPa shift of the WRT boundary to higher pressures relative to that observed in anhydrous Fo_{90} composition. Similar results were obtained by *Inoue et al.* [2001] for the Fo_{80-100} system with 1 wt.% H_2O, although the exact value for the shifts of the WRT boundary was not reported.

Katsura and Ito [1989] determined the wadsleyite-ring-woodite loop of ~0.9 GPa (24 km) for anhydrous Fo_{90} at 1473-1873 K, whereas *Frost* [2003] calculated that it may be reduced to ~0.7 GPa (20 km) by presence of garnet in the pyrolite composition. *Inoue et al.* [1997, 1998] observed a narrower wadsleyite-ringwoodite interval (<0.5 GPa) in hydrous Fo_{90} composition.

Reasons for the observed displacement of the wadsleyite/ ringwoodite phase boundary are not clear. It might be connected to the lower water solubility in ringwoodite relative to wadsleyite, but this suggestion should be verified by data on partitioning of water between the coexisting phases, because the maximum water solubility in ringwoodite is comparable with that in wadsleyite (~2.6 and 3.3 wt% respectively) [e.g. *Kohlstedt et al.*, 1996; *Inoue et al.*, 1998; *Chen et al.*, 2002].

3.3. Post-Spinel Transformation

A significant shift of the PST boundary to higher pressure is expected from the differences in water solubility between ringwoodite and Mg-perovskite and ferroperi-clase. Ringwoodite can incorporate up to 2.6 wt.% H_2O at 1100°C and 20 GPa [*Kohlstedt et al.*, 1996; *Inoue et al.*, 1998], whereas the water solubility in Mg-perovskite and ferropericlase is very limited (<30 and <100 ppm, respectively) [*Bolfan-Casanova et al.*, 2000; 2002; 2003; *Litasov and Ohtani*, 2006].

Recently, we reported *in situ* X-ray diffraction determination of PST boundary in anhydrous and hydrous pyrolite composition [*Litasov et al.*, 2005a,b]. The phase relations were determined at 20-26 GPa and temperatures up to 2300 K. In the anhydrous pyrolite system we observed effective nucleation of Mg-perovskite and ferropericlase from a ring-woodite-bearing assemblage in the temperature range of 1600–2200 K (Fig.4). The obtained PST boundary (appearance line of Mg-perovskite) can be expressed as P (GPa) = -0.0005 T (K) + 23.54 using pressures based on the AT-03 pressure scale and P (GPa) = -0.0008 T (K) + 24.42 using the MS-01 pressure scale.

In the hydrous pyrolite runs approaching the phase boundary of the ringwoodite stability field we observed the appearance of Mg-perovskite peaks, and a decrease in the intensity of the ringwoodite peaks at 23.4-23.6 GPa and 1423–1473 K (Fig.4). The reaction proceeds rapidly and ringwoodite disappears within one hour at an overpressure of only 0.5 GPa. At 1300-1350 K the transformation from ringwoodite to perovskite and backward proceeds with resulted in a pressure jump about 1 GPa, therefore we could not determine

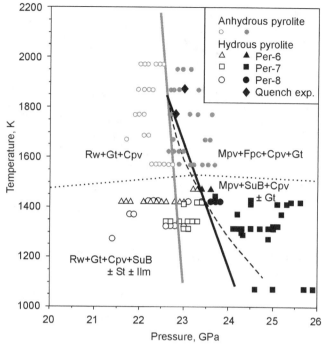

Fig.4. Post-spinel transformation boundary in anhydrous and hydrous pyrolite determined by *in situ* X-ray diffraction experiments [*Litasov et al.*, 2005a,b] using pressures calculated from Au EOS by *Tsuchiya* [2003]. *Open symbols*, ringwoodite, *filled symbols*, Mg-perovskite+ferropericlase. *Quench exp.*, experiments with Pt/Re capsule. Appearance of Mg-perovskite in hydrous (*bold line*) and anhydrous pyrolite (*grey line*) is shown. *Dashed line* shows possible deviation of transformation boundary from linear trend. *Dotted line* shows stability field of superhydrous phase B after *Litasov and Ohtani* [2003]. Rw: ringwoodite; Gt: garnet; Cpv: Ca-perovskite; Mpv: Mg-perovskite; Fpc: ferropericlase; SuB: superhydrous phase B; St: stishovite; Ilm: ilmenite-$MgSiO_3$ phase (akimotoite). Note that a fluid phase is present in hydrous assemblages at or above saturation.

the phase change boundary precisely (Fig.4). We have also performed two runs at higher temperature (1773–1873 K) using a Pt/Re capsule, which is not transparent for X-ray, and the Au+MgO pressure marker placed outside the capsule. In the recovered samples we observed minor amounts of Mg-perovskite at 1773 K and 22.8 GPa. This pressure is close to the post-spinel phase transformation boundary obtained for anhydrous pyrolite (Fig.4). Our data indicate that the post-spinel transformation boundary in hydrous pyrolite shifts to higher pressure by about 0.6 GPa relative to anhydrous pyrolite at 1473 K, whereas there is no obvious shift of this boundary at higher temperatures (1773–1873 K, Fig.4). The resulting linear equation for appearance of Mg-perovskite may be expressed as P (GPa) = -0.002 T (K) + 26.3 and applicable for the temperature range 1000-1800 K. However, we should note a possibility that the transformation boundary deviates from a linear trend and the phase boundary can be shifted by more than 2.5 GPa at 1000 K (Fig.4). It should be noted also that superhydrous phase B coexists with both ringwoodite and Mg-perovskite in all hydrous experiments below 1500 K.

We have compared present data with recent studies of PST in Mg_2SiO_4 and anhydrous pyrolite using both multi-anvil and diamond anvil cell experiments. Two parameters should be useful for this comparison, which are as follows: (a) pressure of the PST, which can be estimated via relation to $d660$ at 1850 K (temperature proposed for the average mantle by *Akaogi et al.* [1989]); and (b) the Clapeyron slope of the transformation boundary.

It is unfortunate for high-pressure experimentalists that most studies made by *in situ* multianvil technique yield the PST pressure that is significantly lower than that expected at $d660$ (Fig.5) and this discrepancy is most likely attributed to inaccuracies in the pressure scales. However, we should note that possible corrections of pressure scales may be related mostly to thermal pressure. The Clapeyron slopes of updated transformation boundaries (e.g. PST) should be more positive relative to those estimated using the present pressure calibration scales, and in the case of the PST boundary, these corrections work against consistency of the Clapeyron slope with the observed topography of $d660$ (see also section 4.3).

First *in situ* data [*Irifune et al.*, 1998] for PST in Mg_2SiO_4 using the Au pressure scale by Anderson et al. [1989] lead to confusion because PST was located at about 2 GPa lower than the pressure at 660 km in seismological models [e.g. *Dziewonski and Anderson*, 1981; *Kennet et al.*, 1995]. Subsequent diamond-anvil cell (DAC) studies [*Chudinovskih and Boehler*, 2001; *Shim et al.*, 2001] and multianvil with an improved technique [*Katsura et al.*, 2003; *Fei et al.*, 2004a; *Litasov et al.*, 2005a] shifted this boundary to higher pressures (Fig.5).

The PST boundary for Mg_2SiO_4 reported by *Fei et al.* [2004a] using the MS-01 scale is closest to 660 km: ΔP_{660} = ($P_{660\ km} - P_{measured}$ at 1850 K) = 0.6 GPa. *Litasov et al.* [2005a] obtained ΔP_{660} = 1.0 using AT-03 scale for anhydrous pyrolite. The results obtained by *Chudinovskikh and Boehler* [2001] for Mg_2SiO_4 using DAC are consistent with the data by *Fei et al.* [2004a]. However, the data obtained by *Shim et al.* [2001] using DAC show a higher pressure of PST (Fig.5), which is almost ideally consistent with the 660-km seismic discontinuity and exceed the pressure by *Fei et al.* [2004a] and *Chudinovskikh and Boehler* [2001]. The results obtained by DAC experiments are however subject to large temperature (100-200 K) and pressure uncertainties. DAC results also exhibit more scatter in the estimated Clapeyron slope of the PST.

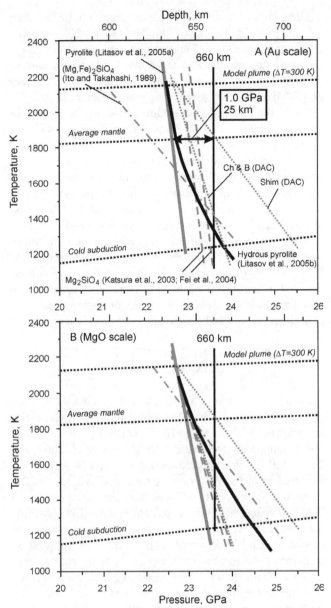

Fig.5. Comparison of post-spinel transformation boundaries obtained in different studies using (A) Au pressure scale by *Tsuchiya* [2003] and (B) MgO pressure scale by Speziale et al. (2001). Average mantle geotherm is after *Akaogi et al.* [1989]. Cold subduction geotherm is after *Kirby et al.* [1996]. DAC, diamond anvil cell experiments after *Chudinovskin and Boehler* [2001] (Ch & B) and *Shim et al.* [2001] (Shim). Calibration of DAC experiments is different from other experiments (see original works for details). *Bold lines*, data for pyrolite. *Dashed lines*, data for Mg_2SiO_4 and Fo_{90}. The discrepancy between experimental data and observed depth of seismic discontinuity is shown.

The pressure interval of coexisting ringwoodite and Mg-perovskite + ferropericlase varies in different experimental studies. *Ito and Takahashi* [1989] reported it to be less than 0.1 GPa for $(Mg_{0.9}Fe_{0.1})_2SiO_4$ (Fig.1). *Wood* [2000] noted that chemographic relations of the coexisting phases near 22–23 GPa also suggest that PST takes place over a very narrow pressure interval. However, *Hirose* [2002] and *Nishiyama et al.* [2004] reported relatively wide pressure intervals (0.5–0.7 GPa) for the PST in pyrolite. Our *in situ* results for PST in the both anhydrous and hydrous pyrolite suggest a width of coexistence of 0.1-0.5 GPa.

Based on the results of *in situ* measurements of PST in anhydrous and hydrous pyrolite we may conclude that the difference in chemical composition between pyrolite and Mg_2SiO_4 do not strongly influence the negative Clapeyron slope of the PST, but it may vary between -0.4 and -0.8 MPa/K using both the Au and MgO equations of state, whereas addition of water may shift PST boundary to higher pressures by ~0.6 GPa (15 km) at 1473 K. At temperatures proposed for *d*660 (~1850 K) there is no obvious shift of the phase boundary by water. Therefore, a major discrepancy between experimental data and seismological observation remains.

3.4. POST-GARNET TRANSFORMATION IN ECLOGITE (MORB)

Litasov and Ohtani [2005] observed ~1 GPa shift of the post-garnet transformation boundary to lower pressure with the addition of 2 wt.% of H_2O to MORB glass using the quench multianvil technique. Then, *Litasov et al.* [2004] and *Sano et al.* [2006] reported results of *in situ* X-ray diffraction studies of the PGT in anhydrous and hydrous MORB respectively. The phase relations were determined at 20-30 GPa and temperature up to 2400 K (Fig.6). In anhydrous MORB, phase transformation began with the appearance of metastable Ca-Mg-perovskite at about 1273 K, followed by transformation to Mg- and Ca-perovskite-bearing assemblages at about 1500 K. Appearance of garnet was observed with increasing temperature in all runs except one at the highest pressures. The experiments revealed that the post-garnet phase boundary is given as P (GPa) = 0.0046 T(K) + 18.40 using AT-03. The pressure interval of coexistence of garnet and Mg-perovskite is found to be very narrow, less than 0.5 GPa.

In hydrous MORB (5 or 10 wt.% H_2O) a Ca-Mg-perovskite-bearing assemblage was observed at temperatures below 1400 K. At 1400-1500 K we observed transformation of Mg-perovskite to garnet and backward at about 22.9–23.1 GPa (Fig.6). We have also performed two runs at high temperature (1673-1873 K) for the MORB starting material with 2 wt.% H_2O using a Pt capsule, which is not transparent to X-ray, and the Au pressure marker placed outside the capsule. In the recovered samples we observed Mg-perovskite,

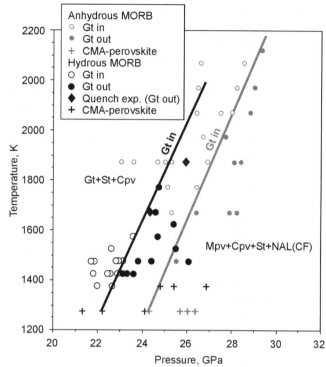

Fig.6. Comparison of garnet-perovskite phase boundaries in the anhydrous (*grey line*) and hydrous (*black line*) MORB after *Litasov et al.* [2004] and *Sano et al.* [2006] determined by *in situ* X-ray diffraction experiments using pressures calculated from Au EOS by *Tsuchiya* [2003]. Gt: garnet; St: stishovite; Cpv: Ca-perovskite; Mpv: Mg-perovskite; NAL: Na-Al hexagonal phase [*Akaogi et al.*, 1999; *Gasparik et al.*, 2000; *Miura et al.*, 2000]; CF: aluminous orthorhombic phase with Ca-ferrite structure [*Akaogi et al.*, 2002b]. Note that a fluid phase is present in hydrous assemblages at or above saturation.

which is consistent with a shift of PGT boundary to lower pressure by ~2 GPa (Fig.6). The resulting equation for PGT in hydrous MORB can be expressed as P (GPa) = 0.0049 T(K) + 15.942 using AT-03.

The shift of the PGT boundary to lower pressure cannot be explained by difference in water solubility between the phases. To the contrary, majorite garnet may accommodate significant amount of water (1130–1250 ppm H_2O, [*Katayama et al.*, 2003]) relative to Mg-perovskite (<30 ppm [*Litasov and Ohtani*, 2006]), thus shift of the phase transformation boundary to higher pressure may be expected. The effects of the other phases like Al-rich NAL phase [*Akaogi et al.*, 1999; *Gasparik et al.*, 2000; *Miura et al.*, 2000], Ca-perovskite and stishovite are not clear. Stishovite may contain significant amount of water (up to 850 ppm as reported by *Chung and Kagi*, 2002), but it exists as a part of both the garnet- and perovskite-bearing assemblage. Although the water solubility in NAL phase is unknown, its influence on the PGT may be significant because NAL is more stable in hydrous MORB relative to the anhydrous system, where it is replaced by Al-rich CF phase [*Akaogi et al.*, 2002b] at only ~0.5 GPa higher pressure relative to Mg-perovskite appearance [*Litasov and Ohtani*, 2005].

Litasov and Ohtani [2005] suggested differences in the oxidation of garnet and Mg-perovskite by hydrous fluid phases as a reason for a shift of the PGT to lower pressure. Aluminous Mg-perovskite contains significantly higher amount of Fe^{3+} than garnet [*McCammon*, 1997] and thus may oxidize garnet-bearing lithology and enhance formation of perovskite. Another explanation is that the fluid can dissolve variable amounts of major components such as Mg, Si and Fe [e.g. *Kawamoto et al.*, 2004], which would affect the composition of remaining solid phases and subsequently influence the pressure of the PGT boundary [*Sano et al.*, 2006].

4. IMPLICATION FOR SEISMIC DISCONTINUITIES AND MANTLE DYNAMICS

For comparison between the experimental boundaries and seismic observations we consider the following three major points: (1) the pressure/depth of discontinuities and mantle temperature; (2) application of the experimental Clapeyron slopes to understand observed topography of the discontinuities; and (3) sharpness of the discontinuities. Other factors such as velocity jumps at the discontinuities, influence of latent heat of transformation and diffusivity of hydrogen will not be considered in detail.

4.1. 410-km Discontinuity

It is generally accepted that $d410$ is caused by the transformation from olivine to wadsleyite in peridotitic mantle. Global averages for the depth of the discontinuity are 411 km [*Gu et al.*, 1998] and 418 km [*Flanagan and Shearer*, 1998, 1999]. Using these estimates *Katsura et al.* [2004] suggested the average temperature at $d410$ is 1760 ± 45 K for a pyrolite mantle. According to our data for anhydrous pyrolite, the temperature at $d410$ would be 1640 K if we use AT-03 as a pressure scale and 1720 K if we use MS-01. These temperatures are generally consistent with widely accepted model for the average mantle geotherm proposed by *Akaogi et al.* [1989] (Fig.7). Along this mantle geotherm, the OWT is shifted insignificantly by water and the width of the discontinuity will not be notably changed at 1760 K. Existence of a minor amount of water (~500 ppm) in the normal mantle (~1760 K) may produce an elevation of a few kilometer in $d410$ depth.

Global topography of $d410$ indicates amplitude of the depth variations of ±10 to ±20 km (*Gu and Dzievonski*, [2002] and *Flanagan and Shearer* [1998], respectively). Regional studies

indicate stronger depth variations. For example, maximum elevations beneath subduction zones are 60-70 km, which is consistent with up to 1000 K temperature anomaly if we apply the Clapeyron slope of OWT +3–4 MPa/K [e.g. *Collier et al.*, 2001; Tibi and Wiens, 2005]. Since the influence of ~2-3 wt.% water on OWT is comparable with that of ~600 K variations in temperature, we suggest that the topography of $d410$ in some regions may be explained by water in the slabs.

Estimates for the width of $d410$ vary between 4 and 35 km [*Shearer*, 2000] while most estimates suggest 6-10 km width. Data for Fo_{90} indicate the width of 25 km at 1473 K and 14 km at 1873 K [*Katsura and Ito*, 1989]. *Frost* [2003] reviewed the recent data on OWT along with the new thermodynamic modeling and argued that 6 km width of the discontinuity is consistent with OWT in anhydrous pyrolite containing garnet and pyroxenes (which makes OWT sharper due to iron partitioning) and corresponds to the minimum of the range of seismic estimates for the width of $d410$. Our results from *in situ* measurement of the OWT in anhydrous pyrolite are in agreement with the thermodynamic calculations by *Frost* [2003].

There are several factors making the OWT at $d410$ broader than 6 km. We would expect a broader transformation in colder, garnet-poor or FeO-rich regions in the mantle. The presence of water, however, has perhaps the most dramatic effect on the discontinuity width. *Wood* [1995] calculated that 100 ppm H_2O in olivine (with assumed olivine/wadsleyite partitioning of 1:10) would broaden the discontinuity by approximately 3 km. Though increasing water content up to or below saturation does appear to broaden the OWT [*Smyth and Frost* 2002; *this study*], both *Chen et al.* [2002] and *Hirschmann et al.* (this volume) suggested that formation of a fluid phase upon saturation would result in a narrower transformation interval. On the other hand, *Smyth and Frost* [2002] suggested that hydrogen diffusion and gravitational stratification narrow the phase transformation interval between olivine and wadsleyite.

Recent measurements and calibrations of water solubilities in olivine and wadsleyite indicate that H_2O partition coefficients ($D_{Wd/Ol}$) are much lower than those estimated by *Wood* [1995] and *Kohlstedt et al.* [1996]. *Chen et al.* [2002] reported results of ion probe measurements of $D_{Wd/Ol}$ and suggested it is 4.6–5.1 at 1473 K. Our estimations of $D_{Wd/Ol}$ from FTIR measurements [*Litasov et al.*, in prep.] suggest $D_{Wd/Ol} = 3.9$ at 1473 K, 2.4 at 1673 K and 1.5 at 1773 K if the calibration of *Bell et al.* [2003] is used for olivine spectra and the calibration of *Paterson* [1982] is used for wadsleyite spectra (see discussions in *Hirschmann et al.*, [2005] and *Litasov and Ohtani* [2006]).

Recent observations that $d410$ is between 20 and 35 km thick beneath some locations in the Mediterranean region [*Van der Meijde et al.*, 2003] have been explained by high H_2O contents in this region. If the transformation is 35 km

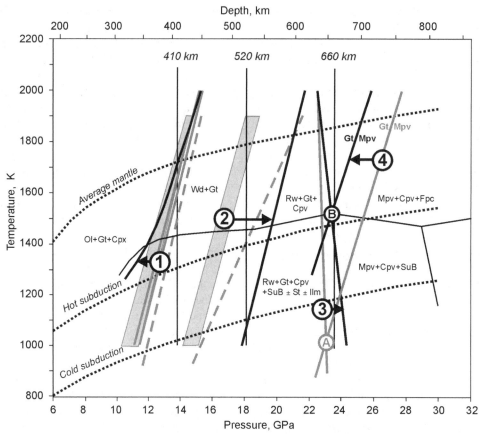

Fig.7. Shift of phase transition boundaries by adding water and comparison with mantle geotherms. *Bold black lines*: phase transitions under hydrous conditions (~2-3 wt.% H_2O); *Grey lines*, anhydrous transitions (*bold lines*, pyrolite; *fields*, Fo_{90}; and *dashed lines*, Mg_2SiO_4 (Fo)). (1) Olivine-wadsleyite transition after *Katsura et al.* [2004] and this work; (2) Wadsleyite-ringwoodite transition *Katsura and Ito* [1989], *Suzuki et al.* [2000] and *Kawamoto* [2004]; (3) Post-spinel transition in pyrolite after *Litasov et al.* [2005a,b]; (4) Post-garnet transition in MORB (Gt/Mpv) after *Litasov et al.* [2004] and *Sano et al.* [2006]. A and B, the points of equal density of basalt and pyrolite under the anhydrous and hydrous conditions, respectively. Thin black lines outline stability field of dense magnesium silicates. Stability of superhydrous phase B (SuB) is after *Litasov and Ohtani* [2003] and *Ohtani et al.* [2003]. Average mantle geotherm is after *Akaogi et al.* [1989]. Hot and cold subduction geotherms are after *Kirby et al.* [1996]. See Fig.2 and 4 for abbreviations.

thick, the H_2O content of olivine and wadsleyite would be 0.21 and 0.84 wt.% respectively according to the model by *Wood* [1995] and $D_{Wd/Ol}$=4. These estimations are supported by the measurements of electrical conductivity near 410 km in this region, which is consistent with 0.1-0.15 wt.% H_2O in olivine [*Tarits et al.*, 2004].

4.2. 520-km Discontinuity

The *d*520 is generally believed to be caused by the transformation from wadsleyite to ringwoodite [e.g. *Rigden et al.*, 1991]. The most consistent seismic evidence for *d*520 observed on a global scale comes from the analysis of long-period S waves using SS precursors [e.g. *Shearer*, 1990,

1996]. However, some studies claim that *d*520 can be observed only in certain regions [*Gossler and Kind*, 1996; *Gu et al.*, 1998]. The absence of reflection at *d*520 in high-frequency studies has been used to suggest a broad transition interval of 10-50 km for this discontinuity [*Benz and Vidale*, 1993; see also *Akaogi et al.*, 1989], which is consistent with the broad transformation interval (20-25 km) observed experimentally in dry systems (see section 3.2). A narrow wadsleyite-ringwoodite transformation interval in hydrous Fo_{90} [*Inoue et al.*, 1997; 1998] may indicate that *d*520 is better recognized in hydrated regions of the mantle.

Systematic searches for *d*520 indicate that its depth varies, and splits in some regions [*Deuss and Woodhouse*, 2001]. Complex structures with double reflections at 500-515 and

550–570 km cannot be explained by the wadsleyite-ringwoodite transformation alone, and require another phase or compositional boundary, such as garnet/clinopyroxene to Ca-perovskite transformation at about 20 GPa (~570 km). However, depth variations of a single 520-km discontinuity in the range from 500 to 560 km might be consistent with hydrous WRT, or at least partially with variations in Mg# number of minerals. Precise estimations are not possible at present because there are no *in situ* X-ray diffraction or detail quench multianvil data for WRT in pyrolite or Fo_{90} compositions in both anhydrous and hydrous systems. The Clapeyron slope of the WRT boundary varies in different studies from +5 MPa/K in Fo_{90} [*Katsura and Ito*, 1989] to +6.9 MPa/K in Mg_2SiO_4 [*Suzuki et al.*, 2000].

4.3. 660-km Discontinuity

The seismic discontinuity with a global average depth of 660 km has been attributed to PST [e.g. *Ito and Takahashi*, 1989; *Flanagan and Shearer*, 1998]. The average depth estimation for the $d660$ varies in the range of 650–670 km. However, maximum amplitude of the discontinuity is 640–710 km [e.g. *Shearer*, 2000]. *Flanagan and Shearer* [1998] reported the average depth of 647 km by using SS travel times, however, after making corrections to topography, crustal structure, lateral S velocity variations, and some other factors they obtained 660 km global average. Recent estimates by *Gu and Dziewonski* [2002] indicate that the global average for the 660-km discontinuity is 654 km, and this value is similar to that proposed earlier (e.g. 650-km by *Anderson and Bass*, [1986]). If we accept 654 km as a global average, then the pressure discrepancy between PST observed in the experiments and that in seismic studies (~1 GPa or 20–25 km) decreases by 0.2-0.3 GPa. Although we could not find significant shift of PST by water at 1773–1873 K, it might be shifted by 0.2-0.3 GPa as reported by some other studies [*Higo et al.*, 2001; *Inoue et al.*, 2001]. *Katsura et al.* [2003] also suggested 0.6 GPa shift of the post-spinel transition boundary to higher pressures in hydrous Mg_2SiO_4 based on their preliminary results at 1663 K. Since the Clapeyron slope of PST is very gentle (about -0.5 MPa/K) we speculate that in dry regions of the mantle the depth of the discontinuity may be 640–650 km, whereas depression of the discontinuity below 650–660 km may indicate presence of water. Taking water into account and possible decrease in temperature at $d410$ (see below), the discrepancy between PST observed in the experiments and that in seismic studies may become negligible.

The relation between the depth of $d660$ and experimental data for PST and that between OWT and $d410$ are used for estimating the average upper mantle geotherm (Fig.7) [e.g.

Akaogi et al., 1989; *Bina and Helffrich*, 1994]. Calculations of the density and seismic velocity profiles of a model pyrolite composition along this geotherm [e.g. *Murakami and Yoshioka*, 2001; *Akaogi et al.*, 2002a] suggest consistency with seismological models, such as PREM [*Dziewonski and Anderson*, 1981] or ak135 [*Kennet et al.*, 1995]. However, it may not be possible to resolve a ~100 K difference in the mantle geotherms comparing the calculated density, velocity profiles and seismological models because of the uncertainties in the modal proportions and thermoelastic parameters of constituent minerals. For instance, the value of density and velocity jump at 660 km varies from 3.5% to 6–7% in different models [e.g. *Murakami and Yoshioka*, 2001; *Akaogi et al.*, 2002a]. A majority of the mantle geotherms calculated using thermoelastic parameters of minerals in the upper and lower mantle [e.g. *Jackson*, 1998; *Deschamps and Trampert*, 2004] are consistent with the geotherm by *Akaogi et al.* [1989]. This geotherm is also consistent with the observed heat flow and plate velocities at shallow levels of the mantle [*Kaula*, 1983]. However, if we presume existence of water in the transition zone and low-velocity zone of the upper mantle, the geotherm could be lower [*Hofmeister*, 1999].

Mapping of the global topography of the 660 km discontinuity shows variations in the depth of $d660$ from about +20 km, for example, in the northern Pacific and Atlantic Ocean and South Africa to -20 km in the western Pacific and South America [*Flanagan and Shearer*, 1998; *Gu and Dziewonski*, 2002; Gu et al., 2003]. *Lebedev et al.* [2003] proposed smaller variations in the range of ±15 km. Regional studies of seismic discontinuities near subduction zones show slab-induced depression of 20-30 km on the $d660$ beneath Tonga [*Niu and Kawakatsu*, 1995] and up to a 50-km depression beneath Izu-Bonin [*Wick and Richards*, 1993 *Castle and Creager*, 1998; *Collier et al.*, 2001]. *Vidale and Benz* [1992] obtained depth perturbations of up to 30 km beneath the Tonga, Izu-Bonin, Marianas, and South American subduction zones. Several regional studies showed that $d660$ elevates by 10-20 km in the areas related to Iceland [*Shen et al.*, 1998], Hawaii [*Li et al.*, 2000], and South Pacific [*Niu et al.*, 2000] due to the effect of plume related upwelling.

Global tomography studies may indicate that the temperature difference between the hot mantle plumes and an average mantle is almost similar to the difference between the subduction zones and an average mantle. Assuming a maximum plume temperature below the solidus of dry pyrolite, an average mantle temperature at $d660$ should be 1650-1750 K, which is 100-200 K lower than generally accepted geotherm (Fig.7). Moreover, placing the mantle geotherm 100-200 K lower, we have also more consistency with our results for PST in anhydrous pyrolite (see section 3.3.). However, regional seismological studies show that depressions of $d660$ in the subduction zones are about two times larger than elevations in the hot regions. This is consistent with the widely accepted temperature model for the upper mantle (Fig.7).

If we assume an average mantle temperature 1850 K at $d660$, we may explain only minor variations in the depth of $d660$ using the experimental data for anhydrous pyrolite (about +10 for hot and -20 km for cold regions of the mantle). Therefore, an additional explanation for larger variations is needed. This may include a delay of the phase transformation due to kinetics, or the influence of minor components such as volatiles. Study of kinetics of PST in Mg_2SiO_4 indicates that with 1 GPa overpressure PST proceeds rapidly (completed within 10^4 years) at temperatures above 1000 K. Only for very cold slabs with temperatures below 1000 K at the PST pressure can the reaction not be completed within $10^6–10^8$ years [Kubo et al., 2002a]. Thus, we can expect a delay of transformation and shift of PST due to kinetics by about 1 GPa (corresponding to 20-25 km) only in very cold slabs.

A more attractive hypothesis might be the presence of water. Displacement of the PST boundary by 0.6 GPa with the addition of ~2 wt.% water corresponds to about 15 km in the depth scale and may account for half of the 30-40 km depressions of the $d660$ in subduction zones at a temperature of about 1473 K. If the transformation boundary is not linear, this effect should be much stronger at lower temperatures (Fig.4-5). Since the Clapeyron slope of the post-spinel transformation boundary in anhydrous pyrolite is very gentle, the temperature effect alone cannot account for the 660 km discontinuity depressions. Thus, if our data for hydrous pyrolite are correct, they can be considered as strong evidence for water in the transition zone [Litasov et al., 2005a,b].

Seismological data show that the $d660$ is sharp in most cases and does not exceed 5 km (0.2 GPa) [e.g. Benz and Vidale, 1993; Stixrude, 1997]. This is consistent with the experimental results, which show a narrow (0.1-0.5 GPa) interval of PST (see section 3.3).

4.4. Basalt-Peridotite Relations Near 660 km

Interaction between the subducted lithosphere and surrounding mantle near the $d660$ is one of the key issues in deep Earth dynamics. The most likely scenario of subduction was suggested by Ringwood [1994] that mature thick and cold slabs with high thermal inertia may penetrate into the lower mantle, whereas young relatively hot and thin slabs are supposed to be deflected above $d660$ and retained in the transition zone. This scenario is supported by tomographic imaging [e.g. Grand et al., 1997; Fukao et al., 2001; Shen and Blum, 2003]

It has been suggested that a density crossover between peridotite and basaltic components can lead to detachment of the basaltic crust from peridotite near 660 km and formation of a garnetite-bearing (former basalt) layer at the base of the transition zone [Anderson and Bass, 1986; Ringwood, 1994; Karato, 1997]. Hirose et al. [1999] concluded that the transformation of MORB to perovskite lithology occurs near 720 km and the density crossover between basalt and peridotite is too narrow for this detachment. Therefore, the basaltic crust may gravitationally sink into the lower mantle. New data on the kinetics of the post-garnet transformation in MORB indicate a very slow reaction rate allowing metastable garnet to survive for a long time (on the order of 10 Ma) after intersection with $d660$ [Kubo et al., 2002b]. Therefore, the problem of basaltic crust separation from the slab near this boundary is not completely clarified.

Our recent in situ experiments [Litasov et al., 2004; Sano et al., 2006] demonstrated that at low temperatures the density crossover between peridotite and basalt might be entirely absent (Fig.7-8), especially under the hydrous conditions of subduction. The pressure and temperature conditions of the crossover of the two phase boundaries lies below the temperature path of a cold subducting slab in anhydrous eclogite (point A in Fig.7) and lies above the temperature path of a hot subducting slab in hydrous eclogite (point B in Fig.7). Therefore, if the slab passes through $d660$, our results provide direct evidence for penetration of the oceanic crust component into the lower mantle without gravitational separation from the peridotite body of the slab at least under

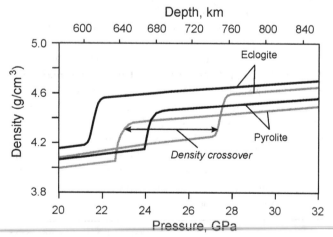

Fig.8. Density profiles of anhydrous and hydrous eclogite (MORB) and pyrolite illustrating the possible density crossover near 660 km. Density calculations were carried out along normal mantle geotherm (grey lines) for anhydrous systems and cold subduction geotherm (dashed lines) for hydrous systems using a third order Birch-Murnaghan equation of state and the set of thermoelastic parameters after Litasov and Ohtani [2005].

the hydrous conditions of subduction. The seismic reflectors with a positive density jump [e.g. *Niu et al.*, 2003] at depths of 1000–1200 km may suggest that the basaltic crust layers with the post-garnet lithology penetrated into the lower mantle.

5. CONCLUDING REMARKS

In this chapter we summarized recent results on the effect of water on major phase transformations in the Earth mantle with implications for seismic discontinuity structure and mantle dynamics. Experimental data are based largely on quench multianvil and *in situ* X-ray diffraction experiments at high pressure and temperature. Differences in water solubility between (a) olivine and wadsleyite and (b) ringwoodite and Mg-perovskite and ferropericlase may cause displacement of these phase transition boundaries that are believed to be responsible for the 410 and 660-km discontinuities, respectively. Experimental results show that water increases the binary loop of olivine-wadsleyite transformation interval, i.e. water expands the stability field of wadsleyite to lower pressures, at least at or below saturation. This interval is ~0.3 GPa (or ~6 km) at 1700 K in the anhydrous pyrolite, whereas it may be ~1.2 GPa at 1500 K (or 33 km) in pyrolite with 2–3 wt.% of H_2O. Broadening of 410-km discontinuity observed in some regions of the mantle is consistent with enrichment by water. Significant shift of the wadsleyite to ringwoodite phase transformation to the higher pressure caused by addition of water in the peridotite system may also be responsible for variations in the depth of 520-km discontinuity. However, the cause of a significant displacement of the wadsleyite/ringwoodite phase boundary remains unclear, and should be addressed in future studies.

In situ X-ray diffraction studies of the post-spinel transformation in hydrous pyrolite indicates that the phase boundary is shifted to higher pressures by 0.6 GPa relative to anhydrous pyrolite at 1473 K, whereas it shows no obvious shift at higher temperature around 1873 K. Displacement of the post-spinel phase transformation boundary in hydrous pyrolite corresponds to about 15 km on the depth scale and may account for half of 30–40 km depressions of the 660-km discontinuity in the subduction zones at the temperature around 1473 K. This effect should be much stronger at lower temperatures. Since the Clapeyron slope of the post-spinel transformation boundary in anhydrous pyrolite is gentle (about -0.5 MPa/K), depressions of the 660 km discontinuity likely cannot be explained by temperature variations alone. A large depression of the 660 km discontinuity is consistent with existence of water in the transition zone.

In situ X-ray diffraction studies of the post-garnet transformation in anhydrous and hydrous MORB show that the phase boundary is shifted to lower pressures by ~2 GPa by addition of water. This observation demonstrates that at moderate slab temperatures (1300-1400 K) the density crossover between peridotite and basalt near 660 km might be entirely absent, especially under hydrous conditions, and would prevent the separation of peridotite and basalitc components at the 660 km discontinuity.

Acknowledgements. We thank S.D. Jacobsen, S. Keshav, S. van der Lee and one anonymous reviewer for thorough reviews of the manuscript and A. Suzuki, H. Terasaki, K. Funakoshi, T. Kubo, S. Kato, K. Tsuno, and R. Ando for collaboration during the experiments at 'SPring-8'. This work was supported by the grants in Aid for Scientific Researches from the Ministry of Education, Culture, Sports, Science and Technology, Japan (No 14102009 and 16075202), to E. Ohtani. This work was conducted as a part of the 21th Century Center-of-Excellence program 'Advanced Science and Technology Center for the Dynamic Earth' at Tohoku University.

REFERENCES

Agee, C. B. (1998) Phase transformations and seismic structure in the upper mantle and transition zone, *Rev. Mineral.*, 37, 165–203.

Akaogi, M., E. Ito, and A. Navrotsky (1989) Olivine-modified spinel-spinel transitions in the system Mg_2SiO_4-Fe_2SiO_4: calorimetric measurements, thermochemical calculation, and geophysical application, *J. Geophys. Res.*, 94, 15671–15685.

Akaogi, M., Y. Hamada, T. Suzuki, M. Kobayashi, and M. Okada (1999) High pressure transitions in the system $MgAl_2O_4$-$CaAl_2O_4$: a new hexagonal aluminous phase with implication for the lower mantle, *Phys. Earth Planet. Inter.*, 115, 67–77.

Akaogi, M., A. Tanaka, and E. Ito (2002a) Garnet-ilmenite-perovskite transitions in the system $Mg_4Si_4O_{12}$-$Mg_3Al_2Si_3O_{12}$ at high pressures and high temperatures: phase equilibria, calorimetry and implications for mantle structure, *Phys. Earth Planet. Inter.*, 132, 303–324.

Akaogi, M., A. Tanaka, M. Kobayashi, N. Fukushima, and T. Suzuki (2002b) High-pressure transformations in $NaAlSiO_4$ and thermodynamic properties of jadeite, nepheline, and calcium ferrite-type phase, *Phys. Earth. Planet. Int.*, 130, 49–58.

Anderson, D. L., and J. D. Bass (1986) Transition region of the Earth's upper mantle, *Nature*, 320, 321-328.

Anderson, O. L., D. G. Issak, and S. Yamamoto (1989) Anharmonicity and the equation state for gold, *J. Appl. Phys.*, 6, 1534–1543.

Bell, D. R., and G. R. Rossman (1992) Water in Earth's mantle: the role of nominally anhydrous minerals, *Science*, 255, 1391–1397.

Bell, D. R., G. R. Rossman, J. Maldener, D. Endisch, and F. Rauch (2003) Hydroxide in olivine: A quantitative determination of the absolute amount and calibration of the IR spectrum, *J. Geophys. Res.*, 108, doi: 10.1029/2001JB000679.

Benz, H. M., and J. E. Vidale (1993) Sharpness of upper-mantle discontinuities determined from high-frequency reflections, *Nature*, 365, 147–150.

Bina, C. R., and G. R. Helffrich (1994) Phase transition Clapeyron slopes and transition zone seismic discontinuity topography, *J. Geophys. Res.*, 99, 15853–15860.

Bolfan-Casanova, N., H. Keppler, and D. C. Rubie (2000) Water partitioning between nominally anhydrous minerals in the MgO-SiO$_2$-H$_2$O system up to 24 GPa: implications for the distribution of water in the Earth's mantle, *Earth Planet. Sci. Lett.*, 182, 209–221.

Bolfan-Casanova, N., H. Keppler, and D. C. Rubie (2003) Water partitioning at 660 km depth and evidence for very low water solubility in magnesium silicate perovskite, *Geophys. Res. Lett.* 30, doi: 10.1029/2003GL017182.

Bolfan-Casanova, N., S. Mackwell, H. Keppler, C. McCammon, and D. Rubie (2002) Pressure dependence of H solubility in magnesiowüstite up to 25 GPa: Implications for the storage of water in the Earth's lower mantle, *Geophys. Res. Lett.*, 29, doi: 10.1029/2001GL014457.

Castle, J. C., and K. C. Creager (1998) Topography of the 660-km seismic discontinuity beneath Izu-Bonin: Implications for tectonic history and slab deformation, *J. Geophys. Res.*, 103, 12511–12527.

Chen J., T. Inoue, H. Yurimoto, and D. J. Weidner (2002) Effect of water on olivine-wadsleyite phase boundary in the (Mg, Fe)$_2$SiO$_4$ system, *Geophys. Res. Lett.*, 29, 1875, doi:10.1029/2001GL014429.

Chen, J., T. Inoue, D. J. Weidner, and Y. Wu (1998) Strength and water weakening of mantle minerals, α, β, and γ Mg$_2$SiO$_4$, *Geophys. Res. Lett.*, 25, 575–578.

Chudinovskikh, L., and R. Boehler (2001) High-pressure polymorphs of olivine and the 660-km seismic discontinuity, *Nature*, 411, 574–577.

Chung, J. I., and H. Kagi (2002) High concentration of water in stishovite in the MORB system, *Geophys. Res. Lett.*, 29, doi:10.1029/202GL015579.

Collier, J. D., G. R. Helffrich, and B. J. Wood (2001) Seismic discontinuities and subduction zones, *Phys. Earth Planet. Inter.*, 127, 35–49.

Deschamps, F., and J. Trampert (2004) Towards a lower mantle reference temperature and composition, *Earth Planet. Sci. Lett.*, 222, 161–175.

Deuss, A., and J. H. Woodhouse (2001) Seismic observations of splitting of the mid transition zone discontinuity, *Science, 294*, 345–357.

Dziewonski, A. M., and D. L. Anderson (1981) Preliminary Reference Earth Model, *Phys. Earth Planet. Int.*, 25, 297–356.

Fei, Y., J. Van Orman, J. Li, W. van Westrenen, C. Sanloup, W. Minarik, K. Hirose, T. Komabayashi, M. Walter, and K. Funakoshi (2004a) Experimentally determined postspinel transformation boundary in Mg$_2$SiO$_4$ using MgO as an internal pressure standard and its geophysical implications, *J. Geophys. Res.*, 109, doi:10.1029/2003JB002562.

Fei, Y., J. Li, K. Hirose, W. Minarik, J. Van Orman, C. Sanloup, W. van Westrenen, T. Komabayashi, and K. Funakoshi (2004b) A critical evaluation of pressure scales at high temperatures by *in situ* X-ray diffraction measurements, *Phys. Earth Planet. Inter.*, 143–144, 515-526.

Flanagan, M. P., and P. M. Shearer (1998) Global mapping of topography on transition zone velocity discontinuities by stacking SS precursors, *J. Geophys. Res.*, 103, 2673–2692.

Flanagan, M. P., and P. M. Shearer (1999) A map of topography on the 410-km discontinuity from PP precursors, *Geophys. Res. Lett.*, 26, 549–552.

Frost, D. J. (1999) The stability of dense hydrous magnesium silicates in Earth's transition zone and lower mantle, *in Mantle Petrology: Field Observations and High Pressure Experimentation: A tribute to Boyd, F.R.*, Geochem. Soc. Special Publ., vol. 6, edited by Y. Fei, C. M. Bertka, and B. O. Mysen, pp. 283–296.

Frost, D. J. (2003) The structure and sharpness of (Mg,Fe)$_2$SiO$_4$ phase transformations in the transition zone, *Earth Planet. Sci. Lett.*, 216, 313–328.

Fukao, Y., S. Widiyantoro, and M. Obayashi (2001) Stagnant slabs in the upper and lower mantle transition zone, *Rev. Geophys.*, 39, 291–323.

Fukao, Y., T. Koyama, M. Obayashi, and H. Utada (2004) Trans-Pacific temperature field in the mantle transition zone derived from seismic and electromagnetic tomography, *Earth Planet. Sci. Lett.*, 217, 425–434.

Gasparik, T., A. Tripathi, and J. B. Parise (2000) Structure of a new Al-rich phase, [K,Na]$_{0.9}$[Mg,Fe]$_2$[Mg,Fe,Al,Si]$_6$O$_{12}$, synthesized at 24 GPa, *Amer. Mineral.*, 85, 613–618.

Gossler, J., and R. Kind (1996) Seismic evidence for very deep roots of continents, *Earth Planet. Sci. Lett.*, 138, 1–13.

Grand, S. P., R. D. Van der Hilst, and S. Widiyantoro (1997) Global seismic tomography: a snapshot of convection in the Earth, *GSA Today, 7*, 1–7.

Griggs, D. T. and J. D. Blacic (1965) Quartz: Anomalous weakness of synthetic crystals, *Science, 147*, 292–295.

Gu, Y. J., A. M. Dziewonski, and C. B. Agee (1998) Global decorrelation of the topography of transition zone discontinuities, *Earth Planet. Sci. Lett.*, 157, 57–67.

Gu, Y. J., A. M. Dziewonski, and G. Ekström (2003) Simultaneous inversion for mantle shear velocity and topography of transition zone discontinuities, *Geophys. J. Int.*, 154, 559–583.

Gu, Y. J. and A. M. Dziewonski (2002) Global variability of transition zone thickness, *J. Geophys. Res.*, 107, doi: 10/1029/2001JB000489.

Helffrich, G. (2000) Topography of the transition zone seismic discontinuities, *Rev. Geophys.*, 38, 141–158.

Higo, Y., T. Inoue, T. Irifune, and H. Yurimoto (2001) Effect of water on the spinel-postspinel transformation in Mg$_2$SiO$_4$, *Geophys. Res. Lett.*, 28, 3505–3508.

Hirose, K. (2002) Phase transitions in pyrolitic mantle around 670-km depth: implications for upwelling of plumes from the lower mantle, *J. Geophys. Res.*, 107, doi: 10.1029/2001JB000597.

Hirose, K., Y. Fei, Y. Ma, and H. K. Mao (1999) The fate of subducted basaltic crust in the Earth's lower mantle, *Nature, 397*, 53–56.

Hirschmann, M. M., C. Aubaud, and A. C. Withers, (2005) Storage capacity of H$_2$O in nominally anhydrous minerals in the upper mantle, *Earth Planet. Sci. Lett.*, 236, 167-181.

Hirth, G., and D. L. Kohlstedt (1996) Water in the oceanic upper mantle: Implications for rheology, melt extraction and the evolution of the lithosphere, *Earth Planet. Sci. Lett., 144,* 93–108.

Hofmeister, A. M. (1999) Mantle values of thermal conductivity and the geotherm from phonon lifetimes, *Science, 283,* 1699–1706.

Ingrin, J., and H. Skogby (2000) Hydrogen in nominally anhydrous upper-mantle minerals: concentration levels and implications, *Eur. J. Mineral., 12,* 543–570.

Inoue, T. (1994) Effect of water on melting phase relations and melt compositions in the system Mg_2SiO_4-$MgSiO_3$-H_2O up to 15 GPa, *Phys. Earth Planet. Inter., 85,* 237–263.

Inoue, T., J. Chen, D. J. Weidner, H. Kagi, H. Yurimoto, P. A. Northrup, and J. B. Parise (1997) Phase boundaries between olivine (α), hydrous wadsleyite (β) and hydrous ringwoodite (γ) and their elastic properties, in *Abst. of IRAPT-16 and HPCJ-38 Joint Conf., Rev. High Pressure Sci. Tech.,* 6, p.489, Kyoto, Japan.

Inoue, T., D. J. Weidner, P. A. Northrup, and J. B. Parise (1998) Elastic properties of hydrous ringwoodite (γ-phase) in Mg_2SiO_4, *Earth Planet. Sci. Lett., 160,* 107–113.

Inoue, T., Y. Higo, T. Ueda, Y. Tanimoto, and T. Irifune (2001) The effect of water on the high-pressure phase boundaries in the system Mg_2SiO_4-Fe_2SiO_4, in *Conference Abst. of Transport of Materials in the Dynamic Earth,* edited by E. Ito, pp.128–129, Kurayoshi, Japan.

Inoue, T., Y. Tanimoto, T. Irifune, T. Suzuki, H. Fukui and O. Ohtaka (2004) Thermal expansion of wadsleyite, ringwoodite, hydrous wadsleyite and hydrous ringwoodite, *Phys. Earth Planet. Inter., 143-144,* 279–290.

Irifune, T., N. Nishiyama, K. Kuroda, T. Inoue, M. Isshiki, W. Utsumi, K. Funakoshi, S. Urakawa, T. Uchida, T. Katsura, and O. Ohtaka (1998) The postspinel phase boundary in Mg_2SiO_4 determined by *in situ* X-ray diffraction, *Science, 279,* 1698–1700.

Ito, E. and E. Takahashi (1989) Post-spinel transformation in the system Mg_2SiO_4-Fe_2SiO_4 and some geophysical implications, *J. Geophys. Res., 94,* 10637–10646.

Jackson, I. (1998) Elasticity, composition and temperature of the Earth's lower mantle: a reappraisal, *Geophys. J. Int., 134,* 291–311.

Jacobsen, S. D., J. R. Smyth, H. Spetzler, C. M. Holl, and D. J. Frost (2004) Sound velocities and elastic constants of iron-bearing hydrous ringwoodite, *Phys. Earth Planet. Inter., 143-144,* 47–56.

Jamieson, J. C., J. N. Fritz, and M. H. Manghnani (1982) Pressure measurements at high temperature in X-ray diffraction studies: gold as a primary standard, in *High pressure research in geophysics* edited by S. Akimoto and M. H. Manghnani, pp.27–48, Center for Academic Publications Japan, Tokyo.

Karato, S. (1997) On the separation of crustal component from subducted oceanic lithosphere near the 660 km discontinuity, *Phys. Earth Planet. Inter., 99,* 103–111.

Karato, S. (1990) The role of hydrogen in the electrical conductivity of the upper mantle, *Nature, 347,* 272–273.

Katayama I., K. Hirose, H. Yurimoto, and S. Nakashima (2003) Water solubility in majoritic garnet in subduction oceanic crust, *Geophys. Res. Lett., 30,* doi: 10.1029/2003GL018127.

Katsura, T., and E. Ito (1989) The system Mg_2SiO_4-Fe_2SiO_4 at high pressures and temperatures: Precise determination of stabilities of olivine, modified spinel, and spinel, *J. Geophys. Res., 94,* 15663–15670.

Katsura, T., H. Yamada, O. Nishikawa, M. Song, A. Kubo, T. Shinmei, S. Yokoshi, Y. Aizawa, T. Yoshino, M.J. Walter, E. Ito, and K. Funakoshi (2004) Olivine-wadsleyite transition in the system $(Mg,Fe)_2SiO_4$, *J. Geophys. Res., 109,* B02209 doi:10.1029/2003JB002438.

Katsura, T., H. Yamada, T. Shinmei, A. Kubo, S. Ono, M. Kanzaki, A. Yoneda, M. J. Walter, E. Ito, S. Urakawa, K. Funakoshi, and W. Utsumi (2003) Post-spinel transition in Mg_2SiO_4 determined by high P-T *in situ* X-ray diffraction, *Phys. Earth Planet. Inter., 136,* 11–24.

Kaula, W. M. (1983) Minimal upper mantle temperature variations consistent with observed heat flow and plate velocities, *J. Geophys. Res., 88,* 10323–10332.

Kawamoto, T. (2004) Hydrous phase stability and partial melt chemistry in H_2O-saturated KLB-1 peridotite up to the uppermost lower mantle conditions, *Phys. Earth Planet. Inter., 143–144,* 387–395.

Kawamoto, T., K. Matsukage, K. Mibe, M. Isshiki, K. Nishimura, N. Ishimatsu, and S. Ono, 2004, Mg/Si ratios of aqueous fluids coexisting with forsterite and enstatite based on the phase relations in the Mg_2SiO_4-SiO_2-H_2O system, *Amer. Miner., 89,* 1433–1437.

Kennet, B. L. N., E. R. Engdahl, and R. Buland (1995) Constrains on seismic velocities in the Earth from traveltimes, *Geophys. J. Int., 122,* 108–124.

Kirby, S. H., S. Stein, E. A. Okal, and D. C. Rubie (1996) Metastable mantle phase transformations and deep earthquakes in subducting oceanic lithosphere, *Rev. Geophys., 34,* 261–306.

Kohlstedt, D. L., H. Keppler, and D. C. Rubie (1996) Solubility of water in the α, β, and γ phases of $(Mg,Fe)_2SiO_4$, *Contrib. Mineral. Petrol., 123,* 345–357.

Kubo, T., E. Ohtani, T. Kato, T. Shinmei, and K. Fujino (1998) Effect of water on the α–β transformation kinetics in San Carlos Olivine, *Science, 281,* 85–87.

Kubo, T., E. Ohtani, T. Kato, S. Urakawa, A. Suzuki, Y. Kanbe, K. Funakoshi, W. Utsumi, T. Kikegawa, and K. Fujino (2002a) Mechanism and kinetics of the post-spinel transformation in Mg_2SiO_4, *Phys. Earth Planet. Inter., 129,* 153–171.

Kubo, T., E. Ohtani, T. Kondo, T. Kato, M. Toma, T. Hosoya, A. Sano, T. Kikegawa, and T. Nagase (2002b) Metastable garnet in oceanic crust at the top of the lower mantle, *Nature, 420,* 803–806.

Kushiro, I. (1968) Liquidus relations in the system forsterite-diopside-silica-H_2O, *Carnegie Inst. Washington, Year Book, 67,* 158–161.

Lebedev, S., S. Chevrot, and R. D. van der Hilst (2003) Correlation between the shear-speed structure and thickness of the mantle transition zone, *Phys. Earth Planet. Inter., 136,* 25–40.

Li, X., R. Kind, K. Priestley, S. V. Sobolev, F. Tilmann, X. Yuan, and M. Weber (2000) Mapping the Hawaiian plume conduit with converted seismic waves, *Nature*, *405*, 938-941.

Litasov, K. D., and E. Ohtani (2002) Phase relations and melt compositions in CMAS pyrolite–H_2O system up to 25 GPa, *Phys. Earth Planet. Inter.*, *134*, 105–127.

Litasov, K. D., and E. Ohtani (2003) Stability of hydrous phases in CMAS-pyrolite-H_2O system up to 25 GPa, *Phys. Chem. Mineral.*, *30*, 147–156.

Litasov, K. D., E. Ohtani, A. Suzuki, T. Kawazoe, and K. Funakoshi (2004) Absence of density crossover between basalt and peridotite in the cold slabs passing through 660 km discontinuity, *Geophys. Res. Lett.*, *31*, doi:10.1029/2004GL021306.

Litasov, K. D., and E. Ohtani (2005) Phase relations in hydrous MORB at 18-28 GPa: implications for heterogeneity of the lower mantle, *Phys. Earth Planet. Inter.*, *150*, 239–263.

Litasov, K. D., and E. Ohtani (2006) Effect of water on the phase relations in the Earth's mantle and deep water cycle, *Geol. Soc. Amer. Spec. Paper* (in press).

Litasov, K. D., E. Ohtani, A. Sano, A. Suzuki, and K. Funakoshi (2005a) *In situ* X-ray diffraction study of post-spinel transformation in a peridotite mantle: Implication to the 660-km discontinuity, *Earth Planet. Sci. Lett.*, *238*, 311–328.

Litasov, K. D., E. Ohtani, A. Sano, A. Suzuki, and K. Funakoshi (2005b) Wet subduction versus cold subduction, *Geophys. Res. Lett.*, *32*, L13312, doi:10.1029/2005GL022921.

Litasov, K. D., E. Ohtani, H. Kagi, and S. Ghosh (2006) Influence of water on olivine-wadsleyite phase transformation and water partitioning near 410-km seismic discontinuity (in prep.)

Matsui, M., S. C. Parker, and M. Leslie (2000) The MD simulation of the equation of state of MgO: application as a pressure calibration standard at high temperature and high pressure, *Am. Mineral.*, *85*, 312–316.

McCammon, C. (1997) Perovskite as a possible sink for ferric iron in the lower mantle: *Nature*, *387*, 694-696.

Mei, S., and D. L. Kohlstedt (2000) Influence of water on plastic deformation of olivine aggregates 1. Diffusion creep regime, *J. Geophys. Res.*, *105*, 21457–21470.

Miura, H., Y. Hamada, T. Suzuki, M. Akaogi, N. Miyajima, and K. Fujino (2000) Crystal structure of $CaMg_2Al_6O_{12}$ a new Al-rich high pressure form, *Amer. Mineral.*, *85*, 1799–1803.

Morishima, H., T. Kato, M. Suto, E. Ohtani, S. Urakawa, W. Utsumi, O. Shimomura, and T. Kikegawa (1994) The phase boundary between α- and β-Mg_2SiO_4 determined by *in situ* X-ray observation, *Science*, *265*, 1202–1203.

Murakami, T., and S. Yoshioka (2001) The relationship between the physical properties of the assumed pyrolite composition and depth distributions of seismic velocities in the upper mantle, *Phys. Earth Planet. Inter.*, *125*, 1–17.

Mysen, B. O. (1973) Melting in a hydrous mantle: Phase relations of mantle peridotite with controlled water and oxygen fugacities, *Carnegie Inst. Washington, Year Book, 72*, 467–478.

Mysen, B. O., and A. L. Boettcher (1975) Melting of a hydrous mantle: I. Phase relations of natural peridotite at high pressure

and high temperature with controlled activities of water, carbon dioxide and oxygen, *J. Petrol., 16*, 520–548.

Nishiyama, N., T. Irifune, T. Inoue, J. Ando, and K. Funakoshi (2004) Precise determination of phase relations in pyrolite across the 660 km seismic discontinuity by *in situ* X-ray diffraction and quench experiments, *Phys. Earth Planet Inter.*, *143–144*, 185–199.

Niu, F., and H. Kawakatsu (1995) Direct evidence for the undulation of the 660-km discontinuity beneath Tonga: Comparison of Japan and California array data, *Geophys. Res. Lett.*, *22*, 531–534.

Niu, F., H. Kawakatsu, and Y. Fukao (2003) A slightly dipping and strong seismic reflector at mid-mantle depth beneath the Mariana subduction zone, *J. Geophys. Res.*, *108*, doi:10.1029/2002JB002384.

Niu, F., H. Inoue, D. Suegetsu, and K. Kanjo (2000) Seismic evidence for a thinner mantle transition zone beneath the South Pacific Superswell, *Geophys. Res. Lett.*, *27*, 1981–1984.

Ohtani, E., K. D. Litasov, T. Hosoya, T. Kubo, and T. Kondo (2004) Water transport into the deep mantle and formation of a hydrous transition zone, *Phys. Earth Planet. Inter.*, *143-144*, 255–269.

Ohtani, E., M. Toma, T. Kubo, T. Kondo, and T. Kikegawa (2003) *In situ* X-ray observation of decomposition of superhydrous phase B at high pressure and temperature, *Geophys. Res. Lett.*, *30*, doi:10.1029/2002GL015549.

Ohtani, E., M. Touma, K. D. Litasov, T. Kubo, and A. Suzuki (2001) Stability of hydrous phases and water storage capacity in the transitional zone and lower mantle, *Phys. Earth Planet. Inter.*, *124*, 105–117.

Paterson, M. S. (1982) The determination of hydrohyl by infrared absorption in quartz, silicate glasses and similar materials, *Bull. Mineral.*, *105*, 20–29.

Pawley, A., P. F. McMillan, and J. R. Holloway (1993) Hydrogen in stishovite, with implications for mantle water content, *Science*, *261*, 1024–1026.

Peacock, S. M. (1990) Fluid processes in subduction zone, *Science*, *248*, 329–337.

Revenaugh, J., and S. A. Sipkin (1994) Seismic evidence for silicate melt atop the 410-km mantle discontinuity, *Nature*, *369*, 474–476.

Rigden, S. M., G. D. Gwanmesia, J. D. Fitzgerald, I. Jackson, and R. C. Liebermann (1991) Spinel elasticity and seismic structure of the transition zone of the mantle, *Nature*, *354*, 143–145.

Ringwood, A. E. (1994) Role of the transition zone and 660 km discontinuity in mantle dynamics, *Phys. Earth Planet. Inter.*, *86*, 5–24.

Rubie, D. C., and C. R. Ross II (1994) Kinetics of the olivine-spinel transformation in subducting lithosphere: experimental constraints and implications for deep slab processes, *Phys. Earth Planet. Inter.*, *86*, 223–241.

Sano, A., E. Ohtani, K. D. Litasov, T. Kubo, T. Hosoya, K. Funakoshi, and T. Kikegawa (2006) Effect of water on garnet-perovskite transformation in MORB and implications for penetrating slab into the lower mantle, *Phys. Earth Planet. Inter.*, (in press).

Shearer, P. M. (1990) Seismic imaging of upper mantle structure with new evidence for a 520-km discontinuity, *Nature, 344*, 121–126.

Shearer, P. M. (1996) Transition zone velocity gradients and the 520-km discontinuity, *J. Geophys. Res., 101*, 3053–3066.

Shearer, P. M. (2000) Upper mantle seismic discontinuities, in *Earth's Deep Interior: Mineral Physics and Tomography from the Atomic to the Global Scale, AGU Geophysical Monograph vol. 117*, edited by S. Karato, A. M. Forte, R. C. Liebermann, G. Masters, and L. Stixrude, pp. 115–131.

Shen, Y., and J. Blum (2003) Seismic evidence for accumulated oceanic crust above the 660-km discontinuity beneath southern Africa, *Geophys. Res. Lett.*, 30, doi:10.1029/2003GL017991.

Shen, Y., S. C. Solomon, I. T. Bjarnason, and C. J. Wolfe (1998) Seismic evidence for a lower-mantle origin of the Iceland plume, *Nature, 395*, 62–65.

Shim, S.-H., T. S. Duffy, and G. Shen (2001) The post-spinel transformation in Mg_2SiO_4 and its relation to the 660-km seismic discontinuity, *Nature, 411*, 571–574.

Shim, S.-H., T. S. Duffy, and K. Takemura (2002) Equation of state of gold and its application to the phase boundaries near 660 km depth in Earth's mantle, *Earth Planet. Sci. Lett., 203*, 729–739.

Smyth, J., and D. Frost (2002) The effect of water on the 410-km discontinuity: An experimental study, *Geophys. Res. Lett.*, 29, doi:10.1029/2001GL014418.

Song, T. R., D. V. Helmberger, and S. P. Grand (2004) Low-velocity zone atop the 410-km seismic discontinuity in the northwestern United States, *Nature, 427*, 530–533.

Speziale, S., C. S. Zha, T. S. Duffy, R. J. Hemley, and H. K. Mao (2001) Quasi-hydrostatic compression of magnesium oxide to 52 GPa: implication for the pressure-volume-temperature equation of state, *J. Geophys. Res., 106*, 515–528.

Stixrude, L. (1997) Structure and sharpness of phase transitions and mantle discontinuities, *J. Geophys. Res., 102*, 14835–14852.

Suzuki, A., E. Ohtani, H. Morishima, T. Kubo, Y. Kanbe, T. Kondo, T. Okada, H. Terasaki, T. Kato, and T. Kikegawa (2000) *In situ* determination of the phase boundary between wadsleyite and ringwoodite in Mg_2SiO_4, *Geophys. Res. Lett., 27*, 803–806.

Tarits, P., S. Hautot, and F. Perrier (2004) Water in the mantle: Results from electrical conductivity beneath the French Alps, *Geophys. Res. Lett., 31*, doi:10.1029/2003GL019277.

Tibi, R., and D. A. Wiens (2005) Detailed structure and sharpness of upper mantle discontinuities in the Tonga subduction zone from regional broadband arrays, *J. Geophys. Res., 110*, doi: 10/1029/2004JB3433.

Tsuchiya, T. (2003) First-principles prediction of the P-V-T equation of state of gold and the 660-km discontinuity in Earth's mantle, *J. Geophys. Res., 108*, doi: 10.1029/2003JB002446.

Van der Meijde, M., F. Marone, D. Giardini, and S. van der Lee (2003) Seismic evidence for water deep in the Earth's upper mantle, *Science, 300*, 1556–1558.

Vidale, J. E., and H. M. Benz (1992) Upper-mantle seismic discontinuities and the thermal structure of subduction zones, *Nature, 356*, 678–683.

Weidner, D.J., and Y. Wang (2000) Phase transformations: Implications for mantle structure Earth's Deep Interior, in *Earth's Deep Interior: Mineral Physics and Tomography from the Atomic to the Global Scale, AGU Geophysical Monograph vol. 117*, edited by S. Karato, A. M. Forte, R. C. Liebermann, G. Masters, and L. Stixrude, pp. 215–235.

Wick, C. W., and M. A. Richards (1993) A detailed map of the 660-kilometer discontinuity beneath the Izu-Bonin subduction zone, *Science, 261*, 1424–1427.

Williams, Q. and R. J. Hemley (2001) Hydrogen in the deep Earth, *Ann. Rev. Earth Planet. Sci., 29*, 365–418.

Wood, B. J. (1995) The effect of H_2O on the 410-kilometer seismic discontinuity, *Science, 268*, 74–76.

Wood, B. J. (2000) Phase transformations and partitioning relations in peridotite under lower mantle conditions, *Earth Planet. Sci. Lett., 174*, 341-354.

Influence of Hydrogen-Related Defects on the Electrical Conductivity and Plastic Deformation of Mantle Minerals: A Critical Review

Shun-ichiro Karato

Yale University, Department of Geology and Geophysics, New Haven, Connecticut 06520, USA

Our current understanding of the microscopic mechanisms by which hydrogen affects the physical properties of minerals is reviewed with emphasis on transport properties including electrical conductivity and plastic deformation. The influence of hydrogen on these physical properties is mostly indirect via hydrogen-related defects. Interpretation of experimental results on defect-dependent properties requires great care because the concentration and speciation of hydrogen in the sample are not always well-controlled and may differ between ambient and high pressure-temperature conditions. Defects with minor concentrations can play a major role. In most (Mg,Fe)-bearing silicates, hydrogen is dissolved as two protons at a vacant M-site (Mg,Fe-site), $(2H)^{\times}_{M}$, though this type of defect has only minor influence on electrical conductivity and plastic deformation. When the concentration of ionized defects exceeds the concentration of hydrogen-free charged defects, the concentration of defects at other sub-lattices increase (in order to maintain charge neutrality) thereby affecting the motion of defects at silicon or oxygen sub-lattice sites, e.g., plastic deformation. Parameterization of the creep law in terms of hydrogen content and pressure (and temperature) is discussed and applied to estimate the upper mantle viscosity and fabric transitions. It is emphasized that both the fugacity effect and the activation volume need to be determined in order to characterize the influence of hydrogen, and therefore the experimental results from a broad range of pressure (pressures beyond a few GPa) are essential to understand the rheological properties of Earth's mantle.

INTRODUCTION

Hydrogen has a remarkable ability to affect some of the key physical and chemical properties of minerals such as rheological properties and melting relations. The addition of only 0.1 wt% H_2O can reduce the effective viscosity of minerals by a factor of $\sim 10^3$ or more [e.g., *Karato*, 1989; *Paterson*, 1989; *Hirth and Kohlstedt*, 1996; Mei *and Kohlstedt*, 2000b;

Earth's Deep Water Cycle
Geophysical Monograph Series 168
Copyright 2006 by the American Geophysical Union.
10.1029/168GM09

Karato and Jung, 2003]. Water also enhances many kinetic processes such as transformation kinetics [*Kubo, et al.*, 1998] and diffusion [*Farver and Yund*, 1991; *Hier-Majumder, et al.*, 2005]. Although these effects have been well known for a long time, the influence of hydrogen at the atomistic level is poorly understood, and quantitative relationships between hydrogen content and physical properties are only now emerging.

During the last ten years, a number of experimental and theoretical studies have attempted to understand the microscopic mechanisms by which hydrogen affects properties such as plastic deformation and electrical conductivity. It is now well established that hydrogen can be structurally incorporated into nominally anhydrous minerals such as olivine and wadsleyite as "point-defects". However, experimental studies of defect-related properties require special attention to sample characterization and a number of fundamental points are not well understood, especially pertaining to experiments with natural samples. There are two primary reasons why measuring the influence of hydrogen on physical properties is challenging. Firstly, hydrogen (water) is abundant in the laboratory environment, and due to its high mobility hydrogen can enter the sample even when water is not intentionally added. At the same time, hydrogen can quickly diffuse out of the sample or change its position (lattice site) in the crystal; the location and speciation of hydrogen in a sample is therefore not necessarily the same at ambient conditions as at high-pressure and temperature. Secondly, the physics by which hydrogen affects physical properties is complicated. Quite often, it is not the major hydrogen-related defects that control the physical properties. Consequently, a great care needs to be taken in conducting experiments and interpreting experimental observation. The purpose of this chapter is to provide a critical review of our current knowledge of the role of hydrogen and related defects in electrical conductivity and plastic deformation.

HYDROGEN INCORPORATION MECHANISMS

Hydrogen in nominally anhydrous minerals can be considered a "point defect" [e.g., Karato, 1989; Yan, 1992; Kohlstedt, et al., 1996; Mei and Kohlstedt, 2000a]. Following charge balance requirements, five models of hydrogen dissolution in silicates containing two types of cations (e.g., Mg_2SiO_4) are considered (Fig. 1):

(i)   molecular water at an interstitial site
(ii) two protons from one water molecule go to M-site (the lattice site that is usually occupied by Mg or Fe), and Mg (or Fe) from M-site and one oxygen from water goes to the surface to form MO (MgO or FeO)
(iii) four protons from two water molecules go to Si-site, and Si from the Si-site and two oxygen atoms from two water molecules form SiO_2 at the surface
(iv) two protons go to two O-site oxygen atoms, and one oxygen from water molecule goes to an interstitial site
(v)   water reacts with an interstitial oxygen to create an interstitial hydroxyl

For each mechanism, one can write an equation for chemical reaction as:

$$H_2O\left(fluid\right) \Leftrightarrow H_2O_I \qquad \text{for reaction (i)} \quad (1)$$

$$H_2O\left(fluid\right) + M_M^\times \Leftrightarrow \left(2H\right)_M^\times + MO \qquad \text{for reaction (ii)} \quad (2)$$

$$2H_2O\left(fluid\right) + Si_{Si}^\times \Leftrightarrow \left(4H\right)_{Si}^\times + SiO_2 \qquad \text{for reaction (iii)} \quad (3)$$

$$H_2O\left(fluid\right) + 2O_O^\times \Leftrightarrow 2\left(OH\right)_O^\bullet + O_I'' \qquad \text{for reaction (iv)} \quad (4)$$

$$H_2O\left(fluid\right) + O_I'' + V_O^{\bullet\bullet} \Leftrightarrow 2\left(OH\right)_I' + V_O^{\bullet\bullet} \qquad \text{for reaction (v)} \quad (5)$$

where the Kröger-Vink notation of point defects is used, i.e., X_Y^Z means species X (V indicates a vacancy) occupies the Y-site with effective charge Z. The effective charge is defined relative to the perfect lattice and a cross indicates neutral charge, prime a negative charge, and a dot a positive charge.

The law of mass action can be applied to each reaction to yield

$$C_H \propto C_w \propto f_{H_2O}\left(P,T\right)\exp\left(-\frac{E_i + PV_i}{RT}\right) \quad (6)$$

$$C_H \propto C_w \propto f_{H_2O}\left(P,T\right)a_{MO}^{-1}\exp\left(-\frac{E_{ii} + PV_{ii}}{RT}\right) \quad (7)$$

$$C_H \propto C_w \propto f_{H_2O}^2\left(P,T\right)a_{SiO_2}^{-1}\exp\left(-\frac{E_{iii} + PV_{iii}}{RT}\right) \quad (8)$$

$$C_H \propto C_w \propto f_{H_2O}^{\frac{1}{2}}\left(P,T\right)\left[O_I''\right]^{-\frac{1}{2}}\exp\left(-\frac{E_{iv} + PV_{iv}}{RT}\right) \quad (9)$$

$$C_H \propto C_w \propto f_{H_2O}^{\frac{1}{2}}\left(P,T\right)\left[O_I''\right]^{\frac{1}{2}}\exp\left(-\frac{E_v + PV_v}{RT}\right) \quad (10)$$

where C_H (C_w) is the concentration of hydrogen (water) in the mineral, $f_{H_2O}\left(P,T\right)$ is the fugacity of water that changes with pressure and temperature, a_{MO}, a_{SiO_2} are the activity of oxides, E_X is the energy change and E_X is the volume change of a solid associated with each reaction, X. Note that both $f_{H_2O}\left(P,T\right)$ and $\exp\left(-\frac{E_X + PV_X}{RT}\right)$ terms strongly depend on pressure (and temperature) and they change with pressure in opposite ways. Consequently the pressure dependence of hydrogen solubility is controlled by a delicate balance of these two terms. Note also that the concentration of OH⁻ (equations (9) and (10)) depends on the number of oxygen interstitial atoms that in turn depends on thermo-chemical conditions.

Other mechanisms of water dissolution can be envisaged that include partially hydrogen-occupied vacancies. However, experimental observations suggest that the concentrations of these defects are smaller than those of neutral defects. The dependence of these minor defect concentrations on chemical environment will be addressed later in relation to electrical conductivity.

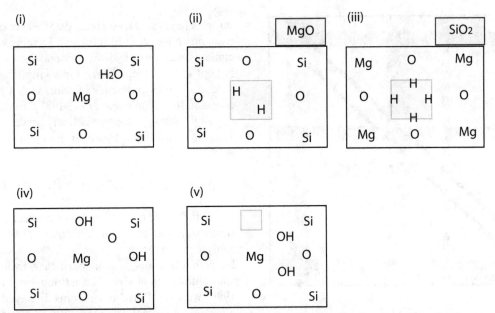

Fig. 1. Mechanisms of dissolution of water (hydrogen) in silicates. Note that mechanisms (ii) and (iii) are associated with an explicit volume change. A volume change also occurs to a lesser extent in mechanisms (i), (iv) and (v).

For olivine, *Zhao, et al.* [2004] and *Kohlstedt et al.* [1996] determined the temperature dependence of hydrogen solubility to get $E_X \sim 50$ kJ/mol and $V_X \sim 10 \times 10^{-6}$ m^3/mol. This value of V_X agrees well with the molar volume of MgO that is consistent with the model (i) above. For wadsleyite, *Williams and Hemley* [2001] proposed that hydrogen solubility decreases with temperature, which was recently confirmed experimentally by *Demouchy et al.* [2005], though the partition coefficient of water between wadsleyite and coexisting melt was roughly independent of temperature at 15 GPa. Using the pressure and temperature dependence of water fugacity, we can estimate $E_X \sim -(50-80)$ kJ/mol and $V_X \sim (11-13) \times 10^{-6}$ m^3/mol for wadsleyite. Again, the estimated volume change, V_X, is roughly consistent with the volume of MgO supporting mechanism (ii). Based on the observation of weak pressure dependence of hydrogen solubility, *Demouchy et al.* [2005] argued that the volume change associated with hydrogen dissolution in wadsleyite is nearly zero, which would be inconsistent with the model (ii). This inference is incorrect. The pressure dependence of water fugacity must be included in estimating V_X. Similarly, the temperature dependence of water fugacity needs to be included in estimating E_X. The large difference in E_X between olivine and wadsleyite likely reflects the difference in crystal chemistry as first noted by *Smyth* [1987] (the wadsleyite structure favors dissolution of hydrogen).

For quartz, mechanism (iii) i.e., the replacement of Si with four protons is the most important mechanism of hydrogen dissolution [e.g., *Doukhan and Paterson*, 1986; *Cordier,*

et al., 1994], whereas in (Mg,Fe)-bearing silicate such as olivine, wadsleyite and orthopyroxene, mechanism (ii), i.e., the replacement of two protons with Mg (or Fe) is the most important mechanism of hydrogen dissolution [e.g., *Kohlstedt, et al.*, 1996]. In the following, I will focus on this particular mechanism since most of minerals we will deal with contain Mg and Fe. Note, however, that the dominant defect type may change when thermo-chemical conditions are changed. It must also be noted that the experimental observations supporting a $(2H)_M^X$ mechanism may not correspond to the true equilibrium mechanism. The kinetics to dissolve hydrogen by a $(4H)_{Si}^X$ mechanism involves diffusion of Si-site vacancies that can be much more sluggish than the kinetics of equilibration with respect to the dissolution by $(2H)_M^X$. It is possible that $(4H)_{Si}^X$ defects are not dominant in short-term laboratory experiments although they can be important under the geologic conditions.

In previous studies, the speciation of hydrogen is typically determined by infrared (IR) spectroscopy on quenched samples [*e.g. Bai and Kohlstedt*, 1993; *Kronenberg*, 1994; *Libowitzky and Beran*, 1995]. However, the site at which hydrogen sits at high pressure and temperature cannot be quenched by any plausible experimental procedure because the characteristic time (τ) for changing the lattice site is too rapid. Using $\tau \approx \frac{d^2}{\pi^2 D}$ (where d is the spacing of lattice sites and D is the diffusion coefficient of hydrogen) and assuming $d \sim 1$ nm and $D \sim 10^{-10}$ m^2/s [*e.g. Mackwell and Kohlstedt*, 1990; *Kohlstedt and Mackwell*, 1998] one gets $\tau \approx 10^{-7}$ s. Consequently, the speciation of lattice site from a quenched

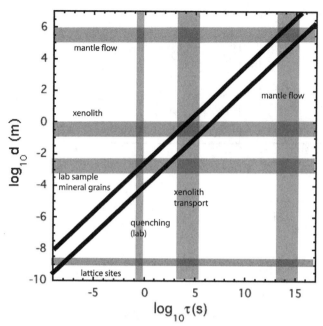

Fig. 2. A diagram showing the conditions where hydrogen-related features (hydrogen content, lattice-site where hydrogen sits) can be preserved in a process with a given time and space-scale. Thick lines show the characteristic time for hydrogen diffusion, τ, for a characteristic distance, d, using $\tau \approx \frac{d^2}{\pi^2 D}$, where D is the diffusion coefficient of hydrogen (a range corresponds to temperature of 1300 to 1800 K). Shown together are some time-scales and length-scales corresponding to laboratory experiments and some geological processes. If (τ, d) for a given process falls below the thick lines, diffusion is efficient and diffusion-loss or change in the speciation will occur, whereas if (τ, d) falls above the lines, these hydrogen-related features will be preserved.

sample can be misleading, although the total amount of dissolved hydrogen is likely quenched in many cases (when d is a sample dimension, ~1 mm, then $\tau \approx 10^3$ s; Fig. 2). That is, the dominant hydrogen-defect can be identified by a "quenching" method from the dependence of total hydrogen content on thermodynamic variables, but information on the speciation of minor defects must be made *in-situ*.

During cooling and depressurization, whether in an experiment or during transport of rock to the surface, hydrogen will diffuse and speciation may change. Cooling and depressurization usually reduces the solubility of hydrogen. Consequently, hydrogen may escape from a sample or hydrous minerals and/or H_2O-rich fluids may precipitate within a crystal. Diffusion loss is rather slow (depending on the sample size and temperature), and in most cases precipitation occurs (Karato, unpublished experimental observation, 1984). Assuming the quenching time of ~ 1 s for typical multianvil experiment, one gets $d \sim 10\text{-}30\ \mu m$. In fact, fluid inclusions have been reported in some experimental

samples [*e.g. Mackwell et al.*, 1985] where quenching times were much longer (~10 minutes). In some cases, these precipitates may form hydrous mineral inclusions if a sample is kept at low temperature for a long time. During slow ascent of a rock in Earth's interior, such a process is likely common [*e.g. Kitamura, et al.*, 1987]. In case of xenoliths, the total hydrogen content is likely preserved but for small xenoliths significant hydrogen loss can occur [*Demouchy et al.*, 2006].

ELECTRICAL CONDUCTIVITY

The influence of hydrogen on electrical conductivity was first discussed by *Duba and Heard* [1980]. Quantitative analyses were made by [*Karato*, 1990] based on the Nernst-Einstein relation using available data on hydrogen diffusivity and solubility in olivine. The assumptions made by [*Karato*, 1990] were (i) all hydrogen atoms dissolved in olivine (and other minerals) contribute equally to electrical conductivity and (ii) hydrogen diffusivity reported by *Mackwell and Kohlstedt* [1990] can be directly applied to estimate electrical conductivity. Both of these assumptions can be questioned because the diffusion coefficient determined by *Mackwell and Kohlstedt* [1990] was chemical diffusion coefficient and not the self-diffusion coefficient. In general, self-diffusion coefficients can be larger than chemical diffusion coefficients (however, for olivine *Kohlstedt and Mackwell* [1998] showed that the diffusion coefficient determined by *Mackwell and Kohlstedt* [1990] was close to the self-diffusion coefficient). Subsequent studies on the electrical conductivity of olivine, wadsleyite and ringwoodite [*Huang et al.*, 2005; *Wang et al.*, 2006] showed that the assumption (i) above was not correct. Electrical conductivity in these minerals (with high hydrogen content) is significantly higher than the values calculated from Karato model using diffusion data (Fig. 3a). When the results are fitted to the following equation

$$\sigma_{cond}^{wet} = A \cdot C_w^r \exp\left(-\frac{H_{cond}^*}{RT}\right) \qquad (11)$$

where C_w is water content (wt%), H_{cond}^* is the activation enthalpy, then Karato model would predict $r=1$ and the activation enthalpy is the same as that of diffusion. However, the actual results summarized in Table 1 show that both r and H_{cond} are smaller than those predicted by the Karato model. To understand the significance of this observation, the dependence of electrical conductivity on chemical environment will be discussed. Electrical conductivity in silicates is sensitive to various activity parameters influencing the defect concentration:

$$\sigma_{cond} \propto [defect] \propto f_{H2O}^r f_{O2}^q a_{MO}^s \qquad (12)$$

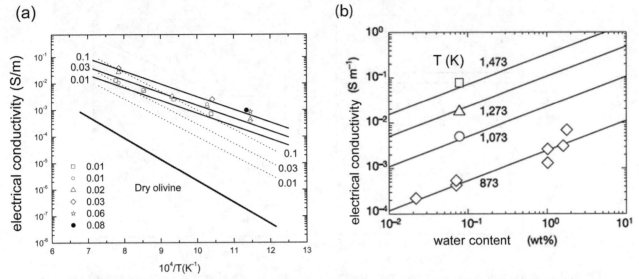

Fig. 3. (a) Electrical conductivity in olivine as a function of inverse temperature from various water contents shown by numbers (wt%). Solid lines with numbers are the least square fit results from the experimental study by [*Wang, et al.*, 2006], broken lines are values calculated from diffusion data using the Karato model. (b) Electrical conductivity in wadsleyite as a function of water content (after [*Huang, et al.*, 2005]). Numbers indicate temperatures.

where [*defect*] is defect concentration, f_{O2} is oxygen fugacity and r, q, and s are the parameters characteristic to each defect. Given the well-accepted model for hydrogen dissolution discussed above (i.e., $(2H)_M^\times$), $C_w \propto f_{H2O}$ and both f_{O2} and a_{MO} are fixed in these experiments. Therefore the model of *Karato* [1990] would predict $\sigma_{cond} \propto C_w$. This is not consistent with the observation shown in Fig. 3. To explain this discrepancy, consider point defect reactions in more detail. The dominant defect, $(2H)_M^\times$, is a neutral defect (neutral with respect to the perfect crystal) in which two protons are trapped at an M-site. However, one of the M-site protons can become a free proton, $H^\cdot$, leaving a single proton at M-site, H_M' through an ionization reaction,

$$(2H)_M^\times \Leftrightarrow H_M' + H^\cdot. \qquad (13)$$

The activation energy for this reaction is on the order of $\frac{e^2}{\varepsilon_0 r}$ where ε_0 is static dielectric constant, e is the charge of electron, and r is the radius of $(2H)_M^\times$ defect. This energy is much smaller than the formation energy of a vacancy at M, O or Si site, $\approx \frac{Z^2 e^2}{\varepsilon_0 r}$ where Z is the valence of the ion involved. Therefore this ionization reaction is comparable to the vacancy formation. In fact, H_{cond}^* is about 1/3–1/4 of activation energies for atomic diffusion in these minerals. If the mobility of free proton is much higher than that of a more abundant defect, $(2H)_M^\times$, then the free proton would determine electrical conductivity. In such a case, electrical conductivity cannot be calculated directly from the solubility and diffusivity of hydrogen as was done by *Karato* [1990]. The high proton mobility leads to a low activation enthalpy of electrical conductivity. Consequently, electrical conductivity is not very sensitive to temperature, but highly sensitive to hydrogen content (Table 1). This makes electrical conductivity a useful means to infer hydrogen content in Earth's mantle.

One immediate consequence of this ionization reaction is that once a large amount of charged defects such as H_M' are created, then the charge balance of a crystal is modified, and the concentration of defects at other sub-lattice sites will

Table 1. Electrical conductivity in wadsleyite and ringwoodite. Parameters in equation (11) are shown. The parameters A, r and H_{cond} are defined by $\sigma_{cond} = A \cdot C_w^r \exp\left(-\frac{H_{cond}^*}{RT}\right)$. Data from *Huang et al.* [2005] and *Wang et al.* [2006].

mineral	$\log_{10} A$ (S/m)	r	H_{cond}^* (kJ/mol)
wadsleyite	2.6±0.2	0.66±0.05	88±3
ringwoodite	3.6±0.1	0.69±0.03	104±2
olivine	3.0±0.4	0.62±0.15	87±5

(a)   (b)

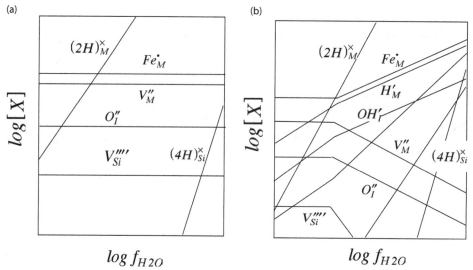

$$\log f_{H2O} \qquad\qquad \log f_{H2O}$$

Fig. 4. Point defect concentration versus water fugacity diagram for (a) a high ionization potential and (b) a low ionization potential. The Kröger-Vink notation of point defects is used. In case (a), all hydrogen-related point defects are neutral, and the dissolution of hydrogen in one sub-lattice (say M-sub-lattice) does not affect point defect concentration in other sub-lattice. In contrast, when ionization potential is low, then dissolution of hydrogen at M-sub-lattice changes the population of charged defects and hence modifies the concentration of point defects in other sub-lattices.

also be affected. The dominant changed defects in olivine (and similar silicates) at water-free conditions are ferric iron, $Fe^{\bullet}_M$, and M-site vacancies, V''_M [e.g. Karato, 1974]. Now, for a reasonable range of conditions, $[(2H)^{\times}_M]/[V''_M] \sim 10^2 - 10^3$ for olivine or wadsleyite, and in such a case, if more than ~0.1-1 % of $(2H)^{\times}_M$ is ionized by the reaction (13), then the charge balance will be modified to $[H'_M] = Fe^{\bullet}_M$. Dependence of concentration of point defects on chemical environment can be calculated for this charge balance equation (see Table 1). From this one can conclude that the charge carrier in wadsleyite under this condition is the free proton, H^g, and this model predicts,

$$\sigma_{cond} \propto f_{H2O}^{\frac{3}{4}} f_{O2}^{-\frac{1}{8}} a_{MO}^{-\frac{1}{2}} \qquad (14)$$

A similar conclusion is obtained for olivine [*Wang* et al., 2006].

Electrical conductivity may also be enhanced by the enhanced diffusion of Mg and Fe. In this case,

$$\sigma_{cond} \propto f_{H2O} a_{MO}^{-1} \qquad (15)$$

Hier-Majumder et al. [2005] studied Mg-Fe diffusion in olivine, and concluded that diffusion is enhanced by hydrogen, but the resultant electrical conductivity is less than that due to proton conduction alone. A relation such as (11) or (14) can be used to infer water fugacity or water content in the mantle [*e.g. Huang* et al., 2005; *Wang* et al., 2006].

In the regime where $[H'_M] = Fe^{\bullet}_M$, concentrations of all other defects are affected by the dissolution of hydrogen.

Consequently, the concentration of all other defects in this regime depend on the hydrogen content in the crystal. Therefore plastic deformation that is likely dependent on the diffusion in Si- or O-sub-lattice is also enhanced by the presence of hydrogen. Fig. 4 illustrates the dependence of point defect concentrations on chemical environment in these two cases.

As previously noted, a certain amount of hydrogen can be dissolved in a sample even though it is not intentionally added, especially when the affinity for hydrogen is very high, such as the case for wadsleyite. A remarkable example is shown in Fig. 5, in which the electrical conductivity of (presumably dry) wadsleyite was reported to be much higher than that of olivine [*Xu* et al., 1998]. It was later found that the observed contrast in electrical conductivity between wadsleyite and olivine was due almost entirely to the difference in hydrogen content and the electrical conductivity of olivine and wadsleyite with similar hydrogen concentrations is nearly identical. The observed large increase in electrical conductivity across the 410-km discontinuity [*e.g. Utada* et al., 2003] may therefore be related to changes in hydrogen concentration. Geodynamic implication of this observation is discussed by Karato et al. [*this volume*].

Hydrogen is also known to enhance rates of Mg-Fe diffusion [*Chakraborty* et al., 1999] and grain-growth kinetics [*Nishihara* et al., 2006a]. In both cases, kinetics are faster in wadsleyite than in olivine due mainly to a higher hydrogen content (the case for hydrogen-enhanced Mg-Fe diffusion has not been demonstrated for wadsleyite, but based on results for olivine

[*Hier-Majumder et al.*, 2005] the high diffusivity of Mg-Fe in wadsleyite is likely due to high hydrogen content). The amount of hydrogen in a sample coming from the pressure medium may also increase at high P-T conditions. This effect will counteract the intrinsic effect of pressure to reduce the rate of mass transport, and could potentially result in a misleading conclusion that pressure has small effects on transport properties (such as diffusion and plastic deformation). The hydrogen content of samples must therefore always be measured before and after experiments on defect-sensitive properties.

PLASTIC DEFORMATION

General Introduction: Needs for High-Pressure Experiments (P>1 GPa)

Hydrogen-related defects also influence plastic flow because deformation at high-temperatures involves movement of point defects either directly (diffusional creep) or indirectly (dislocation creep). In many experimental studies on the mechanisms of hydrogen-weakening effects, the relation between strain-rate and thermo-chemical environment is described as

$$\dot{\varepsilon} \propto f_{H2O}^{r}(P,T) \cdot f_{O2}^{q}(P,T) \cdot a_{MO}^{s}(P,T) \cdot \exp\left(-\frac{E_1^* + PV_1^*}{RT}\right) \quad (16)$$

where E_1^* and V_1^* are the activation energy and volume for defect formation and motion associated with creep (note that $f_{H2O}^{r} f_{O2}^{q} a_{MO}^{s}$ term also depends on pressure and temperature and the net activation energy and volume of creep are different from E_1^* and V_1^*). The water fugacity exponent (r) is often used to identify the atomistic mechanisms of hydrogen-weakening. In order to determine r, the water fugacity is usually controlled by changing the pressure, and strain-rate is measured as a function of pressure [*e.g. Mei and Kohlstedt, 2000a*]. In most cases, the variation of $f_{O2}^{q}(P,T) a_{MO}^{s}(P,T)$ with pressure is small, but the variation of $\exp\left(-\frac{E_1^* + PV_1^*}{RT}\right)$ and $f_{H2O}(P,T)$ can be significant. Consequently, in order to determine r, the activation volume V^* must also be determined, which is difficult in low-pressure (<1 GPa) experiments. To emphasize this point, note that for a typical activation volume in silicates and oxides, $V^*=10 \times 10^{-6}$ m³/mol, the difference in $\exp\left(-\frac{PV^*}{RT}\right)$ over a pressure difference of 0.3 GPa is only a factor of ~1.2, whereas for the pressure difference of 5 (10) GPa, the difference in $\exp\left(-\frac{PV^*}{RT}\right)$ is a factor of ~43 (~1840) at T=1600 K. Consequently, even a small error of any other parameters can cause a large error in the activation volume from low-pressure experiments, whereas even relatively crude measurements at high-pressures (P>5 GPa) provide strong constraints on the activation volume and therefore better constraints on the atomic mechanisms

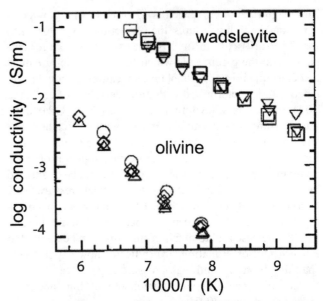

Fig. 5. Electrical conductivity, σ, of wadsleyite and olivine reported by *Xu, et al.* [1998]. The observed systematic difference in conductivity between olivine and wadsleyite turned out to be due mainly to the difference in hydrogen content. Compared at the same water content, temperature (and pressure), the electrical conductivity of these minerals is nearly identical.

of hydrogen weakening. For example, *Mei* [1999] did not take into account the effect of $\exp\left(-\frac{E_1^* + PV_1^*}{RT}\right)$ (i.e., effectively assumed $V^*=0$ m³/mol) and obtained r values indicative of hydrogen weakening mechanisms different from later studies.

An equation for r can be obtained from equation (16):

$$r = r^* + \frac{V^*}{RT} \frac{1}{\left(\frac{\partial \log f_{H2O}}{\partial P}\right)_T} \quad (17)$$

with $r^* \equiv \frac{d \log \dot{\varepsilon}}{d \log f_{H2O}}$ and $r \equiv \left(\frac{\partial \log \dot{\varepsilon}}{\partial \log f_{H2O}}\right)_{P,T}$. Here we used an approximation that oxygen fugacity and oxide activity do not change much with pressure compared to water fugacity. Thus, in an experiment where water fugacity is controlled by changing the pressure, the value that one gets is apparent r, i.e., $r^* \equiv \frac{d \log \dot{\varepsilon}}{d \log f_{H2O}}$. The second term in equation (17) represents a correction term due to the $\exp\left(-\frac{E_1^* + PV_1^*}{RT}\right)$ term. Water is nearly an ideal gas at low pressure (<0.5 GPa) and high-temperature (T>1400 K), and hence $\left(\frac{\partial \log f_{H2O}}{\partial P}\right)_T \approx \frac{1}{P}$, whereas at high-pressure (P>1 GPa), it becomes highly non-ideal gas, and at P=5-20 GPa, $\left(\frac{\partial \log f_{H2O}}{\partial P}\right)_T \approx \frac{\bar{V}^*}{RT}$ with $\bar{V}^* \approx 12 \times 10^{-6}$ m³/mol. Therefore

$$r \approx r^* + \frac{PV^*}{RT} \quad \text{at low pressures (P<0.5 GPa)} \quad (18a)$$

$$r \approx r^* + \frac{V^*}{\bar{V}^*} \quad \text{at high pressures (5-20 GPa).} \quad (18b)$$

The magnitude of correction is $\frac{PV^*}{RT} \approx 0.3$ (0.6) at low pressures, and at high pressures, the correction is $\sim \frac{V}{\bar{V}^*} \approx 0.9$ (1.8) for V^*=10 (20)×10^{-6} m^3/mol. So the correction is large compared with the precision at which the magnitude of r must be determined (in order to obtain a meaningful conclusion, r needs to be determined with an uncertainty of less than ~0.1). In summary, low-pressure studies (P<0.5 GPa) alone are not suited to determine r and V^* with sufficient precision to constrain the hydrogen-related species responsible for hydrogen weakening. Therefore rheological properties under mantle conditions below ~20 km cannot be quantified by low-pressure (<0.5 GPa) data. This critical pressure (~0.5 GPa) is the pressure at which the behavior of water becomes highly non-ideal. Consequently, experimental studies beyond ~1 GPa are essential to obtain results that can be applied to Earth's interior. Equation (18) indicates how experiments spanning both low and high-pressure regimes can be used to determine r and V^* simultaneously with high accuracy [*Karato and Jung, 2003*].

The importance of exploring a broad range of pressure in studying plastic flow is illustrated in Fig. 6 where $\dot{\varepsilon} \propto f_{H2O}^r (P,T) \cdot \exp\left(-\frac{PV^*}{RT}\right)$ (normalized by zero-pressure strain-rate) is plotted as a function of pressure. Strain-rate changes non-monotonically with pressure as first pointed out by *Karato* [1989] and for a reasonable range of V^*, the influence of $\exp\left(-\frac{E_1^* + PV_1^*}{RT}\right)$ is large beyond ~1 GPa. This means that an experimental study beyond ~1 GPa is needed to place strong constraints on V^*, and that the extrapolation of results below ~1 GPa to higher pressure suffers from large uncertainties in inferred strain-rate (effective viscosity). Quantitative experimental studies at pressures beyond a few GPa are essential to obtain any meaningful results on plastic deformation that can be extrapolated to Earth's interior deeper than ~ 20 km.

Diffusional Creep

The mechanisms by which hydrogen affects diffusional creep are straightforward assuming that point-defect mobility does not change with chemical environment and that $\dot{\varepsilon} \propto [defect]$. However, not much is known about point defects at grain-boundaries. If grain-boundary diffusion plays an important role in plastic deformation under hydrous conditions, then application of the previous analysis based on the observations on defects inside the crystal may not be valid. Furthermore, the rate-controlling diffusing species must be known. In olivine, $(Mg,Fe)_2SiO_4$, there are three species ((Mg,Fe), Si and O) that could control the rate of diffusional mass transport. Based on analogy with $(Mg,Fe)O$ and Al_2O_3 [*e.g., Gordon*, 1973] and comparison with a theoretical model, *Karato et al.* [1986] suggested that the bulk diffusion of Mg and Fe controls diffusional creep in fine-grained

olivine. Recent experimental results by *Hier-Majumder et al.* [2005] agree well with the experimental results by [*Karato, et al.*, 1986] and [*Mei and Kohlstedt*, 2000a] in that the diffusion rate is enhanced by water fugacity as $D \propto f_{H2O}$ (this implies that diffusion of Mg-Fe occurs through the motion of $(2H)_M^\times$ and not of V_M''). In contrast, *Mei and Kohlstedt* [2000a] proposed that Si interstitial diffusion (along grain-boundaries) controls diffusional creep in olivine (they assumed that the point defect chemistry at grain-boundaries is the same as that inside of grains). Both models give the same water fugacity dependence and cannot be distinguished from these experimental results alone. This issue will be settled if the dependence of creep rate on grain size, oxygen fugacity and/or oxide activity is determined (see Table 1).

An important parameter for geophysical application is the activation volume, V^*. If V^* for diffusional creep in olivine is as high as 15 ×10^{-6} m^3/mol as *Mei and Kohlstedt* [2000a] reported, then there will not be a transition from dislocation to diffusional creep in the deep upper mantle as proposed by *Karato* [1992]. However, this value of activation volume is not well constrained because only a small percentage of pressure (0.1 to 0.3 GPa) was used in their study. In fact, the activation volume for diffusion of silicon reported by *Mei and Kohlstedt* [2000a] is much larger than that for Si diffusion in olivine reported by *Béjina, et al.* [1997] ($V^*\sim$

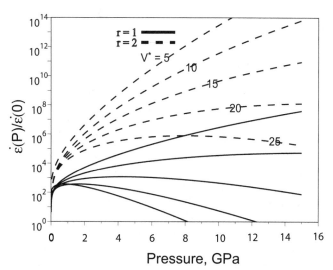

Fig. 6. Influence of pressure on the rate of deformation under hydrogen-saturated conditions. The pressure dependence of rate of deformation is controlled both by the water fugacity and the activation volume as $\dot{\varepsilon} \propto f_{H2O}^r (P,T) \exp\left(-\frac{PV^*}{RT}\right)$ and hence strain-rate changes with pressure in a non-monotonic way. Uncertainties in activation volume, V^*, cause large uncertainties in estimated strain-rate (effective viscosity) above ~1-2 GPa. In other words, a robust constraint on activation volume and r can only be obtained by the experiments above ~1-2 GPa.

$$\dot{\varepsilon}(C_w) = \rho b^2 n_i(C_w) v_i(C_w) \qquad (20)$$

First consider dislocation velocity, $v_i(C_w)$. The physical mechanisms that may control dislocation velocity (mobility) depend on (i) the nature of chemical bonding, (ii) crystal structure, and (iii) the concentration of impurity atoms or secondary phase particles. In most silicates and oxides, the nature of chemical bonding is between ionic and covalent. Si-O is strongly covalent (particularly when Si is coordinated to four oxygen), whereas Mg-O is more ionic. Consequently, the energy required to break chemical bonds is higher in minerals with high SiO_2 content. In these materials, high-temperature steady-state deformation is controlled, in most cases, either by dislocation climb that is in turn controlled by diffusion of the slowest moving atomic species or the glide motion of dislocations through the nucleation and migration of double-kinks. In these cases, hydrogen may affect dislocation velocity through two different effects: (i) through the enhancement of diffusion of the slowest diffusing species, or (ii) through the enhancement of dislocation glide by reducing the energy for nucleation and/or migration of kinks.

Regarding the term $n_i(C_w)$, in a material with strong chemical bonding the density of jogs or kinks is often controlled by the energy of formation of these steps and the chemical environment. The dependence of density of these steps on chemical environment can be analyzed by considering the following chemical reaction where a dislocation line reacts with some chemically active species such as electron holes or water (hydrogen):

$$h^{\cdot}, H_2O + dislocation \Leftrightarrow charged \atop or\ hydrated\ kinks\ (jogs) \qquad (21)$$

Applying the law of mass action to this reaction,

$$n_i(C_w) \propto \left[h^{\cdot}(C_w) \right]^{r_h} exp\left(-\frac{\Delta G_i(C_w, \sigma)}{RT} \right)$$
$$or\ \left[H_2O \right] exp\left(-\frac{\Delta G_i(C_w, \sigma)}{RT} \right) \qquad (22)$$

where $\Delta G_i(\sigma, C_w)$ is the formation free energy of a step and r_h and $rH_2 0$ are the constants that depend on the number of holes or protons at the "steps". Heggie and Jones [1986] showed that the energy of formation and migration of kinks on dislocation line in quartz is reduced by hydrogen and hence the kink density (and mobility) will be higher at hydrogen-rich conditions.

Experimental observations supporting both cases have been reported. For olivine polycrystals deformed in the dislocation creep regime at high temperatures and low stresses, the experimental observations can be interpreted by the dislocation climb-controlled model,

$$\dot{\varepsilon}_{wet} \propto n_i(C_w) v_i(C_w) \propto h^{\cdot} D \propto h^{\cdot} Si_I^{gggg} \propto f_{H2O}^{\frac{5}{4}} \qquad (23)$$

-2 $\times 10^{-6}$ m³/mol). Also the activation volume for Mg-Fe diffusion in olivine [Farber, et al., 2000] is much smaller (~ 6 $\times 10^{-6}$ m³/mol) than this value. The experiments by Hier-Majumder et al. [2005] were conducted to 6 GPa with careful analysis of water content and hence V^* is well constrained. Based on homologous temperature scaling or elastic strain energy model [e.g. Karato, 1981] the activation volume for Si diffusion in olivine should be similar for Mg-Fe diffusion, $V^* \approx \frac{E^*}{K}\left(2\gamma - \frac{2}{3}\right) \sim 5 \times 10^{-6}$ m³/mol (γ: Grüneisen parameter, K: bulk modulus). Although the activation volume for diffusional creep in olivine is not yet well constrained, it is likely smaller than that for dislocation creep (~14 $\times 10^{-6}$ m³/mol) [Karato and Rubie, 1997; Karato and Jung, 2003]).

Dislocation Creep

Dislocations play an important role in plastic deformation, seismic wave attenuation (hence seismic wave velocities) as well as in other kinetic processes such as phase transformations. Currently, the best knowledge of the influence of hydrogen on dislocation mobility comes from experimental studies of dislocation creep combined with theoretical studies on the interaction of hydrogen with dislocations [Karato, 1989; Paterson, 1989]. Karato [1989] predicted a non-monotonic effect of pressure under hydrogen-saturated conditions for a point defect model in olivine. Paterson [1989] reviewed experimental observations and microscopic models of hydrogen weakening of quartz. Here the influence of hydrogen on plastic deformation in the dislocation creep regime is reviewed with emphasis on the influence of hydrogen.

Experimental observations in many silicates including quartz, olivine and pyroxenes indicate that plastic deformation due to dislocations is enhanced by the presence of hydrogen. The degree of weakening is also positively correlated with the silica content. In quartz, this effect is very strong, whereas the weakening effect in olivine is modest.

The rate of deformation by dislocation motion is given by:

$$\dot{\varepsilon} = \rho b v = \rho b^2 n_i v_i \qquad (19)$$

where $\dot{\varepsilon}$ is strain-rate, ρ is dislocation density, b is the length of the Burgers vector, v is dislocation velocity, n_i is the concentration (number per unit length) of "steps" (kinks or jogs) and v_i is their mobility. Dislocation motion occurs in a step-wise fashion through the motion of individual 'steps' such as kinks or jogs. At steady-state, the dislocation density is determined by the balance of the applied stress and the internal stress, and therefore is nearly independent of water content. Therefore the dependence of deformation rate on hydrogen at steady state comes from the dependence of a number of "steps" and of velocity of these steps on water content:

where I assumed $r_h = 1$ and D is the diffusion coefficient of rate-controlling species [e.g., *Mei and Kohlstedt*, 2000b; *Karato and Jung*, 2003]. Therefore the net strain-rate in this case is given by

$$\dot{\varepsilon} \approx \dot{\varepsilon}_{wet} \qquad (24)$$

with

$$\dot{\varepsilon}_{wet} = A \cdot f_{H2O}^{\frac{5}{4}} \propto C_w^{\frac{5}{4}}. \qquad (25)$$

This interpretation is not unique, however. Other models such as $\dot{\varepsilon} \propto [(3H)'_{Si}]$ can also explain the observed dependence on water fugacity. Additional observations such as the dependence of strain-rate on oxide activity and on oxygen fugacity need to be used to better constrain the model.

For quartz, *Griggs* [1974] proposed that creep is controlled by dislocation glide and that hydrogen reduces the energy for nucleation (and growth) of kinks on dislocation lines, thereby enhancing the dislocation glide mobility. Such a model will also apply to other minerals such as olivine at relatively low temperatures and/or high stresses. In this case, the creep rate is proportional to the density of kinks and their mobility and hence (using equation (22)),

$$\dot{\varepsilon}_{wet} \propto [\,H_2O\,]^{rH_2O} exp\left(-\frac{\Delta G_w(\sigma)}{RT}\right) \propto C_w^{rH_2O} exp\left(-\frac{\Delta G_w(\sigma)}{RT}\right) \quad (26)$$

where ΔG_w is the free energy for nucleation and migration of a hydrated double kink. This form of flow law best explains the experimental observations on olivine at relatively high stress levels [*Katayama and Karato*, 2006].

To illustrate the influence of hydrogen on rheological properties of Earth's upper mantle, I have calculated the effective viscosity, $\eta_{eff} \equiv \frac{\sigma}{2\dot{\varepsilon}}$, corresponding to the power-law rheology based on olivine rheology. I used the flow law of the type (23) and (24), namely,

$$\dot{\varepsilon}_{dry} = A_{dry} \cdot exp\left(-\frac{E_{dry}^* + PV_{dry}^*}{RT}\right) \cdot \sigma^n \qquad (27a)$$

$$\dot{\varepsilon}_{wet} = A_{wet} \cdot exp\left(-\frac{E_{wet}^* + PV_{wet}^*}{RT}\right) \cdot \sigma^n \qquad (27b)$$

with parameters given in Table 2. These parameters, particularly activation volumes, were determined by the experiments to P~2 GPa that covered both low-pressure and high-pressure regimes as defined by equation (18) [*Karato and Jung*, 2003] in combination with the low-pressure, high-resolution mechanical data [*Mei and Kohlstedt*, 2000b] (Fig. 7a, b, c).

Note that the activation energy and volume for "wet" conditions include those for solubility of water. The difference in activation volume between "wet" and "dry" conditions is $\sim 10 \times 10^{-6}$ m³/mol, which is close to the volume of incorporation of water in olivine [*Kohlstedt, et al.*, 1996]. This observation is consistent with the present model indicating that the activation energy and volume for creep include those for dissolution of hydrogen-related species and those corresponding to the motion of defects that rate-controls dislocation creep. A very similar result was observed for Mg-Fe diffusion in olivine, in which $V^* \sim 6 \times 10^{-6}$ m³/mol [*Farber, et al.*, 2000] at "dry" condition whereas $V^* \sim 16 \times 10^{-6}$ m³/mol at "wet" condition, a difference in two activation volumes being $\sim 10 \times 10^{-6}$ m³/mol [*Hier-Majumder, et al.*, 2005].

Table 2. The relation between parameters r, q, and s and the point defect concentration in $(Mg,Fe)_2SiO_4$. These parameters are defined by $[defect] \propto f_{H2O}^r f_{O2}^q a_{MO}^s$ (see equation (12) of text). The dominant charged defects are assumed to be H'_M (a proton trapped at M-site) and Fe_M^g (ferric iron).

	r	q	s		r	q	s
$(2H)_M^\times$	1	0	-1	$Si_I^{\bullet\bullet\bullet\bullet}$	1	$-\frac{1}{2}$	-3
H'_M	$\frac{1}{4}$	$\frac{1}{8}$	$-\frac{1}{2}$	$H^\bullet$	$\frac{3}{4}$	$-\frac{1}{8}$	$-\frac{1}{2}$
$Fe_M^\bullet$	$\frac{1}{4}$	$\frac{1}{8}$	$-\frac{1}{2}$	H'''_{Si}	$-\frac{1}{4}$	$\frac{3}{8}$	$\frac{5}{2}$
V''_M	$-\frac{1}{2}$	$\frac{1}{4}$	0	$(2H)''_{Si}$	$\frac{1}{2}$	$\frac{1}{4}$	3
$M_I^{\bullet\bullet}$	$\frac{1}{2}$	$-\frac{1}{4}$	0	$(3H)'_{Si}$	$\frac{5}{4}$	$\frac{1}{8}$	$\frac{5}{2}$
$V_O^{\bullet\bullet}$	$\frac{1}{2}$	$-\frac{1}{4}$	-1	$(4H)_{Si}^\times$	2	0	2
O''_I	$-\frac{1}{2}$	$\frac{1}{4}$	1	$(OH)'_I$	$\frac{1}{4}$	$\frac{1}{8}$	$\frac{1}{2}$
$V'''^{'}_{Si}$	-1	$\frac{1}{2}$	3	$(H_2O)_I^\times$	1	0	0
e'	$-\frac{1}{4}$	$-\frac{1}{8}$	$\frac{1}{2}$	$h^\bullet$	$\frac{1}{4}$	$\frac{1}{8}$	$-\frac{1}{2}$

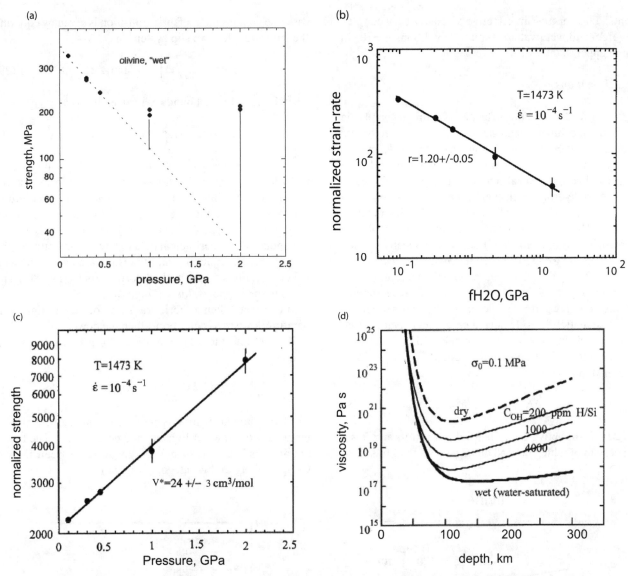

Fig. 7. Effects of pressure on dislocation creep in olivine for water (hydrogen)-saturated conditions (after [*Karato and Jung*, 2003]). (a) A relation between pressure and strength. Data at low-pressures (<0.5 GPa) are from [*Mei and Kohlstedt*, 2000b] and those at higher pressures are from [*Karato and Jung*, 2003]. The broken line shows (schematically) the influence of water fugacity where the effect of activation volume is small. The discrepancy between the extrapolated values from [*Mei and Kohlstedt*, 2000b] assuming zero activation volume (V^*) and the results by [*Karato and Jung*, 2003] is illustrated by solid lines, which gives a measure of the activation volume. A large discrepancy shown in this figure indicates that the activation volume is well-constrained by the comparison of these two data set. (b) and (c) The results of simultaneous inversion of the creep data using the relationship $\dot{\varepsilon} = A\sigma^n f_{H2O}^r \exp\left(-\frac{PV^*}{RT}\right)$ to determine r and V^*. (d) The depth variation of viscosity (for a constant stress) in the upper mantle for a range of water content (bottom) together with the assumed geotherm corresponding to the oceanic environment (age=100 my) and the pressure-depth relationship (top).

Flow law parameters corresponding to "wet" conditions are those for water-saturated conditions. For a case where water content is below saturation, the relation (27b) must be modified with a fixed water content, C_W (ppm H/Si)

$$\dot{\varepsilon}_{fixed} = A_{fixed} \cdot C_W^r \cdot \exp\left(-\frac{E_{fixed}^* + PV_{fixed}^*}{RT}\right) \cdot \sigma^n \qquad (27c)$$

In most of Earth's mantle, the system is closed with respect to water content (see Fig. 2), and the equation (27c) is likely a reasonable approximation. (All the data used in this calculation are summarized in Table 3.)

Fig. 7d shows results of calculation of effective viscosity for a constant stress (with depth) for a range of water contents. The contrast in effective viscosity between dry (water-free) and wet (water-saturated) conditions is ~10^4 in the deep upper mantle.

Fabric Transitions

So far I have reviewed the influence of hydrogen on plastic flow of isotropic mineral aggregates. Dislocation creep occurs by the motion of crystal dislocations. By its very definition, the motion of dislocations is anisotropic. Consequently, the influence of hydrogen can also be anisotropic. In such a case, the addition of hydrogen will change the relative contribution from different slip systems to the total strain, leading to fabric transitions. Hydrogen-induced fabric transitions in olivine were predicted by *Karato* [1995] and systematically studied [*Karato*, 1995; *Jung and Karato*, 2001; *Katayama, et al.*, 2004; *Jung, et al.*, 2006; *Katayama and Karato*, 2006] (Fig. 8).

When a fabric transition occurs at relatively low temperatures, the flow laws are best described by an exponential form, i.e., equation (26). In such a case, the activation free energy for deformation is stress dependent, and in the simplest case, the activation enthalpy, $\Delta H (\Delta G = \Delta H - T\Delta S)$, can be written as

$$\Delta H = \Delta H^o - B\sigma \qquad (28)$$

where ΔH^o is the activation enthalpy at zero stress and B is a constant both of which depend on the slip system. In such a case, the condition for a fabric transition is given by equating the strain-rate for two slip systems as

$$A_1 \exp\left(-\frac{\Delta H_1^o - B_1\sigma}{RT}\right) = A_2 \exp\left(-\frac{\Delta H_2^o - B_2\sigma}{RT}\right) \qquad (29)$$

Therefore the fabric boundary is given by

$$RT \log\frac{A_1}{A_2} = \left(\Delta H_1^o - \Delta H_2^o\right) - \left(B_1 - B_2\right)\sigma \qquad (30)$$

The fabric boundary defined by this equation is insensitive to hydrogen content (as far as a small amount of hydrogen in present), but is sensitive to stress and temperature. This is consistent with the observed fabric transitions from B-type to C-, E-type fabrics (Fig. 8a). This relation is used to extrapolate the experimental data to lower temperatures (the B- to C-type transition) and the results are compared with the observations on naturally deformed rocks (Fig. 8b). The comparison supports this scaling law.

In contrast, when a fabric transition occurs at high temperature/low stress conditions, then the power-law flow law will apply and then the conditions for a fabric transition will be

$$RT \log\frac{C_1 + D_1 f_{H2O}}{C_2 + D_2 f_{H2O}} = \Delta H_1^o - \Delta H_2^o. \qquad (31)$$

The fabric boundary for this case is controlled essentially by temperature and water fugacity and not by stress. The fabric boundaries between A- and E-type, and between E- and C-type belong to this category (Fig. 8a).

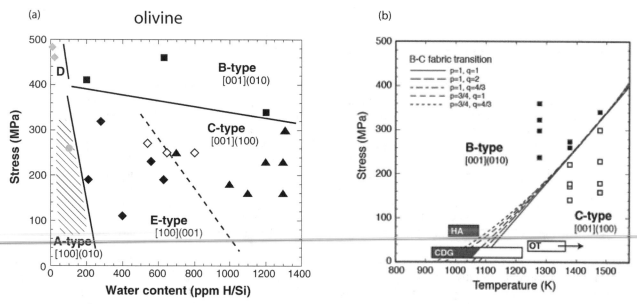

Fig. 8. Fabric diagrams for olivine (a) on a stress-water content space at T~1500 K [*Katayama, et al.*, 2004] and (b) on a stress-temperature space at Cw~1000 ppmH/Si [*Katayama and Karato*, 2006]. Data shown by squares (e.g., HA, CDG, OT) are from naturally deformed peridotites.

The recognition that hydrogen plays an important role in deformation fabrics has provided a new horizon in interpreting deformation fabrics and seismic anisotropy in terms of flow geometry. In particular, the deformation fabrics of rocks from convergent boundaries and the pattern of seismic anisotropy in these regions can be explained in this model by the regional variation in physical/chemical factors (temperature, water content) rather than the complicated flow patterns [e.g. Karato, 2003]. Kneller, et al. [2005] performed a numerical modeling to see how a dominant fabric type is distributed in a realistic subduction zone environment, and indeed showed that olivine B-type fabric dominates near the trench whereas olivine A-, C- or E-type fabric will dominate away from the trench. These studies suggest that an olivine fabric such as B-type may play an important role in certain regions in Earth. If olivine B-type fabric controls seismic anisotropy, the direction of the polarization of fast S-wave will be normal to the flow direction as opposed to cases for A-, C- or E-type fabrics in which it is parallel to the flow direction. The regional distribution of olivine fabrics predicted by our model (B-type near the trench, A- or C- or E-type away from the trench) is consistent with the regional variation of seismic anisotropy in many subduction zones.

SOME SPECULATIONS ON THE INFLUENCE OF HYDROGEN ON PLASTIC DEFORMATION IN DEEP MANTLE MINERALS

Solid-state convection in Earth's ~3000 km thick mantle is largely controlled by plastic deformation. One of the most challenging issues in studying the influence of hydrogen on rheological properties is to conduct experimental studies under deep Earth conditions. I have shown that even for the upper mantle minerals it is critical to conduct quantitative deformation experiments above ~1-2 GPa in order to obtain any meaningful data that can be extrapolated to Earth's interior below a few 10s of km. Although a gas-apparatus such as the one developed by Paterson [1990] provides a high resolution mechanical data and has been used in many laboratories [e.g. Mei and Kohlstedt, 2000a, b], the maximum pressure of operation of that apparatus is ~0.5 GPa (water fugacity ~0.55 GPa). The data obtained under these conditions are limited in their ability to interpret properties in the majority of Earth's mantle.

Experimental studies above ~0.5 GPa are difficult because there is no commercially available apparatus to conduct quantitative deformation experiments above ~0.5 GPa. Quantitative experimental studies beyond this pressure range are limited. Two types of high-pressure deformation apparatus have been developed: D-DIA (deformation

DIA) [Wang et al., 2003] and RDA (rotational Drickamer apparatus) [Yamazaki and Karato, 2001; Xu et al., 2005; Nishihara et al., 2006b]. In both cases, controlled deformation of a sample is possible under high pressure and temperature conditions, and the rheological measurements can be performed using a synchrotron X-ray facility. To date, D-DIA has been operated to ~10 GPa and ~1600 K [e.g. Li et al., 2004] and RDA has been operated to ~16 GPa and ~1800 K [Nishihara et al., 2006b]. However, quantitative experiments on rheological properties under controlled hydrogen content (water fugacity) have only been obtained to 2 GPa (water fugacity ~13 GPa) [Karato and Jung, 2003]. Consequently, the influence of hydrogen on rheological properties of minerals in the deep mantle is not well constrained at this time. However, for wadsleyite, we now have a good idea about the atomistic mechanisms of hydrogen dissolution including the nature of minor charged defects. Processes related to point defects (such as electrical conductivity) in wadsleyite (and ringwoodite) are enhanced by the dissolution of hydrogen. Grain-growth kinetics involving grain-boundary processes in wadsleyite is enhanced by hydrogen [Nishihara et al., 2006a]. Therefore it is almost certain that plastic deformation in wadsleyite and ringwoodite (either by diffusion of point defects or by dislocation motion) is enhanced by hydrogen. Indirect evidence is available suggesting the enhancement of dislocation motion by hydrogen in wadsleyite [e.g. Kubo et al., 1998]. However, there have not been any quantitative experimental studies on the influence of hydrogen on plastic deformation of wadsleyite. Quantitative studies on plastic deformation in wadsleyite (and other deep mantle minerals) are underway in my laboratory. Our preliminary results show a high strength of wadsleyite at water-poor conditions and weakening effects by hydrogen [Nishihara, et al., 2006b].

Much less is known on the influence of hydrogen in lower mantle minerals. For (Mg,Fe)O, dominant hydrogen-related defect is either H'_M or $(2H)^{\times}_M$. Deformation of this material occurs either by diffusion of point defects or by dislocation motion, but deformation by dislocation motion is also controlled by diffusion of point defects (kink formation and migration and jog formation are easy in this material because of a small Peierls stress due to weak chemical bonding). Consequently, the influence of hydrogen on deformation of this material can be understood from the results given in Fig. 4 (the results in Fig. 4 can be applied to (Mg,Fe)O ignoring the defects associated with Si as far as the charge neutrality condition under hydrogen-rich condition is $[H'_M] = Fe^{\bullet}_M$. The following points can be noted: (i) in (Mg,Fe)O, dislocation creep is controlled by dislocation climb that is controlled by diffusion of oxygen. Diffusion of oxygen in (Mg,Fe)O

Table 3. Flow law parameters of olivine for dislocation creep (after *Karato and Jung* [2003])

	$A(s^{-1} (MPa)^{-r-n})$	r	$E^* (kJ/mol)$	$V^* (\times 10^{-6} m^3/mol)$
dry	$10^{6.1 \pm 0.2}$	-	510 ± 30	14 ± 2
wet (water saturated)	$10^{2.9 \pm 0.1}$	1.20 ± 0.05	470 ± 40	24 ± 3
fixed*	$10^{0.56 \pm 0.02}$*	1.20 ± 0.05	410 ± 40	11 ± 3

*: the dimension of A for the fixed water content case is $s^{-1} (MPa)^{-n}$.
"Dry" means water-free condition, "wet" means water-saturated condition, and "fixed" means a case where water content
is independent of pressure and temperature.

occurs through the diffusion of oxygen vacancies or through the diffusion of a neutral defect pair $V_M'' - V_O^{\bullet\bullet}$ (see e.g., *Ando* [1989]). In the former case, the rate of deformation is proportional to $f_{H_2O}^0$, whereas in the latter, it is insensitive to hydrogen content ($V_M'' - V_O^{\bullet\bullet} \propto f_{H_2O}^0$). In either case, the influence of hydrogen on plastic deformation by dislocation creep is likely weaker for (Mg,Fe)O than olivine or other silicates (I also note that the change in activation enthalpy by hydrogen will be smaller in (Mg,Fe)O than other silicate minerals because of a weaker chemical bonding in (Mg,Fe)O than those in silicates). (ii) Diffusion creep in (Mg,Fe)O may be rate-controlled by the volume diffusion of oxygen or by the volume diffusion of Mg (Fe). The latter occurs due to the selected enhancement of oxygen diffusion along grain-boundaries, making the total effective diffusion coefficient of Mg less than that of oxygen [*Gordon*, 1973]. In this case, the creep rate will be proportional to $f_{H2O} (C_w)$ if $(2H)_M^\times$ is the dominant defect. If one accepts the well-documented notion that the volume diffusion of oxygen in (Mg,Fe)O occurs through the diffusion of $V_M'' - V_O^{\bullet\bullet}$ and the diffusion creep is rate-controlled by the volume diffusion of Mg (Fe), then the conclusions on the hydrogen effects are (i) hydrogen has no effects on dislocation creep in (Mg,Fe)O, but (ii) hydrogen has effects similar to olivine for diffusion creep (i.e., $\dot{\varepsilon} \propto f_{H2O}$).

Virtually nothing is known on the influence of hydrogen on physical properties of silicate perovskites. However, based on Mössbauer spectroscopy, the dominant point defects in an Al-containing (Mg,Fe)SiO₃ perovskite are known to be Al_{Si}' and $Fe_M^\bullet$ [*McCammon*, 1997]. The chemical reaction associated with these defects is

$$\tfrac{1}{2} Al_2O_3 + \tfrac{1}{4} O_2 + Fe_M^\times + Si_{Si}^\times \Leftrightarrow Al_{Si}' + Fe_M^\bullet + SiO_2 \quad (33)$$

If the two defects Al_{Si}' and Fe_M^g are the dominant charged defects, then (note that the activity of SiO₂ is fixed by the co-existence of perovskite and magnesiowüstite)

$$[Al_{Si}'] = Fe_M^\bullet \propto f_{O_2}^{\tfrac{1}{8}} a_{Al_2O_3}^{\tfrac{1}{4}} \quad (34)$$

The concentrations of these defects are large (more than 10%; e.g., [*McCammon*, 1997]) and hydrogen solubility

in perovskite is small (~1 ppm wt or less; e.g. *Bolfan-Casanova, et al.* [2000]). Consequently, the incorporation of hydrogen will not change the concentrations of defects at sites other than the one in which hydrogen is incorporated. If analogy with other silicates is a guide, likely sites at which hydrogen is dissolved in (Mg,Fe)SiO₃ perovskite will be either M- or Si-sites (i.e., $(2H)_M^\times$ or $(4H)_{Si}^\times$). Therefore using the model developed in the previous section (the results of Fig. 4 can be applied to (Mg,Fe)SiO₃), I conclude that the rate of deformation will be related to water fugacity as $\dot{\varepsilon} - \dot{\varepsilon}_{dry} \propto f_{H_2O}^q \propto C_w^q$ with $q=1$ if defects at M-site control the rate of deformation and $q=2$ if defects at Si-site control the rate of deformation. In these cases, the influence of hydrogen will be large if plastic deformation under dry conditions is very difficult. The degree to which hydrogen enhances deformation in these cases is primarily controlled by the ratio of concentration of point defects at these sites under dry conditions to the concentration of hydrogen-related defects. Judging from the high solubility of trace elements in CaSiO₃ perovskite [*Corgne and Wood*, 2002; *Hirose et al.*, 2004], it is likely that a very small amount of hydrogen (less than 0.1 ppm wt) goes to (Mg,Fe)SiO₃ perovskite in Earth's lower mantle (a case analogous to hydrogen depletion by partial melting proposed by *Karato* [1986]). Therefore the influence of hydrogen on plastic deformation in (Mg,Fe)SiO₃ perovskite is likely not very large, unless the concentration of defects at dry conditions is significantly smaller than 1 ppm.

In evaluating the role of hydrogen in plastic deformation of lower mantle phases, it is important to estimate the partitioning of hydrogen as well as the magnitude of the hydrogen effect in each phase. The partitioning of hydrogen can be estimated from the experimental results of *Bolfan-Casanova et al.* [2000, 2002], suggesting that more hydrogen is dissolved in (Mg,Fe)O than (Mg,Fe)SiO₃ perovskite. However, I note that *Bolfan-Casanova et al.* [2002] probably underestimated the hydrogen solubility in (Mg,Fe)O due to the effect of unquenchable hydrogen in their samples. Therefore I conclude that a true partitioning coefficient of hydrogen between (Mg,Fe)O and (Mg,Fe)SiO₃ perovskite is yet to be determined.

SUMMARY AND PERSPECTIVES

Our understanding of how hydrogen affects physical properties of minerals has made major progress during the last ten years or so based primarily on experimental studies. However, complications arise due to the difficulties in controlling hydrogen content in a sample, which is subject to "unintentional" hydration but also change in hydration state between ambient high *P-T* conditions. Therefore, great care needs to be exercised to conduct experimental studies and interpret experimental observations on hydrogen-related properties. Based on careful analysis of hydrogen content, the relationship between hydrogen content and electrical conductivity has been determined for wadsleyite, ringwoodite and olivine and the experimental results are compared with the geophysical inference of electrical conductivity to infer the hydrogen (water) content in the transition zone [*Huang et al.*, 2005; *Wang et al.*, 2006]. The influence of hydrogen on the plastic flow of olivine was determined by an experimental work to P=2 GPa [*Karato and Jung*, 2003]. The results provide a robust estimate of effective viscosity of the upper mantle as a function of water content. Hydrogen also changes the deformation fabrics (of olivine). Extensive laboratory studies show a complex relationship between olivine fabrics and physical and chemical conditions [*Jung and Karato*, 2001; *Jung et al.*, 2006; *Katayama et al.*, 2004; *Katayama and Karato*, 2006]. The nature of such a fabric diagram can be understood based on the models of dislocation dynamics under hydrous conditions. Such results provide a new way to interpret complicated patterns of seismic anisotropy in the subduction zones.

This review highlights some areas in which further progress is critical. First, hydrogen effects are highly pressure sensitive and experimental studies need to be conducted at high pressures comparable to those in Earth's interior. For some physical properties (such as electrical conductivity and grain-growth), direct experimental studies have been carried out under deep mantle conditions equivalent to the transition zone or the lower mantle conditions, but high-pressure studies on rheological properties have been limited. I have emphasized that low-pressure studies (P<0.5 GPa) cannot be extrapolated even to ~30 km depth or so, and the conclusions on the atomistic processes from such low-pressure data have large uncertainties. Quantitative deformation experiments at pressures exceeding ~0.5 GPa are not trivial, but a few new apparatus have been developed to explore rheological properties of materials throughout the mantle of Earth [*e.g. Yamazaki and Karato*, 2001; *Wang et al.*, 2003; *Xu et al.*, 2005]. Second, the influence of hydrogen on grain-boundary processes and its atomistic mechanisms are poorly understood. Finally, although

its importance was suggested [*Karato*, 2003] based on pioneering work by *Jackson et al.* [1992], the influence of hydrogen on seismic wave attenuation has not been studied in any detail. There is an urgent need for quantitative study of hydrogen effects on seismic wave attenuation. Future experimental studies on hydrogen-sensitive properties must measure the hydrogen content of the sample both before and after each experiment. Through careful experimental studies at mantle pressures in concert with sound theoretical interpretation progress is now being made toward understanding how hydrogen affects physical properties of minerals and hence the dynamics and evolution of Earth and other terrestrial planets.

Acknowledgments. This work has been supported by National Science Foundation. The review by Greg Hirth and an anonymous reviewer, and the editorial suggestions by Steve Jacobsen and Susan van der Lee helped improve the presentation.

REFERENCES

Ando, K., Self-diffusion in oxides, in Rheology of Solids and of the Earth, edited by S. Karato and M. Toriumi, pp. 57–82, Oxford University Press, Oxford, 1989.

Bai, Q., and D. L. Kohlstedt, Effects of chemical environment on the solubility and incorporation mechanism for hydrogen in olivine, Phys. Chem. Min., 19, 460–471, 1993.

Béjina, F., P. Ratteron, J. Zhang, O. Jaoul, and R. C. Liebermann, Activation volume of silicon diffusion in San Carlos olivine, Geophys. Res. Lett., 24, 2597–2600, 1997.

Bolfan-Casanova, N., H. Keppler, and D. C. Rubie, Water partitioning between nominally anhydrous minerals in the MgO-SiO$_2$-H$_2$O system up to 24 GPa: implications for the distribution of water in the Earth's mantle, Earth Planet. Sci. Lett., 182, 209–221, 2000.

Bolfan-Casanova, N., S. J. Mackwell, H. Keppler, C. McCammon, and D. C. Rubie, Pressure dependence of H solubility in magnesiowüstite up to 25 GPa: implications for the storage of water in the Earth's lower mantle, Geophys. Res. Lett., 29, 89-81/89-84, 2002.

Chakraborty, S., R. Knoche, H. Schulze, D. C. Rubie, D. Dobson, N. L. Ross, and R. J. Angel, Enhancement of cation diffusion rates across the 410-kilometer discontinuity in Earth's mantle, Science, 283, 362–365, 1999.

Cordier, P., J. A. Weil, D. F. Howarth, and J-C. Doukhan, Influence of the (4H)$_{Si}$ defect on dislocation motion in crystalline quartz, Eur. J. Min., 6, 17–22, 1994.

Corgne, A., and B. J. Wood, CaSiO$_3$ and CaTiO$_3$ perovskite-melt partitioning of trace elements: Implications for gross mantle differentiation, Geophys. Res. Lett., 29, 10.1029/2001GL014398, 2002.

Demouchy, S., E. Deloule, D. J. Frost, and H. Keppler, Pressure and temperature-dependence of water solubility in iron-free wadsleyite, Amer. Min., 90, 1084–1091, 2005.

Demouchy, S., S. D. Jacobsen, F. Gaillard, and C. R. Stern, Rapid magma ascent recorded by water diffusion profiles in mantle olivine, *Geology, in press*, 2006.

Doukhan, J.-C., and M. S. Paterson, Solubility of water in quartz, *Bull. Min.*, *109*, 193–198, 1986.

Duba, A., and H. C. Heard, Effect of hydration on the electrical conductivity of olivine, Trans. Amer. Geophys. Union, 61, 404, 1980.

Farber, D. L., D. Williams, and F. J. Ryerson. Divalentcation diffusion in Mg_2SiO_4 spinel (ringwoodile), β phase (wadsleyite) and olivine: implications for electrical conductivity of the mantle, *J Geophys. Res.*, 105, 513–529, 2000.

Farver, J. R., and R. A. Yund, Oxygen diffusion in quartz: dependence on temperature and water fugacity, Chem. Geol., 90, 55–70, 1991.

Gordon, R. S., Mass transport in the diffusional creep of ionic solids, J. Amer. Cer. Soc., 65, 147–152, 1973.

Griggs, D. T., A model of hydrolytic weakening in quartz, J. Geophys. Res., 79, 1653–1661, 1974.

Heggie, M., and R. Jones, Models of hydrolytic weakening in quartz, Phil. Mag., A., 53, L65–L70, 1986.

Hier-Majumder, S., I. M. Anderson, and D. L. Kohlstedt, Influence of protons on Fe-Mg interdiffusion in olivine, J. Geophys. Res., 110, 10.1029/2004JB003292, 2005.

Hirose, K., N. Shimizu, W. van Westrenen, and Y. Fei, Trace element partitioning in Earth's lower mantle and implications for geochemical consequences of partial melting at the core-mantle boundary, Phys. Earth Planet. Inter., 146, 249–260, 2004.

Hirth, G., and D. L. Kohlstedt, Water in the oceanic upper mantle—implications for rheology, melt extraction and the evolution of the lithosphere, Earth Planet. Sci. Lett., 144, 93-108, 1996.

Huang, X., Y. Xu, and S. Karato, Water content of the mantle transition zone from the electrical conductivity of wadsleyite and ringwoodite, Nature, 434, 746-749, 2005.

Jackson, I., M. S. Paterson, J. D. Fitz Gerald, Seismic wave dispersion and attenuation in Åheim dunite, Geophys. J. Inter., 108, 517–534, 1992.

Jung, H., and S. Karato, Water-induced fabric transitions in olivine, Science, 293, 1460–1463, 2001.

Jung, H., I. Katayama, Z. Jiang, T. Hiraga, and S. Karato, Effects of water and stress on the lattice preferred orientation in olivine, Tectonophysics, 421, 1–22.

Karato, S., Point Defects and Transport Properties of Olivine, MSc thesis, 56 pp, University of Tokyo, Tokyo, 1974.

Karato, S., Pressure dependence of diffusion in ionic solids, Phys. Earth Planet. Inter., 25, 38–51, 1981.

Karato, S., Does partial melting reduce the creep strength of the upper mantle?, Nature, 319, 309–310, 1986.

Karato, S., Defects and plastic deformation in olivine, in Rheology of Solids and of the Earth, edited by S. Karato and M. Toriumi, pp. 176–208, Oxford University Press, Oxford, 1989.

Karato, S., The role of hydrogen in the electrical conductivity of the upper mantle, Nature, 347, 272–273, 1990.

Karato, S., On the Lehmann discontinuity, Geophys. Res. Lett., 19, 2255–2258, 1992.

Karato, S., Effects of water on seismic wave velocities in the upper mantle, Proc. Japan Acad., 71, 61–66, 1995.

Karato, S., Mapping water content in Earth's upper mantle, in Inside the Subduction Factory, edited by J. E. Eiler, pp. 135–152, Amer. Geophys. Union, Washington DC, 2003.

Karato, S., and H. Jung, Effects of pressure on high-temperature dislocation creep in olivine polycrystals, Phil. Mag., A., 83, 401–414, 2003.

Karato, S., J. D. Fitz Gerald, and M. S. Paterson, Rheology of synthetic olivine aggregates: influence of grain-size and water, J. Geophys. Res., 91, 8151–8176, 1986.

Karato, S., and D. C. Rubie, Toward experimental study of plastic deformation under deep mantle conditions: a new multianvil sample assembly for deformation experiments under high pressures and temperatures, J. Geophys. Res., 102, 20111–20122, 1997.

Katayama, I., H. Jung, and S. Karato, New type of olivine fabric at modest water content and low stress, Geology, 32, 1045–1048, 2004.

Katayama, I., and S. Karato, Effects of temperature on the B- to C-type fabric transition in olivine, Phys. Earth Planet. Inter., 157, 35–45, 2006.

Kitamura, M., S. Kondoh, N. Morimoto, G. H. Miller, G. R. Rossman, and A. Putnis, Planar OH-bearing defects in mantle olivine, Nature, 328, 143–145, 1987.

Kneller, E. A., P. E. van Keken, S. Karato, and J. Park, B-type olivine fabric in the mantle wedge: Insights from high-resolution non-Newtonian subduction zone models, Earth Planet. Sci. Lett., 237, 781–797, 2005.

Kohlstedt, D. L., H. Keppler, and D. C. Rubie, Solubility of water in the α, β and γ phases of $(Mg,Fe)_2SiO_4$, Contrib. Mineral. Petrol., 123, 345–357, 1996.

Kohlstedt, D. L., and S. J. Mackwell, Diffusion of hydrogen and intrinsic point defects in olivine, Zeit. Phis. Chem., 207, 147–162, 1998.

Kronenberg, A. K., Hydrogen speciation and chemical weakening of quartz, in Silica: Physical Behavior, Geochemistry, and Materials Applications, edited by P. J. Heaney, et al., pp. 123–176, Min. Soc. Amer., Washington, DC, 1994.

Kubo, T., E. Ohtani, T. Kato, T. Shinmei, and K. Fujino, Effects of water on the α–β transformation kinetics in San Carlos olivine, Science, 281, 85–87, 1998.

Li, L., D. J. Weidner, P. Ratteron, J. Chen, and M. T. Vaughan, Stress measurements of deforming olivine at high pressure, Phys. Earth Planet. Inter., 143/144, 357–367, 2004.

Libowitzky, E., and A. Beran, OH defects in forsterite, Physics and Chemistry of Minerals, 22, 387–392, 1995.

Mackwell, S. J., and D. L. Kohlstedt, Diffusion of hydrogen in olivine: implications for water in the mantle, J. Geophys. Res., 95, 5079–5088, 1990.

Mackwell, S. J., D. L. Kohlstedt, and M. S. Paterson, The role of water in the deformation of olivine single crystals, J. Geophys. Res., 90, 11319–11333, 1985.

McCammon, C., Perovskite as a possible sink for ferric iron in the lower mantle, Nature, 387, 694–696, 1997.

Mei, S., The effects of water on the plastic deformation of olivine and olivine-basalt aggregates, Ph D thesis, 142 pp, University of Minnesota, Minneapolis, 1999.

Mei, S., and D. L. Kohlstedt, Influence of water on plastic deformation of olivine aggregates, 1. Diffusion creep regime, J. Geophys. Res., 105, 21457–21469, 2000a.

Mei, S., and D. L. Kohlstedt, Influence of water on plastic deformation of olivine aggregates, 2. Dislocation creep regime, J. Geophys. Res., 105, 21471–21481, 2000b.

Nishihara, Y., T. Shinmei, and S. Karato, Grain-growth kinetics in wadsleyite: effects of chemical environment, Phys. Earth Planet. Inter., 154, 30–43, 2006a.

Nishihara, Y., D. Tinker, Y. Xu, Z. Jing, K. N. Matsukage, and S. Karato, Plastic deformation of wadsleyite and olivine at high-pressures and high-temperatures using a rotational Drickamer apparatus (RDA), J. Geophys. Res., submitted, 2006b.

Paterson, M. S., The interaction of water with quartz and its influence in dislocation flow—an overview, in Rheology of Solids and of the Earth, edited by S. Karato and M. Toriumi, pp. 107–142, Oxford University Press, Oxford, 1989.

Paterson, M. S., Rock deformation experimentation, in The Brittle-Ductile Transition in Rocks: The Heard Volume, edited by A. G. Duba, et al., pp. 187–194, Amer. Geophys. Union, Washington DC, 1990.

Smyth, J. R., β–Mg_2SiO_4: a potential host for water in the mantle?, Amer. Min., 75, 1051–1055, 1987.

Utada, H., T. Koyama, H. Shimizu, and A. D. Chave, A semi-global reference model for electrical conductivity in the mid-mantle beneath the north Pacific region, Geophys. Res. Lett., 30, 10.1029/2002GL016092, 2003.

Wang, D., Y. Xu, and S. Karato, The effect of hydrogen on the electrical conductivity in olivine, Nature, in press, 2006.

Wang, Y., W. B. Durham, I. C. Getting, and D. J. Weidner, The deformation-DIA: A new apparatus for high temperature triaxial deformation to pressures up to 15 GPa, Review of Scientific Instruments, 74, 3002–3011, 2003.

Williams, Q., and R. J. Hemley, Hydrogen in the deep Earth, Ann. Rev. Earth Planet. Sci., 29, 365–418, 2001.

Xu, Y., Y. Nishihara, and S. Karato, Development of a rotational Drickamer apparatus for large-strain deformation experiments under deep Earth conditions, in Frontiers in High-Pressure Research: Applications to Geophysics, edited by J. Chen, et al., pp. 167–182, Elsevier, Amsterdam, 2005.

Xu, Y., B. T. Poe, T. J. Shankland, and D. C. Rubie, Electrical conductivity of olivine, wadsleyite, and ringwoodite under upper-mantle conditions, Science, 280, 1415–1418, 1998.

Yamazaki, D., and S. Karato, High pressure rotational deformation apparatus to 15 GPa, Rev. Sci. Instrum., 72, 4207–4211, 2001.

Yan, H., Dislocation Recovery in Olivine, Master of Science thesis, 93 pp, University of Minnesota, Minneapolis, 1992.

Zhao, Y.-H., S. B. Ginsberg, and D. C. Kohlstedt, Solubility of hydrogen in olivine: dependence on temperature and iron content, Contrib. Min. Petrol., 147, 155–161, 2004.

Effect of Water on the Sound Velocities of
Ringwoodite in the Transition Zone

Steven D. Jacobsen

Department of Geological Sciences, Northwestern University, Evanston, IL 60208

Joseph R. Smyth

Department of Geological Sciences, University of Colorado, Boulder, CO 80309

High-pressure elasticity studies will play a central role in efforts to constrain the potential hydration state of the Earth's mantle from seismic observations. Here we report the effects of 1 wt% H_2O (as structurally bound OH) on the sound velocities and elastic moduli of single-crystal ringwoodite of Fo_{90} composition, thought to be the dominant phase in the deeper part of the transition zone between 520 and 660-km depth. The experiments were made possible through development of a GHz-ultrasonic interferometer used to monitor P and S-wave travel times through micro-samples (30–50 μm thickness) under hydrostatic compression in the diamond-anvil cell. The velocity data to ~10 GPa indicate that hydrous ringwoodite supports 1–2% lower shear-wave velocities than anhydrous ringwoodite at transition zone pressures, though elevated pressure derivatives ($K' = 5.3 \pm 0.4$ and $G' = 2.0 \pm 0.2$) bring calculated hydrous P-velocities close to anhydrous values within their mutual uncertainties above ~12 GPa. Corresponding V_P/V_S ratios are elevated by ~2.3% and not strongly dependent on pressure. Velocities for hydrous ringwoodite are calculated along a 1673 K adiabat using finite-strain theory and compared with existing data on anhydrous ringwoodite and various radial seismic models. It may be possible to distinguish hydration from temperature anomalies by low S-velocities associated with "normal" P-velocities and accompanying high V_P/V_S ratios. The presence of a broadened and elevated 410-km discontinuity, together with depressed 660-km discontinuities and intervening low S-wave anomalies along with high V_P/V_S ratios are the most seismologically diagnostic features of hydration considering the available information from mineral physics.

1. INTRODUCTION

Based on the composition of primitive meteorites and the volatile content of igneous rocks, *Ringwood* [1966] recognized that the bulk Earth may contain three to five times the amount of hydrogen currently present in the atmosphere, hydrosphere, and sediments. Wadsleyite and ringwoodite (β, and γ-Mg_2SiO_4) display an unusual affinity for hydrogen, and together may be storing more H_2O by mass in the mantle transition zone (at 410–660 km depth) than is present in all the oceans combined [e.g. *Smyth*, 1987, 1994; *Inoue et al.*, 1995; *Kohlstedt et al.*, 1996; *Bolfan-Casanova et al.*, 2000]. The transition zone may therefore play a critical role in maintaining a global deep-Earth water cycle [*Kawamoto*

Earth's Deep Water Cycle
Geophysical Monograph Series 168
Copyright 2006 by the American Geophysical Union.
10.1029/168GM09

et al., 1996; Bercovici and Karato, 2003; Ohtani et al., 2004; Hirschmann et al., 2005; Karato et al., this volume]. Since the transition zone partly controls heat and mass transfer between the lower and upper mantle, the presence of water in this region may influence conductive and radiative transport properties [*Karato, 1990; Huang et al., 2005; Hofmeister, 2004; Keppler and Smyth, 2005*], rheology [*Karato et al., 1986; Mei and Kohlstdet 2000; Kavner, 2003*] and mantle convection [*Richard et al., 2002; Bercovici and Karato, 2003*].

The colloquial use of "water" or "hydration" of mantle minerals refers to the incorporation of hydrogen as structurally bound hydroxyl (OH), typically associated with point defects such as cation vacancies, coupled substitutions involving trivalent cations (such as aluminum), or reduction of ferric iron (Fe^{3+}) [*Kudoh et al., 1996; Smyth et al., 2003; McCammon et al., 2004; Blanchard et al., 2005*]. Because hydration leads to modified (i.e. defect) structures, an understanding of the influence of water on the elastic properties and sound velocities of nominally anhydrous mantle phases is needed to interpret seismological observation in regions of a potentially hydrous transition zone [e.g. *van der Meijde et al., 2003; Song et al., 2004; Blum and Shen, 2004*]. Equation of state parameters used in comparing laboratory mineral physics data with seismology include the initial density (ρ_0), thermal expansivity (α_V), isothermal and adiabatic bulk moduli (K_T, K_S) and the shear modulus (G, sometimes written μ). In addition, pressure and temperature derivatives of the moduli $K' = (dK/dP)_T$, $G' = (dG/dP)_T$, $(dK/dT)_P$ and $(dG/dT)_P$ are required to extrapolate compressional and shear-wave velocities (V_P, V_S) to deeper mantle conditions using finite-strain equations of state [e.g. *Duffy and Anderson, 1989*].

Ringwoodite is the high-pressure polymorph of olivine thought to be the dominant phase in the lower part of the transition zone (TZ) between about 520 and 660-km depth. Ringwoodite is known to incorporate 1.0–2.8 wt% H_2O at TZ conditions [*Kohlstedt et al., 1996; Inoue et al., 1998; Smyth et al., 2003*], though hydration mechanisms remain uncertain. Infrared (IR) spectra are characterized by a broad absorption band at 2700–3800 cm^{-1}, typical of OH stretching. A strong band at ~1600 cm^{-1}, which could result from molecular H_2O are present in both hydrous and anhydrous samples and has therefore been interpreted as an overtone of the strong 800 cm^{-1} Si-O stretching [*Bolfan-Casanova et al., 2000*]. Single-crystal X-ray structure refinements of hydrous ringwoodite show predominantly octahedral-site vacancies and some Mg-Si disorder [*Kudoh et al., 2000; Smyth et al., 2003*], but classical atomistic simulations indicate that hydrogen associated with tetrahedral vacancies should be energetically favorable over other negatively charged point defects [*Blanchard et al., 2005*]. Hydration leads to a positive

volume change, lowering the density of ringwoodite (with $X_{Fe} \approx 0.10$) from 3700 (± 5) kg/m^3 [*Sinogeikin et al., 2003*] to 3650 (± 5) kg/m^3 with ~1 wt% H_2O [*Jacobsen et al., 2004*], representing about 1.4% density decrease upon hydration at 300 K and room pressure. Thermal expansion data from high-temperature X-ray diffraction studies at room pressure are available for pure-Mg ringwoodite [*Suzuki et al., 1979*] and hydrous pure-Mg ringwoodite [*Inoue et al., 2004*]. Thermal expansivity at high pressure has been obtained by fitting P-V-T X-ray diffraction data for anhydrous pure-Mg ringwoodite [*Meng et al., 1994; Katsura et al., 2004*] and anhydrous iron-bearing ringwoodite [*Nishihara et al., 2004*], but not hydrous ringwoodite. The comparative thermal expansion study of *Inoue et al.* [2004] indicates that α_V is ~10% lower in hydrous ringwoodite compared with anhydrous ringwoodite. Static compression studies at 300 K have been carried out on γ-Mg_2SiO_4, γ-$(Mg_{0.6}Fe_{0.4})_2SiO_4$, γ-$(Mg_{0.4}Fe_{0.6})_2SiO_4$, γ-$(Mg_{0.2}Fe_{0.8})_2SiO_4$ and γ-Fe_2SiO_4 spinels [*Hazen, 1993; Zerr et al., 1993*]. For comparison, compressibility studies of hydrous pure-Mg ringwoodite [*Yusa et al., 2000*] and hydrous Fe-bearing ringwoodite [*Smyth et al., 2004; Manghnani et al., 2005*] indicate that K_{T0} is reduced by 8–10% upon hydration, meaning that hydrous ringwoodite is more compressible than anhydrous ringwoodite. While the various P-V-T datasets obtain α_V, K_{T0}, K' and dK/dT needed to estimate mineral densities in the transition zone, they do not provide direct information on the adiabatic elastic wave velocities, V_P and V_S.

The elastic properties of ringwoodite have been studied by various techniques including Brillouin spectroscopy [e.g. *Sinogeikin et al., 2003*], ultrasonic interferometry [e.g. *Li, 2003*], the resonant sphere technique [*Mayama et al., 2005*], radial X-ray diffraction [*Kavner, 2003; Nishiyama et al., 2005*] and computational methods [e.g. *Kiefer et al., 1997*]. Room pressure Brillouin studies have determined the single-crystal elastic constants (C_{ij}) of anhydrous pure-Mg ringwoodite [*Weidner et al., 1984*], anhydrous Fe-bearing ringwoodite [*Sinogeikin et al., 1998*], and hydrous pure-Mg ringwoodite [*Inoue et al., 1998; Wang et al., 2003*]. The C_{ij} are symmetry reduced elements of the fourth-rank tensor (C_{ijkl}) relating stress and strain in Hooke's law [*Nye, 1972*] and are useful in calculating elastic anisotropy as well as K_S and G from polycrystalline averaging schemes [*Hashin and Strikman, 1962*]. The C_{ij} of Fe-bearing hydrous ringwoodite were determined ultrasonically at room P-T conditions [*Jacobsen et al., 2004*]. Together, these studies indicate adding ~10% iron to pure-Mg ringwoodite leaves K_S and G unchanged within experimental uncertainties, but the addition of 1–2 wt% H_2O to either pure-Mg or Fe-bearing ringwoodite reduces K_S by 6–9% and G by 9–13%. Temperature derivatives of the elastic moduli for pure-Mg and Fe-bearing anhydrous ringwoodite have been obtained by Brillouin spectroscopy [*Jackson et al.,* 2000; *Sinogeikin et al.,* 2003] and the resonant sphere techniques [*Mayama et al.,* 2005]. Temperature derivatives of the elastic moduli for hydrous ringwoodite are not yet available. Pressure derivatives of the elastic moduli for anhydrous pure-Mg ringwoodite have been obtained ultrasonically [*Li,* 2003] and for anhydrous Fe-bearing ringwoodite by Brillouin spectroscopy [*Sinogeikin et al.,* 2003].

Though structurally bound hydroxyl is readily detectible in laboratory samples using IR spectroscopy, can hydration of the mantle transition zone be detected seismically in the real Earth? In order to address this question, efforts are underway to quantify the effects of hydration on the physical properties of the major nominally anhydrous mantle phases. In this study, we have determined the pressure dependence of the C_{ij}, aggregate moduli (K_S, G) and sound velocities of hydrous ringwoodite with Fo_{90} composition containing about 1 wt% of H_2O to 9.2 GPa. Finite strain theory is used to calculate compressional and shear-wave velocities at transition zone pressures and 300 K. Using some assumptions about temperature derivatives of the elastic moduli for hydrous ringwoodite, we also estimate monomineralic velocities along a 1673 K adiabat for comparison with anhydrous ringwoodite and various radial seismic models.

2. EXPERIMENT

2.1 Sample Synthesis and Characterization

Hydrous ringwoodite was synthesized at 20 GPa and 1400°C in the 5000-ton multianvil press at Bayerisches Geoinstitut. The sample corresponds to run SZ9901 in our previous X-ray diffraction studies [*Smyth et al.,* 2003; *Smyth et al.,* 2004]. A large 18/11 assembly [*Frost et al.,* 2004] consisting of an 18-mm edge-length MgO octahedron for 54-mm carbide cubes with 11-mm truncations was used to compress a 2.0-mm welded-Pt capsule fitted with a stepped lanthanum chromate heater. The large capsule size in this assembly may facilitate the growth of large crystals, which measured up to 800 μm in this experiment. Starting materials consisted of natural Fo_{90} olivine, En_{90} orthopyroxene, hematite, and silica, with ~3 wt% water was added as brucite. Heating duration was five hours. The crystals are deep-blue in color and have approximate composition γ-$(Mg_{0.85}Fe_{0.11})_2H_{0.16}SiO_4$ from microprobe analysis [*Smyth et al.,* 2003]. Mössbauer and Electron Energy Loss Spectroscopy using TEM [*Smyth et al.,* 2003] indicate that approximately 10% of the iron is ferric (i.e. $Fe^{3+}/\Sigma Fe = 0.10$). Single-crystal samples were oriented on crystallographic vectors [100], [110], and [111] with a four-circle X-ray diffractometer and parallel polished into platelets measuring between 30 and 50 μm in thickness.

IR-spectroscopy was carried out on the ultrasonic samples showing approximately 0.96–1.01 wt% H_2O [*Jacobsen et al.,* 2004] using the calibration of *Paterson* [1982]. Unlike olivine [*Bell et al.,* 2003], an absolute calibration for water is not available for ringwoodite, so the accuracy of the measured water content is not known and can only be directly compared with other studies using infrared spectroscopy. Single-crystal X-ray diffraction of the ultrasonic samples gives a cubic unit-cell volume of 530.80(5) Å^3 for a calculated density of $\rho_0 = 3.651(5)$ g/cm^3.

2.2 Gigahertz-Ultrasonic Interferometry

High-pressure ultrasonic measurements were carried out in a diamond-anvil cell using GHz-ultrasonic interferometry [*Spetzler et al.,* 1993; *Jacobsen et al.,* 2005]. Diamond cells typically require samples less than ~100 μm thick in order to achieve pressures above ~5 GPa. The wavelength of strain waves is given by $\lambda = V/f$, where V is the velocity and f is the frequency. In mantle minerals where V_P ~10 km/s and V_S ~5 km/s, acoustic wavelengths at 10–100 MHz with commercial transducers are typically 0.1 to 1.0 mm, being much too long for interferometric techniques with diamond-cell samples. By driving ultrasonic transducers at 0.5 to 2 GHz, acoustic wavelengths are reduced to ~5–10 μm, suitable for ultrasonics in the diamond cell. *P*-waves are generated by thin-film, zinc-oxide transducers sputtered directly onto an acoustic buffer rod. In separate experiments, *S*-waves are generated by *P*-to-*S* conversion inside a single-crystal acoustic buffer rod before transmission into the diamond cell. Shear waves produced in this way are purely polarized, given by the polarization direction of the incident *P*-wave. Details of the new GHz-ultrasonic technique with both *P*- and *S*-wave capabilities are given elsewhere [*Jacobsen et al.,* 2004; *Jacobsen et al.,* 2005]. Hydrostatic pressures were maintained to the maximum pressure of 9.2 GPa using a 16:3:1 methalol:ethanol:water pressure transmitting medium.

2.3 Data Analysis and Equation of State Fitting Procedures

Ringwoodite is cubic (space group $Fd3m$, spinel structure) and has three unique single-crystal elastic constants (C_{11}, C_{44}, C_{12}). The elastic constants were determined from *P*- and *S*-wave travel times (t_P, t_S) along the crystallographic vectors [100], [110], and [111]. Four different pure-mode velocities were used to determine the C_{ij} as follows:

$$\rho(V_P^{[100]})^2 = C_{11} \tag{1}$$

$$\rho(V_S^{[100]})^2 = C_{44} \tag{2}$$

$$\rho(V_P^{[110]})^2 = {}^1\!/_2(C_{11} + 2C_{44} + C_{12}) \qquad (3)$$

$$\rho(V_S^{[111]})^2 = {}^1\!/_3(C_{11} + C_{44} - C_{12}) \qquad (4)$$

In equations (1) through (4), the superscript [uvw] to each velocity gives the direct-space vector of wave propagation, which is parallel to the wave polarization for V_P. For $V_S^{[100]}$ and $V_S^{[111]}$, the shear-wave polarization is degenerate. Velocities were calculated from the measured travel times using the relation $V_{P,S} = (2l/t_{P,S})$, where l is the sample length, initially determined by micrometer measurements to ±1 µm [*Jacobsen et al., 2004*]. The change in sample length with pressure was calculated using third-order Birch Murnaghan equation of state parameters $K_{T0} = 175(3)$ GPa and $K' = 6.2(6)$, determined by separate static compression experiments on hydrous ringwoodite to 45 GPa [*Manghnani et al., 2005*]. The choice of K' in the length calculation is not critical over the experimental pressure range. The difference in calculated length at 10 GPa using $K' = 6.2$ versus $K' = 4.0$ is only 0.06–0.10% (for 30–50 µm initial length). The resulting difference in velocity at 10 GPa is only about 5 m/s, which is considerably smaller than the typical uncertainty in velocity due to standard deviation in the travel times (a few parts in 10^3). Thus, errors in the reported velocities and elastic constants (Tables 1–4) are dominated by errors in the measured travel times. We distinguish absolute accuracy in the room pressure values, dominated by uncertainty in the length [*Jacobsen et al., 2004*], with relative uncertainties at high pressure, dominated by uncertainty in the travel-times and change in length and used for the purpose of equation of state fitting. Pressures were obtained with the ruby-fluorescence scale [*Mao et al., 1986*] and carry a nominal uncertainty of ±0.05 GPa.

Compressional and shear-wave velocities in various crystallographic directions of hydrous ringwoodite are plotted in Figure 1. The velocities were used to calculate C_{ij} using equations (1)–(4). Both C_{11} and C_{44} were determined directly by $V_P^{[100]}$ and $V_S^{[100]}$ respectively. C_{12} was determined in two

Table 1. [100] P-velocities for hydrous Fo_{90} ringwoodite.

P (GPa)	ρ (kg/m³)	$V_P^{[100]}$ (m/s)	$\rho V^2 = C_{11}$ (GPa)
0	3651(5)	9036(62)	298(4)
1.15	3675	9182(82)	310(6)
2.36	3699	9256(62)	317(4)
3.71	3725	9428(45)	331(3)
4.65	3742	9524(46)	339(3)
5.48	3757	9576(54)	345(4)
6.28	3772	9661(60)	352(4)
7.04	3785	9720(68)	358(5)
7.71	3797	9771(59)	362(4)
8.50	3810	9834(57)	368(4)

Table 2. [100] S-velocities for hydrous Fo_{90} ringwoodite.

P (GPa)	ρ (kg/m³)	$V_S^{[100]}$ (m/s)	$\rho V^2 = C_{44}$ (GPa)
0	3651(5)	5536(20)	111.9(8)
2.32	3698	5585(20)	115.3(8)
2.98	3711	5591(24)	116.0(9)
3.48	3720	5606(12)	116.9(5)
4.17	3733	5620(21)	117.9(9)
4.81	3745	5635(10)	118.9(4)
5.92	3765	5652(13)	120.3(5)
6.60	3777	5669(19)	121.4(8)
7.16	3787	5679(12)	122.1(5)
7.75	3797	5685(23)	122.7(9)
8.37	3808	5691(19)	123.3(8)
9.00	3819	5703(17)	124.2(7)

separate experiments using $V_P^{[110]}$ and $V_S^{[111]}$. These velocities give a linear combination of elastic constants shown in equations (3)–(4), which were solved independently for C_{12} at their respective experimental pressures using linear fits to C_{11} and C_{44} given in Table 5. Therefore, equations (3) and (4) provide an independent cross-check of C_{12} since it is determined by two separate experiments involving two different crystallographic directions, in one case with P-waves and in the other case S-waves. Variation of the elastic constants with pressure for Fe-bearing hydrous ringwoodite is plotted in Figure 2, along with linear fits to C_{ij} for nominally anhydrous Fe-bearing ringwoodite measured by *Sinogeikin et al.* [2003].

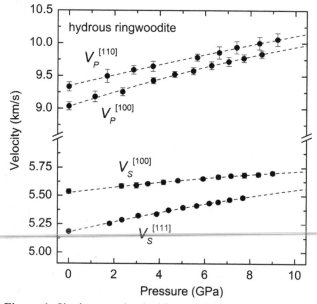

Figure 1. Single-crystal velocities in various crystallographic directions of hydrous ringwoodite (Fo_{90} composition) as a function of pressure measured by GHz-ultrasonic interferometry.

Table 3. [110] *P*-velocities for hydrous Fo$_{90}$ ringwoodite.

Pressure (GPa)	ρ (kg/m^3)	$V_P^{[110]}$ (m/s)	$\rho V^2 =$ ½(C_{11}+2C_{44}+C_{12}) (GPa)	C_{12}* (GPa)
0	3651(5)	9334(70)	318(5)	113(2)
1.68	3685	9496(101)	332(7)	123(3)
2.84	3708	9593(66)	341(5)	128(2)
3.69	3724	9647(78)	347(6)	129(2)
5.65	3760	9783(54)	360(4)	134(1)
6.62	3778	9857(94)	367(7)	138(3)
7.39	3791	9938(96)	374(7)	144(3)
8.40	3808	10003(97)	381(7)	146(3)
9.20	3822	10061(99)	387(8)	149(3)

*calculated using $V_P^{[110]}$ and fits to C_{11} and C_{44} at these pressures.

Aggregate elastic moduli and acoustic velocities were determined using the Voigt-Reuss-Hill (VRH) averaging scheme [*Hill*, 1952]:

$$C_S = \frac{(C_{11} - C_{12})}{2} \qquad (5)$$

$$G_R = \frac{(5C_S C_{44})}{(2C_{44} + 3C_S)} \qquad (6)$$

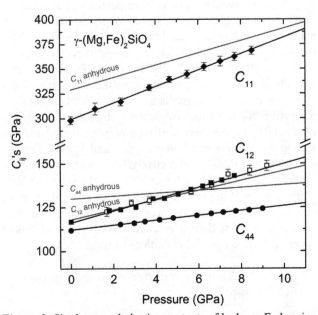

Figure 2. Single-crystal elastic constants of hydrous Fe-bearing ringwoodite as a function of pressure. C_{11} (filled diamonds) and C_{44} (filled circles) were determined by $V_P^{[100]}$ and $V_S^{[100]}$ directly using equations (1) and (2) in the text. C_{12} was determined in two separate experiments using $V_S^{[111]}$ (filled squares) and $V_P^{[110]}$ (open squares). Solid curves are linear fits (Table 5) to all the data. Variation of C_{ij} with pressure for anhydrous Fe-bearing ringwoodite from Brillouin spectroscopy [*Sinogeikin et al.*, 2003] is shown by grey curves.

$$G_V = \frac{(2C_S + 3C_{44})}{5} \qquad (7)$$

$$G_{VRH} = \frac{G_R + G_V}{2} \cong G = \rho(V_S)^2 \qquad (8)$$

$$K_S = \frac{(C_{11} + 2C_{12})}{3} = \rho(V_P)^2 - (^4/_3)G \qquad (9)$$

While the approach of *Hashin and Strikman* [1962] has been shown to place tighter bounds on the shear modulus, in this case the difference between the Voigt bound (G_V) and Reuss bound (G_R) was typically 0.5% or less, so the simpler VRH approach was employed. At each pressure where C_{44} was determined directly, K_S, G, and the acoustic velocities were determined using linear fits to C_{11} and C_{12} and equations (8) and (9). The aggregate acoustic velocities at high pressure (Table 6) were fitted to a third-order finite strain equation of state [*Davies and Dziewonski*, 1975; *Duffy and Anderson*, 1989] using the following formulation:

$$\rho(V_P)^2 = (1 - 2\varepsilon)^{5/2}(L_1 + L_2\varepsilon) \qquad (10)$$

$$\rho(V_S)^2 = (1 - 2\varepsilon)^{5/2}(M_1 + M_2\varepsilon) \qquad (11)$$

$$P = -(1 - 2\varepsilon)^{5/2}(C_1\varepsilon + \frac{C_2(\varepsilon^2)}{2}) \qquad (12)$$

where strain (ε) is $^1/_2[1 - (\rho/\rho_0)^{2/3}]$. Finally, equation of state parameters K_0, G_0, K' and G' were determined from the finite strain fitting parameters using:

$$M_1 = G_0 \qquad (13)$$

$$L_1 = K_0 + (^4/_3)G_0 \qquad (14)$$

$$C_1 = 3L_1 - 4M_1 \qquad (15)$$

Table 4. [111] S-velocities for hydrous Fo90 ringwoodite.

Pressure (GPa)	ρ (kg/m^3)	$V_S{}^{[111]}$ (m/s)	$\rho V^2 = \frac{1}{3}(C_{11}+C_{44}-C_{12})$ (GPa)	C_{12}* (GPa)
0	3651(5)	5184(13)	98.1(5)	116.8(6)
1.8	3688	5254(15)	101.8(6)	123.2(7)
2.34	3698	5287(11)	103.4(4)	123.7(5)
3.06	3712	5324(15)	105.2(6)	125.1(7)
3.87	3728	5338(10)	106.2(4)	130.0(5)
4.40	3738	5373(12)	107.9(5)	130.1(6)
5.04	3749	5396(14)	109.2(6)	132.5(7)
5.64	3760	5413(11)	110.2(5)	135.2(6)
6.20	3770	5435(11)	111.4(5)	137.1(6)
6.59	3777	5445(14)	112.0(6)	138.9(7)
7.10	3786	5466(13)	113.1(6)	140.5(7)
7.68	3796	5484(16)	114.1(7)	143.0(8)

*calculated using $V_S{}^{[111]}$ and fits to C_{11} and C_{44} at these pressures.

$$M_2 = 5G_0 - 3K_0 G' \tag{16}$$

$$L_2 = 5(K_0 + (^4/_3)G_0) - 3K_0(K' + (^4/_3)G') \tag{17}$$

$$C_2 = 3L_2 - 4M_2 + 7C_1 \tag{18}$$

3. RESULTS AND DISCUSSION

3.1 Elastic Moduli of Hydrous Iron-Bearing Ringwoodite

Jacobsen et al. [2004] determined the single-crystal elastic constants and aggregate moduli of hydrous ringwoodite (Fo$_{90}$ composition) at room P-T conditions using the same samples measured in the current high-pressure study. They reported C_{11} = 298(13), C_{44} = 112(6), C_{12} = 115(6) GPa, where the errors in parentheses reflect the absolute accuracy of the measurements, dominated by uncertainty in sample thicknesses. These values are in agreement with C_{ij} obtained by linear fits (Table 5) to the variation of C_{ij} with pressure in the current study; C_{11} = 299(3), C_{44} = 112.2(5), and C_{12} = 116.2(4) GPa, where errors in parentheses reflect error-weighted regressions dominated by the measured travel times and change in sample length as described above. VRH averaged moduli from *Jacobsen et al.* [2004] are K_{S0} = 176(7), and G_0 = 103(5), in agreement with the current high-pressure study where third-order finite strain fitting to the ρV^2 data (Table 6) yields K_{S0} = 177(4) and G_0 = 103.10(9) GPa, with K' = 5.3(4) and G' = 2.0(2). Errors in the equation of state parameters were calculated from errors in the fitting parameters L_1, L_2, M_1, M_2 (Eq. 10–11) from experimental error in travel times, density, and length change.

Upon hydration to ~1 wt% H$_2$O, all of the single-crystal moduli are reduced by variable amounts compared with anhydrous Fe-bearing ringwoodite measured by Brillouin spectroscopy [*Sinogeikin et al.*, 2003], with C_{11} = 329(3), C_{44} = 130(1), and C_{12} = 118(2). From this comparison, C_{11} and C_{44} are reduced by 9.4% and 13.8% respectively, whereas the C_{12} is lowered by only about 2.5%. Aggregate moduli from the anhydrous Fe-bearing ringwoodite study are K_{S0} = 188(3) and G_0 = 120(2) GPa. Thus, hydration to 1 wt% H$_2$O lowers K_{S0} and G_0 by about 6% and 13%, respectively. The reduction in moduli upon hydration was used to estimate the effect of water on aggregate velocities [*Jacobsen et al.*, 2004], resulting in −42 m/s and −36 m/s (V_P and V_S respectively) for every 0.1 wt% H$_2$O (1000 ppm wt.% H$_2$O) added to Fe-bearing ringwoodite at room pressure. To illustrate the magnitude of this effect, we note that in consideration of measured temperature derivatives for anhydrous ringwoodite in Table 7 [*Sinogeikin et al.*, 2003], the addition of ~1 wt% of water is equivalent to raising the temperature, in terms of V_P and V_S reduction, by about 600 and 1000 °C respectively. At low pressure, hydration of ringwoodite has a larger effect on the velocities than temperature within possible ranges of these parameters in the mantle. In this study, we incorporate the effects of water on measured pressure derivatives, and will revisit the effects of water on V_P and V_S at TZ conditions in section 3.4.

3.2 Elastic Moduli of Hydrous Iron-Bearing Ringwoodite at High Pressure

Pressure derivatives of all the C_{ij} are markedly higher upon hydration (Table 5, Figure 2) in comparison with anhydrous Fe-bearing ringwoodite [*Sinogeikin et al.*, 2003]. The pressure derivative of C_{11} is 8.3(5) for hydrous ringwoodite compared with 6.2(2) for anhydrous ringwoodite. The pressure derivative of C_{44} is 1.37(8), compared with 0.8(2) for anhydrous ringwoodite, and dC_{12}/dP is 3.38(8), compared with

Table 5. Single-crystal elastic properties of ringwoodite.

		X_{Fe}	H_2O (wt%)	C_{11} (GPa)	dC_{11}/dP	C_{44} (GPa)	dC_{44}/dP	C_{12} (GPa)	dC_{12}/dP	Ref.
ringwoodite										
	γ-Mg_2SiO_4	0	0	327(3)		131(2)		114(2)		1
	γ-Mg_2SiO_4	0	0	348	6.32	129	0.82	112	3.18	2
hydrous ringwoodite										
	γ-$Mg_{1.89}Si_{0.97}H_{0.33}O_4$	0	2.2	281(6)		117(4)		92(5)		3
	γ-$Mg_{1.85}Si_{0.985}H_{0.356}O_4$	0	2.3	290.6(7)		118.4(3)		104.0(6)		4
Fo$_{90}$ ringwoodite										
	γ-$(Mg_{0.91}Fe_{0.09})_2SiO_4$	0.09	0	329(2)	6.2(2)	130(2)	0.8(2)	118(3)	2.8(3)	5
Fo$_{90}$ hydrous ringwoodite										
	γ-$(Mg_{0.85}Fe_{0.11})_2SiH_{0.16}O_4$	0.11	1.0	299(4)	8.3(5)	112.2(5)	1.37(8)	116.2(4)	3.38(8)	6

[1][*Jackson et al.*, 2000]; Brillouin scattering, 1 atm.
[2][*Kiefer et al.*, 1997]; Pseudopotential calculation.
[3][*Inoue et al.*, 1998]; Brillouin Scattering, 1 atm.
[4][*Wang et al.*, 2003]; Brillouin Scattering, 1 atm.
[5][*Sinogeikin et al.*, 2003]; Brillouin Scattering (Pmax = 16 GPa).
[6]This study. C_{ij} and C_{ij}' from fits to the experimental data (Pmax = 9 GPa).

2.8(3) for anhydrous ringwoodite. For anhydrous ringwoodite $C_{12} < C_{44}$ at room pressure, but because $dC_{12}/dP > dC_{44}/dP$ the term C_{12} crosses C_{44} at approximately 6 GPa (Figure 2). Upon hydration, the dramatic lowering of C_{44} from 130 to 112 GPa results in C_{12} being ~4 GPa higher than C_{44}. In these experiments, equations (3) and (4) represent an important redundancy check of C_{12} since it was determined in two separate experiments using the same direct measurements of C_{11} and C_{44} and from both V_P and V_S in different crystallographic directions. In fact, that $C_{12} > C_{44}$ at room pressure

is typical for oxide spinels [*Yoneda* 1990; *Reichmann and Jacobsen* 2005], so it might be considered unusual that $C_{12} < C_{44}$ for anhydrous ringwoodite. In simple structures, the relative values of C_{44} and C_{12} may provide some insight into the nature of interatomic bonding [*e.g. Weidner and Price*, 1988]. When interatomic forces are parallel to the relative position between atom pairs, the elastic constants satisfy an identity known as the Cauchy condition:

$$\tfrac{1}{2}(C_{12} - C_{44}) = P \qquad (19)$$

Table 6. Aggregate elastic moduli and velocities of hydrous Fo$_{90}$ ringwoodite used for equation of state fitting.

Pressure* (GPa)	ρ (kg/m³)	K_S (GPa)	G (GPa)	V_P (m/s)	V_S (m/s)	$\nu^\dagger$
0	3651(5)	177.1(20)	103.2(11)	9285(58)	5317(32)	0.256(3)
2.32	3698	188.8	107.7	9480	5396	0.260
2.98	3711	192.1	108.7	9531	5413	0.262
3.48	3720	194.6	109.8	9574	5433	0.263
4.17	3733	198.1	111.1	9630	5456	0.264
4.81	3745	201.3	112.4	9682	5478	0.265
5.92	3765	206.9	114.3	9768	5510	0.267
6.60	3777	210.3	115.7	9823	5534	0.268
7.16	3787	213.1	116.7	9867	5551	0.268
7.75	3797	216.0	177.7	9910	5567	0.270
8.37	3808	219.2	118.7	9955	5582	0.271
9.00	3819	222.3	119.8	10003	5602	0.272

Numbers in parenthesis represent one standard deviation in the last place. Uncertainties are estimated to be ~1% in aggregate moduli and ~0.5% in velocity, based upon uncertainties in C_{ij}.
*Pressures where C_{44} was determined directly. Precision is ~0.05 GPa from ruby fluorescence.
†Poisson's ratio.

Table 7. Thermoelastic properties of γ-(Mg,Fe)$_2$Si$_2$O$_4$.

	ρ_0 (kg/m^3)	V_P (km/s)	V_S (km/s)	K_{S0} (GPa)	K' $(dK_S/dP)_T$	$(dK_S/dT)_P$ GPa K^{-1}	G_0 (GPa)	G' $(dG/dP)_T$	$(dG/dT)_P$ GPa K^{-1}	Ref.
ringwoodite										
γ-Mg$_2$SiO$_4$	3515(8)	9.86(3)	5.78(2)	185(2)	4.5(2)	−0.024(3)	120(1)	1.5(1)	−0.015(2)	[1,2]
Fo$_{90}$ ringwoodite										
γ-(Mg$_{0.91}$Fe$_{0.09}$)$_2$SiO$_4$	3701(5)	9.69(2)	5.68(1)	188(3)	4.1(3)	−0.021(2)	120(2)	1.3(2)	−0.016(2)	[3]
Fo$_{90}$ hydrous ringwoodite										
γ-(Mg$_{0.85}$Fe$_{0.11}$)$_2$SiH$_{0.16}$O$_4$	3651(5)	9.29(6)	5.32(3)	177(4)	5.3(4)		103.1(9)	2.0(2)		[4]

[1][*Li*, 2004]; Ultrasonic interferometry, Pmax = 12 GPa
[2][*Jackson et al.*, 2000]; Brillouin scattering, Tmax = 873 K
[3][*Sinogeikin et al.*, 2003]; Brillouin scattering, Pmax = 16 GPa, Tmax = 923 K
[4][*Jacobsen et al.*, 2004] and this study; GHz-ultrasonic interferometry, Pmax = 9 GPa

where P is pressure. At room pressure, ½($C_{12} - C_{44}$) = 2 GPa for hydrous ringwoodite, being closer and of positive sign (like the other spinel structures) compared with that reported for anhydrous ringwoodite at room pressure [*Sinogeikin et al.*, 2003], where ½($C_{12} - C_{44}$) = −6 GPa. Deviation from the Cauchy condition of about +2 and −6 GPa for hydrous and anhydrous ringwoodite respectively are maintained over the experimental pressure ranges.

The single crystal elastic constants of hydrous ringwoodite were used to calculate bulk and shear moduli as a function of pressure using equations (5) through (9), and are plotted together with results on anhydrous Fe-bearing ringwoodite and anhydrous pure-Mg ringwoodite in Figure 3. A third-order finite strain equation of state was fitted to the ρV^2 data in Table 6 resulting in pressure derivatives K' = 5.3(4) and G' = 2.0(2), compared with K' = 4.1(3) and G' = 1.3(2) for anhydrous Fe-bearing ringwoodite [*Sinogeikin et al.*, 2003]. Reduced moduli, and elevated pressure derivatives measured ultrasonically in this study are consistent with the emerging trends for hydrous phases (of K_{T0} and K_T') determined in static compression studies with X-ray diffraction. For example in olivine, K_{T0} is reduced from ~129 GPa to 120 GPa with ~0.8 wt% H$_2$O, while K' ~ 7 compared with K' ~ 5 for dry olivine [*Smyth et al.*, 2005]. Similarly for wadsleyite,

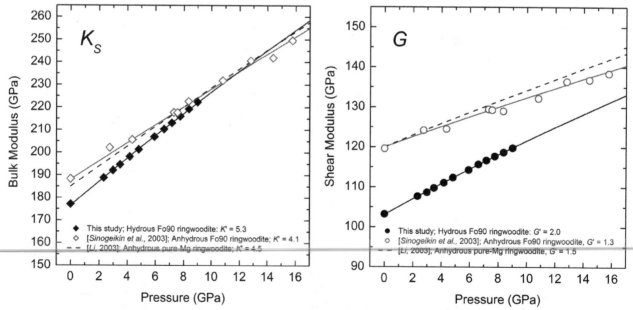

Figure 3. Variation of the bulk (K_S) and shear (G) moduli with pressure for hydrous Fe-bearing ringwoodite (this study, filled symbols and solid lines), anhydrous Fe-bearing ringwoodite from *Sinogeikin et al.* [2003] (open symbols and grey solid lines), and anhydrous pure-Mg ringwoodite from *Li* [2003] (dashed lines).

anhydrous samples give K_{T0} = 174(4) GPa with K' = 3.9(11), and when hydrated to ~1 wt% H_2O, K_{T0} = 152(6) GPa with K' = 6.5(20) [*Holl et al., 2003*]. For ringwoodite, the current ultrasonic values K_{S0} = 177(4) GPa with K' = 5.3(4) are comparable to those obtained by static compression of hydrous ringwoodite (of Fo_{90} composition) from X-ray diffraction with K_{T0} = 175(3) GPa with K' = 6.2(6) [*Manghnani et al., 2005*]. The difference between measured adiabatic and isothermal values of K for hydrous Fe-bearing ringwoodite is consistent with isothermal-adiabatic transformations represented by the Grüneisen relation, $K_S = K_T(1 + \alpha\gamma_{th}T)$, for values of α = 2.7–3.0 x10^{-5} K^{-1} [*Inoue et al., 2004*] and thermal Grüneisen the parameter (γ_{th}) of 1.2-1.5 [*Poirier, 2000*]. Though the adiabatic and isothermal K obtained ultrasonically and by static compression are in excellent agreement, the ultrasonic K' is about 15% lower. The reason for this discrepancy is unknown, but since K' is a third-order fitting parameter in the static compression studies, we consider the ultrasonic K' to be more accurate (i.e. since K is obtained at each pressure).

3.3 Velocities and Anisotropy at High Pressure and 300 K

Is it possible to detect the presence of water or variations in hydration at the level of ~1 wt% H_2O in the transition zone seismically? In order to evaluate the effect of water on aggregate sound velocities of ringwoodite at high pressures, third-order equation of state parameters (Table 7) and equations (10) and (11) were used to compare velocities of pure-Mg anhydrous ringwoodite [*Li, 2003*], anhydrous Fe-bearing ringwoodite [*Sinogeikin et al., 2003*], and hydrous Fe-bearing ringwoodite from the current study at TZ pressures (Figure 4). Although initially (P_0, 300 K) P-wave velocities are about 400(±60) m/s (~4.0%) slower in hydrous ringwoodite, due to the elevated pressure derivatives of the moduli, V_P for hydrous ringwoodite is comparable within uncertainty to anhydrous ringwoodite above ~12 GPa. Therefore, in the depth range where ringwoodite is the stable phase (~18–24 GPa), we expect that the 300 K P-wave velocities for hydrous and anhydrous ringwoodite are indistinguishable (Figure 4).

Shear-wave velocities are initially (P_0, 300 K) about 360(±60) m/s (~6.7%) slower in hydrous Fe-bearing ringwoodite (Figure 4). Despite the elevated G', and unlike V_P, the pronounced slowing of S-wave velocities upon hydration means that hydrous shear velocities remain lower than anhydrous V_S throughout TZ pressures. At 20 GPa, the calculated hydrous V_S is still 1–2% slower than anhydrous V_S.

Elastic wave anisotropy is a critical feature of seismic data typically associated with mantle flow [*e.g. Karato, 1998*]. Although evidence for anisotropy in the TZ is somewhat sparse, there are reports of observed anisotropy in the TZ associated with remnant subducted oceanic lithosphere beneath Fiji-Tonga [*Chen and Brudzinski, 2003*]. One way to

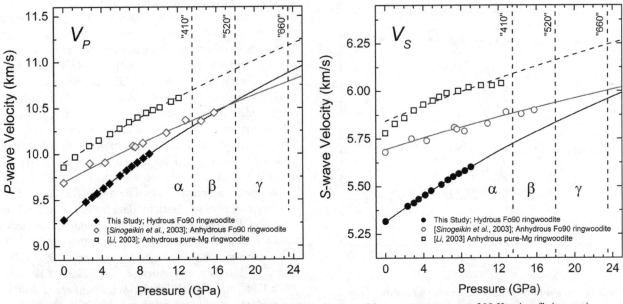

Figure 4. Compressional and shear-wave velocities projected to transition zone pressures at 300 K using finite-strain equations of state for hydrous Fe-bearing ringwoodite (this study, filled symbols and solid lines), anhydrous Fe-bearing ringwoodite [*Sinogeikin et al., 2003*] (grey open symbols, grey solid lines) and anhydrous pure-Mg ringwoodite [*Li, 2003*] (black open squares, dashed lines).

quantify elastic anisotropy in cubic phases such as spinel is given by the anisotropy factor [*Karki et al.*, 1999]:

$$A = \left[\frac{2C_{44} + C_{12}}{C_{11}} \right] - 1 \qquad (20)$$

where $A = 0$ for an elastically isotropic crystal. Despite the dramatic reduction of C_{44} compared with C_{11} and C_{12} upon hydration, the anisotropy factor of anhydrous Fe-bearing ringwoodite, $A = 0.15$ [*Sinogeikin et al.*, 2003], is indistinguishable with $A = 0.14(2)$ for hydrous Fe-bearing ringwoodite within error. Compared with other cubic phases like MgO ($A = 0.36(2)$ [*Jacobsen et al.*, 2002]), the anisotropy factor of ringwoodite is quite low anyway. The variation of A with pressure for anhydrous and hydrous ringwoodite is plotted together with MgO for comparison in Figure 5. Although the effects of temperature are neglected here, we suggest that anisotropy is not likely a useful indicator of hydration because A is so similar for anhydrous and hydrous ringwoodite, but also because both decrease to near zero at 16–25 GPa (Figure 5).

The most diagnostic difference between anhydrous and hydrous ringwoodite with respect to seismological observation is the V_P/V_S ratio and related Poisson's ratio (ν):

$$\nu = \frac{3K_S - 2G}{2(3K_S + G)} = \frac{1}{2} \frac{(V_P/V_S)^2 - 2}{(V_P/V_S)^2 - 1} \qquad (21)$$

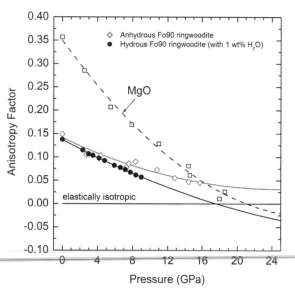

Figure 5. Elastic anisotropy factor A (see text) for hydrous Fe-bearing ringwoodite (this study, filled circles, solid line), anhydrous Fe-bearing ringwoodite from *Sinogeikin et al.* [2003] (open diamonds, grey solid line) and for comparison MgO from *Sinogeikin and Bass* [2000] (open squares, dashed line).

The V_P/V_S ratio of ringwoodite is elevated from 1.706(7) [*Sinogeikin et al.*, 2003] to 1.75(2) upon hydration, or about 2.3% higher. Elevated V_P/V_S ratios persist at high pressure, where above 14 GPa (and 300 K) the calculated V_P/V_S ratio of hydrous ringwoodite is still ~2.3% higher (Figure 6). Poisson's ratio of hydrous ringwoodite is 0.256(3), or ~8% higher than anhydrous ringwoodite with ν = 0.237(1). Above 14 GPa, the calculated Poisson's ratio of hydrous ringwoodite is still ~4.5% higher than anhydrous ringwoodite.

Hydration has a much stronger influence on V_P/V_S than temperature or variation in Fe-content within their respective possible variations in the transition zone. A high-temperature Brillouin study of γ-$(Mg_{0.91}Fe_{0.09})_2SiO_4$ indicates that V_P/V_S increases from 1.706 at 300 K to 1.712 at 923 K, a variation of just 0.4% over 628 K [*Sinogeikin et al.*, 2003]. The variation in V_P/V_S with Fe-content is indistinguishable within experimental uncertainties, with $V_P/V_S = 1.71(1)$ for pure-Mg ringwoodite [*Li*, 2003] compared with 1.706(7) for γ-$(Mg_{0.91}Fe_{0.09})_2SiO_4$ [*Sinogeikin et al.*, 2003]. The ~2% increase in V_P/V_S on hydration is maintained at high-pressure conditions, and is therefore the best seismic parameter to consider, especially where elevated V_P/V_S is associated with "normal" P-velocities and reduced S-velocities since elevated temperatures will reduce both V_P and V_S.

3.4 Calculation of Hydrous Velocities in the Transition Zone

In order to evaluate the effects of water on seismic velocities of ringwoodite in the transition zone, we calculated monomineralic V_P and V_S along a 1673 K adiabat using finite strain theory for comparison with anhydrous ringwoodite and seismic models such as PREM [*Dziewonski and Anderson*, 1981] and IASP91 [*Kennett and Engdahl*, 1991]. Temperature derivatives of the moduli for hydrous ringwoodite are not available, so for the purpose of this calculation we chose to examine two different cases. In the first, we simply used the same temperature derivatives as for anhydrous ringwoodite [*Sinogeikin et al.*, 2003], given in Table 7. In the second case, we reduced dK/dT and dG/dT by an amount proportional to the product αK_T. Examination of a large number of elasticity data for oxide minerals indicates that the magnitude of temperature derivatives tends to increase by an amount proportional to this product [*Duffy and Anderson*, 1989]. Taking anhydrous $\alpha_V = 3.07 \times 10^{-5}$ K^{-1} with $K_T = 186$ GPa, and hydrous $\alpha_V = 2.73 \times 10^{-5}$ K^{-1} with $K_T = 175$ GPa [*Inoue et al.*, 2004; *Manghnani et al.*, 2005], the product αK_T is reduced by about 15% in the hydrous case. Therefore, in the reduced derivative calculation we used $dK_S/dT = -0.024$ and $dG/dT = -0.018$ GPaK^{-1}. Densities at high temperature were calculated using thermal expansivities of

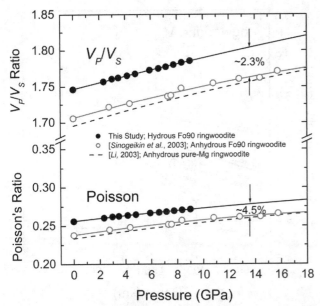

Figure 6. Plot of the V_P/V_S ratio and Poisson's ratio for hydrous ringwoodite (this study, filled circles, solid lines), anhydrous ringwoodite from *Sinogeikin et al.* [2003] (open grey circles, grey solid lines), and anhydrous pure-Mg ringwoodite from *Li* [2003] (dashed lines). Hydration has a much larger effect of V_P/V_S than does Fe-content or temperature (not shown), and is the most diagnostic feature of hydrous ringwoodite.

Inoue et al. [2004] given above. Moduli were then calculated at room pressure at the foot of a 1673 K adiabat using temperature derivatives for the two different cases. Finally, the high-temperature velocities were projected adiabatically into the transition zone using equations (10)–(12) with model parameters given in Table 7.

Monomineralic velocities for anhydrous and hydrous ringwoodite are plotted together with two seismic models in Figure 7. In the case where we have assumed the same temperature derivatives for anhydrous and hydrous ringwoodite, P-wave velocities for anhydrous and hydrous ringwoodite are within about 50 m/s (~0.5%) of each other between 520 and 660 km depth, and therefore are considered to be equal within their mutual uncertainties, and lie about 80 m/s (~0.8%) above PREM. When the temperature derivatives of hydrous (or anhydrous) ringwoodite are reduced by 15% as described above, P-velocities match PREM very well between ~550 and 660 km depth. Hydrous S-wave velocities remain well below anhydrous V_S throughout the lower part of the transition zone and show a good fit to PREM. At 520 km-depth hydrous V_S is about 160 m/s (~2.9%) below anhydrous V_S and at 660 km depth hydrous V_S is about 100 m/s (~1.8%) below anhydrous V_S. When the temperature derivative of hydrous ringwoodite is varied by 15% as described

above, hydrous S-velocities bound both PREM and IASP91 models whereas anhydrous S-velocities lie about 135 m/s (~2.5%) above these models in the lower part of the transition zone. To summarize, hydrous and anhydrous P-velocities are indistinguishable at TZ conditions and are ~1% faster than PREM, whereas hydrous S-velocities are 2–3% slower than anhydrous S-velocities and match radial seismic models very well (Figure 7).

The V_P/V_S ratio of PREM at 600 km depth is 1.84, compared with 1.85 for hydrous ringwoodite and 1.81 for anhydrous ringwoodite in the current calculation, another indication that hydrous (monomineralic) ringwoodite velocities provide a better fit to radial seismic models. However, there are many uncertainties in the current comparison. Though our intention here is only illustrate the magnitude of effects of water on monomineralic velocities, the lower part of the transition zone is unlikely to be pure ringwoodite. The $(Mg,Fe)SiO_3$ phase (majorite) is also likely to be present in various modal proportions depending on the mineralogical model [e.g. *Ringwood,* 1975; *Bass and Anderson,* 1984]. Majorite velocities are lower than PREM [see e.g. *Smyth and Jacobsen,* this volume] and could therefore also explain the discrepancy between (high) anhydrous V_P and V_S velocities and PREM. The comparison with PREM and IASP91 also involves radially averaged seismic models, which do not reflect local variation in temperature or composition, both of which should vary to some extent locally in the transition zone. Water, if present, may not be evenly distributed throughout the transition zone. Fractional melting (due to water) is yet another key parameter not treated here but expected to influence bulk sound velocities. Furthermore, we note that the velocities measured here at GHz frequencies are fully unrelaxed and do not reflect high-temperature dispersion or attenuation [e.g. *Jackson et al.,* 2005], and in particular the possible effects of water on high-temperature viscoelastic relaxation, which may act to reduce observed seismic (Hz) velocities further [e.g. *Karato and Jung,* 1998]. An additional uncertainty arises from the potential influence of second (pressure) derivatives, K'' and G'', which currently are implied values from the third-order truncation. Data spanning a greater pressure range will be required to ascertain to what extent the observed elevated K' and G' may be offset at higher pressure due to increasing $|K''|$ and $|G''|$. However, we note that the crossing of hydrous K_S with anhydrous K_S (Figure 3) is only 2–3 GPa beyond the current measured data and therefore this unusual observation is fairly robust. We emphasize that the model is not a rigorous test of water in the transition zone, and is only meant to highlight the most important features of hydration for monomineralic ringwoodite velocities at high pressure.

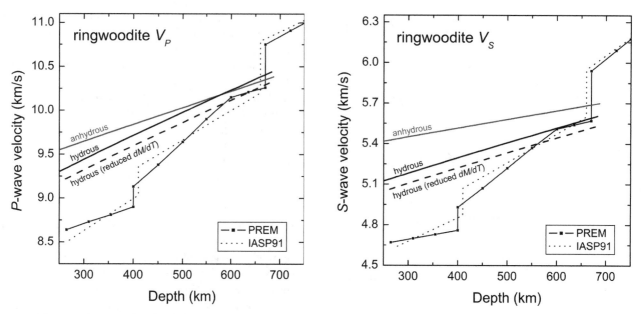

Figure 7. Compressional and shear-wave velocities of ringwoodite projected along an adiabat (1673 K foot temperature) for anhydrous ringwoodite from *Sinogeikin et al.* [2003] (grey solid line) and hydrous ringwoodite (this study, solid black line) where dM/dT is assumed to be the same. The dashed line shows hydrous velocities when dM/dT is reduced by 15% (see text). Seismic models are PREM [*Dziewonski and Anderson,* 1981] and IASP91 [*Kennett and Engdahl,* 1991].

4. SEISMIC STRUCTURE OF A HYDROUS TRANSITION ZONE

Identifying hydration, or lateral variation of hydration in the TZ will require several types of seismic data together with a comprehensive view of regional mantle structure. Here we have shown how hydration of ringwoodite could manifest itself in the lower part of the transition zone as "*normal*" *P*-velocities and reduced *S*-wave velocities accompanied by elevated V_P/V_S ratios at the level of several percent. The same is expected to be true for wadsleyite in the upper part of the TZ, however, at the time of writing sound velocity data (even at room pressure) are not yet available for hydrous wadsleyite. Temperature derivatives of the elastic moduli, for both hydrous wadsleyite and hydrous ringwoodite are also needed.

Water is also expected to influence the width (depth interval) and absolute depth of the "410" and "660" seismic discontinuities. A discussion of the effects of water on the "410" structure is presented elsewhere [see *Hirschmann et al.,* this volume]. Briefly, the width of the two-phase loop between olivine and wadsleyite increases with increasing water content [*Wood,* 1995; *Smyth and Frost,* 2002], at least up to the point of saturation where a free fluid phase (melt) forms and the loop narrows again [*Chen et al.,* 2002; *Hirschmann et al.,* 2005], favoring a sharper seismic discontinuity. Despite the higher water storage capacity of olivine at conditions of ~410 km depth relative to shallow mantle conditions [e.g. *Chen et al.,* 2002; *Smyth et al.,* 2005], the partition coefficient of water between olivine and wadsleyite is still expected to be greater than unity, probably around 5 [*Hirschmann et al,* this volume]. Similarly, the storage capacity of water in ringwoodite is much higher than in the silicate perovskite at "660", probably by around 1000 times [*Bolfan-Casanova et al.,* 2003]. Displacement of the "410" upwards and the "660" downwards due to the presence of water would therefore create a thick transition zone, i.e. with a greater depth interval between "410" and "660" compared with a nominally dry region. Thus, seismic studies for water should include high-resolution local velocity structure in candidate locations, preferably away from subduction zones where thermal anomalies are likely minimal. In the same regions, accurate transition zone structure (discontinuity depths and TZ thickness) will be needed. A locally thick transition zone could also occur due to an anomalously low temperature relative to the surrounding mantle, however, if intervening (especially *S*-wave) velocities are low and V_P/V_S ratios high, the best compositional explanation appears to be H_2O. The presence of a broadened and elevated "410" discontinuity, together with depressed "660" discontinuity with intervening low *S*-wave anomalies and high V_P/V_S ratios would be highly compatible with hydration considering the available information from mineral physics.

Acknowledgments. Support during various stages of this study was provided by Fellowships from the Alexander von Humboldt Foundation and the Carnegie Institution of Washington, as well as National Science Foundation grants EAR-0440112 (to SDJ) and EAR-0337611 (to JRS), the Carnegie/DOE Alliance Center, and the Bayerisches Geoinstitut Visitors Program.

REFERENCES

Bass, J.D., and D.L. Anderson, Composition of the upper mantle: Geophysical tests of two petrological models, *Geophys. Res. Lett., 11,* 237–240, 1984.

Bell, D., G. Rossman, J. Maldener, D. Endisch, and F. Rauch, Hydroxide in olivine: a quantitative determination of the absolute amount and calibration of the IR spectrum: *J. Geophys. Res., 108,* 2105, doi:10.1029/2001JB000679, 2003.

Bercovici, D., and S.-i. Karato, Whole-mantle convection and the transition-zone water filter, *Nature, 425,* 39–44, 2003.

Blanchard, M., K. Wright, and J.D. Gale, A computer simulation of OH defects in Mg_2SiO_4 and Mg_2GeO_4 spinels, *Phys. Chem. Minerals, 32,* 585–593, 2005.

Blum, J., and Y. Shen, Thermal, hydrous, and mechanical states of the mantle transition zone beneath southern Africa, *Earth Planet. Sci. Lett., 217,* 367–378, 2004.

Bolfan-Casanova, N., H. Keppler, and D.C. Rubie, Water partitioning between nominally anhydrous minerals in the $MgO-SiO_2$-H_2O system up to 24 GPa: implications for the distribution of water in the Earth's mantle, *Earth Planet. Sci. Lett., 182,* 209–221, 2000.

Bolfan-Casanova, H. Keppler, and D.C. Rubie, Water partitioning at 660 km depth and evidence for very low water solubility in magnesium silicate perovskite, *Geophys. Res. Lett., 30,* 1905, doi:10.1029/2003GL017182, 2003.

Chen, W.P., and M.R. Brudzinski, Seismic anisotropy in the mantle transition zone beneath Fiji-Tonga, *Geophys. Res. Lett., 30,* 1682, doi:10.1029/2002GL016330, 2003.

Chen, J., T. Inoue, H. Yurimoto, and D.J. Weidner, Effect of water on olivine-wadsleyite phase boundary in the $(Mg,Fe)_2SiO_4$ system, *Geophys. Res. Lett., 29,* 1875, doi:1810.1029/2001GL014429, 2002.

Davies, G.F., and A.M. Dziewonski, Homogeneity and constitution of the Earth's lower mantle and outer core, *Phys. Earth Planet. Int., 10,* 336–343, 1975.

Duffy, T.S. and D.L. Anderson, Seismic velocities in mantle minerals and the mineralogy of the upper mantle, *J. Geophys. Res., 94,* 1895–1912, 1989.

Dziewonski, A.M. and D.L. Anderson, Preliminary reference earth model, *Phys. Earth Planet. Int., 25,* 297–356, 1981.

Frost, D.J., B.T. Poe, R.G. Trønnes, C. Liebske, A. Duba, and D. C. Rubie, A new large-volume multianvil system, *Phys. Earth Planet. Int., 143–144,* 507–514, 2004.

Hashin, Z. and A. Shtrikman, A variational approach to the theory of the elastic behavior of polycrystals, *J. Mech. Phys. Solids, 29,* 81–95, 1962.

Hazen, R.M., Comparative crystal chemistry of silicate spinels: anomalous behavior of $(Mg,Fe)_2SiO_4$, *Science, 259,* 206–209, 1993.

Hill, R., The elastic behavior of a crystalline aggregate, *Proc. Phys. Soc. London, 64,* 349–354, 1952.

Hirschmann, M.M., C. Aubaud and A.C. Withers, Storage capacity of H_2O in nominally anhydrous minerals in the upper mantle, *Earth Planet. Sci. Lett., 236,* 167–181, 2005.

Hofmeister, A.M., Enhancement of radiative transfer in the upper mantle by OH^- in minerals, *Phys. Earth Planet. Int., 146,* 483–495, 2004.

Holl, C.M., J.R. Smyth, and S.D. Jacobsen, The effect of water on the compressibility of wadsleyite, *Eos Trans., 84(47),* Am. *Geophys. Union,* Abstract T11C-0414.

Huang, X., Y. Xu, and S. Karato, Water content of the mantle transition zone from the electrical conductivity of wadsleyite and ringwoodite, *Nature, 434,* 746–749, 2005.

Inoue, T., Y. Tanimoto, T. Irifune, T. Suzuki, H. Fukui, and O. Ohtaka, Thermal expansion of wadsleyite, ringwoodite, hydrous wadsleyite and hydrous ringwoodite, *Phys. Earth Planet. Int., 143–144,* 279–290, 2004.

Inoue, T., D.J. Weidner, P.A. Northrup, and J.B. Parise, Elastic properties of hydrous ringwoodite (γ-phase) in Mg_2SiO_4, *Earth Planet. Sci. Lett., 160,* 107–113, 1998.

Inoue, T., Y. Yurimoto, and T. Kudoh, Hydrous modified spinel, $Mg_{1.75}SiH_{0.5}O_4$: a new water reservoir in the mantle transition region, *Geophys. Res. Lett., 22,* 117–120, 1995.

Jackson, J.M., S.V. Sinogeikin, and J.D. Bass, Sound velocities and elastic properties of γ-Mg_2SiO_4 to 873 K by Brillouin spectroscopy, *Am. Mineral., 85,* 296–303, 2000.

Jackson, I., S. Webb, L. Weston, and D. Boness, Frequency dependence of elastic wave speeds at high temperature: a direct experimental demonstration, *Phys. Earth Planet. Int., 148,* 85–96, 2005.

Jacobsen, S.D., H.J. Reichmann, H.A. Spetzler, S.J. Mackwell, J.R. Smyth, R.J. Angel, and C.A. McCammon, Structure and elasticity of single-crystal (Mg,Fe)O and a new method of generating shear waves for gigahertz ultrasonic interferometry, *J. Geophys. Res., 107,* doi:10.1029/2001JB000490, 2002.

Jacobsen, S.D., J.R. Smyth, H. Spetzler, C.M. Holl, and D.J. Frost, Sound velocities and elastic constants of iron-bearing hydrous ringwoodite, *Phys. Earth Planet Int., 143–144,* 47–56, 2004.

Jacobsen, S.D., H.J. Reichmann, A. Kantor, and H. Spetzler, A gigahertz ultrasonic interferometer for the diamond-anvil cell and high-pressure elasticity of some iron-oxide minerals. *In:* J Chen et al., (Eds.) *Advances in High-Pressure Technology for Geophysical Applications,* Elsevier, Amsterdam, pp. 25–48, 2005.

Karato, S., The role of hydrogen in the electrical conductivity of the upper mantle, *Nature, 347,* 272–273, 1990.

Karato, S., Seismic anisotropy in the deep mantle, boundary layers and the geometry of mantle convection, *Pure Appl. Geophys., 151,* 565–587, 1998.

Karato, S., and H. Jung, Water, partial melting and the origin of the seismic low velocity and high attenuation zone in the upper mantle, *Earth Planet. Sci. Lett., 157,* 193–207, 1998.

Karato, S., M.S. Paterson, and J.D. Fitz Gerald, Rheology of synthetic olivine aggregates – influence of grain-size and water, *J. Geophys. Res., 91,* 8151–8176, 1986.

Karki, B.B., R.M. Wentzcovitch, S. de Gironcoli, and S. Baroni, First-principles determination of elastic anisotropy and wave velocities of MgO at lower mantle conditions, *Science, 286,* 1701–1705, 1999.

Katsura, T., S. Yokoshi, M. Song, K. Kawabe, T. Tsujimura, A. Kubo, E. Ito, Y. Tange, N. Tomioka, K. Saito, A. Nozawa, and K.-i. Funakoshi, Thermal expansion of Mg_2SiO_4 ringwoodite at high pressure, *J. Geophys. Res., 109,* B12209, doi:10.1029/2004JB003094, 2004.

Kavner, A., Elasticity and strength of hydrous ringwoodite at high pressure, *Earth Planet. Sci. Lett., 214,* 645–654, 2003.

Kawamoto, T., R.L. Hervig, and J.R. Holloway, Experimental evidence for a hydrous transition zone in the early Earth's mantle, *Earth Planet. Sci. Lett., 142,* 587–592, 1996.

Kennett, B.L.N., and E.R. Engdahl, Travel times for global earthquake location and phase identification. *Geophys. J. Int., 105,* 429–465, 1991.

Keppler, H., and J.R. Smyth, Optical and near infrared spectra of ringwoodite to 21.5 GPa: Implications for radiative heat transport in the mantle, *Am. Mineral., 90,* 1209–1212, 2005.

Kiefer, B., L. Stixrude, and R.M. Wentzcovitch, Calculated elastic constants and anisotropy of Mg_2SiO_4 spinel at high pressure, *Geophys. Res. Lett., 24,* 2841–2844, 1997.

Kohlstedt, D.L., H. Keppler, and D.C. Rubie, Solubility of water in the α, β, and γ phases of $(Mg,Fe)_2SiO_4$, *Contr. Min. Pet., 123,* 345–357, 1996.

Kudoh, Y., T. Inoue, and H. Arashi, Structure and crystal chemistry of hydrous wadsleyite, $Mg_{1.75}SiH_{0.5}O_4$: possible hydrous magnesium silicate in the mantle transition zone, *Phys. Chem. Minerals, 23,* 461–469, 1996.

Kudoh, Y., T. Kuribayashi, H. Mizobata, and E. Ohtani, Structure and cation disorder of hydrous ringwoodite, $γ-Mg_{1.89}Si_{0.98}H_{0.30}O_4$, *Phys. Chem. Minerals, 27,* 474–479, 2000.

Li, B., Compressional and shear wave velocities of ringwoodite $γ-Mg_2SiO_4$ to 12 GPa, *Am. Mineral, 88,* 1312–1317, 2003.

Manghnani, M.H., G. Amulele, J.R. Smyth, C.M. Holl, G. Chen, V. Prakapenka, and D.J. Frost, Equation of state of hydrous Fo_{90} ringwoodite to 45 GPa by synchrotron powder diffraction, *Min. Mag., 69,* 317–323, 2005.

Mayama, N., I. Suzuki, T. Saito, I. Ohno, T. Katsura and A. Yoneda, Temperature dependence of the elastic moduli of ringwoodite, *Phys. Earth Planet. Int., 148,* 353–359, 2005.

Mao, H.K., J. Xu, and P.M. Bell, Calibration of the ruby pressure gauge to 800 kbar under quasi-hydrostatic conditions, *J. Geophys. Res., 91,* 4673–4676, 1986.

McCammon, C.A., D.J. Frost, J.R. Smyth, H.M.S. Lausten, T. Kawamoto, N.L. Ross, and P.A. Van Aken, Oxidation state of iron in hydrous mantle phases: implications for subduction and mantle oxygen fugacity, *Phys. Earth Planet. Int., 143–144,* 157–169, 2004.

Mei, S., and D.L. Kohlstedt, Influence of water on plastic deformation of olivine aggregates: 1. Diffusion creep regime, *J. Geophys. Res., 105,* 21457–21469, 2000.

Meng, Y., Y. Fei, D.J. Weidner, G.D. Gwanmesia, and J. Hu, Hydrostatic compression of $γ-Mg_2SiO_4$ to mantle pressures and 700 K: Thermal equation of state and related thermoelastic properties, *Phys. Chem. Minerals, 21,* 407–412, 1994.

Nye, J.F., *Physical Properties of Crystals,* Oxford University Press, London, pp. 322, 1972.

Nishihara, Y.E., K. Takahashi, K. Matsukage, T. Iguchi, K. Nakayama, and K. Funakoshi, Thermal equation of state of $(Mg_{0.91}Fe_{0.09})_2SiO_4$ ringwoodite, *Phys. Earth Planet. Int., 143–144,* 33–46, 2004.

Nishiyama, N., Y. Wang, T. Uchida, T. Irifune, M.L. Rivers, and S.R. Sutton, Pressure and strain dependence of the strength of sintered polycrystalline Mg_2SiO_4 ringwoodite, *Geophys. Res. Lett., 32,* doi: 10.1029/2004GL022141, 2005.

Ohtani, E., K. Litasov, T. Hosoya, T. Kubo and T. Kondo, Water transport into the deep mantle and formation of a hydrous transition zone, *Phys. Earth Planet. Int., 143–144,* 255–269, 2004.

Paterson, M.S., The determination of hydroxyl by infrared absorption in quartz, silicate glasses and similar materials, *Bull. Mineral., 105,* 20–29, 1982.

Poirier, J.P., *Introduction to the Physics of the Earth's Interior* (2nd edition), Cambridge University Press, New York, pp. 312, 2000.

Reichmann, H.J., and S.D. Jacobsen, Sound wave velocities and elastic constants of $ZnAl_2O_4$ spinel and implications for spinel-elasticity systematics, *Am. Mineral., 91,* 1049–1054, 2006.

Richard, G., M. Monnerau, and J. Ingrin, Is the transition zone an empty water reservoir? Inferences from numerical model of mantle dynamics, *Earth Planet. Sci. Lett., 205,* 37–51, 2002.

Ringwood, A.E., Composition and origin of the Earth, *In:* Hurley, P.M. (Ed.) *Advances in Earth Science,* M.I.T. Press, Cambridge, pp. 287–356, 1966.

Ringwood, A.E., *Composition and Petrology of the Earth's Mantle,* McGraw-Hill, New York, pp. 618, 1975.

Sinogeikin, S.V., and J.D. Bass, Single-crystal elasticity of pyrope and MgO to 20 GPa by Brillouin scattering in the diamond cell, *Phys. Earth Planet. Inter., 120,* 43–62, 2000.

Sinogeikin, S.V., J.D. Bass, and T. Katsura, Single-crystal elasticity of ringwoodite to high pressures and high temperatures: implications for 520 km seismic discontinuity, *Phys. Earth Planet. Int., 136,* 41–66, 2003.

Sinogeikin, S.V., T. Katsura, and J.D. Bass, Sound velocities and elastic properties of Fe-bearing wadsleyite and ringwoodite, *J. Geophys. Res., 103,* 20819–20825, 1998.

Smyth, J.R., $β-Mg_2SiO_4$: a potential host for water in the mantle? *Am. Mineral., 72,* 1051–1055, 1987.

Smyth, J.R., A crystallographic model for hydrous wadsleyite $(β-Mg_2SiO_4)$: An ocean in the Earth's interior? *Am. Mineral., 79,* 1021–1024, 1994.

Smyth, J.R., and D.J. Frost, Effect of water on olivine-wadsleyite phase boundary in the (Mg,Fe)$_2$SiO$_4$ system, *Geophys. Res. Lett., 29,* 1875, doi:10.1029/2001GL014429, 2002.

Smyth, J.R., D.J. Frost, and F. Nestola, Hydration of olivine and the Earth's deep water cycle, *Geochim Cosmochim Acta, 69,* A746, 2005.

Smyth, J.R., C.M. Holl, D.J. Frost, and S.D. Jacobsen, High pressure crystal chemistry of hydrous ringwoodite and water in the Earth's interior, *Phys. Earth Planet. Int., 143–144,* 271–278, 2004.

Smyth, J.R., C.M. Holl, D.J. Frost, S.D. Jacobsen, F. Langenhorst, and C.A. McCammon, Structural systematics of hydrous ringwoodite and water in Earth's interior, *Am. Mineral., 88,* 1402–1407, 2003.

Song, T.-R.A., D.V. Helmberger, and S. Grand, Low velocity zone atop the 410 seismic discontinuity beneath the northwestern US, *Nature, 427,* 530–533, 2004.

Spetzler, H.A., G. Chen, S. Whitehead, and I.C. Getting, A new ultrasonic interferometer for the determination of equation of state parameters of sub-millimeter single crystals, *Pure Appl. Geophys., 141,* 341–377, 1993.

Suzuki, I., E. Ohtani, and M. Kumazawa, Thermal expansion of γ-Mg$_2$SiO$_4$, *J. Phys. Earth, 27,* 53–71, 1979.

van der Meijde, M., F. Marone, D. Giardini, and S. van der Lee, Seismic evidence for water deep in Earth's upper mantle, *Science, 300,* 1556–1558, 2003.

Wang, J., S.V. Sinogeikin, T. Inoue, and J.D. Bass, Elastic properties of hydrous ringwoodite, *Am. Mineral., 88,* 1608–1611, 2003.

Weidner, D.J., and G.D. Price, The effect of many-body forces on the elastic properties of simple oxides and olivine, *Phys. Chem. Minerals, 16,* 42–50, 1988.

Weidner, D.J., H. Sawamoto, S. Sasaki, and M. Kumazawa, Single-crystal elastic properties of the spinel phase of Mg$_2$SiO$_4$, *J. Geophys. Res., 89,* 7852–7860, 1984.

Wood, B.J., The effect of H$_2$O on the 410-kilometer seismic discontinuity, *Science, 268,* 74–76, 1995.

Yoneda, A., Pressure derivatives of elastic constants of single crystal MgO and MgAl$_2$O$_4$, *J. Phys. Earth, 38,* 19–55, 1990.

Yusa, H., T. Inoue, and Y. Ohishi, Isothermal compressibility of hydrous ringwoodite and its relation to the mantle discontinuities, *Geophys. Res. Lett., 27,* 413–416, 2000.

Zerr, A., H.J. Reichmann, H. Euler, and R. Boehler, hydrostatic compression of γ-(Mg$_{0.6}$Fe$_{0.4}$)$_2$SiO$_4$ to 50 GPa, *Phys. Chem. Minerals, 19,* 507–509, 1993.

High-Pressure and High-Temperature Stability and Equation of State of Superhydrous Phase B

Toru Inoue[1], Takayuki Ueda[1], Yuji Higo[1],
Akihiro Yamada[1], Tetsuo Irifune[1], and Ken-ichi Funakoshi[2]

We have determined the stability and *P-V-T* equation of state of superhydrous phase B, $Mg_{10}Si_3O_{18}H_4$ (hereafter referred to as Shy-B) by in situ X-ray diffraction measurements at SPring-8 at pressure between 13 and 23 GPa and temperatures up to 1400°C. High temperature X-ray diffraction measurements using a conventional X-ray source were also conducted to determine the thermal expansion coefficient at atmospheric pressure. Shy-B was found to melt incongruently into assemblies of wadsleyite (β) + liquid, or ringwoodite (γ) + liquid, depending on pressure at temperatures of 1000-1250°C. The isothermal bulk modulus (K_{T0}) and temperature derivative (*dK/dT*) of Shy-B are 132(1) GPa and $-2.5(1) \times 10^{-2}$ GPa/K, respectively, when we adopted an equation of state of Au proposed by Anderson et al. [1989] and also assumed its pressure derivative is 5.8 proposed by Crichton et al. [1999], while the thermal expansion coefficient (α_0) is $37.6(1) \times 10^{-6}$ K^{-1}. Phase relations imply that Shy-B may be stable in relatively low temperature regions of the mantle transition zone, such as within older subducted slabs, but not in the surrounding mantle. The density difference between model subducted hydrous slabs and the surrounding mantle was calculated at 20 GPa as functions of water content and temperature in the slab using the present results, which demonstrates that the Shy-B bearing slabs with H_2O contents of 0.5 wt%, 1.0 wt % and 1.5 wt% are denser than the surrounding mantle at temperatures below 1200°C, 900°C and 600°C, respectively. This implies that the possible penetration of a hydrous slab into the deeper mantle depends critically on the water content and temperature profile of the slab.

1. INTRODUCTION

Water is one of the main volatile components in the Earth. The influence of water on phase relations, physical properties and melting temperatures of minerals plays a key role in Earth's potential deep water cycle [*e.g. Karato et al., 1986;*

Inoue, 1994; Kawamoto et al., 1996; Yusa and Inoue, 1997; Irifune et al, 1998; Inoue et al, 1998; Chen et al., 1998;, Komabayashi et al., 2004; Ohtani et al, 2004; Huang et al., 2005]. Water can be transported into the Earth's interior via subducted slabs in the form of dense hydrous magnesium silicates (DHMSs), such as phase A, phase B, phase D, phase E, Shy-B, hydrous wadsleyite and hydrous ringwoodite [*e.g. Ringwood and Major*, 1967; *Liu*, 1986, 1987; *Kanzaki*, 1991; *Gasparik*, 1993; *Inoue et al.*, 1995; *Kohlstedt et al.*, 1996; *Ohtani et al.*, 1997; *Inoue et al.*, 1998; *Irifune et al.*, 1998; *Frost and Fei*, 1998]. Shy-B ($Mg_{10}Si_3H_4O_{18}$) is one of the most important DHMS, which may exist stably in low temperature regions of the subducted slabs at pressures equivalent to the mantle transition zone [*e.g. Gasparik,*

[1]Geodynamics Research Center, Ehime University, Matsuyama, Ehime, Japan
[2]Japan Synchrotron Radiation Research Institute, Sayo, Hyogo, Japan

Earth's Deep Water Cycle
Geophysical Monograph Series 168
Copyright 2006 by the American Geophysical Union.
10.1029/168GM11

1993; *Irifune et al.*, 1998; *Ohtani et al.*, 2001]. Shy-B has a zero-pressure density of 3.327 g/cm³ and contains 5.8 wt% water [*Pacalo and Parise*, 1992]. The stability of Shy-B was determined by conventional quench experiments [*e.g. Gasparik*, 1993; *Ohtani et al.*, 2001], and the high pressure decomposition of Shy-B was reported by in situ X-ray observation [*Ohtani et al.*, 2003], but no in situ X-ray experiments within the high temperature stability field of Shy-B have been conducted due to the difficulty in confining water in the system. Metal capsules normally used in quench experiments, such as Pt and Au, significantly reduce the signal to noise ratio of the X-ray diffraction profile. At room temperature, the compressibility of Shy-B was determined by *Crichton et al.* [1999] to 7.7 GPa, but the thermal expansivity and temperature derivative of bulk modulus have not been clarified to date.

In this study, we introduced a thin AgPd capsule for in situ X-ray diffraction measurement of the hydrous phase at high pressure and temperature, as this metal is highly transparent to X-rays and is also easily sealed by welding. Using this technique, *Sano et al.* [2004] has already reported the decomposition of hydrous aluminum silicate AlSiO₃OH by in situ X-ray experiments. Here, we investigated the stability field and some thermoelastic properties, such as bulk modulus, its temperature derivative, and the thermal expansion coefficient, of Shy-B. Combining the present results with the previously reported physical properties of ringwoodite and majorite garnet, we calculated the density differences between model hydrous slab peridotites containing Shy-B and the surrounding mantle as functions of water content and temperature. On the basis of these estimated density contrasts, we discuss the role of Shy-B in subduction behaviors of the model hydrous slabs in the deep mantle.

2. EXPERIMENTS

Shy-B was synthesized using the MA-8 type high pressure apparatus (Orange-2000) at GRC, Ehime University. A mixture of MgO, Mg(OH)$_2$ and SiO$_2$ reagents in a molar ratio of 8:2:3 was used as the starting material to obtain stoichiometry identical to that of Shy-B (Mg$_{10}$Si$_3$O$_{18}$H$_4$). The sample was sealed in a Pt capsule, and held at 18 GPa and 1000°C for 90 minutes. The recovered sample was examined by powder X-ray diffraction and found to be a pure Shy-B, which was used as the starting materials for in situ X-ray diffraction measurements.

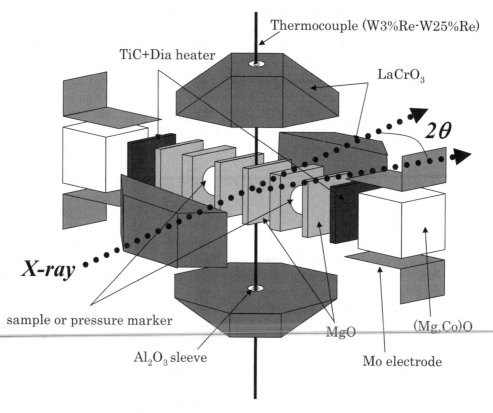

Figure 1. Schematic illustration of the cell assembly used for the present in situ X-ray experiments. X-ray path is shown by dotted arrows.

In situ high pressure and high temperature X-ray experiments were performed using SPEED-1500, installed at the BL04B1 beamline of SPring-8 [*Utsumi et al.*, 1998]. A beam of white synchrotron X-ray radiation with dimensions of 50 x 200 μm was directed to the sample via horizontal and vertical slits through the anvil gap of second stage anvils, and the diffracted beam was collimated by a 50 μm horizontal slit and detected with a pure Ge solid state detector, together with a CCD camera for radiographic imaging of the sample. A 4096 channel analyzer was used to acquire photons in a range of 30-150 keV. The diffraction angle was fixed to about 6 degrees, which was calibrated using diffraction peaks of Au and MgO under ambient conditions. The uncertainty of the diffraction angle was typically ±0.002 degree in each experiment.

Fig. 1 shows the cell assembly used in the present experiment. Tungsten carbide (WC) cubes of 26 mm edge length and the truncations of 3.0 mm (TEL=3.0 mm) were used as second stage anvils. We placed the hot junction of a $W_{97\%}Re_{3\%}$-$W_{75\%}Re_{25\%}$ thermocouple in contact with the sample and a pressure marker with thin MgO sheets. An Al_2O_3 sleeve was used to prevent the thermocouple wires from contacting the surrounding $LaCrO_3$ pressure medium, and the coils of thinner wires of $W_{97\%}Re_{3\%}$ and $W_{75\%}Re_{25\%}$ were used to avoid the thermocouple failure at the boundary between pressure medium and pyrophyllite gasket under compression. The temperature fluctuation was kept within ±1°C throughout heating under pressure.

Mo foils of 50 μm thickness were used as leads for electric power supply to twin TiC plus diamond (1:1 by weight) sheet heaters. The pressure medium was made of semi-sintered $LaCrO_3$ and MgO doped with 13 wt% of CoO. We used pyrophyllite gaskets of 2.9 mm thickness and 4 mm width with cemented amorphous boron rods for X-ray windows.

Pt or AuPd are normally used as capsules for hydrous experiment because they are inert and do not react with most mineral samples under investigation. These metals, however, have large atomic weights and X-ray absorption edges at around 80 keV, which hinders X-ray diffraction requiring sufficient quality and intensity from the encapsuled sample for equation of state studies. In the present experiment, we adopted an Ag70%Pd30% capsule with a wall thickness of 50 μm and outer diameter of 2.0 mm for in situ X-ray diffraction measurements, because it is easily sealed by welding and its atomic weight is substantially smaller than those of Pt or AuPd. In addition, the absorption edges of AgPd are located around 25 keV, which would not interfere the major diffraction signals between 40-100 keV in our experiments. Fig. 2 shows a CCD image of the AgPd capsule and adjacent thermocouple junction through a circular boron window. The sample inside the AgPd capsule can be imaged by the bright synchrotron

Figure 2. Transmission X-ray image of the AgPd capsule (left) and thermocouple wire (middle) imaged through a cylindrical boron window in pyrophyllite gasket before compression. The square shadow area on the right is the pressure marker of Au + MgO powders. The sample in the capsule can be seen by both X-ray imaging and diffraction.

X-ray beam, which also enables us to obtain X-ray diffraction from the sample through the capsule in situ.

We used a mixture of Au and MgO (1:20 by weight) powders as pressure markers, which were directly enclosed in the pressure medium next to the sample (Fig. 1). The pressure was calculated using an equation of state of Au proposed by *Anderson et al.* [1989]. Although there have been considerable debates on the accuracy of pressure scales and it is suggested that the Anderson's scale of gold may underestimate the pressures under the *P-T* conditions of the mantle transition region [*e.g. Matsui and Nishiyama*, 2002; *Shim et al.*, 2002; *Tsuchiya*, 2003; *Fei et al.*, 2004], we used this scale for consistency with earlier *P-V-T* studies on some major high pressure phases [*e.g. Funamori et al.*, 1996; *Wang et al.*, 1996]. Moreover such a possible underestimation of pressure is not very significant at pressures below 20 GPa and temperatures below 1200°C, where most of the present in situ observations were conducted. Nevertheless, we listed unit-cell volume data of Au for future re-evaluation of the pressures using more appropriate equations of state, and the pressure calculated based on another equation of state of Au proposed by *Shim et al.* (2002) was also shown for comparison. In addition, we also listed unit-cell volume data of MgO and the calculated pressures based on equations of state of MgO proposed by *Jamieson et al.* (1982) and *Matsui et al.* (2000) for comparison. It should be also noted that the pressures based on the Anderson scale of Au are lower than those calculated with the latest MgO scale of *Matsui et al.* [2000] by 0.5-1.5 GPa for the P, T conditions of the present study (Table. 2). The same kinds of discussion have been made in *Matsui and Nishiyama* (2002), *Shim et al.* (2002), *Tsuchiya* (2003), *Fei et al.* (2004) and *Inoue et al.* (2006).

We estimated temperature gradient using the Au pressure maker, when we assumed that the pressure is the same inside the cell assembly. The estimated temperature gradient was about 100°C /500 μm. We collected the diffractions of the pressure marker and the sample just close to thermocouple junction, so this temperature gradient does not mean the measured temperature uncertainty.

In the present in situ X-ray experiments, pressure was applied first by increasing the press load to a target value, and then temperature was increased keeping the press load constant. X-ray diffraction data were acquired every 100°C for both the sample and the pressure marker from 700°C to 1400°C for about 600 seconds. The phases present were identified based on the obtained X-ray diffraction profiles in order to determine the stability field of Shy-B.

Table 1. Experimental conditions and the results of the phase stability runs

Run	Pressure (GPa)$	Temperature (°C)	Result#
S610	12.9 (1)	1100	phB+β+melt (+Shy-B)
S608	16.8 (1)	1400	β+melt (+Shy-B)
S631	17.8 (1)	1200	γ+melt (+Shy-B)
S629	18.5 (1)	1400	γ+melt (+Shy-B)
S580*	18.7 (3)	1400	γ+melt (+Shy-B)
S667	20.0 (2)	1300	γ+melt (+Shy-B)

*Temperature was estimated by power supply because of thermocouple failure.

$The pressure at the highest temperature in each run. The number in parentheses shows 1σ of each pressure.

#Shy-B was found in the low temperature portion of the sample charge in all run. Shy-B : superhydrous phaseB, β: wadsleyite γ: ringwoodite, phB : phaseB ($Mg_{12}Si_4O_{21}H_2$)

Two additional experiments (S692 and S722) were performed to study the equation of state of Shy-B at pressures to 23 GPa and temperatures to 900°C. After the desired press load was achieved, temperature was increased to 900°C to reduce deviatric stress within the sample, and the X-ray diffraction data was acquired every 200°C from 900°C down to 100°C and finally at room temperature (27°C). All data acquisitions were made inside the stability field of Shy-B. Nine clear diffraction peaks of Shy-B were indexed as (132), (051), (142), (221), (133), (231), (241), (091) and (303), were used to calculate the lattice constants at high pressure and high temperature.

The thermal expansion coefficient of Shy-B was also determined by high temperature X-ray diffraction measurements at atmospheric pressure, using Rigaku PINT-2000 instruments equipped with a high temperature furnace at The University of Kitakyushu. Cu K α_1 radiation (1.5405Å) was used as the X-ray source, and no inert gas such as Ar was used in the present experiments. The powdered sample was mixed with a powder of α-Al_2O_3 (corundum) in the same volume proportions as the sample, which was used as an internal standard to calibrate the X-ray optical system. Thermal expansion data for corundum of *Aldebert and Traverse* [1984] were used as references, and six diffraction lines ((104), (110), (113), (024), (030) and (124)) were used to calculate the lattice parameters of corundum.

Temperature was measured by Pt-Rh thermocouple, which was calibrated on the basis of the melting temperatures of some reference materials such as In (156.6°C), Sn (232.0°C) and Pb (327.2°C). Temperature uncertainty was less than 1°C in the present X-ray diffraction measurements. Twenty-five diffractions of Shy-B were used to calculate the lattice constants at high temperature and at the ambient pressure.

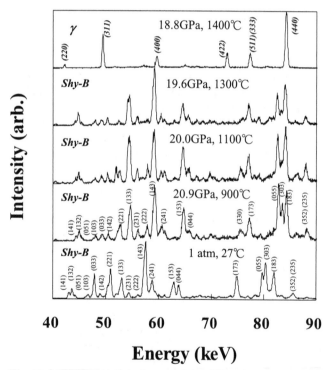

Figure 3. Representative change of the diffraction pattern with increasing temperature at around 19–20 GPa. The hkl indexes of Shy-B are shown in normal letters in parenthesis, and the hkl indexes of ringwoodite are shown in bold and italic letters in parenthesis. No AgPd diffractions were observed because the diffraction area (~1.0 mm) was much smaller than sample diameter (1.9 mm in initial diameter, which is the inner diameter of AgPd capsule). γ, ringwoodite; Shy-B, superhydrous phase B.

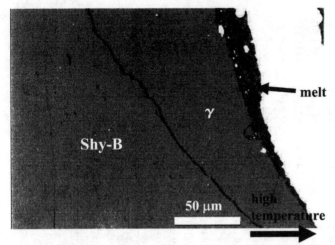

Figure 4. Back scattered electron image of the sample from run S629 (18.5GPa,1400°C). γ, ringwoodite; Shy-B, superhydrous phase B.

3. RESULTS AND DISCUSSION

Six in situ X-ray diffraction measurements were conducted at pressures of 13–20 GPa at temperatures up to 1400°C to determine the stability of Shy-B as shown in Table 1. The results in this table represent the phases present at the final stage of the heating under pressure, immediately before quenching. A representative variation of the diffraction profiles with increasing temperature is shown in Fig. 3. Below 1300°C at 19.6 GPa, only the diffraction peaks of Shy-B were identified, whereas the peaks of Shy-B completely disappeared and those of ringwoodite were observed at 1400°C and 18.8 GPa. After confirming the disappearance of Shy-B, the charge was quenched, because it was often difficult to see the ideal powder diffraction pattern of the run products after a few tens of minutes due to crystal growth at such a high temperature.

An example of the recovered run product is shown in Fig. 4. On the high temperature side of the sample near the thermocouple hot junction, where the X-ray diffraction data were acquired, crystals of ringwoodite were recognized with a small amount of melt, now converted to quench crystals. Because of the temperature gradient in the capsule, however, only Shy-B was identified on the lower temperature side, suggesting that this phase melts incongruently into ringwoodite and liquid at pressures around 18.5 GPa.

Fig. 5 summarizes the results of the present in situ X-ray diffraction measurements in a P-T diagram. At 13 GPa, Shy-B melted incongruently to an assemblage of phase B plus liquid and then to that of wadsleyite plus liquid. The identification of phase B was difficult by only X-ray observation because of similarity of crystal structure with Shy-B. Nevertheless, we observed phase B in recovered sample of

run S610, so we judged that phase B existed in 1100°C at around 13 GPa. Further studies should be needed to clarify the stability region of phase B in this system.

The solidus temperature was around 1000°C at 13 GPa. In contrast, Shy-B melted incongruently to form ringwoodite and liquid at around 1200–1300°C at 18–20 GPa. Thus the solidus of Shy-B should have a positive Clapeyron slope of dP/dT ~0.03 GPa/K at a pressure range between 13–20 GPa.

Based on the present results on the stability field of Shy-B combined with those of an earlier study [*Gasparik*, 1993], two in situ X-ray experiments were performed to determine the equation of state of Shy-B within its stability field. The obtained lattice constants and the unit cell volume at each pressure and temperature are listed in Table 2. We also used same data obtained in the phase stability runs at 700°C and 900°C, because the diffraction peaks were as sharp as those observed in the P-V-T runs (S692 and S722) at these temperatures, suggesting that the effect of the deviatric stress is insignificant under these conditions.

The unit cell volume data were fitted to the Birch-Murnaghan equation of state at room temperature (27°C) to calculate the bulk modulus of Shy-B. The results of the least-square fitting are shown in Fig. 6, where the previous data by *Crichton et al.* [1999] using diamond anvil cell are also shown for comparison. We fixed the pressure derivative of bulk modulus as 5.8 according to *Crichton et al.* [1999]. The obtained bulk modulus was 132.2(9) GPa, which was slightly lower than that of *Crichton et al.* [1999] (K=142.6(8) GPa). This difference may come from the selected pressure standards and the EOS of both studies, because *Crichton et al.* [1999] used the equation of state of quartz reported by *Angel et al.* [1997] to calculate the pressure.

The measurement of the unit-cell volume change in Shy-B was conducted at high temperatures up to 500°C at

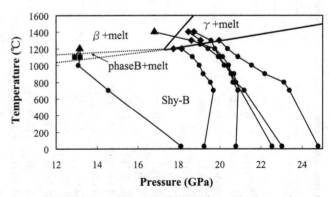

Figure 5. High temperature stability of superhydrous phase B at pressures of 13–20 GPa determined by in situ X-ray diffraction. β, wadsleyite; γ, ringwoodite; Shy-B, superhydrous phase B. circle, Shy-B; square, phase B+melt; triangle, β+melt; diamond, γ+melt.

Table 2. Lattice constants and the unit-cell volumes of superhydrous phase B at each P-T condition

Run No.	Au V(Å³)	P_Anderson (GPa)	P_Shim (GPa)	MgO V(Å³)	P_Jamieson (GPa)	P_Mastui (GPa)	Temp. (°C)	a-axis (Å)	b-axis (Å)	c-axis (Å)	Volume (Å³)
S580006	67.85(1)	1atm	1atm	74.78(4)	1atm	1atm	27	5.102(2)	13.992(4)	8.748(8)	624.4(7)
S581002	67.85(1)	1atm	1atm	74.78(2)	1atm	1atm	27	5.100(2)	13.998(3)	8.735(7)	623.6(6)
S608004	67.85(3)	1atm	1atm	74.78(5)	1atm	1atm	27	5.107(1)	13.995(2)	8.733(5)	624.2(4)
S610002	67.85(3)	1atm	1atm	74.78(4)	1atm	1atm	27	5.108(1)	14.000(2)	8.722(4)	623.9(3)
S629002	67.85(1)	1atm	1atm	74.78(2)	1atm	1atm	27	5.110(1)	13.999(3)	8.727(7)	624.1(6)
S631002	67.85(1)	1atm	1atm	74.78(3)	1atm	1atm	27	5.107(2)	14.004(4)	8.745(10)	625.4(8)
S667002	67.85(4)	1atm	1atm	74.78(4)	1atm	1atm	27	5.106(1)	14.000(3)	8.752(7)	625.7(6)
S692002	67.85(2)	1atm	1atm	74.78(4)	1atm	1atm	27	5.110(1)	13.995(3)	8.739(5)	624.9(4)
S722010	67.85(3)	1atm	1atm	74.78(4)	1atm	1atm	27	5.102(1)	13.996(2)	8.743(5)	624.3(4)
S749002	67.85(2)	1atm	1atm	74.78(3)	1atm	1atm	27	5.104(2)	14.004(4)	8.734(9)	624.2(7)
S722072	63.53(6)	13.13(24)	12.95	69.32(3)	13.65(10)	14.22	27	4.961(1)	13.632(3)	8.517(6)	576.0(5)
S692021	63.04(2)	14.98(7)	14.76	68.91(4)	15.23(11)	15.52	27	4.955(1)	13.623(3)	8.504(6)	574.1(5)
S722028	62.94(5)	15.36(19)	15.14	68.64(5)	16.08(15)	16.40	27	4.947(1)	13.593(3)	8.496(7)	571.3(5)
S722070	62.91(6)	15.49(23)	15.26	68.57(3)	16.33(9)	16.63	27	4.943(1)	13.586(3)	8.486(6)	569.9(5)
S722055	62.43(6)	17.44(24)	17.13	67.95(3)	18.09(11)	18.72	27	4.928(1)	13.547(4)	8.466(7)	565.2(6)
S722030	62.10(4)	18.80(18)	18.46	67.72(4)	19.21(13)	19.52	27	4.927(1)	13.535(3)	8.429(9)	562.1(6)
S692018	63.15(1)	15.03(4)	14.90	69.01(3)	15.21(8)	15.57	100	4.956(1)	13.622(1)	8.509(2)	574.4(1)
S722025	63.03(4)	15.49(16)	15.35	68.76(5)	15.99(17)	16.37	100	4.949(3)	13.599(5)	8.489(10)	571.3(8)
S722067	62.97(4)	15.70(17)	15.57	68.64(4)	16.40(11)	16.76	100	4.943(2)	13.586(4)	8.492(7)	570.2(6)
S722054	62.43(5)	17.86(19)	17.67	68.01(3)	18.51(11)	18.87	100	4.929(1)	13.547(3)	8.463(6)	565.1(4)
S692017	63.51(7)	14.92(25)	15.04	69.19(4)	15.63(13)	16.10	300	4.955(2)	13.624(3)	8.502(7)	574.0(5)
S722024	63.22(4)	16.01(14)	16.10	68.93(4)	16.44(13)	16.92	300	4.951(1)	13.604(2)	8.498(4)	572.4(3)
S722051	62.57(4)	18.56(16)	18.57	68.17(4)	18.93(13)	19.41	300	4.931(1)	13.559(2)	8.469(4)	566.2(3)
S722021	63.36(3)	16.77(12)	17.02	69.14(4)	16.86(12)	17.44	500	4.951(1)	13.613(3)	8.505(6)	573.2(5)
S722050	62.69(4)	19.29(16)	19.54	68.28(3)	19.64(10)	20.22	500	4.934(1)	13.560(2)	8.465(5)	566.4(4)
S722020	63.60(3)	17.17(10)	17.60	69.39(2)	17.22(6)	17.89	700	4.957(2)	13.620(4)	8.509(8)	574.5(6)
S722014	63.27(3)	18.37(10)	18.79	68.97(5)	18.49(14)	19.21	700	4.946(1)	13.605(3)	8.500(5)	572.0(4)
S631006	62.91(4)	19.71(14)	20.13	68.56(1)	19.81(5)	20.52	700	4.933(2)	13.574(4)	8.460(8)	566.5(8)
S722047	62.83(3)	20.03(10)	20.44	68.45(3)	20.20(11)	20.88	700	4.936(2)	13.569(3)	8.462(7)	566.8(5)
S608014	62.62(4)	20.81(17)	21.25	68.15(4)	20.91(13)	21.88	700	4.924(1)	13.554(2)	8.473(5)	565.5(4)
S629006	62.53(3)	21.19(14)	21.60	68.01(5)	21.69(17)	22.35	700	4.921(2)	13.543(4)	8.449(8)	563.1(6)
S722017	63.78(2)	17.83(7)	18.39	69.58(4)	17.79(11)	18.58	900	4.957(1)	13.631(2)	8.505(5)	574.6(4)
S722059	63.51(6)	18.76(23)	19.35	69.26(7)	18.77(22)	19.56	900	4.950(2)	13.613(4)	8.484(7)	571.7(6)
S631010	63.39(4)	19.19(15)	19.78	69.16(2)	19.07(4)	19.87	900	4.943(1)	13.586(2)	8.485(5)	569.8(4)
S722046	63.04(1)	20.46(5)	21.07	68.67(3)	20.49(12)	21.43	900	4.940(1)	13.575(3)	8.470(6)	568.0(4)
S608018	63.02(3)	20.57(11)	21.15	68.62(3)	20.33(13)	21.59	900	4.922(2)	13.586(3)	8.482(6)	567.2(5)
S629010	62.98(3)	20.71(10)	21.30	68.43(13)	21.41(50)	22.21	900	4.932(2)	13.563(3)	8.471(6)	566.6(5)
S722031	62.89(2)	21.03(9)	21.64	68.52(3)	21.05(11)	21.92	900	4.934(2)	13.568(4)	8.461(8)	566.4(6)

( ) : standard deviation

Table 3. Lattice constants and the unit-cell volumes of superhydrous phase B at high temperature and ambient pressure

Run No.	Temp. (°C)	a-axis (Å)	b-axis (Å)	c-axis (Å)	Volume (Å^3)
Shy-B-27	27	5.1033 (5)	13.9975 (11)	8.7253 (7)	623.28 (9)
Shy-B-50	50	5.1056 (6)	14.0041 (11)	8.7251 (7)	623.84 (10)
Shy-B-100	100	5.1086 (6)	14.0100 (13)	8.7314 (8)	624.92 (11)
Shy-B-150	150	5.1120 (5)	14.0180 (11)	8.7358 (7)	626.01 (9)
Shy-B-200	200	5.1160 (5)	14.0274 (11)	8.7396 (7)	627.19 (10)
Shy-B-250	250	5.1207 (6)	14.0346 (11)	8.7464 (7)	628.57 (10)
Shy-B-300	300	5.1230 (5)	14.0463 (10)	8.7513 (6)	629.74 (9)
Shy-B-350	350	5.1270 (4)	14.0563 (9)	8.7547 (6)	630.93 (8)
Shy-B-400	400	5.1294 (5)	14.0660 (10)	8.7607 (7)	632.08 (9)
Shy-B-450	450	5.1333 (6)	14.0754 (11)	8.7650 (7)	633.29 (10)
Shy-B-500	500	5.1368 (6)	14.0849 (13)	8.7696 (8)	634.50 (11)

( ): standard deviation

atmospheric pressure, and the results are shown in Table 3 and Fig. 7. Shy-B persisted at least up to 500°C without any notable anomalous changes in the lattice parameters. The thermal expansion coefficient at atmospheric pressure is α_0 = 37.6(1) x 10^{-6} /K, which is significantly larger than those of wadsleyite (34.0(5) x 10^{-6} /K) and ringwoodite (30.7(6) x 10^{-6} /K) [*Inoue et al.*, 2004].

We also determined the bulk modulus of Shy-B at high temperature using the Birch-Murnaghan equation of state, and the results are shown in Fig. 8. Fig. 9 shows the variation of K_0 as a function of temperature, yielding a temperature derivative of bulk modulus dK/dT of -0.025(1) GPa/K. This

value is quite close to those of wadsleyite (-0.027 GPa/K) and ringwoodite (-0.028 GPa/K) [*Meng et al.*, 1993].

4. GEOPHYSICAL IMPLICATIONS

We demonstrated that Shy-B was stable at temperatures below 1200°C and at pressures corresponding to the mantle transition zone. Because the temperatures of the transition zone is believed to be in the range of 1400–1600°C [*e.g. Brown and Shankland*, 1981; *Katsura and Ito*, 1989; *Akaogi et al.*, 1989; *Ito and Takahashi*, 1989], Shy-B may exist only in lower temperature regions of the mantle, such as within subducting slabs.

Using the present P-V-T equation of state of Shy-B and existing thermochemical data on other mantle minerals, we

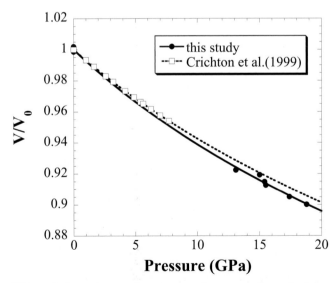

Figure 6. Room-temperature compression curve of superhydrous phase B up to 19 GPa based on the present unit-cell volume data (solid line). The volume data (open squares) and the compression curve (broken line) reported by Crichton et al. (1999) is also shown for comparison.

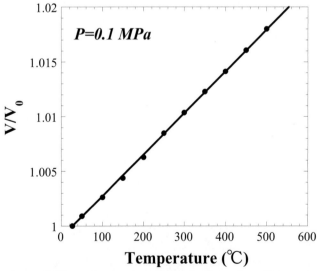

Figure 7. Thermal expansion of superhydrous phase B at atmospheric pressure.

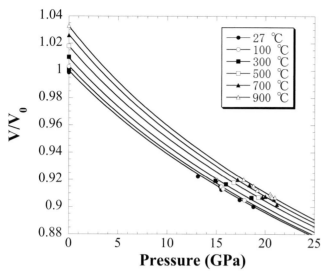

Figure 8. High temperature compression curves of superhydrous phase B. The data at atmospheric pressure are calculated values based on the thermal expansion data shown in Fig. 7.

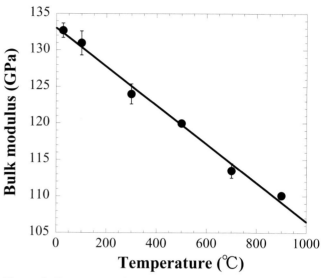

Figure 9. Temperature dependence of bulk modulus of superhydrous phase B.

calculated the density changes in hydrous subducted slabs as functions of water content and temperature at a pressure of 20 GPa, corresponding to the lower part of the transition zone. The obtained density changes were compared with the density of a model mantle composition.

At pressure around 20 GPa, Mg_2SiO_4-rich ringwoodite and $MgSiO_3$-rich garnet are the main constituent minerals of the transition zone [*e.g., Irifune*, 1993]. First, the mantle composition was simplified to consist solely of MgO and SiO_2, where the Mg/Si molar ratio was assumed to be 1.5. In this case, the molar ratio of Mg_2SiO_4-ringwoodite and $MgSiO_3$-majorite garnet becomes 1:1 (58:42 in volume ratio). This is in close agreement with the ratio of ringwoodite and garnet (~6:4 in volume) in a pyrolite mantle with a more complex chemical composition [*Irifune*, 1993].

Next, it was assumed that the chemical composition of the subducted slab was identical to a surrounding mantle except for the H_2O contents, i.e. Mg/Si ratio is 1.5. The addition of H_2O was assumed to yield only Shy-B as a hydrous phase according to the following reaction.

$$7\ Mg_2SiO_4\ +\ 2\ H_2O\ \rightarrow\ Mg_{10}Si_3O_{18}H_4\ +\ 4\ MgSiO_3$$

ringwoodite   fluid   Shy-B   majorite

This assumption should be reasonable because recent experiments of peridotite-2 wt% water system [*Litasov et al.*, 2003; *Ohtani et al.*, 2004] and Schreinemakers analysis in the system MgO-SiO_2-H_2O [*Komabayashi et al.*, 2004; *Komabayashi*, this volume] show that other hydrous phases except for Shy-B and hydrous ringwoodite

should not exist in hydous peridotite at around 20 GPa. It is reported that ringwoodite can contain significant amounts of water (up to 2.8 wt%) in its crystal structure [*Inoue et al.*, 1998; *Yusa et al.*, 2000]. The surrounding mantle in the mantle transition zone may be also hydrated as hydrous wadsleyite and hydrous ringwoodite [e.g. *Inoue et al.*, 2004; *Ohtani et al.*, 2004]. So in the

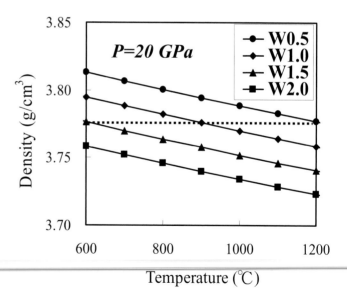

Figure 10. A comparison of the densities of model hydrous subducted slabs and that of the surrounding mantle as functions of temperature and water content at 20 GPa. Horizontal broken line indicates the density of a model anhydrous surrounding mantle at 1600°C. W0.5 means the water (H_2O) content of 0.5 wt% in subducted slab.

Table 4. Parameters used in density calculations of the model mantle and slabs.

	ρ (g/cm^3)	K (GPa)	dK/dP	dK/dT(x10^{-2}GPa/K)	α (x10^{-6}/K)
superhydrous phase B	3.327[*1]	132[*2]	5.8[*3]	-2.5[*2]	37.6[*2]
ringwoodite	3.563[*4]	185[*5]	4.5[*5]	-2.8[*6]	30.7[*7]
majorite	3.513[*8]	167[*9]	6.7[*9]	-2.0[*10]	25[*10]

[*1] Pacalo&Parise (1992)
[*2] This study
[*3] Crichton et al. (1999)
[*4] Sasaki et al. (1982)
[*5] Li (2003)
[*6] Meng et al. (1993)
[*7] Inoue et al. (2004)
[*8] Angel et al. (1989)
[*9] Gwanmesia et al. (1998)
[*10] Wang et al. (1998)

present calculation, we only judge the density difference between Shy-B bearing suducted slab and the anhydrous surrounding mantle. If the surrounding mantle contains some water, the Shy-B bearing slab can penetrate more easily depending on the water content.

With increasing water content, ringwoodite will be consumed to form Shy-B and majorite garnet. Since the Mg/Si molar ratio is assumed to be 1.5, the total water content in this model slab composition may not exceed 2 wt %, provided that all water is retained in Shy-B.

The parameters used for the present calculation are shown in Table 4. The density under high temperature and high pressure was calculated using the 3rd order Birch-Murnaghan equation of state with corrections using thermal expansion coefficients. The result is shown in Fig. 10, where the density changes in the model hydrous slabs with changing temperature are depicted as a function of bulk water content between 0.5 and 2.0 wt%. The density of the anhydrous model mantle at a typical mantle temperature (1600°C) in this region of the transition zone is also shown in Fig. 10

It is seen that addition of a small amount of water to slab material significantly reduces the bulk density, by up to ~0.08 g/cm^3 for the maximum water content of 2.0 wt%. The hydrous model slab may be denser than the surrounding mantle for only limited water contents of less than ~1.5 wt% and temperatures significantly lower than ~1200°C, ~900°C, and ~600°C, for the water contents of 0.5, 1.0 and 1.5 wt%, respectively. Thus the presence of DHMS, such a Shy-B, may significantly affect the behavior of the subducting slab in the deep mantle.

The present results indicate that whether slab can penetrate into the deeper mantle or not strongly depends on the water content and temperature profile of the slab. If the water content of the subducted slab is high enough (1.5~2.0 wt%), such a slab can not be denser than the surrounding mantle at any reasonable slab temperatures at ~20 GPa and would stagnate in the mantle transition zone.

Acknowledgments. The authors are thankful to T. Sanehira, Y. Sueda and Y. Tanimoto, T. Kawahara, K. Ochi, A. Kurio and Y. Okajima for help in the in situ X-ray observations at SPring-8 (No: 2001A0385-ND-np, 2001B0443-CD-np). We thank T. Suzuki for help in the thermal expansion measurement in The University of Kitakyushu. We also thank J. Ando, W. Utsumi and O. Shimomura for advice and encouragement during this study. Constructive comments by two anonymous reviewers and the editor, S. Jacobsen, were greatly helpful to improve the manuscript. The present study is partially supported by the grand-in-aid for Scientific Research of the Ministry of Education, Science, Sport, and Culture of the Japanese government (no:13740270) to T. Inoue.

REFERENCES

Akaogi, M., E. Ito, and A. Navrotsky, Olivine-modified spinel-spinel transitions in the system Mg$_2$SiO$_4$-Fe$_2$SiO$_4$: Calorimetric measurements, thermochemical calculation, and geophysical application, *J. Geophys. Res.*, 94, 15671–15685, 1989.

Aldebert, P., and J. P. Traverse, α-Al$_2$O$_3$: A high-temperature thermal expansion standard, *High Temp.-High Press.*, 16, 127–135, 1984.

Anderson, O.L., D. G. Isaak, and S. Yamamoto, Anharmonicity and the equation of state for gold, *J. Appl. Phys.*, 65, 1535–1543, 1989.

Angel, R. J., L. W. Finger, R. M. Hazen, M. Kanzaki, D. J. Weidner, R. C. Liebermann, and D. R. Veblen, Structure and twinning of single-crystal MgSiO$_3$ garnet synthesized at 17 GPa and 1800°C, *Am. Mineral.*, 74, 509–512, 1989.

Angel, R. J., D. R. Allan, R. Militech, and L. W. Finger, The use of quartz as an internal pressure standard in high-pressure crystallography, *J. Appl. Crystallogr.*, 30, 461–466, 1997.

Brown, J. M., and T. J. Shankland, Thermodynamic parameters in the Earth as determined from seismic profiles, *Geophys. J.*, 66, 579–596, 1981.

Chen, J., T. Inoue, D.J. Weidner, Y. Wu, and T. Vaughan, Strength and water weaking of mantle minerals, olivine, wadsleyite and ringwoodite, *Geophys. Res. Lett.*, 25, 575–578, 1998.

Crichton, W. A., N. L. Ross, and T. Gasparik, Equation of state of magnesium silicates anhydrous B and superhydrous B, *Phys. Chem. Minerals*, 26, 570–575, 1999.

Fei, Y., J. Li, K. Hirose, W. Minarik, J. V. Orman, C. Sanloup, W. V. Westrenen, T. Komabayashi, and K. Funakoshi, A critical evaluation of pressure scales at high temperatures by in situ X-ray diffraction measurements, *Phys. Earth Planet. Inter.*, 143–144, 515–526,. 2004.

Frost, D., and Y. Fei, Stability of phase D at high pressure and high temperature, *J. Geophys. Res., 103*, 7463–7474, 1998.

Funamori, N., T. Uchida, W. Utsumi, T. Kondo, and T. Yagi, Thermoelastic properties of MgSiO₃ perovskite determined by in situ X-ray observations up to 30 GPa and 2000 K, *J. Geophys. Res., 101*, 8257–8269, 1996.

Gasparik, T., The role of volatiles in the transition zone, *J. Geophys. Res., 98*, 4287–4299, 1993.

Gwanmesia, G. D., G. Chen, and R. C. Liebermann, Sound velocities in MgSiO₃-garnet to 8 GPa, *Geophys. Res. Lett., 25*, 4553–4556, 1998.

Huang, X., Y. Xu, and S. Karato, Water content in the transition zone from electrical conductivity of wadsleyite and ringwoodite, *Nature, 434*, 746–749, 2005.

Inoue, T., Effect of water on melting phase relations and melt composition in the system Mg₂SiO₄-MgSiO₃-H₂O up to 15 GPa, *Phys. Earth Planet. Inter., 85*, 237–263, 1994.

Inoue, T., H. Yurimoto, and Y. Kudoh, Hydrous modified spinel, Mg₁.₇₅SiH₀.₅O₄: a new water reservoir in the mantle transition zone, *Geophys. Res. Lett., 22*, 117–120, 1995.

Inoue, T., D.J. Weidner, P. A. Northrup, and J. B. Parise, Elastic properties of hydrous ringwoodite (γ-phase) in Mg₂SiO₄, *Earth Planet. Sci. Lett., 160*, 107–113, 1998.

Inoue, T., Y. Tanimoto, T. Irifune, T. Suzuki, H. Fukui, and O. Ohtaka, Thermal expansion of wadsleyite, hydrous wadsleyite, ringwoodite and hydrous ringwoodite, *Phys. Earth Planet. Inter., 143–144*, 279–290, 2004.

Inoue, T, T. Irifune, Y. Higo, T. Sanehira, Y. Sueda, A. Yamada, T. Shinmei, D. Yamazaki, J. Ando, K. Funakoshi, and W. Utsumi, The phase boundary between wadsleyite and ringwoodite in Mg₂SiO₄ determined by in situ X-ray diffraction, *Phys. Chem. Minerals*, doi 10.1007/s00269-005-0053-y, 2006.

Jamieson J. C., J. N. Fritz, and M. H. Manghnani, Pressure measurement at high temperature in X-ray diffraction studies: gold as a primary standard, *In: Akimoto S., Manghnani M.H. (eds) High pressure research in geophysics.* Center for Academic Publishing, Tokyo, pp 27–48, 1982.

Irifune, T., Phase transformations in the Earth's mantle and subducting slabs: Implications for their compositions, seismic velocity and density structures and dynamics, *The Island Arc, 2*, 55–71, 1993.

Irifune, T., N. Kubo, M. Isshiki, and Y. Yamazaki, Phase transformations in serpentine and transportation of water into the lower mantle, *Geophys. Res. Lett., 25*, 203–206, 1998.

Ito, E., and E. Takahashi, Postspinel transformations in the syatem Mg₂SiO₄-Fe₂SiO₄ and some geophysical implications, *J. Geophys. Res., 94*, 10637–10646, 1989.

Kanzaki, M., Stability of hydrous magnesium silicates in the mantle transition zone, *Phys. Earth Planet. Inter., 66*, 307–312, 1991.

Karato, S., M. S. Paterson, and J. D. FitzGerald, Rheology of synthetic olivine aggregates: influence of grain size and water, *J. Geophys. Res., 91*, 8151–8176, 1986.

Katsura, T., and E. Ito, The system Mg₂SiO₄-Fe₂SiO₄ at high pressures and temperatures: Precise determination of stabilities

of olivine, modified spinel, and spinel, *J. Geophys. Res., 94*, 15663–15670, 1989.

Kawamoto, T., R. L. Hervig, and J. R. Holloway, Experimental evidence for a hydrous transition zone in the early Earth's mantle, *Earth Planet. Sci. Lett., 142*, 587–592, 1996.

Kohlstedt, D. L., H. Keppler, and D. C. Rubie, Solubility of water in the α, β and γ phases of (Mg,Fe)₂SiO₄, *Contrib. Mineral. Petrol., 123*, 345–357, 1996.

Komabayashi, T., S. Omori, and S. Maruyama, Petrogenetic grid in the system MgO-SiO₂-H₂O up to 30 GPa, 1600°C : Applications to hydrous peridotite subducting into the Earth's deep interior, *J. Geophys. Res., 109*, B03206, doi: 10.1029/2003JB002651, 2004.

Li, B., Compressional and shear wave velocities of ringwoodite γ-Mg₂SiO₄ to 12 GPa, *Am. Mineral., 88,* 1312–1317, 2003.

Litasov, K.D., and E. Ohtani, Stability of various hydrous phases in CMAS pyrolite–H₂O system up to 25 GPa, *Phys. Chem. Mineral, 30*, 147–156, 2003.

Liu, L.-G., Phase transformations in serpentine at high pressures and temperatures and implications for subducting lithosphere, *Phys. Earth Planet. Inter., 42*, 255–262, 1986.

Liu, L.-G., Effect of H₂O on the phase behavior of the forsterite-enstatite system at high pressures and temperatures and implications for the Eearth, *Phys. Earth Planet. Inter., 49*, 142–167, 1987.

Matsui, M., S. C. Parker, and M. Leslie, The MD simulation of the equation of state of MgO: Application as a pressure calibration standard at high temperature and high pressure, *Am. Mineral., 85*, 312–316, 2000.

Matsui, M., and N. Nishiyama, Comparison between the Au and MgO pressure calibration standards at high temperature, *Geophys. Res. Lett., 29*(10), pp.6–1, 2002.

Meng, Y., D. J. Weidner, G. D. Gwanmesia, R. C. Liebermann, M. T. Vaughan, Y. Wang, K. Leinenweber, R. E. Pacalo, A. Yeganeh-Haeri, and Y. Zhao, In situ high P-T X-ray diffraction studies on three polymorphs (α, β, γ) of Mg₂SiO₄, *J. Geophys. Res., 98*, 22, 22199–22207, 1993.

Ohtani, E., H. Mizobata, Y. Kudoh, T. Nagase, H. Arashi, H. Yurimoto, and I. Miyagi, A new hydrous silicate, a new water reservoir in the upper part of the lower mantle, *Geophys. Res. Lett., 24*, 1047–1050, 1997.

Ohtani, E., M. Toma, K. Litasov, T. Kubo, and A. Suzuki, Stability of dense hydrous magnesium silicate phases and water storage capacity in the transition zone and lower mantle, *Phys. Earth Planet. Inter., 124*, 105–117, 2001.

Ohtani, E., M. Toma, T. Kubo, T. Kondo, and T. Kikegawa, In situ X-ray observation of decomposition of superhydrous phase B at high pressure and temperature, *Geophys. Res. Lett., 30*(2), pp. 1–1, 2003.

Ohtani, E, K. Litasov, T. Hosoya, T. Kubo, and T. Kondo, Water transport into the deep mantle and formation of a hydrous transition zone, *Phys. Earth Planet. Inter., 143–144*, 255–269, 2004.

Pacalo, R. E. G., and J. B. Parise, Crystal structure of superhydrous B, a hydrous magnesium silicate synthesized at 1400°C and 20 GPa, *Am. Mineral., 77*, 681–684, 1992.

Ringwood, A. E., and A. Major, High-pressure reconnaissance investigastions in the system Mg$_2$SiO$_4$-MgO-H$_2$O, *Earth Planet. Sci. Lett., 2*, 130–133, 1967.

Sano, A., E. Ohtani, T. Kubo, and K. Funakoshi, In situ X-ray observation of decomposition of hydrous aluminum silicate AlSiO$_3$OH and aluminum oxide hydrate δ-AlOOH at high pressure and temperature, *J. Phys. Chem. Solids, 65(8–9)*, 1547–1554, 2004.

Sasaki, S., C. T. Prewitt, S. Sato, and E. Ito, Single crystal X-ray studies of γ-Mg$_2$SiO$_4$ spinel, *J. Geophys. Res., 87*, 7829–7832, 1982.

Shim, S., T. S. Duffy, and K. Takemura, Equation of state of gold and its application to the phase boundaries near 660 km depth in the Earth's mantle, *Earth Planet. Sci. Lett., 203*, 729–739, 2002.

Tuchiya, T., First-principles prediction of the P-V-T equation of state of gold and the 660-km discontinuity in Earth's mantle, *J. Geophys. Res., 108*, 2462, doi:10.1029/2003JB002446, 2003.

Utsumi, W., K. Funakoshi, S. Urakawa, M. Yamakata, K. Tsuji, H. Konishi, and O. Shimomura, SPring-8 beamline for high pressure science with multianvil apparatus, *Rev. High Press. Sci. Tech., 7*, 1484–1486, 1998.

Wang, Y., D. J. Weidner, and F. Guyot, Thermal equation of state of CaSiO$_3$ perovskite, *J. Geophys. Res., 101*, 661–672, 1996.

Wang, Y., D. J. Weidner, J. Zhang, G. D. Gwanmesia, and L. C. Liebrmann, Thermal equation of state of garnets along the pyrope-majorite join, *Phys. Earth Planet. Inter., 105*, 59–71, 1998.

Yusa, H., and T. Inoue, Compressibility of hydrous wadsleyite (β-phase) in Mg$_2$SiO$_4$ by high pressure X-ray diffraction, *Geophys. Res. Lett., 24*, 1831–1834, 1997.

Yusa, H., T. Inoue, and Y. Ohishi, Isothermal compressibility of hydrous ringwoodite and its relation to the mantle discontinuities, *Geophys. Res. Lett., 27*, 413–416, 2000.

_____

Yuji Higo, Toru Inoue, Tetsuo Irifune, Takayuki Ueda, Akihiro Yamada, Geodynamics Research Center, Ehime University, 2–5 Bunkyo-cho, Matsuyama 790-8577, Ehime, Japan, (inoue@sci.ehime-u.ac.jp)

Ken-ichi Funakoshi, Japan Synchrotron Radiation Research Institute, 1-1-1 Kouto, Sayo 679-5198, Hyogo, Japan

Phase Diagram and Physical Properties of H_2O at High Pressures and Temperatures: Applications to Planetary Interiors

Jung-Fu Lin, Eric Schwegler, and Choong-Shik Yoo

Lawrence Livermore National Laboratory, 7000 East Avenue, Livermore, CA 94550

Here we discuss the phase diagram and physical properties of H_2O under pressure-temperature conditions relevant to planetary interiors. Recent studies show that the melting curve of H_2O increases rapidly above a recently discovered triple point at approximately 35 to 47 GPa and 1000 K, indicating a large increase in $\Delta V/\Delta S$ (volume versus entropy change) and associated changes in the physical properties of H_2O at high pressures and temperatures. Existence of the triple point is thought to be associated with the formation of a superionic phase, dynamically-disordered ice VII, or extension of the ice VII-ice X phase boundary; although the precise pressure and temperature of the triple point, curvature of the melting line, and nature of the solid-solid transition below the triple point all remain to be further explored. The steep increase in the melting curve of H_2O at high pressures and temperatures has important implications on our understanding of planetary interiors. Depending on its curvature, the melting line of H_2O may intersect the isentropes of Neptune and Uranus as well as the geotherm of Earth's lower mantle. Furthermore, if the triple point is due to the occurrence of the theoretically predicted superionic phase, besides leading to significant ionic conductivity, fast proton diffusion would cause enhanced chemical reactivity and formation of complex compounds in these planets. For example, reaction of H_2O with iron and other metals to form metal hydrides such as FeH_x could provide a mechanism for incorporation of hydrogen as a light element into Earth's core. The equation of state of water is also presented as it pertains to the properties of hydrous fluid and melt phases in the mantle.

INTRODUCTION

Due to a triple point that occurs near the Earth's mean surface temperature and pressure, H_2O is present in three different forms within the biosphere; water, vapor, and solid ice-Ih. H_2O is also believed to be a major component of the intermediate ice layers in the interiors of Uranus and Neptune [*Nellis et al.*, 1988; *Hubbard et al.*, 1991],

and water vapor plumes have been detected near the south polar regions of Saturn's icy moon, Enceladus [*Hansen et al.*, 2006]. Water is also a major component of fluid and melt phases forming in the Earth's mantle on dehydration. Based upon the bulk water content of ordinary chondritic meteorites, two-thirds of Earth's original H_2O may have either been lost to space or is present in the interior as trace quatities of $(OH)^-$ in nominally anhydrous mantle minerals [e.g. *Smyth*, 1987; *Bell and Rossman*, 1992; *Williams and Hemley*, 2001; *see Smyth and Jacobsen*, 2006, *this volume for a review*]. Deep reservoirs of H_2O may also present in dense magnesium silicates [e.g. *Angel et al.*, 2001; also see

Earth's Deep Water Cycle
Geophysical Monograph Series 168
Copyright 2006 by the American Geophysical Union.
10.1029/168GM12

Komabayashi et al., (2006) of this volume] or even as ice-VII in cold subducting slabs [*Bina and Navrotsky*, 2000]. Therefore, the physical and chemical properties of H_2O at high pressures and temperatures (*P-T*) play an important role in planetary science. The behavior and phase diagram of H_2O at high *P-T* are also of fundamental interest in physics, chemistry, and biological sciences because the flexibility of hydrogen bonding gives rise to a myriad of crystalline, amorphous, and liquid phases with unique physical and chemical properties [*Petrenko and Whitworth*, 1999].

Shock-wave experiments have served as the main tool to characterize liquid water under high *P-T*. The results have been used to derive and to constrain a variety of equation of state (EOS) models [*Homles et al.*, 1985; *Nellis et al.*, 1988]. Shock wave studies have also been used to determine additional properties of water at high pressure such as Raman spectra [*Holmes et al.*, 1985] and electrical conductivity [*David et al.*, 1960; *Hamann et al.*, 1969; *Holzapfel*, 1969; *Mitchell et al.*, 1982; *Chau et al.*, 2001; *Celliers et al.*, 2004]. These experiments suggest that high *P-T* conditions generate highly mobile charge carriers through molecular dissociation leading to an increase in ionic conductivity [*David et al.*, 1960; *Hamann et al.*, 1969; *Holzapfel*, 1969; *Mitchell et al.*, 1982; *Chau et al.*, 2001] and eventually to the onset of electronic conduction in liquid water [*Celliers et al.*, 2004].

In addition to shockwave and theoretical studies, static measurements of the melting curve, phase behavior, and physical properties of H_2O have recently been performed under *P-T* conditions that are relevant to the interior of planets, such as Earth. These studies have found strong evidence for a rapid increase in the melting temperature of H_2O above 35 to 47 GPa. Here we discuss recent developments in high-pressure experimental and theoretical studies on the behavior and physical properties of H_2O in terms of possible implications for planetary interiors.

HIGH *P-T* PHASES OF H_2O

Our current understanding of the phase diagram of H_2O is summarized in Figure 1. In the following, we will focus primarily on the high *PT* regions that are most relevant to planetary interiors (the liquid and solid phases above 2 GPa and 300 K). As illustrated in Figure 2a, Ice VII is a molecular crystal consisting of oxygen atoms arranged on a body-centered cubic (*bcc*) lattice and hydrogen atoms arranged in random (disordered) positions that satisfy the ice rules. The overall structure is often described as two interpenetrating hydrogen bond networks of cubic ice disconnected in the sense that they do not share any hydrogen bonds in common. In addition, there is experimental evidence that in the range of 2.2 to 25 GPa, ice VII exhibits spatially modulated

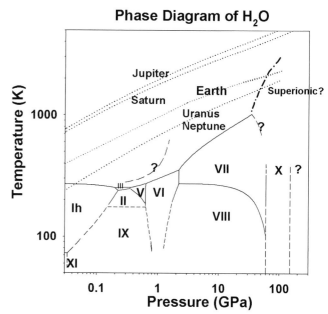

Figure 1. High *P-T* phase diagram of H_2O. The Roman numerals are various forms of solid ices, whereas ice Ih represents ice in the hexagonal crystal structure. Black solid line, experimentally determined phase boundaries [after *Petrenko and Whitworth*, 1999; *Lin et al.*, 2005]; dash line: schematic phase boundaries; dotted lines: isentropes of Jupiter, Saturn, Uranus, and Neptune [*de Pater and Lissauer*, 2001], respectively, and geotherm of the Earth [*Brown and Shankland*, 1981]. The dash line in the liquid region represents a proposed liquid-liquid transition by *Kawamoto et al.* [2004]. At pressures near 150 GPa additional structural phase transitions have been proposed based on both theory [*Demontis et al.*, 1988; *Benoit et al.*, 1996] and experiment [*Loubeyre et al.*, 1999]. Depending on the curvature of the extrapolated melting curve above 35 GPa (dash-dotted line), the melting curve may intersect proposed isentropes of Neptune and Uranus at ~50 GPa and geotherm at ~60 GPa [*Lin et al.*, 2005].

(incommensurate) phases; a modulation in the periodicity of the structure [*Loubeyre et al.*, 1999].

All of the ice phases observed below 60 GPa have structures that closely follow Pauling's ice rules [*Pauling*, 1935]: 1) molecular crystals composed of water molecules with "gas-phase" like geometries, 2) each water molecule is oriented so that it participates in four hydrogen bonds with neighboring water molecules in tetrahedral coordination, and 3) at most one hydrogen atom is located between any two neighboring oxygen atoms. However, at higher pressures the ice rules are no longer satisfied as symmetrization of the hydrogen bond leads to protons residing midway between adjacent oxygen atoms, resulting in stable forms of ice that are non-molecular [*Holzapfel*, 1972; *Polian et al.*, 1984; *Benoit et al.*, 1996; *Aoki et al.*, 1996; *Goncharov et al.*, 1996,

1999; *Loubeyre et al.*, 1999]. As illustrated in Figure 2b, with increasing pressure, nearest neighbor oxygen-oxygen distances in ice VII decrease, and eventually in the pressure range of 40 to 60 GPa the hydrogen atoms are located at the midway point between neighboring oxygen atoms leading to a transition from a molecular to an atomic crystal called ice X. At even higher pressures, near 150 GPa, additional structural phase transitions have been proposed based on both theory [*Demontis et al.*, 1988; *Benoit et al.*, 1996] and experiment [*Loubeyre et al.*, 1999] where the *bcc* oxygen sublattice in ice X undergoes a transformation to an anti-fluorite or a hexagonal-close packed (*hcp*) structure.

In addition to the intriguing high *P*, low *T* behavior of ice, *ab initio* simulations indicate that the high *P-T* behavior of H_2O may hold surprises as well. For instance, it has been predicted that many of the high *P-T* phases of ice have regions characterized by fast protonic diffusion with stable oxygen sublattices, which could lead to significant increases in ionic conductivity while still in the solid phase [*Demontis et al.*, 1988; *Benoit et al.*, 1996; *Cavazzoni et al.*, 1999; *Goldman et al.*, 2005]. These high *P-T* domains of fast proton diffusion in ice are often referred to as a superionic phase [*Cavazzoni et al.*, 1999; *Goldman et al.*, 2005].

The liquid phase of H_2O also undergoes significant changes with increasing *P-T*. In particular, the local structure of liquid H_2O changes from the open four-fold hydrogen bonded structure at ambient *P-T* conditions to a nearest-neighbor coordination shell of up to 13 at pressures of 10 GPa (Figure 2c) [*Schwegler et al.*, 2000; *Strässle et al.*, 2006]. This change from low to high density H_2O appears gradually based on neutron diffraction studies up to 400 MPa [*Soper et al.*, 2000]. However, there is also indirect evidence from

Raman spectroscopy that the transition is abrupt and possibly first-order [*Kawamoto et al.*, 2004]. It is perhaps surprising to note that below 15 GPa and at temperatures near the melting curve, the large increase in coordination number from four-fold bonded to higher nearest-neighbor coordination is not accompanied by a corresponding decrease in the average nearest neighbor oxygen-oxygen distance or an appreciable change in the number of hydrogen bonds [*Schwegler et al.*, 2000; *Strässle et al.*, 2006]. Although this process is typically referred to as a simple collapse of the second coordination shell down on the first [*Schwegler et al.*, 2000; *Soper et al.*, 2000], one could speculate that the collapse occurs primarily between H_2O molecules that are not connected via "hydrogen bond wires", and results in a set of interpenetrating hydrogen bond networks in the high-density liquid that closely resembles structures of the ice phases (*e.g.* ice VII) found at corresponding pressures (Figure 2).

Shock wave studies have shown that the conductivity of liquid water increases rapidly along the primary Hugoniot (the locus of states reached by passage of a shock wave through a material initially at ambient *P-T* conditions) above 10 GPa and eventually levels off to values typical of molten ionic salt above 20 GPa [*Nellis et al.*, 1988]. The increased conductivity of liquid H_2O at high *P-T* is commonly interpreted as being caused by a rapid increase in the fraction of dissociated H_2O molecules in the liquid [*Holzapfel*, 1969]. As illustrated in Figure 2c, a similar onset of molecular dissociation in water is found in *ab initio* MD simulations [*Schwegler et al.*, 2001] and estimates of the ionic conductivity along the planetary isentrope of Uranus agree well between simulation [*Cavazzoni et al.*, 1999] and reverberating shock measurements [*Chau et al.*, 2001].

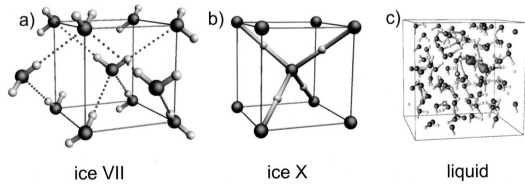

a)   ice VII   b)   ice X   c)   liquid

Figure 2. Schematic representations of ice VII (**a**), ice X (**b**), and a snapshot from an *ab initio* MD simulation of liquid water at 27 GPa and 1400 K (**c**) [*Schwegler et al.*, 2001]. The large black spheres represent oxygen atoms, the small white spheres are hydrogen atoms, and the black dashed/dotted lines in **Fig. 2a** are hydrogen bonds. The isosurfaces in **Fig. 2c** are the orbitals involved in a typical dissociation event in the liquid as a proton is shuttled between neighboring oxygen atoms.

EXPERIMENTAL AND THEORETICAL METHODS AT HIGH *P-T*

A variety of techniques have been used to study the behavior of H_2O under extreme conditions. Here we focus on static and theoretical methods capable of reaching *P-T* conditions relevant to planetary interiors. In particular, externally-heated diamond anvil cells (EHDAC) and laser-heated diamond anvil cells (LHDAC) are two common high-pressure techniques used to reach to static high *P-T* conditions [*Bassett et al., 1993; Fei et al., 1993; Lin et al., 2004a*]. Because commonly used pressure calibrants such as samarium-doped yttrium-aluminum-garnet (YAG) and ruby have a tendency to dissolve in water above 600 K (*Dachi et al., 2002*), secondary holes near the sample chamber have been used to hold the pressure calibrant in optical spectroscopic studies [*Lin et al., 2004a*]. In order to avoid reaction between H_2O and metal-based gasket materials, various other metal gaskets and gasket inserts, such as Re, Au, Ir, W, and Pt, have been tested and shown to successfully confine H_2O at high *P-T* [*Lin et al., 2005*]. Inert pressure calibrants such as Au or Ta has also been used in thermal EOS studies with X-ray diffraction [*Lin et al., 2005*]. The EHDAC apparatus can be readily interfaced with a variety of experimental probes, such as *in situ* Raman spectroscopy, X-ray diffraction, and/or visual observations for investigating the phase diagram of H_2O.

LHDACs have also proven to be a useful tool for studying H_2O at *in-situ* high *P-T* conditions. Because H_2O is transparent to the 1 µm infrared laser often used in these experiments, metal foils (Pt, Pt-Ir alloys, Ir, Re, or W) are used as a laser absorber, both with and without a small hole of 10–20 µm in diameter in the center of the foil [*Schwager et al., 2004; Goncharov et al., 2005; Lin et al., 2004b*]. The temperature of the laser-heated sample can be determined from either the surface temperature of the foil [*Schwager et al., 2004*] or from the intensity ratios of the Stokes-to-anti-Stokes Raman peaks [*Goncharov et al., 2005*]. In these experiments, chemical reactions with Fe, Re, Pt, and W foils to form metal hydrides have been observed at high temperature [*Schwager et al., 2004; Lin et al., 2005; Ohtani et al., 2005*], while Ir foils apparently react with H_2O only if the system is heated to several hundred degrees above the melting curve [*Schwager et al., 2004*], and Pt-Ir (20% Ir) foils do not show any evidence of reactions with H_2O up to temperatures of 1500 K [*Goncharov et al., 2005*]. Visual observation of changes in the laser-speckle pattern (*Schwager et al., 2004*) or *in situ* Raman spectroscopy (*Lin et al., 2004b; Goncharov et al., 2005*) have recently been employed as criteria for identifying melting and structural phase transitions.

In addition to the recent developments in DAC technologies, first-principles theoretical approaches have proven to be a useful tool for investigating the high *P-T* properties of H_2O. Ever since the first classical molecular dynamics (MD) simulation of liquid H_2O [*Rahman et al., 1971*], a great deal of effort has gone into the development of improved models. In fact, for modest *P-T* conditions, a variety of different classical potentials are known to accurately reproduce the properties of both the liquid and solid phases of H_2O [*Mahoney et al., 2000; Sanz et al., 2004*]. However, as one considers the extreme conditions that are relevant to planetary interiors, many of these simple empirical models are known to break down. This general failure is partly due to the overall lack of reliable experimental data to fit the potentials to, but more importantly, as higher *P-T* are considered, molecular dissociation begins to play a dominant role. For instance, at pressures above 15 GPa, the Hugoniot of water obtained from commonly used empirical potentials deviate significantly from experiment due to the empirical potentials' inability to describe molecular dissociation effects. It is in these regimes of high *P-T* where first-principles based methods can be used to go beyond the limitations of a typical classical model. In addition to providing enough predictive power to reliably investigate regions of phase space where there is little or no existing experimental data, first-principles methods are appropriate for investigating extreme *P-T* conditions, since bond making and breaking processes are taken into account in a quantitative fashion.

There are several notable exceptions to the general failure of empirical models (such as theoretical models with classical potentials) to describe H_2O under extreme conditions. For instance, a rather simple model based on Morse potentials was first used to successfully argue that hydrogen bond symmetrization should occur in ice VII as it is compressed, and eventually result in a transition to ice X [*Holzapfel, 1972*]. Subsequent calculations with increasingly sophisticated levels of theory such as *ab initio* MD and path integral sampling were used to further characterize the transition to ice X in much greater detail [*Bernasconi et al., 1998; Benoit et al., 1998*] and to propose additional structural phase transitions at higher pressures [*Demontis et al., 1988; Benoit et al., 1996*].

THE MELTING CURVES OF H_2O

Precise determination of the melting curve of H_2O is essential for modeling planetary interiors, understanding how numerous chemical reactions may affect the stability field of mineral assemblages at extreme conditions, and verifying theoretical predictions. Recently, conflicting reports on the melting curve of H_2O below ~30 GPa have been resolved using angle-dispersive synchrotron X-ray diffraction, Raman spectroscopy, and visual observation as melting criteria [*Lin et al.*, 2004a; 2005] (Figures 3, 4 and Plate 1). The disappearance of diffraction peaks from the oxygen lattice of solid H_2O phases in an EHDAC has been used as an indication of melting in previous studies [*Fei et al.*, 1993; *Dubrovinskaia and Dubrovinsky*, 2003a,b; *Frank et al.*, 2004]. However, recrystallization of ice VII in different orientations at *P-T* conditions close to the melting line makes it difficult to identify the onset of the melting in such experiments [*Datchi et al.*, 2000; *Lin et al.*, 2004a]. Use of the very intense synchrotron X-ray source now makes it possible to detect diffuse X-ray scattering characteristic of water, providing a very reliable melting criterion (Figure 3). Future advances in modeling diffuse X-ray scattering patterns will further provide knowledge of the local structure of water at high *P-T*.

A change of Raman-active OH-stretching bands and the appearance of the translational modes (specific modes of

Raman Spectra of H₂O

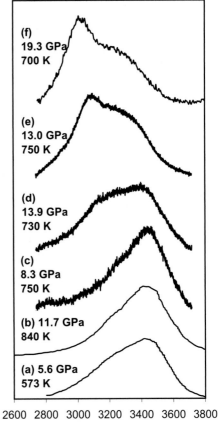

Figure 4. Representative Raman spectra in liquid and solid H_2O in shock wave (pattern **a**) (*Homles et al.*, 1985) and static DAC (pattern **b**: *Kawamoto et al.*, 2004; patterns **c,d,e,f**: *Lin et al.*, 2004a; 2005) studies. **a,b,c,d**; liquid water; **e,f**: solid ice. The OH-stretching modes change significantly across melting; the low-frequency A_{1g} mode is the dominant band in ice VII while the high frequency mode dominates in liquid water. The change of the Raman-active OH-stretching bands can be used to detect melting [*Lin et al.*, 2004a; 2005].

X-ray Diffraction of H₂O

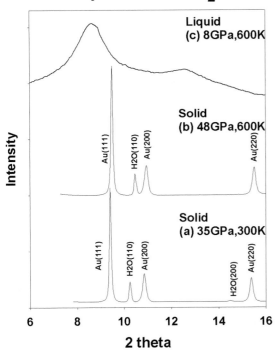

Figure 3. Angle-dispersive X-ray diffraction and diffuse scattering patterns of the solid (**a,b**) and liquid (**c**) H_2O at high *P-T*. A monochromatic beam (wavelength = 0.4157 Å) was used as the X-ray source and the diffracted X-rays were collected by a CCD (MARCCD). The diffraction patterns **a** and **b** showed diffraction peaks of the oxygen atoms in a bcc sublattice, indicating the existence of the solid ice phase. The observation of the diffuse scatterings (pattern **c**) showed clearly the occurrence of the liquid water under high *P-T*.

the lattice vibrations) have also been used to reliably detect melting in ice VII (Figure 4) (*Lin et al.*, 2004a; *Lin et al.*, 2005); the low-frequency A_{1g} mode is the dominant band in ice VII while the high frequency mode dominates in liquid water. The observed OH-stretching bands of liquid water at static high pressures are very similar to those obtained in shock-wave Raman measurements reaching to 26 GPa and 1700 K [*Holmes et al.*, 1985]. The intensity increase in the high frequency band obtained in the shock-wave experiments has been interpreted as an indication of an increase in monomeric water molecule concentration due to the breaking of intermolecular hydrogen bonding [*Holmes et al.*, 1985]. In both static and shock wave Raman spectra measurements there is a lack of broad band centered at 2900 to 3000 cm^{-1} that is characteristic of the hydronium ion (H_3O^+) at low pressures [*Holmes et al.*, 1985; *Kawamoto et al.*, 2004; *Lin et al.*, 2004a; 2005]. This has led to the suggestion that the dissociation of water under high pressure conditions results in the formation of free H^+ and OH^- ions, which in turn are responsible for the observed conductivity increase in water for pressures between 3 to 30 GPa [*Holmes et al.*, 1985; *Chau et al.*, 2001]. However, a series of *ab initio* MD simulations at similar thermodynamic conditions found evidence that the dissociation of water still occurs through a bimolecular process similar to what is found at ambient conditions, leading to the formation of short-lived OH^- and H_3O^+ ions [*Schwegler et al.*, 2000; 2001]. For pressures above ~25 GPa, the OH-stretching modes begin to overlap with the second order Raman signal from the diamond anvils, making it difficult to use these modes to detect melting. Nevertheless, subtraction of the second-order Raman scattering of the diamond anvils from the Raman spectrum taken from a LHDAC has been used to measure the phase diagram of H_2O up to 56 GPa and 1500 K [*Goncharov et al.*, 2005].

Plate 1 shows a comparison of recent experimental and theoretical melting lines [*Cavazzoni et al.*, 1999; *Dubrovinskaia et al.*, 2003a,b; *Frank et al.*, 2004; *Lin et al.*, 2004a; *Schwager et al.*, 2004; *Goldman et al.*, 2005; *Goncharov et al.*, 2005; *Lin et al.*, 2005]. Although significant discrepancies occur in these reported melting curves, these studies, in general, point to a discontinuous increase along the melting line approximately between 35 GPa to 47 GPa. MD simulations from 30 to 300 GPa first indicated a discontinuous increase in the melting curve at high *P-T* relative to the extrapolated experimental melting curve of ice VII from lower *P-T* conditions [*Cavazzoni et al.*, 1999]. According to these and subsequent MD simulations, the change in slope is likely due to the appearance of a superionic phase below the melting curve [*Cavazzoni et al.*, 1999; *Goncharov et al.*, 2005; *Goldman et al.*, 2005]. On the experimental side, visual observations of the laser-speckle pattern in a LHDAC first reported a distinct change in melting slope at 43 GPa and 1600 K, and it was speculated that this change was due to a first-order transformation from ice VII to ice X [*Schwager et al.*, 2004]. *In situ* Raman spectroscopy in a LHDAC confirmed the discontinuous melting curve and indicated the presence of a triple point at about 47 GPa and 1000 K [Goncharov *et al.*, 2005], which is accompanied by a substantial decrease in the intensity of the O-H stretch band in the Raman spectra. Similar trends in the power spectra obtained in *ab initio* MD simulations

indicate that the observed change in the melting curve may be due to the occurrence of a superionic phase at high pressure [*Goncharov et al.*, 2005]. EHDAC experiments with *in-situ* X-ray diffraction, Raman spectroscopy, and reliable *P-T* determinations also showed a discontinuous change in the melting curve at approximately 35 GPa and 1040 K, although the melting temperature of H_2O above 40 GPa is beyond the temperature capability of the EHDAC, leading to a large uncertainty in the curvature of the extrapolated melting curve [*Lin et al.*, 2005]. Inconsistencies between reported melting lines from various experimental techniques highlight the importance of reliable melting criteria and precise *P-T* determinations, as well as the need for expanded high *P-T* capabilities and improved techniques for detecting subtle changes in oxygen-hydrogen bonding in future experimental studies of these unresolved regions of the high *P-T* phase diagram of H_2O.

There are also significant differences between recent theoretical studies of the melting curve of H_2O. For instance, *ab initio* MD calculations have predicted that at 2000 K, H_2O will melt at pressures ranging from 30 GPa [*Cavazzoni et al.*, 1999], to 65 GPa [*Goncharov et al.*, 2005], to 75 GPa [*Goldman et al.*, 2005]. The large discrepancy in the calculated melting pressures at 2000 K is somewhat surprising given the nearly identical levels of theory and simulation protocols used. In addition to issues related to finite size effects and simulation timescales, it is possible that a large fraction of this discrepancy comes from the specific computational approaches used to determine the phase boundary. In all of the previous investigations with *ab initio* MD, the simulations have started with a single phase (either the liquid or the solid) and have proceeded with a "heat-until-it-melts" or a "squeeze-until-it-freezes" strategy for locating the transition between the liquid and the solid phase. The primary objection to these types of approaches is that the observed phase transition does not directly correspond to the melting temperature (or pressure), but instead to conditions of thermal metastability, which for small system sizes and short simulations timescales can be significantly different from the equilibrium melting point. The two main approaches for reliably computing a material's melting temperature is to use either a free-energy based method where the equivalence of the Gibbs free energy of the solid and the liquid is computed [*Alfe, et al.*, 2003], or a two-phase approach where coexistence between the solid and the liquid is directly simulated [*Ogitsu et al.*, 2003]. To date, neither of these computational techniques has been applied within an *ab initio* MD context to examine H_2O, and as such, the accurate computation of the high-pressure melting curve remains an open challenge to theorists.

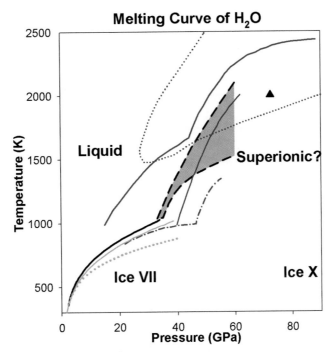

Plate 1. Melting curve of H$_2$O at high *P-T*. Black solid line, melting curve of ice VII determined by Raman spectra, optical observation, and X-ray diffraction [*Lin et al.*, 2004a; 2005]; black dashed line and grey area: upper bound and lower bound of the extrapolated melting curve [*Lin et al.*, 2005; also see Fig. 1]; blue dotted line: occurrence of the theoretically predicted superionic phase [*Cavazzoni et al.*, 1999]; red dash line with dots: melting curve determined by Raman spectroscopy in a LHDAC [*Goncharov et al.*, 2005]; blue solid line: theoretically calculated freezing of H$_2$O [*Goncharov et al.*, 2005]; solid gray line: melting curve determined by angle-dispersive X-ray diffraction [*Dubrovinskaia and Dubrovinsky*, 2003a,b]; gray dotted line: melting curve determined by energy-dispersive X-ray diffraction in an EHDAC [*Frank et al.*, 2004]; red solid line: optical observation in a LHDAC [*Schwager et al.*, 2004]; solid triangle: liquid to superionic transition in *ab initio* MD calculations [*Goldman et al.*, 2005]. The melting line of Datchi *et al.* [2000] is in agreement with that of Lin *et al.* [2004a, 2005].

THERMODYNAMIC PROPERTIES OF LIQUID AND SOLID H_2O PHASES

A detailed understanding of the EOS properties of liquid and solid phases of H_2O under extreme conditions is essential to the development of accurate models of planetary interiors. Information on the thermodynamic properties of H_2O under extreme conditions has come from a variety of sources, depending on the specific *P-T* of interest. For the liquid phase under moderate *P-T* conditions, a large amount of experimental data has been combined into EOS models based on highly parameterized formulations of the Helmoltz free energy [*Wagner et al.*, 2002]. Although this has proven to be quite useful for interpolation of data, the highest *P-T* conditions that are relevant to planetary science fall outside the range of these models, so their direct use would require large extrapolations with questionable levels of accuracy. For this reason, EOS models for liquid water under extreme *P-T* conditions have been based on either shock wave experiments [*Walsh et al.*, 1957] or MD simulations [*Belonoshko et al.*, 1991; *Brodholt et al.*, 1993; *Sakane et al.*, 2001]. In particular, the TIP4P empirical potential [*Jorgensen et al.*, 1983] has been used to tabulate *P-V-T* data for water up to 2500 K and 35 GPa. By adding virial terms to a modified Redlich Kwong-style EOS equation:

$$P = \frac{RT}{V-b} - \frac{a}{T^{1/2}V(V+b)} + \frac{c}{V} + \frac{d}{V^2} + \frac{e}{V^3} + \frac{f}{V^4}$$

$$a = -582468 - 3038.79T - 9.24574 \times 10^{-3}T^2 + 3.02674 \times 10^9/T^2$$
$$b = -3.90463 \times 10^{-2} - 0.991078V$$
$$c = 3.64905 \times 10^4$$
$$d = -1.02451 \times 10^7$$
$$e = -1.79681 \times 10^8$$
$$f = 2.18437 \times 10^9$$

where V is the volume in cm^3/mol, P is in bars and T is in Kelvin, the simulation data can be faithfully reproduced over a wide range of phase-space [*Brodholt et al.*, 1993]. However, some care should be exercised when using empirical potentials like TIP4P for describing water under high-*P* conditions. As mentioned earlier, simple empirical models typically do not allow for intramolecular dissociation reactions, which most certainly readily occur in liquid water for pressures above 15 GPa [*Holzapfel*, 1969].

Most thermodynamic data on ice under pressure have come from room temperature DAC measurements due to the difficulty of simultaneously achieving high *P-T* in a DAC and the rapid rise in *T* characteristic of shock experiments. As shown in Table 1, DAC experiments on ice VII have been reported and used to fit different isothermal models with relatively good levels of agreement for properties such as the bulk modulus

and the equilibrium volume. However, in the development of a complete EOS model for ice VII, some difficulties are encountered due to the fact that ice VII is not recoverable under ambient *P-T* conditions, which could otherwise provide a convenient reference point for determining the Gibbs free energy. To deal with this complication, EOS models for ice VII are usually developed by taking advantage of the fact that along the melting curve, the Gibbs free energy of the liquid and the solid are equal. By starting from an appropriate EOS for the liquid phase (*e.g.* from shock wave measurements or MD simulations) and matching to an accurate measurement of the melting curve, thermodynamically consistent EOS models for both liquid and solid H_2O can be readily constructed [*Frank et al.*, 2004; *Dolan et al.*, 2005].

PLANETARY AND GEOPHYSICAL APPLICATIONS

The discontinuous and rapid increase in the melting curve of H_2O at and above the triple point at ~35 to 47 GPa indicates a large increase in $\Delta V/\Delta S$ (entropy versus volume change), significant changes in the physical properties of H_2O, and possible existence of a solid ice phase at higher *P-T*. The unusual behavior of H_2O has several important implications to understanding planetary interiors, including Earth's. Based on extrapolations of the H_2O melting curve [*Datchi et al.*, 2000], Bina and Navrotsky [2000] suggested that ice VII exists in portions of the coldest subducting slabs after H_2O is liberated from hydrous minerals by successive dehydration processes. Although recent studies on the melting curve of H_2O are consistent with the calculations, an eutectic behavior and melting point depression of H_2O with surrounding materials are expected in a multi-component system such as the Earth's mantle, making the presence of solid ice VII phase in the Earth's mantle unlikely. Depending on the curvature of the extrapolated melting curve above the triple point, the solid-liquid phase boundary may intersect the isentropes of Neptune and Uranus and the geotherm of Earth's lower mantle (i.e., at 60 GPa based on the extrapolated melting curve by *Lin et al.* [2005]) (Figures 1, Plate 1). Thus, H_2O could exist in a solid form at *P-T* conditions between the middle to lowermost mantle and the intermediate layers of Neptune and Uranus. Based on seismic and geodynamic data, a high viscosity layer with strongly suppressed flow-induced deformation and convective mixing has been proposed to exist near 2000 km depth [*Forte and Mitrovica*, 2001]. The intersection between the melting curve of H_2O and the mantle geotherm at approximately 60 GPa suggests significant changes in the physical properties of H_2O and the stability of hydrous minerals within the mid-lower mantle and may provide an additional explanation for its viscosity heterogeneity [*Forte and Mitrovica*, 2001; *Lin et al.*, 2005]. However, if the significant increase in the melting curve

Table 1. Comparison of reported EOS parameters determined for ice VII and liquid water by isothermal compression at 300 K. The transition to ice X (~40–60 GPa) is often neglected in determining the EOS parameters of ice VII. V_0: the zero pressure volume; K_{0T}: isothermal bulk modulus at ambient conditions; K_{0T}': the derivative of the isothermal bulk modulus at ambient conditions; P: the pressure range that the measurements were taken over; EOS: 3rd order Birch-Munaghan (BM), Vinet, or Murnaghan (M).

V_0 (cm³/ mol)	K_{0T} (GPa)	K_{0T}'	P (GPa)	EOS	Authors
Ice VII:					
12.3	23.7	4.15	4.3–128	BM	Hemley *et al.*, 1987
14.52	4.26	7.75	2.2–170	Vinet	Loubeyre *et al.*, 1999
12.3	23.9	4.2	3.16–18.55	BM	Fei *et al.*, 1993
12.22	25.04	3.66	3.16–128	M	Dolan *et al.*, 2005
12.4	21.1	4.4	6.82–60.52	BM	Frank *et al.*, 2004
Liquid:					
18.07	2.21	6.029	0.05–0.8	M	Dolan *et al.*, 2005

is due to the occurrence of a superionic phase [*Cavazzoni et al.*, 1999; *Goncharov et al.*, 2005; *Goldman et al.*, 2005], fast protonic diffusion could enhance chemical reactivity with silicates and oxides and lead to the formation of hydrous silicate and oxide compounds in the Earth's lower mantle. Based on cosmochemical abundances and density profiles of Uranus and Neptune, H_2O, CH_4, and NH_3 are presumed to be major components in the middle layers of these icy planets. Fully disassociated, ionic H_2O and NH_3 (and possible CH_4) would lead to reactions between these components, and the formation of complex oxygen, nitrogen, carbon, and hydrogen compounds. In addition, the reaction of H_2O with iron at high P-T to form iron hydrides (FeH_x) provides a mechanism for the incorporation of hydrogen as a light element into growing Earth's core [*Okuchi*, 1997; *Williams and Hemley*, 2001; *Ohtani et al.*, 2005], but also raises complications regarding possible solid ice phase in the Earth's mantle.

Knowledge of the EOS of liquid H_2O is also needed to understand the behavior of possible hydrous fluids and melts that may be present in the Earth's mantle. In addition to the formation of fluids/melts upon dehydration of hydrous minerals in the subducting slabs, regions deeper in the mantle that may become saturated in H_2O would result in formation of dense melts rich in H_2O, MgO, SiO_2, and other components that fractionate into the melt. Since H_2O is likely the most compressible component in the melt, the presence and concentration of H_2O would have a large influence on the melt density and physical properties at depth [*e.g. Richet and Polian*, 1998; *Matsukage et al.*, 2006]. High-pressure studies have shown that wadsleyite and ringwoodite, two major minerals in the transition zone (410-660 km depth), have anomalously high water solubility on the order of 1 wt% (and as much as 3 wt%) whereas the solubility of water in upper- and lower-mantle minerals is 5 to 10 times lower [*e.g. Hohlstedt*

et al., 1996; *Bolfan-Casanova et al.*, 2000; *Murakami et al.*, 2002]. *Bercovici and Karato* [2003] proposed that a thin melt layer may form at 41-km depth if the wadsleyite in the transition zone contained more hydrogen than is soluble into olivine just above 410-km. Thus, passively upwelling mantle crossing the 410-km discontinuity would experience dehydration-induced partial melting. In order to test this hypothesis, it is necessary to know if such a melt layer at 410-km depth would be denser than the solid material above it. Therefore, the physical properties of water and hydrous melts and fluids must be further understood for geodynamic modeling of Earth's potential deep-water cycle.

Acknowledgments. This work was performed under the auspices of the U.S. Department of Energy at the University of California/Lawrence Livermore National Laboratory under Contract No. W-7405-Eng-48. Support for the study was also provided by the Lawrence Livermore Fellowship to J. F. Lin. We thank S. D. Jacobsen for numerous comments and discussions. We also thank G. Galli, W. J. Evans, Z. Jenei, E. Gregoryanz, M. Somayazulu, B. Militzer, V. V. Struzhkin, S. Gramsch, R. J. Hemley, and H. K. Mao for helpful discussions. J. F. Lin and E. Schwegler contributed equally to the paper.

REFERENCES

Alfe, D (2003), First-principles simulations of direct coexistence of solid and liquid aluminum, *Phys. Rev. B* 68, 064423.

Angel, R. J., D. J. Frost, N. L. Ross, and R. Hemley (2001), Stabilities and equations of state of dense hydrous magnesium silicates, *Phys. Earth Planet. Int.*, 127, 181–196.

Aoki, K., H. Yamawaki, M. Sakashita, and H. Fujihisa (1996), Infrared absorption study of the hydrogen-bond symmetrization in ice to 110 GPa, *Phys. Rev. B*, 54, 15673–15677.

Bassett, W. A., A. H. Shen, M. Bucknum, and I. M. Chou (1993), A new diamond anvil cell for hydrothermal studies to 10 GPa and -190°C to 1100°C, *Rev. Sci. Instrum.*, 64, 2340–2345.

Bell, D., and G. Rossman (1992), Water in Earth's mantle: The role of nominally anhydrous minerals: *Science*, 255, 1391–1397.

Benoit, M., M. Bernasconi, P. Focher, and M. Parrinello (1996), New high-pressure phase of ice, *Phys. Rev. Lett.*, 76, 2934–2936.

Benoit, M., D. Marx, and M. Parrinello (1998), Tunneling and zero-point motion in high-pressure ice, *Nature*, 392, 258–261.

Bercovici, D., and S. Karato (2003), Whole-mantle convection and the transition-zone water filter, *Nature*, 425, 39–44.

Bernasconi, M., P. L. Silvestrelli, and M. Parrinello (1998), *Ab initio* infrared absorption study of the hydrogen-bond symmetrization in ice, *Phys. Rev. Lett.*, 81, 1235–1238.

Bina, C. R., and A. Navrotsky (2000), Possible presence of high-pressure ice in cold subducting slabs, *Nature*, 408, 844–847.

Bolfan-Casanova, N., H. Keppler, and D. Rubie (2000), Water partitioning between nominally anhydrous minerals in the MgO–SiO$_2$–H$_2$O system up to 24GPa: implications for the distribution of water in the earth's mantle, *Earth Planet. Sci. Lett.*, 182, 209–221.

Brodholt, J., and B. Wood (1993), Simulations of the structure and thermodynamic properties of water at high pressures and temperatures, *J. Geophys. Res.*, 98, 519–536.

Brown, J. M., and T. J. Shankland (1981), Thermodynamic parameters in the Earth as determined from seismic profiles, *Geophys. J. R. Astr. Soc.*, 66, 579–596.

Cavazzoni, C., *et al.* (1999), Behavior of ammonia and water at high pressure and temperature: implications for planetary physics, *Science*, 283, 44–46.

Celliers, P. M., *et al.* (2004), Electronic conduction in shock-compressed water, *Phys. Plasmas*, 11, L41–L44.

Chau, R., A. C. Mitchell, R. W. Minich, and W. J. Nellis (2001), Electrical conductivity of water compressed dynamically to pressures of 70–180 GPa (0.7–1.8 Mbar), *J. Chem. Phys.*, 114, 1361–1365.

David, H. G., and S. D. Hamann (1959) The chemical effects of pressure. Part 5. The electrical conductivity of water at high shock pressures, *Trans. Faraday Soc.*, 55, 72–78.

Datchi, F., P. Loubeyre, and R. LeToullec (2000), Extended and accurate determination of the melting curves of argon, helium, ice (H$_2$O), and hydrogen (H$_2$), *Phys Rev. B*, 61, 6535–6546.

Demontis, P., R. LeSar, and M. L. Klein (1988), New high-pressure phases of ice, *Phys. Rev. Lett.*, 60, 2284–2287.

de Pater, I., and J. J. Lissauer (2001), Planetary Sciences, pp. 544, University Cambridge Press.

Dolan, D. H., J. N. Johnson, and Y. M. Gupta (2005), Nanosecond freezing of water under multiple shock wave compression: Continuum modeling and wave profile measurements *J. Chem. Phys.*, 123, 064702.

Dubrovinskaia, N., and L. Dubrovinsky (2003a), Melting curve of water studied in externally heated diamond-anvil cell, *High Pressure Res.*, 23, 307–310.

Dubrovinskaia, N., and L. Dubrovinsky (2003b), Whole-cell heater for the diamond anvil cell, *Rev. Sci. Instrum.*, 74, 3433–3437.

Fei, Y., H. K. Mao, and R. J. Hemley (1993), Thermal expansivity, bulk modulus, and melting curve of H$_2$O-ice VII to 20 GPa, *J. Chem. Phys.*, 99, 5369–5373.

Forte, A. M., and J. X. Mitrovica (2001), Deep-mantle high-viscosity flow and thermochemical structure inferred from seismic and geodynamic data, *Nature*, 410, 1049–1056.

Frank, M.R., Y. Fei, and J. Hu (2004), Constraining the equation of state of fluid H$_2$O to 80 GPa using the melting curve, bulk modulus, and thermal expansivity of Ice VII, *Geochim. Cosmochim. Acta*, 68, 2781–2790.

Frost, D. J., *et al.* (2004), Experimental evidence for the existence of iron-rich metal in the Earth's lower mantle, *Nature*, 428, 409–412.

Goldman, N., L. E. Fried, I.-F. W. Kuo, and C. J. Mundy (2005), Bonding in the superionic phase of water, *Phys. Rev. Lett.*, 94, 217801.

Goncharov, A. F., V. V. Struzhkin, M. Somayazulu, R. J. Hemley, and H. K. Mao (1996), Compression of H$_2$O ice to 210 GPa: evidence for a symmetric hydrogen-bonded phase, *Science*, 273, 218–220.

Goncharov, A. F., V. V. Struzhkin, H. K. Mao, and R. J. Hemley (1999), Raman spectroscopy of dense ice and the transition to symmetric hydrogen bonds, *Phys. Rev. Lett.*, 83, 1998–2001.

Goncharov, A. F., *et al.* (2005), Dynamic ionization of water under extreme conditions, *Phys. Rev. Lett.*, 94, 125508.

Hamann, S. D., and M. Linton (1969), Electrical conductivities of aqueous solutions of KCl, KOH and HCl, and the ionization of water a high shock pressures, *Trans. Faraday Soc.*, 65, 2186–2196.

Hansen, C. J., L. Esposito, A. I. F. Stewart, J. Colwell, A. Hendrix, W. Pryor, D. Shemansky, and R. West. (2006) Enceladus' water vapor plume. Science, 311, 1422–1425.

Hemley, R. J., *et al.* (1987), Static compression of H$_2$O-ice to 128 GPa (1.28 Mbar). *Nature*, 330, 737–740.

Holmes, N. C., W. J. Nellis, W. B. Graham, and G. E. Walrafen (1985), Spontaneous Raman scattering from shocked water, *Phys. Rev. Lett.*, 55, 2433–2436.

Holzapfel, W. B. (1969), Effect of pressure and temperature on the conductivity and ionic dissociation of water up to 100 kbar and 1000°C, *J. Chem. Phys.*, 50, 4424–4428.

Holzapfel, W. B. (1972), On the symmetry of hydrogen bonds in ice VII, *J. Chem. Phys.*, 56, 712–715.

Hubbard, W. B., *et al.* (1991), Interior structure of Neptune: comparison with Uranus, *Science*, 253, 648–651.

Jorgensen, W. A., J. Chandrasekhar, J. D. Madura, R. W. Impley, and M. L. Klein, Comparison of simple potential functions for simulating liquid water, *J. Chem. Phys.*, 79, 926–935.

Katoh, E., H. Yamawaki, H. Fujihisa, M. Sakashita, and K. Aoki (2002), Protonic diffusion in high-pressure ice VII, *Science*, 295, 1264–1266.

Kawamoto, T., S. Ochiai, and H. Kagi (2004), Changes in the structure of water deduced from the pressure dependence of the Raman OH frequency, *J. Chem. Phys.*, 120, 5867–5870.

Kohlstedt, D., H. Keppler, and D. Rubie (1996), Solubility of water in the α, β and γ phases of (Mg, Fe)$_2$SiO$_4$, *Contrib. Mineral. Petrol.*, 123, 345–357.

Komabayashi, T. (2006), Phase relations of hydrous peridotite: implications for water circulation in the Earth's mantle, *this volume*.

Lin, J. F., *et al.* (2004a), High pressure-temperature Raman measurements of H_2O melting to 22 GPa and 900 K, *J. Chem. Phys.,* 121, 8423–8427.

Lin, J. F., M. Santoro, V. V. Struzhkin, H. K. Mao, and R. J. Hemley (2004b), *In situ* high pressure-temperature Raman spectroscopy technique with laser-heated diamond anvil cells, *Rev. Sci. Instrum.,* 75, 3302–3306.

Lin, J. F., *et al.* (2005), Melting behavior of H_2O at high pressures and temperatures, *Geophys. Res. Lett.,* 32, L11306, doi:10.1029/2005GL022499.

Loubeyre, P., R. LeToullec, E. Wolanin, M. Hanfland, and D. Hausermann (1999), Modulated phases and proton centering in ice observed by X-ray diffraction up to 170 GPa, *Nature,* 397, 503–506.

Mahoney, M. W., and W. L. Jorgensen (2000), A five-site model for liquid water and the reproduction of the density anomaly by rigid, nonpolarizable potential functions, *J. Chem. Phys.,* 112, 8910–8922.

Matsukage, K. N., Z. Jing, and S.-i. Karato (2006) Density of hydrous silicate melt at the conditions of Earth's deep upper mantle. *Nature,* 438, 488–491.

Mitchell, A. C., and W. J. Nellis (1982), Equation of state and electrical conductivity of water and ammonia shocked to the 100 GPa (1 Mbar) pressure range, *J. Chem. Phys.* 76, 6273–6281.

Murakami, M., K. Hirose, H. Yurimoto, S. Nakashima, and N. Takafuji (2002), Water in earth's lower mantle, *Science,* 295, 1885–1887.

Nellis, W. J., *et al.* (1988), The nature of the interior of Uranus based on studies of planetary ices at high dynamic pressure, *Science,* 240, 779–781.

Ogitsu, T., E. Schwegler, F. Gygi. and G. Galli (2003), Melting of lithium hydride under pressure, *Phys. Rev. Lett.,* 91, 175502.

Ohtani, E., N. Hirao, T. Kondo, M. Ito, and T. Kikegawa (2005), Iron-water reaction at high pressure and temperature, and hydrogen transport into the core, *Phys. Chem. Miner.,* 32, 77–82.

Okuchi, T. (1997) Hydrogen partitioning into molten iron at high pressure: implications for the Earth's core. *Science,* 278, 1781–1784.

Pauling, L. (1935), Structure and entropy of ice and of other crystals with some randomness of atomic arrangement, *J. Am. Chem. Soc.,* 57, 2680–2684.

Petrenko, V. F., and R. W. Whitworth (1999), Physics of ice, pp. 252–283, Oxford Univ. Press, New York.

Polian, A., and M. Grimsditch, (1984), New high-pressure phase of H_2O: Ice X, *Phys. Rev. Lett.,* 52, 1312–1314.

Rahman, A., and F. H. Stillinger (1971), Molecular dynamics study of liquid water, *J. Chem. Phys.,* 33, 3336–3359.

Richet, P., and A. Polian (1998) Water as a dense icelike component in silicate glass. *Science,* 281, 396–398.

Sakane, S., W. Liu, D. J. Doren, E. L. Shock, and R. H. Wood (2001), Prediction of the Gibbs energies and an improved equation of state for water at extreme conditions from ab initio energies with classical simulations, *Geochim. Cosmochim. Acta,* 65, 4067.

Sanz, E., C. Vega, J. L. F. Abascal, and L. G. MacDowell (2004), Phase diagram of water from computer simulation, *Phys. Rev. Lett.,* 92, 255701.

Schilling, J. G., M. B. Bergeron, and R. Evans (1980), Halogens in the mantle beneath the North Atlantic, *Phil. Trans. R. Soc. London A,* 297, 147–178.

Schwegler, E., G. Galli, and F. Gygi (2000), Water under pressure, *Phys. Rev. Lett.,* 84, 2429–2432.

Schwegler, E., G. Galli, F. Gygi, and R. Q. Hood (2001), Dissociation of water under pressure, *Phys. Rev. Lett.,* 87, 265501.

Schwager, B., L. Chudinovskikh, A. Gavriluk, and R. Boehler (2004), Melting curve of H_2O to 90 GPa measured in a laser-heated diamond cell, *J. Phys. Condens. Matter,* 16, S1177–S1179.

Smyth, J. R. (1987), β-Mg_2SiO_4: a potential host for water in the mantle? *Am. Mineral.,* 72, 1051–1055.

Smyth, J.R. and S.D. Jacobsen (2006), Nominally anhydrous minerals and Earth's deep water cycle, *this volume.*

Strässle, Th., A. M. Saitta, Y. Le Godec, G. Hamel, S. Klotz, J. S. Loveday, and R. J. Nelmes (2006), Structure of dense liquid water by neutron scattering to 6.5 GPa / 670 K, *Phys. Rev. Lett.,* 96, 1067801.

Williams, Q., and R. J. Hemley (2001), Hydrogen in the deep earth, *Annu. Rev. Earth Planet. Sci.,* 29, 365– 418.

Wagner, W., and A. Pruß (2002), The IAPWS formulation 1995 for the thermodynamic properties of ordinary water substance for general and scientific use, *J. Phys. Chem. Ref. Data,* 31, 387–535.

Walsh, J. M., and M. H. Rice (1957), Dynamic compression of liquids from measurements on strong shock waves, *J. Chem. Phys.,* 26, 815–823.

Water Content in the Mantle Transition Zone Beneath the North Pacific Derived From the Electrical Conductivity Anomaly

Takao Koyama[1,2], Hisayoshi Shimizu[1], Hisashi Utada[1], Masahiro Ichiki[2], Eiji Ohtani[3], and Ryota Hae[3]

Fukao et al. (2004) inverted semi-global electromagnetic network data for three-dimensional electrical conductivity structure in the mantle transition zone beneath the north Pacific. In this paper we interpret the electrical conductivity structure in terms of the water distribution in the mantle transition zone, using partial derivatives determined by laboratory experiments on mantle materials. Fukao et al. (2004) explained both electrical conductivity and seismic P-wave velocity anomalies with thermal anomalies because of the overall coincidence of high electrical conductivity with low seismic velocity. However, a significant discrepancy is found beneath the Mariana islands where the seismic tomography would indicate little temperature anomaly, while electromagnetic tomography implies high temperatures. Despite limitations and differences in spatial resolution, this result indicates that this particular feature may not be explained by only a thermal effect. Taking into consideration that this region is well populated by subducted slabs, we further assume that this discrepancy is caused by water dehydrated from those slabs. Under this assumption, by combining the Nernst-Einstein relationship (e.g. Karato, 1990) and the recent result of laboratory measurements of hydrogen diffusivity in wadsleyite (Hae et al., 2006), the water content anomaly was estimated from the electrical conductivity anomalies. We find that the mantle transition zone beneath Mariana islands could contain about 0.3 weight % water.

1. INTRODUCTION

Electrical conductivity is an important physical parameter that elucidates the Earth's deep interior because it varies

[1] Earthquake Research Institute, University of Tokyo, Tokyo 113-0032, Japan
[2] Institute for Research on Earth Evolution, Japan Agency of Marine-Earth Science and Technology, Yokosuka 237-0061, Japan
[3] Institute of Mineralogy, Petrology, and Economic Geology, Tohoku University, Sendai 980-8578, Japan

Earth's Deep Water Cycle
Geophysical Monograph Series 168
Copyright 2006 by the American Geophysical Union.
10.1029/168GM13

by orders of magnitude with environment, and can therefore detect some anomalies in physicochemical state of the Earth. Electromagnetic (EM) induction methods can do as well as seismological methods at estimating Earth's mantle structure from the surface to about 1000 km deep in the mid-mantle (e.g. Yukutake, 1965; Banks, 1969) by using the response of the EM field variation induced in the Earth by fields of external origin.

To estimate the deep and large-scale structure such as the mid-mantle, however, it is necessary to measure the EM fields for long periods, say, 100 days to several years at observatories covering a wide area. Conventional studies used geomagnetic variation data from permanent observatories around the world to estimate the electrical conductivity structure by separating the potential of geomagnetic field variation into external and internal parts. It is easily shown that the ratio of external and internal parts is a function of the electrical conductivity in the Earth (Schuster, 1889; Lahiri and Price, 1939; Rikitake, 1950; Banks, 1969; Olsen, 1999). The EM method that uses only the geomagnetic field is called the geomagnetic depth sounding (GDS) method, in which a ratio of vertical and horizontal component in the frequency domain is used as the so-called induction response functions.

$$Hz(f) = T(f) \, Hh(f), \qquad (1)$$

where $Hz(f)$ and $Hh(f)$ are vertical and horizontal components of geomagnetic field in a frequency f, respectively. $T(f)$ is a GDS response, which includes information on the electrical conductivity structure in the Earth.

This GDS method, however, is useful only in the period range of several days and longer, because the vertical geomagnetic variations due to source field morphology are very small at shorter periods. This means that shallow mantle structure, above the mid-mantle, cannot be estimated only by measurement of the geomagnetic field. Because the electrical field variations do not vanish at shorter periods it is essential to use an EM induction method based on electric field data, such as the MT method (Cagniard, 1953) .

$$Eh(f) = Z(f) \, Hh(f), \qquad (2)$$

where $Eh(f)$ and $Hh(f)$ are horizontal components of geoelectric and geomagnetic field in a frequency f, respectively. $Z(f)$ is a MT response or a MT impedance.

To elucidate the Earth's interior, the electrical conductivity structures estimated by EM induction methods are compared with laboratory measurements of the electrical conductivity of mantle materials (Akimoto and Fujisawa, 1965; Omura, 1991; Shankland et al., 1993; Xu et al., 1998). EM induction methods are very sensitive to highly conductive media such as hot regions and fluids, and can thus detect some anomalies in mantle temperature and composition (Tarits et al. 2004, Ichiki et al. 2006). Regional EM studies can detect anisotropic structure (Lizarralde et al, 1995; Evans et al. 2005; Baba et al. 2006), using theoretical and experimental results of mineral physics (Karato, 1990; Mackwell and Kohlstedt, 1990; Constable et al., 1992; Simpson and Tommasi, 2005).

Utada et al. (2003) applied both the GDS and MT methods to geomagnetic field data from observatories and voltage data from submarine cables, and estimated a one dimensional reference model for the electrical conductivity structure of the mid-mantle beneath the Pacific region. The mid-mantle model of Utada et al. (2003) has two jumps at 400 and 650

km depth and is very similar to the model derived from laboratory measurement of the electrical conductivity of the mantle materials at high pressure and high temperature (Xu et al., 1998, 2000). The χ^2 misfit, however, is greater than one even for the one dimensional reference model, which implies significant lateral heterogeneity in the mid-mantle.

In this paper, heterogeneous structures derived from both the electrical conductivity and a seismic velocity model are compared. We then elucidate the origins of these anomalies. Finally we use the electrical conductivity structure to estimate anomalies in water content in the mid-mantle beneath the north Pacific.

2. THREE-DIMENSIONAL EM TOMOGRAPHY

Koyama (2001) used the same geomagnetic field data from sixteen observatories/stations, voltage data from eight submarine cables in the Pacific region (Fig. 1), and calculated MT and GDS response data as used in Utada et al. (2003) and Fukao et al. (2004). In these studies, the GDS response (eq. 1) is defined as the ratio of the vertical and northward components of the geomagnetic field in the frequency domain, and the MT response (eq. 2) is defined as the ratio of voltage $V(f)$, that is, integrated value of the horizontal geoelectric field along the submarine cable, and the northward component of the geomagnetic field $Hx(f)$ in the frequency domain f.

$$Z(f) = V(f)/Hx(f) = \left(\int -Eh(f)dl \right)/Hx(f) \qquad (3)$$

To calculate the value of the voltage V in numerical modeling in practice, V is evaluated by summing discretized electrical field Eh in each line element of the cable dl, that is, $V(f) = -\Sigma \, Eh(f)dl$.

Utada et al. (2003)'s 1-D reference model for the electrical conductivity beneath the Pacific has a jump of about 1.5 orders of magnitude at 400 km and one of 0.5 orders of magnitude at 650 km. This 1-D model is remarkably similar to the conductivity profile deduced from laboratory measurement by Xu et al. (1998, 2000). On the other hand, this 1-D study suggested significant lateral heterogeneity beneath the Pacific from the joint D^+ analysis for all the data (Parker, 1980).

Koyama (2001) imaged the 3-D electrical conductivity anomalous structures relative to the 1-D reference model of Utada et al. (2003). Koyama (2001) estimated the 3-D heterogeneous conductivity structure of the north Pacific area: longitude 90 to 270 degrees east and latitude 0 to 90 degrees north. Vertically, the region extends from 350 to 850 km depth, because the skin depth at one day period is about 650 km and the electrical conductivity in the upper

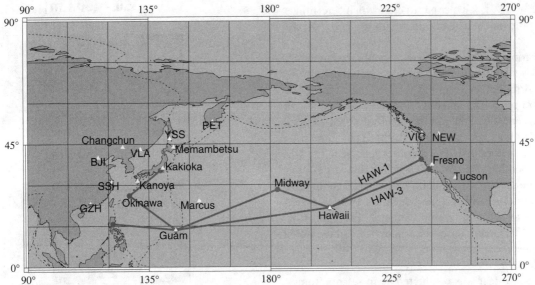

Fig.1. Observatory map: grey lines and white triangles indicate submarine cables and geomagnetic observatories, respectively. The voltages are measured between two ends of each cable. Dashed lines indicate plate boundaries.

mantle is much smaller than that in the transition zone (see Fukao et al., 2004). This 3-D model provides the logarithmic electrical conductivity $\log(\sigma_{3D}/\sigma_{1D})$ in each cell, where σ_{3D} is the estimated three-dimensional heterogeneous structure of the electrical conductivity and σ_{1D} is the electrical conductivity of the reference 1-D model of Utada et al. (2003). The grid size of each cell is 15 degrees horizontally and 100 km vertically. The total number of model parameters is 360 (= 6 by 12 by 5). The other regions of the Earth are fixed to the 1-D reference model of Utada et al. (2003) except for the surface layer including ocean-land contrasts of electrical conductivity.

At the surface, ocean-land contrasts are critical for EM induction problems, known as galvanic distortions (e.g. Jiracek, 1990; Utada and Munekane, 2000). The galvanic distortion is due to contamination by the charge accumulating at the surface topographic contrast and the associated contrast in the electrical conductivity must be taken into account. Therefore, a top layer with a homogeneous thickness of 3000 m was included with laterally heterogeneous conductance values corrected for topography. Topographic data is taken from ETOPO5 data. In inversion processing, the 3-D forward calculations were carried out with 3 degrees grids horizontally and 50 km grids vertically except the surface layer with 3000 m thickness by using a 3-D forward modeling code in the spherical Earth (Koyama et al., 2002). The data are both the real and imaginary parts of MT and GDS responses along with their estimated errors. The MT responses are estimated at eight periods for each of the eight submarine cables. The GDS responses are estimated at eight periods for each of the eight geomagnetic observatories, except at Marcus where GDS responses only at five periods are estimated. The so-called C responses are given at five periods for each of the eight geomagnetic observatories by Fujii and Schultz (2002) and are converted to the GDS responses. The total number of responses, which are complex numbers, is 165, and thus the total number of data is doubled, 330. Hereafter, the geomagnetic variation of the external origin is approximated by a dipole field due to a ring current in the magnetosphere (Banks, 1969).

3. THREE-DIMENSIONAL CONDUCTIVITY STRUCTURE BENEATH THE NORTH PACIFIC

Using the 1-D reference model as the initial model in this 3D inversion, Koyama (2001) inverted the data from submarine cables and geomagnetic observatories for 3-D anomalies of the electrical conductivity beneath the Pacific to minimize the objective function Φ.

$$\Phi = \sum_i \left(\frac{d_i - f_i(m)}{e_i} \right)^2 + \lambda (Lm)^2, \qquad (4)$$

where the first and second terms of the right hand side indicate the total residuals of data parameters and the constraint on model parameters, respectively. d_i and $f_i(m)$ are the EM response data at i-th observatory and the corresponding synthetic response for the model parameter m, respectively. e_i is the data error of d_i. L is the constraint operator for the model parameter m. In this study, horizontal smoothness was adapted for L. λ is a hyper parameter, which determines the weight of the model constraints. In this study, λ is fixed to 5.

Thirty iterations of the quasi-Newton method were made. The data residuals were reduced by 20 %, relative to those for the initial model. A resultant ratio of the term of the model constraints to the objective function Φ in eq. (4) is about ten percent. Fig. 2 shows the resultant three-dimensional model of the electrical conductivity. The model provides logarithmic conductivity anomalies relative to the 1-D reference model, $\log_{10}(\sigma_{3D}/\sigma_{1D})$. Three notable large-scale features can be seen in this model. The electrical conductivity of the mantle transition zone beneath Hawaii is twice as conductive as the 1-D reference model. The transition zone beneath Mariana is thrice as conductive. On the other hand, the uppermost lower mantle beneath Philippine is half as conductive.

A checkerboard test was conducted as in seismic tomographic studies. The result of the test shows that the region beneath the submarine cables through Hawaii, Guam and Philippine has relatively good resolution (Fig.3). However, the electrical conductivity relates non-linearly to MT and GDS responses and, unlike linear seismic tomography, the checkerboard test may not be applied to EM induction problems. The intensity of the anomalies in our 3-D model (Fig.2) is almost as small as the anomaly intensity of the checkerboard test (Fig.3a), allowing us to regard the test's intensity resolution as applicable to the data. We conducted another resolution test by calculating the sensitivity of the electrical conductivity of each model cell on the data set, which is supposed to be the following,

$$S_j(m) = \sum_i \left| \frac{\partial f_i(m)}{e_i \partial m_j} \right|^2 , \qquad (5)$$

where $S_j(m)$ is the total deviation of set of data parameters divided by its error, by slightly changing each model parameter m_j. that is $\log(\sigma_{3D}/\sigma_{1D})$ of the j-th cell, and then can represent the sensitivity of each model parameter on data set. $S_j(m)$ can express the effect only by each model cell unlike the previous checkerboard test which may have inductive interactions between the checkers, and thus this sensitivity test can be complementary to the checkerboard tests. Fig.4 shows $S_j(m)$ in logarithmic scale, and it turned out that there is high sensitivity at the region which can be also well-resolved in the checkerboard test, and then the result of both tests are

Fig.2. The contour map of three-dimensional model of the electrical conductivity inverted from the data of MT and GDS responses: Each model grid size is 15° x 15° (horizontal) x 100 km (vertical). A unit is anomalies from a 1-D reference model (σ_{1D}) by Utada et al. (2003) in logarithmic scale, $\log_{10}(\sigma_{3D}/\sigma_{1D})$.

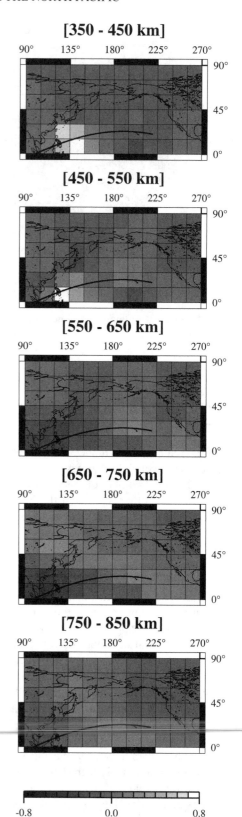

[350 - 450 km]

[450 - 550 km]

[550 - 650 km]

[650 - 750 km]

[750 - 850 km]

-0.8   0.0   0.8

Electrical Conductivity Anomaly, $\log_{10}(\sigma_{3D}/\sigma_{1D})$

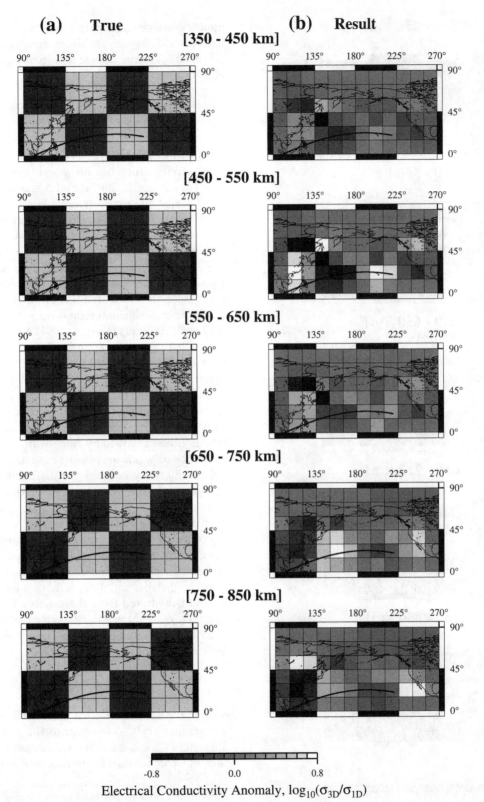

Fig.3. Checkerboard test: (a) checkerboard model given a priori which are ±0.5 unit. (b) result of checkerboard test inverted from synthetic MT and GDS data responses for a checkerboard model (a).

[350 - 450 km]

[450 - 550 km]

[550 - 650 km]

[650 - 750 km]

[750 - 850 km]

$$\log_{10} S_j(\boldsymbol{m})$$

almost consistent. Hereafter, we focus on the presumably high-resolved region. Details on resolution of both seismic and electrical structure are shown and discussed in Fukao et al. (2004).

4. ESTIMATING WATER DISTRIBUTION IN THE MANTLE TRANSITION ZONE

As in Fukao et al. (2004), we compare the electrical conductivity structure with seismic P-wave velocity structure (Fukao et al., 2003). Fig. 5 shows both models. On large scales, both models seem similar, for example, the mid-mantle beneath Hawaii has high conductivity and low velocity, and the uppermost lower mantle beneath Philippine has low conductivity and high velocity. Fukao et al. (2004) assumed both conductivity and P-wave velocity anomalies were simply due to thermal perturbation in the mantle and converted them to thermal anomalies, by using the experimental relationship derived from laboratory measurements (Karato, 1993; Xu et al., 2000). They concluded that these features thermal in origin because the temperatures derived from electrical conductivity and from seismic velocity are similar. A very high conductive region in the mid-mantle beneath Mariana, however, cannot be explained with a thermal origin, because the seismic velocity is not very anomalous. This anomaly may thus be explained by compositional differences. The existence of small ions is effective at enhancing electrical conductivity because the diffusion of small ions is very fast. On the other hand, P wave velocity does not change as much due to existence of small ions, such as H^+ from water (Yusa and Inoue, 1997). Therefore we find hydrogen from water the most preferable explanation for the discrepancies in thermal model derived from seismic and electrical structures. This region is well populated by subducted slabs, and it might have a lot of water carried by and dehydrated from these slabs. The electrical conductivity σ (S/m) is related to concentration of hydrogen c (atom/m^3), that is, water content linearly by the Nernst-Einstein relationship (Karato, 1990)

$$\sigma = cDq^2/kT, \qquad (6)$$

where D, q are diffusivity and electrical charge of hydrogen, respectively. q is equal to the elementary charge, 1.60×10^{-19}(C). k is the Boltzman's constant. T is a temperature, of which we used a reference model in Ito and Katsura

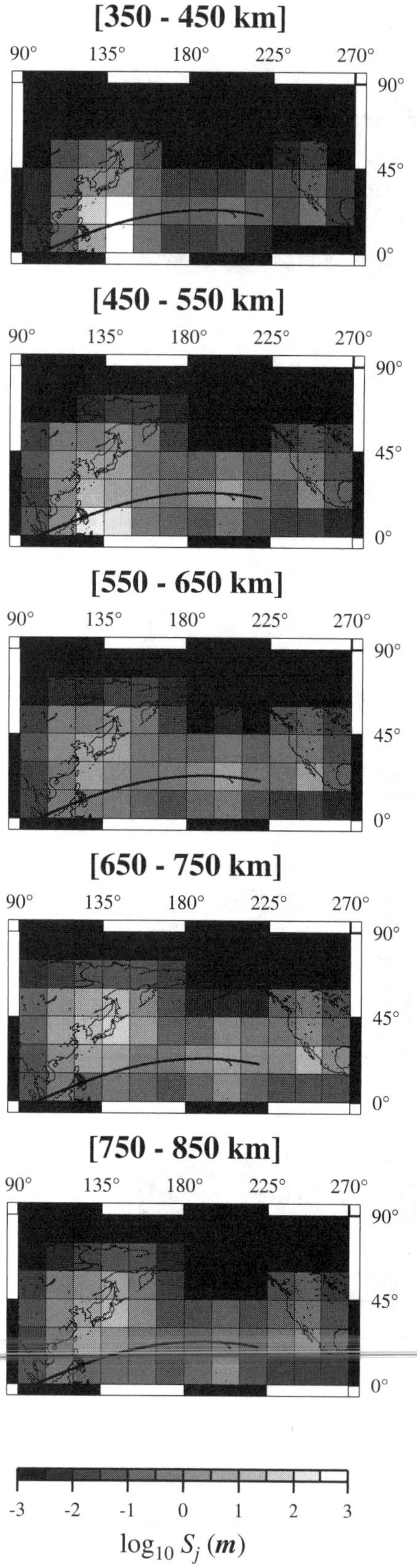

Fig. 4. The value of sensitivity $S_j(m)$ in logarithmic scale is contoured in each model cell, which is defined at eq. (5). It shows how sensitive to used data sets in this analysis each model parameter is.

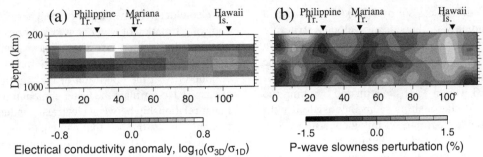

Fig. 5. Profile of anomalies beneath curve through Hawaii, Guam and Philippine shown in maps of Fig 2 as a black curve. (a) electrical conductivity (b) seismic velocity anomaly in Fukao et al. (2004). Two horizontal black lines in each map indicate the upper and lower boundaries of the mantle transition zone.

(1989), which was also used in Xu et al. (2000) for conversion of temperature to the depth profile, and in Fukao et al. for conversion of physical anomaly to temperature anomaly. Water content c_w (weight% H_2O) is estimated by using c (atom/m^3) as $c_w / 100 = (c/N_A/2) * M_{water} / d$, where N_A, M_{water} and d are the Avogadro's number 6.02×10^{23}(atom/mol), a molecular weight of H_2O 18(g/mol) and a density of wadsleyite 4×10^6(g/m^3), respectively. Although we cannot preclude that the Mariana anomaly may be partly affected by a temperature anomaly, we assume that the Mariana anomaly is caused only by hydrogen, for reasons of simplicity, limitations in resolving power, and differences in the resolution length scales between the seismic and conductivity data.

Very recently, Hae et al. (2006) measured the diffusivity of hydrogen in wadsleyite at high pressure (15–16 GPa) and temperature (900–1200 °C) with a Kawai-type multi anvil. The measured diffusivity D is $D = 9.6 \times 10^{-6}$ exp [-123 (kJ mol^{-1}) / RT]. With this value for D by following eq. (6), we estimate the water content in the upper transition zone beneath Mariana to be about 0.3 weight % H_2O (Fig. 6).

Recently, Huang et al. (2005) measured the electrical conductivity of hydrous wadsleyite and ringwoodite at pressure of 14–16 GPa and temperature of 500–1000 °C, and found the relationship between the electrical conductivity σ and water content c_w (weight% H_2O),

$$\sigma = A c_w^r \exp(-H^*/RT), \qquad (7)$$

where H^*(kJ/mol) is activation enthalpy, and A(S/m) and r are experimental constants, 88(kJ/mol), 380(S/m) and 0.66 for wadsleyite, 104(kJ/mol), 4070(S/m) and 0.69 for ringwoodite, respectively. R is the gas constant. T is the temperature, just the same as one in eq. (6). In this paper, just the laboratory measurement was simply accounted and the oxygen coefficient correction was not adapted, although it is assumed by Huang et al. (2005). Unlike eq. (6), r is not unity, and then electrical conductivity is not linearly related

to water content. Huang et al. suggested that it means that not all hydrogen atoms are free and a ratio of free hydrogen atoms decreases as the water content increases. If this relationship by Huang et al. (2005) is adapted to the electrical conductivity structure beneath the Pacific, the water content could be about 2.3 weight %.

5. DISCUSSION

We discuss the water content anomaly in the mid-mantle beneath the Marianas estimated in the previous section. The value of 0.3 weight % H_2O is well below the saturation limit of water in wadsleyite of about 3 weight % H_2O (Kohlstedt et al., 1996).

The water in wadsleyite decreases P wave velocity Vp. The dependency of the bulk modulus on water content in wadsleyite is found to be -2 to -4 (%)/(weight% H_2O) (Yusa and Inoue, 1997). For ringwoodite (Jacobsen et al., 2004), the dependency of Vp on H_2O is (-0.4 (km/s)/(weight % H_2O). Because the anelastic effect of water on Vp is unknown, only the anharmonic effect is included in the above evaluation (Karato, 1993, 2003). According to this water effect on Vp,

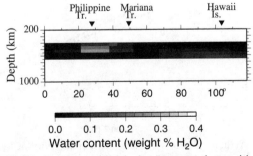

Fig. 6. Water content anomalies in the upper mantle transition zone derived from the electrical conductivity anomalies in Fig 5(a) by using the Nernst-Eisnstein relationship combining with the laboratory measurement of diffusivity of hydrogen in wadsleyite by Hae et al. (2006). A unit is weight % H_2O.

0.3 weight % H_2O decreasess Vp by 0.3–0.6 %, consistent with the small anomaly in the seismic velocity structure (Fig. 5). Thus 0.3 weight % H_2O is feasible to explain both electrical conductivity and seismic velocity anomalies in the mid-mantle beneath Mariana.

However, the 2.3 weight % H_2O estimated by using the Hunag et al (2005) relationships seem inconsistent with the result of seismic tomography. If this inconsistency is a consequence of having ignored temperature anomalies, the temperature should be about 300K lower than the reference model to explain the seismic structure in the presence of this amount of water (see Suetsugu et al., in this volume). It, however, should require much more water to explain the electrical conductivity with lower temperature and water content exceeds its limit solved in wadsleyite. Then thermal effect cannot explain the both anomalies of seismic and electrical structure including water according to the relationship of Huang et al. (2005).

Therefore if a relationship by Huang et al. (2005) is correct, other mechanisms than water and thermal effects must be considered to explain both seismic and electrical conductivity structures in the future work.

Fig.7 shows the temperature dependency of the electrical conductivity inferred from the laboratory measurement by both Hae et al. (2006) and Huang et al. (2005), that is, experimental relationship of eqs. (6) and (7) in which water content is supposed to 0.1 weight % H_2O. This figure indicates the both estimations of the electrical conductivity are not very different in the temperature range where the measurements were conducted in the laboratories. Therefore large discrepancy of estimation of water content may be caused by extrapolation of the temperature dependency. Therefore laboratory measurements of both hydrogen diffusivity and electrical conductivity at the condition of the mantle transition zone are aspired to estimate the water content more precisely. Non linear relationship between electrical conductivity and water content, that is, the fact that r in eq. (7) is not unity may be true, and this coefficient is also required to be measured in the mid-mantle condition.

Combined with other geophysical information such as seismic velocity structure and laboratory studies, electrical conductivity can provide important information on the deep Earth's interior, including the distribution of water. To elucidate further details on the electrical conductivity structure in the whole Earth's mantle, the efforts should be paid to install and maintain the long term measurement of geoelectrical and geomagnetic field on land and on the sea floor, including geomagnetic measurement by long life satellites.

Acknowledgements. We thank D. Suetsugu, S. Ono, T. Hanyu, K. Baba, A. Shito, and T. Yoshino for fruitful discussion. Y. Fukao and M. Obayashi kindly provided us their results of seismic tomography. Critical comments by two anonymous reviewers were very constructive to improve our manuscripts. The research was supported by the MUD project of IFREE/JAMSTEC. Numerical calculations were partly executed by using the Earth Simulator in JAMSTEC.

REFERENCES

Akimoto, S., and H. Fujisawa, Demonstration of the electrical conductivity jump produced by the olivine-spinel transition, *J. Geophys. Res., 70,* 443–449, 1965.

Baba, K., A.D. Chave, R.L. Evans, G. Hirth, and R.L. Mackie, Mantle dynamics beneath the East Pacific Rise at 17°S: Insights from the Mantle Electromagnetic and Tomography (MELT) experiment, *J. Geophys. Res., 111,* doi:10.1029/2004JB003598, 2006.

Banks, R. J., Geomagnetic variations and the electrical conductivity of the upper mantle, *Geophys. J. R. astr. Soc., 17,* 457–487, 1969.

Cagniard, L., Basic theory of the magneto-telluric method of geophysical prospecting, *Geophysics, 45,* 1–16, 1953.

Constable, S., T.J. Shankland, and A. Duba, The electrical conductivity of an isotropic olivine mantle, *J. Geophys. Res., 97,* 3397–3404, 1992.

Evans, R.L., G. Hirth, K. Baba, D. Forsyth, A. Chave, and R. Mackie, Geophysical evidence from the MELT area for compositional controls on oceanic plates, *Nature, 437,* 249–252, 2005.

Fujii, I., and A. Schultz, The three-dimensional electromagnetic response of the Earth to ring current and auroral oval excitation, *Geophys, J. Int., 151,* 689–709, 2002.

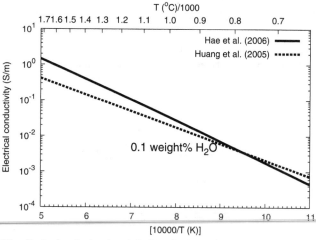

Fig. 7. Arrhenius's plots of electrical conductivity of hydrous wadsleyite inferred from laboratory measurement. Bold line shows temperature dependency of electrical conductivity derived from Hae et al. (2006), that is eq. (6). Dashed line shows one derived from Huang et al. (2005), that is eq. (7). Water content is supposed to 0.1 weight% H_2O.

Fukao, Y., A. To, and M. Obayashi, Whole mantle P-wave tomography using *P* and *PP-P* data, *J. Geophys. Res. 108*, doi:10.1029/2001JB000989, 2003.

Fukao, Y., T. Koyama, M. Obayashi, and H. Utada, Trans-Pacific temperature field in the mantle transition region derived from seismic and electromagnetic tomography, *EPSL, 217*, 425–434, 2004.

Hae, R., E. Ohtani, T. Kubo, T. Koyama, and H. Utada, Hydrogen diffusivity in wadsleyite and water distribution in the mantle transition zone, *EPSL, 243*, 141–148, 2006.

Huang, X., Y. Xu, and S. Karato, Water content in the transition zone from electrical conductivity of wadsleyite and ringwoodite, *Nature, 434*, 746–749, 2005.

Ichiki, M., K. Baba, M. Obayashi, and H. Utada, Water content and geotherm in the upper mantle above the stagnant slab: Interpretation of electrical conductivity and seismic P-wave velocity models, *Phys. Earth Planet. Inter., 155*, 1–15, 2006.

Ito, E. and T. Katsura, A temperature profile of the mantle transition zone, *Geophys. Res. Lett., 16*, 425–428, 1989.

Jacobsen, S.D., J.R. Smyth, H. Spetzler, C.M. Holl, and D.J. Frost, Sound velocity and elastic constants of iron-bearing hydrous ringwoodite, *Phys. Earth Planet. Inter., 143–144*, 47–56, 2004.

Jiracek, G. R., Near-surface and topographic distortion in electromagnetic induction, *Surveys in Geophysics, 11*, 163–203, 1990.

Karato, S., The role of hydrogen in the electrical conductivity of the upper mantle, *Nature, 347*, 272–273, 1990.

Karato, S., Importance of anelasticity in the interpretation of seismic tomography, *Geophys. Res. Lett., 20*, 1623–1626, 1993.

Karato, S., Mapping water content in the upper mantle, *in Inside the Subduction Factory, AGU Monograph, 138*, 135–152, 2003.

Kohlstedt, D.L, H. Keppler, and D.C. Rubie, Solubility of α, β, and γ phases of $(Mg,Fe)_2SiO_4$, *Contrib. Mineral. Petrol., 123*, 345–357, 1996.

Koyama, T., A Study on the Electrical Conductivity of the Mantle by Voltage Measurements of Submarine Cables, *Ph.D. Thesis, University of Tokyo*, 129 pp, 2001.

Koyama, T., H. Shimizu, and H. Utada, Possible effects of lateral heterogeneity in the D" layer on electromagnetic variations of core origin, *Phys. Earth Planet. Inter., 129*, 99–116, 2002.

Lahiri, B.N., and A.T. Price, Electromagnetic induction in non-uniform conductors and the determination of the conductivity of the Earth from terrestrial magnetic variations, *Phil. Trans. Roy. Soc. London A, 237*, 509–540, 1939.

Lizarralde, D., A. Chave, G. Hirth, and A. Schultz, Northeastern Pacific mantle conductivity profile from long-period magnetotelluric sounding using Hawaii-to California submarine cable data, *J. Geophys. Res., 100*, 17,837–17,854, 1995.

Mackwell, S.J. and D.L. Kohlstedt, Diffusion of Hydrogen in Olivine: Implication for Water in the Mantle, *J. Geophys. Res., 95*, 5079–5088, 1990.

Olsen, N., Long–period (30 days–1 year) electromagnetic sounding and the electrical conductivity of the lower mantle beneath Europe, *Geophys. J. Int., 138*, 179–187, 1999.

Omura, K., Change of electrical conductivity of olivine associated with the olivine-spinel transition, *Phys. Earth Planet. Inter., 65*, 292–307, 1991.

Parker, R.L., The inverse problem of electromagnetic induction: existence and construction of solutions based on incomplete data, *J. Geophys. Res., 85*, 4421–4428, 1980.

Rikitake, T., Electromagnetic induction within the Earth and its relation to the electrical state of the Earth's interior. 2., *Bull. Earthquake Res. Inst., Univ. Tokyo, 28*, 263–283, 1950.

Schuster, A., The diurnal variation of terrestrial magnetism, *Phil. Trans. Roy. Soc. London A, 180*, 467–518, 1889.

Shankland, T.J., J. Peyronneau, and J.P. Poirier, Electrical conductivity of the Earth's lower mantle, *Nature, 366*, 453–455, 1993.

Simpson, F., and A. Tommasi, Hydrogen diffusivity and electrical anisotropy of a peridotite mantle, *Geophys. J. Int., 160*, 1092–1102, 2005.

Suetsugu, D., T. Inoue, A. Yamada, D. Zhao, and M. Obayashi, Temperature anomalies and water content in the mantle transition zone from P-wave velocity structure and the 660-km discontinuity depths beneath subduction zones, *AGU monograph, this volume*, 2006.

Tarits, P, S. Hautot, and F. Perrier, Water in the mantle: Results from electrical conductivity beneath the French Alps, *Geophys. Res. Lett., 31*, doi:10.1029/2003GL019277, 2004.

Utada, H., T. Koyama, H. Shimizu, and A. D. Chave, A semi-global reference model for electrical conductivity in the mid-mantle beneath the north Pacific region, *Geophys. Res. Lett., 30*, doi:10.1029/2002GL016092, 2003.

Utada, H., and H. Munekane, On galvanic distortion of regional three-dimensional magnetotelluric impedances, *Geophys. J. Int., 140*, 385–398, 2000.

Xu, Y., B. T. Poe, T. J. Shankland, and D. C. Rubie, Electrical conductivity of olivine, wadsleyite and ringwoodite under upper-mantle conditions, *Science, 280*, 1415–1418, 1998.

Xu, Y., T. J. Shankland, and B. T. Poe, Laboratory-based electrical conductivity in the Earth's mantle, *J. Geophys. Res. 105*, 27,865–27,875, 2000.

Yukutake, T., The solar cycle contribution to the secular change in the geomagnetic field, *J. Geomag. Geoelectr., 17*, 287–309, 1993.

Yusa, H. and T. Inoue, Compressibility of hydrous wadsleyite (β-phase) in Mg_2SiO_4 by high pressure X-ray diffraction, *Geophys. Res. Lett., 24*, 1831-1834, 1997.

Takao Koyama, Earthquake Research Institute, Yayaoi 1-1-1, Bunkyo, 113-0032, Japan (tkoyama@eri.u-tokyo.ac.jp)

A Water-Rich Transition Zone Beneath the Eastern United States and Gulf of Mexico From Multiple ScS Reverberations

Anna M. Courtier and Justin Revenaugh

Department of Geology and Geophysics, University of Minnesota, Minneapolis, Minnesota

We examine mantle discontinuities beneath the United States and Gulf of Mexico using multiple ScS reverberations from earthquakes in Central and South America captured by 65 broadband and long-period seismometers across the United States. The depths of discontinuities and the impedance contrasts across them were estimated using a hierarchical waveform inversion and stacking method. The path-averaged depth of the 410-km discontinuity varies moderately across the study area and is particularly shallow (~395 km) beneath the eastern United States. Topography on the 660-km discontinuity is more subdued and is close to the global mean depth. The 520-km discontinuity is seen consistently across the study area, though both the depth and the impedance contrast of the discontinuity vary significantly. Corridors in the eastern United States and Gulf of Mexico have extremely strong 520-km discontinuities relative to the corresponding 410-km and 660-km discontinuities. We attribute the shallow 410-km and strong 520-km discontinuities beneath the eastern United States and Gulf of Mexico to a locally water-rich transition zone.

INTRODUCTION

Water must be present to some extent in the Earth's mantle [e.g. *Bell and Rossman*, 1992], though the total mass of water and its distribution in the interior are largely unknown [e.g. *Drake and Righter*, 2002; *Hirschmann et al.*, 2005]. The water content of the mantle is largely governed by the amount of hydrogen that partitioned into the accreting core and mantle during Earth formation, the amount of water that subsequently degassed to form the hydrosphere, and the amount that is recycled into the interior through subduction [e.g. *Ahrens*, 1989; *Williams and Hemley*, 2001]. Each of the major mantle minerals has the ability to incorporate at least trace amounts of water in its structure [e.g. *Bell and Rossman*, 1992; *Kohlstedt et al.*, 1996; *Bolfan-Casanova et al.*, 2000; *Murakami et al.*, 2002; *Bolfan-Casanova et al.*, 2003; *Litasov et al.*, 2003; *Bolfan-Casanova*, 2005], allowing for the equivalent of half of the world's oceans in a "dry" mantle [*Hirschmann et al.*, 2005] and potentially much more [*Smyth*, 1987; *Bell and Rossman*, 1992]. Hydrogen may be present as point defects, as structurally bound hydroxyl [(OH)$^-$], or as molecular water (H_2O), each of which is colloquially described as "water" in the literature.

The distribution of water in the mantle can be constrained by considering seismic observations in conjunction with experimental constraints on water in mantle minerals. A number of authors have examined the solubility of water in mantle minerals [e.g. *Bell and Rossman*, 1992; *Kohlstedt et al.*, 1996; *Ingrin and Skogby*, 2000; *Bolfan-Casanova et al.*, 2000; *Bolfan-Casanova et al.*, 2003; *Bolfan-Casanova*, 2005] and the pressure-temperature stability fields of minerals containing water [e.g. *Irifune et al.*, 1998; *Higo et al.*, 2001; *Ohtani et al.*, 2001; *Smyth and Frost*, 2002]. Although all of the most abundant mantle minerals are capable of storing water in their structures, for some the concentration of water may reach only the parts per million level.

Earth's Deep Water Cycle
Geophysical Monograph Series 168
Copyright 2006 by the American Geophysical Union.
10.1029/168GM14

Water solubility measurements indicate that transition zone minerals can incorporate considerably more water than both the remaining primary upper mantle minerals and primary lower mantle minerals [e.g. *Smyth*, 1987; *Kohlstedt et al.*, 1996; *Bolfan-Casanova et al.*, 2000]. While it is highly unlikely that the entire mantle is saturated with water, the transition zone may be a significant reservoir for water due to its enhanced water solubility.

The transition zone seismic discontinuities, herein referred to as the 410-km, 520-km, and 660-km discontinuities, correspond to the phase transition from olivine to wadsleyite (β-Mg$_2$SiO$_4$), the transition from wadsleyite to ringwoodite (γ-Mg$_2$SiO$_4$), and the dissociation of ringwoodite to perovskite-type (Mg,Fe)SiO$_3$ and magnesiowüstite-(Mg,Fe)O, respectively (see *Helffrich* [2000] for a review). These three reactions have pressure-temperature stability fields that vary between hydrous and anhydrous conditions. The 410-km discontinuity migrates to shallower depths in the presence of water [*Smyth and Frost*, 2002; *Chen et al.*, 2002; *Komabayashi et al.*, 2005; *Komabayashi et al.*, this volume], whereas the 660-km discontinuity may deepen [*Higo et al.*, 2001]. The presence of water also can broaden or sharpen mantle velocity transitions [e.g. *Akaogi et al.*, 1989; *Wood*, 1995]. Transition thicknesses under anhydrous conditions are 9-18 km for the 410-km discontinuity [*Akaogi et al.*, 1989], 30 km for the 520-km discontinuity [*Akaogi et al.*, 1989], and 4 km or less for the 660-km discontinuity [*Ito and Takahashi*, 1989]. For under-saturated hydrous conditions, the 410-km and 660-km discontinuities broaden to thicknesses of up to 40 km [*Wood*, 1995; *Helffrich and Wood*, 1996; *Smyth and Frost*, 2002] and 13 km [*Higo et al.*, 2001], respectively. The 520-km discontinuity sharpens in the presence of water, occurring over a thickness of less than 15 km [*Inoue et al.*, 1998].

Seismic methods yield the most direct observations of the transition zone discontinuities. A global study by *Flanagan and Shearer* [1998] reported mean depths of 418 km, 515 km, and 660 km for the three discontinuities. The 520-km discontinuity is seen more frequently in areas where data density is high, and it typically has a much smaller apparent impedance contrast than either the 410-km or 660-km discontinuities. The poor signal-to-noise ratio of data sampling the low-contrast 520-km discontinuity almost certainly adds to the apparent variability in its depth. It is also occasionally observed as a two-part discontinuity, further adding to apparent depth variability [e.g. *Deuss and Woodhouse*, 2001]. Transition thicknesses influence the impedance contrasts measured by seismic methods. As a transition broadens, its effective reflection and conversion coefficients decrease [e.g. *Richards*, 1972], making the discontinuity more difficult to detect and, often, biasing downward the estimated impedance and/or velocity contrast. The contrasts at the 410-km and 660-km discontinuities are large enough that even if water broadens them, they should still be easily detectable with multiple *ScS* reverberations. The 520-km discontinuity, which narrows in the presence of water, may appear stronger and be more easily detected in hydrous environments. If the transition zone is a significant reservoir for water in the mantle, the effect of that water should be apparent in seismic observations.

Several seismic studies have called upon water to explain anomalous observations of the transition zone and the overlying mantle. A layer of silicate partial melt was proposed by *Revenaugh and Sipkin* [1994] to explain an impedance decrease detected at an average of 80 km above the 410-km discontinuity beneath easternmost China and the Sea of Japan. The inferred partial-melt layer was detected with multiple *ScS* reverberations and was seen consistently across the study area, though the thickness of the layer varied from ~50 to 100 km. Volatiles, principally water, were suggested as a catalyst for producing the melt.

Song et al. [2004] reported a very similar feature beneath the northwestern United States. The low velocity zone in this study was detected in triplicated shear-wave arrivals and extended from ~20 to 90 km above the 410-km discontinuity. It too is interpreted as a partial melt layer and linked to hydration from past subduction in the region. *Van der Meijde et al.* [2004] used receiver functions to infer the presence of up to 1000 ppm by weight of water near depths of 400 km beneath the Mediterranean, another region of subduction, which again may be the source of the water. A second receiver function study [*Vinnik et al.*, 2003] reported a ~60-km thick low velocity layer on top of the 410-km discontinuity beneath the Arabian plate. This is interpreted as a water-rich layer underlying the dry continental root of the plate. The transition zone water filter model of *Bercovici and Karato* [2003] predicts occasional melt above the 410-km discontinuity as a byproduct of the difference in water solubilities of transition zone and overlying upper mantle minerals. The model predicts that as upwelling mantle material reaches the water-rich transition zone, it is able to incorporate water on the order of $1-2$ wt% into its mineral structures. When this material passes through the 410-km transition, it becomes at least saturated, if not super-saturated, with water due to the lower solubility of water in olivine than in wadsleyite [e.g. *Kohlstedt et al.*, 1996; *Hirschmann et al.*, 2005]. Water saturation greatly depresses the melting temperature of the olivine and can cause localized partial melting above the transition zone [*Inoue*, 1994]. The resulting melt may be intermediate in density to olivine and wadsleyite [e.g. *Stolper et al.*, 1981; *Ohtani et al.*, 1995] and could pond above the 410-km discontinuity.

Broadening of the 410-km phase transition to thicknesses of $20-35$ km is observed in the Mediterranean region [*Van* der Meijde et al., 2003]. This can be explained by 0.07 wt % H_2O (700 ppm) in the surrounding mantle. The 660-km transition also may broaden in the region, although the receiver function method used may not be able to detect the amount of broadening across the transition interval expected at that depth for the estimated water content. *Van der Meijde et al.* [2005] observed increases in transition zone thickness beneath the same region, which is consistent with either water or low temperature persisting throughout the depth range of the transition zone. They further note that the velocity transition at the 520-km discontinuity may be as sharp as 20 km in the region. Several thousand kilometers to the north-northwest, weak conversions from the 660-km discontinuity beneath the North Sea could be the result of a broadened discontinuity [*Helffrich et al.*, 2003]. A transition occurring over a 13-km interval could cause the observed behavior, consistent with the expected influence of water at 660 km depth [*Higo et al.*, 2001].

Low seismic velocities due to water in the mantle have been reported for compressional and shear waves in regions of both current and past subduction [e.g. *Zhao*, 2001; *Nolet and Zielhuis*, 1994]. In the mantle wedge, at depths to 400 km, a low velocity zone often lies just above the subducting slab. Here the low velocities are attributed to water released from the slab as hydrous minerals dewater [*Zhao*, 2001]. Deeper in the mantle, *Nolet and Zielhuis* [1994] find evidence of water in the upper mantle and transition zone in shear wave tomography of the mantle beneath the Russian platform. There, a low shear wave velocity anomaly extends from 300 to 500 km depth and follows the trend of the ancient Tornquist-Teisseyre subduction zone, which initiated with closure of the Tornquist Sea a maximum of ~450 million years ago [*Bergstrom*, 1990; *Scotese and McKerrow*, 1990]. The authors conclude that 85 million years of subduction have injected a significant amount of water into the transition zone, substantially lowering shear wave velocities.

In this context, we conducted an *ScS* reverberation study examining variability in discontinuity depths and impedance contrasts beneath the United States and Gulf of Mexico. Multiple *ScS* reverberations are useful tools for detecting regional variability that occurs due to changes in temperature or chemistry along mantle discontinuities. Past and present subduction of the Farallon and Juan de Fuca plates lead to slab-dewatering and the possibility for areas of water storage in the mantle beneath the study area.

DATA

We compiled a dataset of 130 long-period and broadband seismograms from fourteen intermediate depth events in Central and South America. Only events with magnitude m_b

≥ 5.9 and depth $z \geq 95$ km were considered. Data with low signal to noise ratios or with excessive apparent source complexity were discarded. The events occurred between 1974 and 2001 and were recorded at 65 stations from a variety of networks across the United States; the Digital World-Wide Standardized Seismograph Network (DWWSSN), United States National Seismic Network (USNSN), High-Gain Long-Period Network (HGLP), Lamont-Doherty Cooperative Seismographic Network (LCSN), IRIS Global Seismograph Network (GSN), Seismic Research Observatory (SRO), Pacific Northwest Regional Seismic Network (PNSN), Leo Brady Network (LB), TERRAscope (Southern California Seismic Network, SCSN), GEOSCOPE, Berkeley Digital Seismograph Network (BDSN), California Transect Network (CT), and ANZA Regional Network (ANZA). See Table 1 for source parameters of the events.

Seismograms were rotated, deconvolved to ground velocity, low-pass filtered, and decimated to a three-second sampling interval following *Revenaugh and Jordan* [1989]. Transverse component data were separated into eight source-receiver paths based on geographic sampling (Figure 1) and considerations of data density. Six paths connect Central America and the United States. The remaining two connect South America and the United States. An example trace is shown in Figure 2.

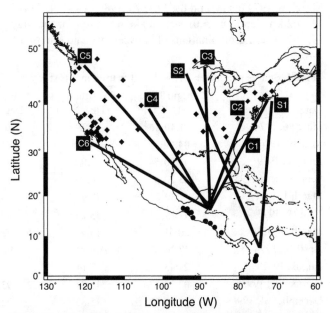

Figure 1. Map of the study area showing earthquakes (circles) and seismic stations (diamonds). Black bars indicate particular geographic paths and are numbered from east to west according to the source region; Central America (C) or South America (S). The bars are schematic; actual sampling is much broader geographically.

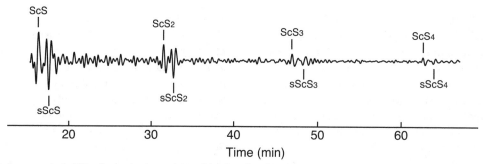

Figure 2. Long period, SH-polarized seismogram of the October 21, 1995 earthquake recorded at station SBC (34.442°N, 119.713° W) with multiple *ScS* phases within the reverberative interval labeled.

METHOD

Zeroth- and first-order *ScS* reverberations (i.e. multiple *ScS* and *sScS* phases and similar arrivals once-reflected from discontinuities within the mantle; Figure 3) were modeled using the hierarchical waveform inversion method of *Revenaugh and Jordan* [1991a]. The only change to their method was the addition of a finite source duration, taken as the half-duration of the event as given by the Harvard Centroid Moment Tensor (CMT) catalog which scales directly with seismic moment. Source parameters for the two events prior to 1977 were taken from nearby events in the CMT catalog. Specifically, the July 18, 1983 (12.67° N, 87.18° W, 86 km depth, m_b 6.0) earthquake was used for the March 6, 1974 event, and the June 25, 1980 (4.44° N, 75.78° W, 162 km depth, m_b 6.0) earthquake was used for the May 19, 1976 event. Combinations of between two and four ScS_n-

$sScS_n$ phase pairs were modeled on each seismogram, with the number depending on the level of ambient noise and interference from major arc arrivals and phases from other earthquakes. Model parameters include the whole-mantle quality factor (Q_{ScS}), crustal thickness, and whole-mantle travel time. Q_{ScS} and the crustal thickness are regarded as "nuisance" parameters due to the geographic length of the corridors and complex variations in crustal structure expected along these primarily continental paths [e.g. *Sipkin and Revenaugh*, 1994].

Modeled zeroth-order reverberations were stripped from the data, leaving a residual signal consisting of first- and higher-order reverberations from discontinuities throughout the mantle [*Revenaugh and Jordan*, 1989], noise, and residual multiple *ScS* energy. An estimate of the mantle radial shear-wave reflection coefficient is obtained by 1D migration of the first-order reverberations as per *Revenaugh and*

Table 1. Source parameters of earthquakes used in this study. See Figure 1 for path locations.

Date	Origin Time (UTC)	Latitude (Deg. N)	Longitude (Deg. W)	Depth (km)	m_b	Paths
March 6, 1974	01:40:26	12.29	86.39	110	6.1	C1 and C5
May 19, 1976	04:07:15	4.46	75.78	157	6.4	S1
June 22, 1979	06:30:54	17.00	94.61	107	6.3	C5
September 16, 1989	23:20:53	16.50	93.67	108	6.6	C3, C5, and C6
March 1, 1991	17:30:26	10.94	84.64	196	6.1	C5
June 12, 1993	11:15:07	13.25	87.53	217	5.9	C3, C5, and C6
March 14, 1994	20:51:24	15.99	92.43	164	6.9	C2, C3, C5, and C6
April 10, 1994	17:36:57	14.72	92.00	100	6.0	C5 and C6
August 19, 1995	21:43:31	5.14	75.58	119	6.7	S1
October 21, 1995	02:38:57	16.84	93.47	159	7.2	C2, C3, C4, C5, and C6
December 31, 1996	12:41:42	15.83	92.97	99	6.4	C4, C5, and C6
September 2, 1997	12:13:22	3.85	75.75	198	6.8	S1
November 9, 1997	22:56:42	13.85	88.81	176	6.5	C1, C2, C3, C4, C5 and C6
December 11, 1997	07.56:28	3.93	75.79	177	6.4	S1 and S2
December 18, 1997	15:02:00	13.84	88.74	182	6.1	C1, C3, C4, C5, and C6

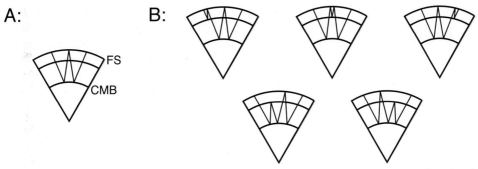

Figure 3. Schematic raypaths for (A) zeroth-order ScS_2 reverberations and (B) first-order ScS_2 reverberations interacting with a hypothetical mantle discontinuity by either top-side (upper panel) or bottom-side (lower panel) reflection. FS = Free Surface, CMB = Core-Mantle Boundary.

Jordan [1991*b*]. The general velocity model of *Revenaugh and Jordan* [1991*a*] was used for the migration of all source-receiver paths following simple velocity scaling to match the whole-mantle travel time specific to each region. The velocity used in migration must match the vertical *ScS* travel time estimated during the zeroth-order reverberation stripping, but is otherwise largely immaterial. For the near-vertical raypaths followed by multiple *ScS* and higher-order reverberations, move-out is minimal and nearly identical for all reasonable mantle models. What differs are the absolute depths, and we include the uncertainty in this mapping in the error bounds on the discontinuity depths. The stacking of all seismograms sampling an individual geographic corridor accomplished by migration greatly increases the signal-to-noise ratio of low-amplitude reflected shear waves. Unfortunately it also introduces artifacts and biases by collapsing 3D structure into a single spatial dimension, a point we address below. The reflection coefficient profiles are modeled using migrated synthetic seismograms in a graduated series of steps, resulting in a preferred reflection coefficient profile that contains the fewest reflectors necessary to accurately replicate the data profile.

There are several effects that can bias our reflection coefficient estimates; chief among these are along-path heterogeneity and anisotropy. Velocity and attenuation heterogeneity and crustal structure variability along-path are not well accounted for in the 1D migration methods used here. The result is discontinuity response functions (synthetic seismograms used as migration matched filters) that differ in waveform detail and arrival time from the data, lowering waveform cross-correlation and biasing downward the estimates of reflection coefficient. For the low-frequency waveforms we examine, these effects are not great, but they are present. Experiments with migration of synthetic data intentionally made to differ from the discontinuity response functions suggest biases on the order of 25% are possible. Along-path variability in the depth of individual discontinuities imposes an additional downward bias that scales with discontinuity topography, but which may also reach 25%. Anisotropy is present beneath much if not all of the study region, splitting the multiple *ScS* reverberations and increasing the waveform complexity of arrivals on the transverse components records we use. Because of inefficient reflection at the free surface and core-mantle boundary, SV-polarized multiple *ScS* phases are not well excited along these paths and are seldom observable in our dataset. (Their amplitude is typically only one-fourth of the amplitude of SH). As a result, the primary effect of anisotropy on the transverse component is a downward amplitude bias that increases roughly linearly with the number of passages through anisotropic depth intervals. This influences our estimates of Q_{ScS} and reduces apparent reflection coefficients of mantle discontinuities. As with along-path heterogeneity, the effect should be near-equal for all discontinuities and, as such, is unlike the effect of discontinuity topography, which is reflector specific.

To minimize the impact of these sources of bias on our interpretations of mantle discontinuity structure, we will focus on reflection coefficient ratios, which cancel out common downward biases. In theory, the only remaining biases are due to variable degrees of discontinuity topography and transition width.

RESULTS

SH reflectivity estimates for each of the eight source-receiver paths are shown in Figure 4. Shallow upper mantle discontinuities are seen sporadically across the study area (Table 2). H (see *Revenaugh and Jordan*, 1991*c* for discontinuity nomenclature) may be present beneath the westernmost paths (C5 and C6) as well as path S2, but reflectivity peaks at this depth can also arise as artifacts of inadequate modeling of crustal impulse response, and these interpretations of H should be treated with caution. This is especially so in the case of path S2, where the putative H discontinuity has a

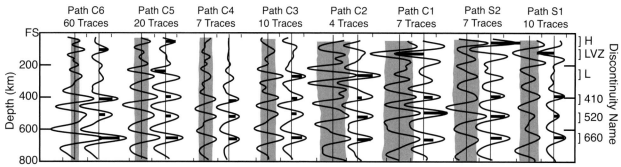

Figure 4. SH-Reflectivity profiles obtained for each of the eight paths. Vertical axis is depth (km) from the free surface (FS); the horizontal axis, R(z), is an estimate of the SH reflection coefficient of the mantle (%). Tick marks to the right and left of each centerline represent 3%. Data (left) are paired with the preferred synthetic profile (right). Peaks exceeding the gray shaded region are significant at the 95% confidence level. Bold horizontal bars on the preferred profile mark the depths and reflection coefficients of discontinuities included in the model. The number of seismograms contributing to each data stack is labeled above each profile. Paths are ordered from west to east and are grouped by source region. Numbers correspond to Figure 1.

particularly high impedance contrast. Several discontinuities are seen in the depth range of L as well, though the impedance contrast is negative below the western United States and positive beneath the eastern United States. The easternmost paths (C1 and S1) are equally well modeled by either a low velocity zone near 115 km depth or an impedance increase in D" above the core-mantle boundary. This seemingly ridiculous ambiguity, a peak in the upper mantle possibly arising from an opposite polarity impedance contrast in the lower mantle, is due to the sensitivity of first-order reverberation travel time to separation of the mantle reflector from the core-mantle boundary and/or the free surface (see *Revenaugh and Jordan*, 1991b) and not directly on reflector depth. Path

C3 also has a discontinuity in D"; modeling an upper-mantle low velocity zone in its place produced a much lower quality of profile fit, but either interpretation is plausible in light of previous work. These paths cross the Caribbean in a region where a D" discontinuity has been detected [e.g. *Garnero and Lay*, 2003], but the paths with a potential low velocity zone also traverse oceanic lithosphere, where such a discontinuity is often observed [*Revenaugh and Jordan*, 1991c]. It is possible that it is some combination of the two that create our results. Two other paths which cross the Caribbean region (C2 and S2) do not require a discontinuity in D". In general, the sporadic appearances of the shallow upper mantle (or D") discontinuities limit further interpretation.

Table 2: Path-averaged discontinuity depth, z (km), and measured impedance contrasts, R(z) (%), of upper mantle discontinuities in the preferred models for the eight paths in the study. Bandpass filter parameters provided are low cut, low corner, high corner, and high cut in mHz. See Figure 1 for path locations.

Path	Bandpass (mHz)	H z,km	H R(z)	G/D" z,km	G/D" R(z)	L z,km	L R(z)	410 z,km	410 R(z)	520 z,km	520 R(z)	660 z,km	660 R(z)
C1a	8 10 40 60			2673	6.48			397	2.34	497	4.93	662	3.73
C1b	8 10 40 60			123	-6.88			397	2.18	496	5.79	667	2.39
C2	12 14 40 60					256	3.86	397	1.19	519	2.68	662	3.39
C3	10 12 40 60			2803	3.71	269	2.57	408	1.41	507	1.28	658	2.68
C4	10 12 40 60							421	1.99	522	1.81	662	2.18
C5	16 18 40 60	54	2.63			241	-2.70	402	1.45	526	1.48	663	3.27
C6	10 12 40 60	105	2.15					416	3.37	516	1.57	663	5.16
S1a	8 10 40 60			2691	5.60			388	2.71	507	1.40	643	2.55
S1b	8 10 40 60			110	-4.16			385	2.65	507	1.38	643	2.67
S2	14 16 40 60	55	7.28					395	2.48	519	3.57	657	2.77

The 410-km, 520-km, and 660-km discontinuities are seen in all profiles and have path-averaged depths and impedance contrasts that vary across the study area (Table 2). The small amount of path-to-path average depth variation (Plate 1) that exists extends beyond the error estimate of 5-7 km for path-averaged discontinuity depth used by *Revenaugh and Jordan* [1989]. However, we increase that error estimate to ±12 km to account for greater lateral heterogeneity along the long source-receiver paths, which places the topography along the 410-km discontinuity (±12 km) at the edge of resolution and depth variability along the 520-km and 660-km discontinuities (±10 km and ±8 km, respectively) below it. On average, the discontinuities are shallower beneath the eastern part of the study area and impedance contrasts for the 410-km and 660-km discontinuities are small throughout. The latter is likely due to heterogeneity along these long and tectonically variable paths. As previously discussed, impedance contrast estimates obtained in this study should be considered in a relative, as opposed to absolute, sense (Plate 2). Of particular note is path C1, which has a very strong 520-km discontinuity, but only weak impedance contrasts for the 410-km and 660-km discontinuities (Table 2). In fact, the 520-km discontinuity is roughly twice the strength of the other two discontinuities. Nearby paths (C2 and S2) also have strong 520-km discontinuities, on the order of that seen at 410 or 660 km. We believe this is evidence of a wet transition zone along the three paths.

DISCUSSION

Wadsleyite and ringwoodite are the major minerals of the transition zone and can incorporate considerable amounts of water into their structures, favoring water storage in the transition zone rather than the upper mantle above [*Kohlstedt et al.*, 1996; *Hirschmann et al.*, 2005]. Water solubility is much lower in the lower mantle [*Bolfan-Casanova et al.*, 2000; *Murakami et al.*, 2002], and the partitioning of water favors incorporation into the transition zone at the 660-km discontinuity. The transition zone thus may act as a key water reservoir in the mantle. The presence of a large amount of water there would affect discontinuities bracketing and inside the transition zone, influencing their depths and apparent impedance contrasts.

The most striking characteristic of anomalous path C1 is the unusually large impedance contrast at the 520-km discontinuity (both relative to the 410-km and 660-km discontinuity and in reference to other studies sampling the 520-km discontinuity over much of the globe [*Shearer*, 1990; *Deuss and Woodhouse*, 2001; *Revenaugh and Jordan*, 1991*b*]). The ratios of the apparent impedance contrasts between the 520-km and 410-km or 660-km discontinuities were the

characteristics used to define the geographic extent of the anomalous corridor, with ratios of either 520:410 or 520:660 > 1.25 considered highly anomalous. Only paths C1, C2, and S2 fall into this category. Although geographically proximal, no data is shared between these paths, strongly suggesting some regional structure is responsible for the unusually energetic reflections from the 520-km discontinuity.

The impedance ratios of the 520-km discontinuity to the 410-km and 660-km discontinuities are sufficiently high, however, as to cast some doubt on the veracity of this result. To test it, a modified jack-knife was applied to each of the anomalous paths to examine the contribution of individual seismograms to the measured impedance contrast at 520 km. Stacked reflectivity profiles were recalculated for each permutation of the dataset created by dropping one seismogram. The resulting impedance ratios showed variability, both upwards and down, and scaling with the number of seismograms in the stack. In the cases where removing a trace lowered the 520:410 or 520:660 impedance contrast ratios, no stations, events, or event depths were common among removed traces. The highly anomalous impedance ratios are not the product of a noise burst or "rogue" seismogram or event, but rather appear to be reliable, albeit noisy, estimates.

The phase transition from wadsleyite to ringwoodite at 520 km depth is expected to occur over a depth range of 30 km [*Akaogi et al*, 1989]. The incorporation of water into the structures of transition zone minerals narrows the range over which the phase transition occurs to as little as 15 km [*Inoue et al.*, 1998]. Similar observations have been made using seismic methods. The 520-km discontinuity is generally observed to occur over a range of up to 50 km [*e.g. Cummins et al.*, 1992]. Due in part to this broad transition interval, the 520-km discontinuity typically has a measured impedance contrast that is only roughly half that at the 410-km discontinuity [*Shearer*, 1990]. However, in the presence of water, the apparent impedance contrast observed at 520 km depth is increased [e.g. *Van der Meijde et al.*, 2005], as is the ratio between the impedance contrasts for the 520-km and 410-km discontinuities.

Under hydrous conditions, the phase change from olivine to wadsleyite at the top of the transition zone occurs at lower pressure conditions and over a broader range of pressures than under anhydrous or saturated conditions [*Wood*, 1995; *Chen et al.*, 2002; *Smyth and Frost*, 2002; *Hirschmann et al.*, 2005]. If water is present, the 410-km discontinuity should be observed at shallower depths [*Wood*, 1995], and the transition interval will depend on the partition coefficient of water between wadsleyite and olivine at those conditions [*Hirschmann et al.*, 2005]. Similarly, the transition across the 660-km discontinuity is also expected to broaden if the

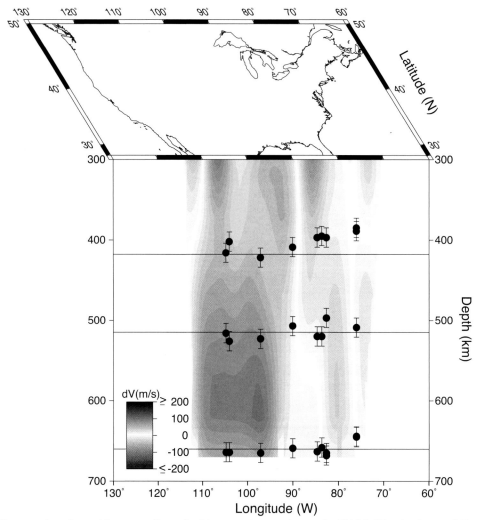

Plate 1. Topography of transition zone discontinuities superimposed over the NA04 (Van der Lee and Frederiksen, 2005) tomography model sliced across 28° N. Horizontal lines are global mean depths for the three discontinuities (*Flanagan and Shearer*, 1998).

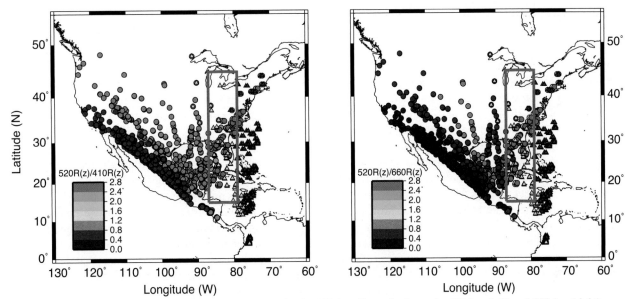

Plate 2. Path-averaged ratios of impedance contrast for the 520-km discontinuity to the 410-km (left) and 660-km (right) discontinuities. Symbols plotted are surface bounce points of multiple *ScS* phases for individual traces. Gray rectangles highlight a swath of elevated 520-km impedance contrasts.

partitioning of water between the transition zone and lower mantle is of the same order of magnitude as the partitioning between the transition zone and the overlying mantle [*Smyth and Frost*, 2002], though the broadening expected at 660 km depth is less than that at 410 km [e.g. *Higo et al.*, 2001; *Smyth and Frost*, 2002].

Broadening of a transition that is centered on the same depth as the original discontinuity will not directly affect the multiple *ScS* phases, unless that broadening is unusually large (>25 to 30 km for most of our paths). However, it can affect apparent discontinuity depth and estimated impedance contrast if the broadening is not symmetric above and below the original transition. Asymmetric broadening of a discontinuity changes its centroid depth; variable broadening induces additional topography along the path and leads to a dampened or lowered impedance contrast obtained from the *ScS* reverberation method. Therefore, while the multiple *ScS* reverberations may not be directly sensitive to the thickness of the phase boundary at a mantle discontinuity, their interactions with that broadened region across a source-receiver path still result in a lowered impedance contrast.

The response of the 410-km and 660-km discontinuities to water results in a broadened transition zone in hydrous regions. Transition zone thicknesses range from 243 to 267 km (± 17 km) in our study, and all are thicker than the global average of 241 km [*Flanagan and Shearer*, 1998]. Paths that show hydrous signatures in impedance contrasts have the thickest transition zones. The broadened transition zones are primarily the result of shallow 410-km discontinuities. The 660-km discontinuity exhibits less depth variation and no topographic correlation, negative or otherwise, with the 410-km discontinuity.

We see no evidence of a partial melt layer on top of the 410-km discontinuity [e.g., *Song and Helmberger*, this volume]. The absence of a melt layer does not necessarily contradict the implication of water in the transition zone. There could be enough water in the transition zone to amplify the measured impedance contrast across the 520-km discontinuity without exceeding the saturation limit (~0.4 wt% H_2O) for the mantle above 410 km depth [*Hirschmann et al.*, 2005]. Alternatively, the water content of the transition zone may exceed the saturation limit of the mantle above but the hydrous material may be localized within the transition zone and not upwelling across the 410-km discontinuity. Lastly, a thin melt layer may be present but not detected. For the long-period shear waves we examine, any melt layer would need to exceed 20 to 30 km to appear separate from the underlying 410-km discontinuity. A thinner layer would act to lower the apparent impedance contrast across the 410-km reflector, augmenting the effect of transition broadening and topography.

Before concluding that water is present in large quantities in the transition zone along paths C1, C2, and S2, we must consider other mechanisms for increasing the measured impedance contrast across the 520-km discontinuity relative to the 410-km and 660-km discontinuities. And in fact, some additional upward driver is needed to achieve the unusually large ratios we observe. Since path-averaged topography along the three transition zone discontinuities is at or below the error estimates, other mechanisms for producing topography are not discussed. The shallow 410-km discontinuity that exists along two paths sampling the eastern United States is interpreted in conjunction with the elevated impedance contrast at 520 km rather than as a separate line of evidence for a wet transition zone beneath the anomalous region.

Significant along-path topography on the 410-km and 660-km discontinuities would lower the measured impedance contrasts across those boundaries. A more-nearly "flat" 520-km discontinuity would then appear larger in a relative sense. Path-averaged depths of the three discontinuities do not exhibit greater variability for the 410-km and 660-km discontinuities. Nor is this behavior seen globally, as seismic studies show the opposite: a more undulatory 520-km discontinuity [e.g. *Deuss and Woodhouse*, 2001]. The inferred reflection coefficient of the 520-km discontinuity along path C1 also is large in an absolute sense, not just in relative terms, implying it is being driven up by some process and is not simply less damped than the 410-km and 660-km discontinuities. As a result, we consider relative topography variability as a possible, perhaps likely, addition to water-induced impedance ratio amplification, but not a substitute. A second possibility is that the multiple *ScS* reverberations are imaging a fast velocity structure confined to the lowermost transition zone and geographically situated beneath the Gulf of Mexico. Under this scenario, what we image as the 520-km discontinuity is a mix of the globally observed phase transition and a seismically fast local thermal and/or chemical anomaly. The latter, however, is not observed in North American tomographic models [*Van der Lee and Frederiksen*, 2005]. A third possibility is that there is a velocity-neutral, high-density feature in the region. The associated mass excess should produce a large excursion in the geoid in the Gulf region, an excursion not seen in the measured geoid [*Lemoine et al.*, 1997]. A final possibility is that another compositional defect has the same effect as hydrogen on discontinuity depths and impedance contrasts across the transition zone discontinuities and also lowers the shear wave velocities of the minerals in the deep upper mantle. With future research, another feasible option may be presented, however a mechanism for introducing that compositional defect into a localized region of the mantle is also required to explain our results. In the absence of a viable alternative, we conclude that the transition zone is unusually wet along paths C1, C2, and S2. The expected behavior of water on transition zone discontinuity depths and impedance contrasts matches most of our observations, and a subducted slab beneath the region provides a mechanism for introducing water to the region.

Previous seismic studies made observations or predictions consistent with a hydrated transition zone beneath the eastern United States. Nominally anhydrous upper mantle and transition zone minerals containing water in some form are expected to have lower seismic velocities [*Karato*, 1995; *Wang et al.*, 2003; *Jacobsen et al.*, 2004; *Jacobsen and Smyth*, this volume]. Low compressional wave velocities measured in rocks beneath the Appalachians in the eastern United States [*Taylor and Toksoz*, 1982] could be an indication of serpentinization and hydration of the mantle rocks [*Christensen*, 1966; *Karato*, 1995]. A low velocity corridor trending along the eastern coast of the United States is found by *Van der Lee and Nolet* [1997] and *Van der Lee et al.* [2005]. The low velocity anomaly is strongest beneath the Gulf of Mexico and extends from 200 km to transition zone depths. The authors interpret it as a water-rich region of the upper mantle. Injection of water into overlying mantle by subduction has been proposed as a mechanism for creating the hydrous region [*Van der Lee and Nolet*, 1997]. *Song et al.* [2004] and *Song and Helmberger* [this volume] observe a low velocity layer atop the transition zone beneath the western United States. This layer is not apparent in our reflection profiles for the same region. However, *Song and Helmberger* [this volume] report a thickness of ~25 km for the layer, which is pushing the detection limit of multiple *ScS* reverberations.

Turning the hypothesis of subduction-related water injection into the upper mantle and transition zone around, one could ask why other areas of active or recent subduction do not show clear signs of hydration. The lack of a seismic signature of water in areas where it may be expected does not necessarily exclude the presence of water. The seismic signatures of water are generally subtle and easily masked by stronger heterogeneities that result from large lateral thermal gradients, melt, and anisotropy. *Williams and Hemley* [2001] estimate that seismic methods generally have a minimum detection limit of more than 1 wt% H_2O in the transition zone and 2 wt% H_2O above it. Water in quantities greater than these limits may be rare, even in areas where there exists an abundant source of water.

CONCLUSIONS

Discontinuity depths and apparent impedance contrasts obtained from multiple *ScS* reverberations sampling the mantle beneath the eastern United States and Gulf of Mexico are consistent with a water-rich transition zone. A similar hydrous signature is not seen beneath the central or western United States. Laboratory studies find that the major mantle minerals are capable of storing large amounts of water, leading to as much as 0.4 wt% above 410 km depth and 1.5 wt% in the transition zone [*Hirschmann et al.*, 2005]. The distribution of water in the mantle has implications for planetary accretion, mantle composition, rheology, and convection, but its seismic signature can be subtle. We believe that multiple *ScS* reverberations are important tools in the search for mantle water.

Acknowledgements. Helpful reviews were received from Emile Okal, an anonymous reviewer, and the book editors. GMT software [*Wessel and Smith*, 1998] was used to prepare some of the figures. This research was partially funded by the National Science Foundation (EAR-0437424). AMC received support from a University of Minnesota Graduate School Fellowship and Department of Geology and Geophysics Mooney and Dennis Graduate Fellowships.

REFERENCES

Ahrens, T.J., Water storage in the mantle, *Nature, 342,* 122–123, 1989.

Akaogi, M., E. Ito, and A. Navrotsky, Olivine-modified spinel-spinel transitions in the system Mg_2SiO_4-Fe_2SiO_4: Calorimetric measurements, thermochemical calculation, and geophysical application, *J. Geophys. Res., 94,* 15671–15685, 1989.

Bell, D.R. and G.R. Rossman, Water in Earth's mantle: The role of nominally anhydrous minerals, *Science, 255,* 1391–1397, 1992.

Bercovici, D. and S.-I. Karato, Whole-mantle convection and the transition-zone water filter, *Nature, 425,* 39–44, 2003.

Bergstrom, S.M., Relations between conodont provincialism and the changing palaeogeography during the Early Palaeozoic, in *Palaeozoic Palaeogeography and Biogeography,* edited by W.S. McKerrow and C.R. Scotese, *Geological Society Memoir, 12,* 105 –121, 1990.

Bolfan-Casanova, N., Water in the Earth's mantle, *Mineralogical Magazine, 69,* 229–257, 2005.

Bolfan-Casanova, N., H. Keppler, and D.C. Rubie, Water partitioning between nominally anhydrous minerals in the MgO-SiO_2-H_2O system up to 24 GPa: Implications for the distribution of water in the Earth's mantle, *Earth Planet Sci. Lett., 182,* 209–221, 2000.

Bolfan-Casanova, N., H. Keppler, and D.C. Rubie, Water partitioning at 660 km depth and evidence for very low water solubility in magnesium silicate perovskite, *Geophys. Res. Lett., 30(17),* 1905, doi:10.1029/2003GL017182, 2003.

Chen, J., T. Inoue, H. Yurimoto, D.J. Weidner, Effect of water on olivine-wadsleyite phase boundary in the $(Mg,Fe)_2SiO_4$ system, *Geophys. Res. Lett., 29(18),* 1875, doi:10.1029/2001GL014429, 2002.

Christensen, N.I., Elasticity of ultrabasic rocks, *J. Geophys. Res., 71*, 5921 –5931, 1966.

Cummins, P.R., B.L.N. Kennett, J.R. Bowman, and M.G. Bostock, The 520-km discontinuity?, *Bull. Seismol. Soc. Am., 82*, 323–336, 1992.

Deuss, A. and J. Woodhouse, Seismic observations of splitting of the mid-transition zone discontinuity in Earth's mantle, *Science, 294*, 354–357, 2001.

Drake, M.J. and K. Righter, Determining the composition of the Earth, *Nature, 416*, 39–44, 2002.

Flanagan, M.P. and P.M. Shearer, Global mapping of topography on transition zone velocity discontinuities by stacking SS precursors, *J. Geophys. Res., 103*, 2673–2692, 1998.

Garnero, E.J. and T. Lay, D" shear velocity heterogeneity, anisotropy and discontinuity structure beneath the Caribbean and Central America, *Phys. Earth Planet. Inter., 140*, 219–242, 2003.

Helffrich, G., Topography of the transition zone seismic discontinuities, *Rev. Geophys., 38*, 141–158, 2000.

Helffrich, G.R. and B.J. Wood, 410 km discontinuity sharpness and the form of the olivine α-β phase diagram: resolution of apparent seismic contradictions, *Geophys. J. Int., 126*, F7–F12, 1996.

Helffrich, G., E. Asencio, J. Knapp, and T. Owens, Transition zone structure in a tectonically inactive area: 410 and 660 km discontinuity properties under the northern North Sea, *Geophys. J. Int., 155*, 193–199, 2003.

Higo, Y., T. Inoue, T. Irifune, and H. Yurimoto, Effect of water on the spinel-postspinel transformation in Mg_2SiO_4, *Geophys. Res. Lett., 28*, 3505–3508, 2001.

Hirschmann, M.M., C. Aubaud, and A.C. Withers, Storage capacity of H_2O in nominally anhydrous minerals in the upper mantle, *Earth Planet. Sci. Lett., 236*, 167–181, 2005.

Ingrin, J. and H. Skogby, Hydrogen in nominally anhydrous upper-mantle minerals: Concentration levels and implications, *European Journal of Mineralogy, 12*, 543–570, 2000.

Inoue, T., Effect of water on melting phase relations and melt composition in the system Mg_2SiO_4-$MgSiO_3$-H_2O up to 15 GPa, *Phys. Earth Planet. Inter., 85*, 237–263, 1994.

Inoue, T., D.J. Weidner, P.A. Northrup, and J.B. Parise, Elastic properties of hydrous ringwoodite (γ-phase) in Mg_2SiO_4, *Earth Planet. Sci. Lett., 160*, 107–113, 1998.

Irifune, T., N. Kubo, M. Isshiki, and Y. Yamasaki, Phase transformation in serpentine and transportation of water in the lower mantle, *Geophys. Res. Lett., 25*, 203–206, 1998.

Ito, E. and E. Takahashi, Post-spinel transformations in the system Mg_2SiO_4-Fe_2SiO_4 and some geophysical implications, *J. Geophys. Res., 94*, 10637–10646, 1989.

Jacobsen, S.D., J.R. Smyth, H. Spetzler, C.M. Holl, and D.J. Frost, Sound velocities and elastic constants of iron-bearing hydrous ringwoodite, *Phys. Earth Planet. Inter., 143–144*, 47–56, 2004.

Karato, S-I., Effects of water on seismic wave velocities in the upper mantle, *Proc. Japan Acad., 71*, 61–66, 1995.

Kohlstedt, D.L., H. Keppler, and D.C. Rubie, Solubility of water in the α, β and γ phases of $(Mg,Fe)_2SiO_4$, *Contrib. Mineral. Petrol., 123*, 345–357, 1996.

Komabayashi, T., S. Omori, and S. Maruyama, Experimental and theoretical study of stability of dense hydrous magnesium silicates in the deep upper mantle, *Phys. Earth Planet. Inter., 153*, 191–209.

Lemoine, F.G., D.E. Smith, L. Kunz, R. Smith, E.C. Pavlis, N.K. Pavlis, S.M. Klosko, D.S. Chinn, M.H. Torrence, R.G. Williamson, C.M. Cox, K.E. Rachlin, Y.M. Wang, S.C. Kenyon, R. Salman, R. Trimmer, R.H. Rapp, and R.S. Nerem, The development of the NASA GSFC and NIMA joint geopotential model, *International Association of Geodesy Symposia, 117*, 461–469, 1997.

Litasov, K., E. Ohtani, F. Langenhorst, H. Yurimoto, T. Kubo, and T. Kando, Water solubility in Mg-perovskites and water storage capacity in the lower mantle, *Earth Planet. Sci. Lett., 211*, 189–203, 2003.

Murakami, M., K. Hirose, H. Yurimoto, S. Nakashima, and N. Takafuji, Water in Earth's lower mantle, *Science, 295*, 1885–1887, 2002.

Nolet, G. and A. Zielhuis, Low S velocities under the Tornquist-Teisseyre zone: Evidence for water injection into the transition zone by subduction, *J. Geophys. Res., 99*, 15813–15820, 1994.

Ohtani, E., Y. Nagata, A. Suzuki, and T. Kato, Melting relations of peridotite and the density crossover in planetary mantles, *Chemical Geology, 120*, 207–221, 1995.

Ohtani, E., M. Toma, K. Litasov, T. Kubo, and A. Suzuki, Stability of dense hydrous magnesium silicate phases and water storage capacity in the transition zone and lower mantle, *Phys. Earth Planet. Inter., 124*, 105–117, 2001.

Revenaugh, J. and T.H. Jordan, A study of mantle layering beneath the western Pacific, *J. Geophys. Res., 94*, 5787–5813, 1989.

Revenaugh, J. and T.H. Jordan, Mantle layering from ScS reverberations, 1, Waveform inversion of zeroth-order reverberations, *J. Geophys. Res., 96*, 19749–19762, 1991a.

Revenaugh, J. and T.H. Jordan, Mantle layering from ScS reverberations, 2, The transition zone, *J. Geophys. Res., 96*, 19763–19780, 1991b.

Revenaugh, J. and T.H. Jordan, Mantle layering from ScS reverberations, 3, The upper mantle, *J. Geophys. Res., 96*, 19780–19810, 1991c.

Revenaugh, J. and S.A. Sipkin, Seismic evidence for silicate melt atop the 410-km mantle discontinuity, *Nature, 369*, 474–476, 1994.

Richards, P.G., Seismic waves reflected from velocity gradient anomalies within the Earth's upper mantle, *J. Geophys., 38*, 517–527, 1972.

Scotese, C.R. and W.S. McKerrow, Revised world maps and introduction, in *Palaeozoic Palaeogeography and Biogeography*, edited by W.S. McKerrow and C.R. Scotese, *Geological Society Memoir, 12*, 1–21, 1990.

Shearer, P.M., Seismic imaging of upper-mantle structure with new evidence for a 520-km discontinuity, *Nature, 344*, 121–126, 1990.

Sipkin, S.A. and J. Revenaugh, Regional variation of attenuation and travel time in China from analysis of multiple-ScS phases, *J. Geophys. Res., 99*, 2687–2699, 1994.

Smyth, J.R., β-Mg$_2$SiO$_4$: A potential host for water in the mantle?, *Amer. Mineral., 72,* 1051–1055, 1987.

Smyth, J.R. and D.J. Frost, The effect of water on the 410-km discontinuity: An experimental study, *Geophys. Res. Lett., 29(10),* 1485, doi:10.1029/2001GL014418, 2002.

Song, T.A., D.V. Helmberger, and S.P. Grand, Low-velocity zone atop the 410-km seismic discontinuity in the northwestern United States, *Nature, 427,* 530–533, 2004.

Stolper, E., D. Walker, B. Hager, and J. Hays, Melt segregation from partially molten source regions: The importance of melt density and source region size, *J. Geophys. Res., 86,* 6261–6271, 1981.

Taylor, S.R. and M.N. Toksoz, Crust and upper-mantle velocity structure in the Appalachian orogenic belt: Implications for tectonic evolution, *Geol. Soc. Am. Bull., 93,* 315–319, 1982.

Van der Lee, S. and G. Nolet, Upper mantle S-velocity structure of North America, *J. Geophys. Res., 102,* 22815–22838, 1997.

Van der Lee, S. and A. Frederiksen, Surface wave tomography applied to the North American upper mantle, in *Seismic Earth: Array analysis of broadband seismograms, Geophys. Monogr. Ser., 157,* edited by A. Levander and G. Nolet, 67–80, 2005.

Van der Lee, S., K. Regenauer-Lieb, and D. Yuen, The role of water in connecting past and future episodes of subduction, *EOS Trans. AGU, 86(52), Fall Meeting Supplement,* Abstract DI43A-05, 2005.

Van der Meijde, M., F. Marone, D. Giardini, and S. van der Lee, Seismic evidence for water deep in Earth's upper mantle, *Science, 300,* 1556–1558, 2003.

Van der Meijde, M., F. Marone, and S. van der Lee, Seismic evidence for water atop the Mediterranean transition zone, *EOS Trans. AGU, 85(47),* Fall Meeting Supplement, Abstract T31F-03, 2004.

Van der Meijde, M., S. van der Lee, and D. Giardini, Seismic discontinuities in the Mediterranean mantle, *Phys. Earth Planet. Inter., 148,* 233–250, 2005.

Vinnik, L., M. Ravi Kumar, R. Kind, and V. Farra, Super-deep low-velocity layer beneath the Arabian plate, *Geophys. Res. Lett., 30(7),* 1415, doi:10.1029/2002GL016590, 2003.

Wang, J., S.V. Sinogeiken, T. Inoue, and J.D. Bass, Elastic properties of hydrous ringwoodite, *Amer. Mineral., 88,* 1608–1611, 2003.

Wessel, P. and W.H.F. Smith, New, improved version of the Generic Mapping Tools released, *EOS Trans. AGU, 79,* 579, 1998.

Williams, Q. and R.J. Hemley, Hydrogen in the deep Earth, *Annu. Rev. Earth Planet. Sci., 29,* 365–418, 2001.

Wood, B., The effect of H$_2$O on the 410-kilometer seismic discontinuity, *Science, 268,* 74–76, 1995.

Zhao, D., Seismological structure of subduction zones and its implications for arc magmatism and dynamics, *Phys. Earth Planet. Inter., 127,* 197–214, 2001.

A.M. Courtier and J. Revenaugh, Dept. of Geology and Geophysics, University of Minnesota, 108 Pillsbury Hall, 310 Pillsbury Drive SE, Minneapolis, MN 55455. (e-mail: cour0090@umn.edu; justinr@umn.edu)

Low Velocity Zone Atop the Transition Zone in the Western US From S Waveform Triplication

Teh-Ru Alex Song and Don V. Helmberger

Division of Geological and Planetary Sciences, Seismo Lab, Caltech, Pasadena, USA

Song et al. [2004] modeled regional S wave triplications in the northwestern US and found a low velocity zone atop the 410 seismic discontinuity. Strong azimuthal variation in waveforms associated with paths sampling the western edge of this structure are observed on the TriNet array for several events. Here, we model this data with a new 3D simulation technique which combines 2D finite-difference with Kirchhoff diffraction operators to include responses off the great circle. To reconcile such sharp changes in waveforms requires a sharp western edge less than 100 km across a boundary with a change of 3–5% in velocity. Though the geometry of the LVZ is not unique due to limited data analyzed, the sharp edge of the LVZ is robustly constrained with available array data. Such a LVZ is consistent with the existence of water in Earth's transition zone, at least locally.

1. INTRODUCTION

Melt segregation is one of the most important processes involved in understanding the chemical evolution of the Earth's mantle [*Stolper et al.*, 1981]. High pressure experiments indicate that dry basaltic melt produced by partial melt of peridotite is neutrally buoyant at conditions of the 410-km discontinuity [*Rigden et al.*, 1984; *Agee,* 1998; *Chen et al.*, 2002]. However, it is not clear if hydrous silicate melt also behaves in the same manner. One of the most recent studies indicates that the effect of water on the density of silicate melt under high pressure is small enough such that the melt can be trapped in the deep upper mantle [*Matsukage et al.*, 2005], if the water content is less than 2 wt%. To form such a molten layer near the 410, dehydration melting across the 410 can generate the melt due to the difference in solubility between the deep upper mantle (Olivine-rich) and the transition zone (Wadsleyite-rich) [*Kohlstedt et al.*, 1996].

Furthermore, *Hirschmann et al.* [2005] suggested that hydrous melting can continue to generate small amounts of melt as the material ascends through the upper mantle, while its depth extent would depend on the storage capacity profile of the upper mantle. In any case, the presence of such molten layers provides evidence for a dissolved water component in the transition zone [*Kawamoto et al.*, 1996; *Bercovici and Karato,* 2003], if not globally. Therefore, it is important to seismically constrain the geometry of such layers.

The structure near the 410 is known to be complicated beneath western US from receiver function analysis [*Gilbert et al.*, 2003]. These receiver functions are constructed by stacking multiple profiles from various azimuths. Normally, a clear positive peak near 410 km and 660 km appears, indicating the expected polarity of a P to S conversion (fast to slow velocity jump). However, some abnormal results occur where a strong negative pulse appears, suggesting a low velocity zone [*Gilbert et al.*, 2003]. These samples essentially establish locations of low-velocity zones (LVZ) from vertically traveling paths. As demonstrated in *Song et al.* [2004], the observed abnormal samples also occur where there are

Earth's Deep Water Cycle
Geophysical Monograph Series 168
Copyright 2006 by the American Geophysical Union.
10.1029/168GM15

abnormal wide-angle reflections or triplication shifts. In order to sample the deep upper mantle with complementary horizontal paths, we analyze body waves propagating at distances of 14°–24°, which form the upper mantle triplications and have great sensitivity near the 410. A more complete review of this type of data follows here.

Upper mantle triplications were one of the first observations used to constrain the upper mantle seismic discontinuity structure [*Johnson,* 1967; *Helmberger and Wiggins,* 1971]. Because of the relatively sharp increase in velocity, the seismic wavefield bifurcates into multiple branches (see Plate. 1A, Plate. 1B). The 410 seismic discontinuity produces thesecondary arrival before 17°, which are refraction/reflections from the 410, named CD and BC branches, respectively, while the waves traveling shallower in the upper mantle are called the AB branch. While most current studies focus on travel time tomography and receiver function studies, it is a useful complement to also analyze seismic triplications, which are very sensitive to mantle structure where seismic waves travel horizontally. Introducing a low velocity zone not only increases the separation of AB and CD branches, but also extends the AB branches beyond 22°. We illustrate an example of an upper mantle triplication at 16°–23° to display the usefulness of these data in constraining the velocity structure near the 410 (Plate. 1C). Two relevant branches will be discussed here, AB (slow) and CD (fast). The separation of these two branches is very useful to constrain the velocity while their relative amplitude is helpful in determining the velocity gradient. The crossover of AB and CD branches is used to determined the depth and size of the discontinuity. The back-branch AB is particularly useful in determining the velocity and its gradient near the 410 [*Song et al.,* 2004; *Walck,* 1984a, b]. The move out and amplitude of the forward CD branches beyond the crossover of AB and CD indicates the velocity and its gradient below the 410 seismic discontinuity.

One important aspect of the AB branch is the position of triplication point B since it is relevant to where the velocity gradient starts to decrease or whenever the low velocity zone exists atop the 410. It was observed that the AB branch crosses over with the CD branch and appears from 19°–20°. It disappears near 22° [*Walck, 1984b*] because the velocity gradient is relatively steep, forcing shallow turning depths. However, if the AB branch crosses over with the CD branch at larger distances and emerges at 22°, it indicates a decrease in velocity gradient occurring at depth, which provides the best evidence of a low velocity zone atop the 410, along with the timing of CD and AB branches. In other words, we can potentially determine the thickness of the low velocity zone. With the help of the timing between AB and CD branches, we can determine the velocity reduction of the LVZ and its thickness although there are trade-offs. Note that the uncertainty of the depth of shallow earthquakes produce relatively small differences in the frequency band of 0.01–0.2 Hz [*Song,* 2006] and does not obscure the delayed AB branch. In addition, the source mechanisms for the events we used are primarily strike-slip events and the radiation pattern is favorable for observing SH waveforms even if minor differences in source mechanisms are allowed. However, observing these relatively subtle differences requires dense arrays with station spacing less than 30 km.

PASSCAL broadband seismic experiments have been greatly increasing during the past decades. Mostly these data are used to study regional crust and mantle structure, with the help of receiver functions [*Wilson et al.,* 2005] and travel time tomography [*Gao et al.,* 2004]. Sometimes, surface wave tomography is also performed to understand the regional tectonics [*West et al.,* 2004]. In most cases, the spacing of the arrays is much shorter than global and permanent regional networks. Such arrays in fact provide dense sampling in the regions of interest. We have first analyzed broadband waveform data for the permanent TriNet seismic broadband array in southern California and several PASSCAL arrays, including La Ristra Array, CDROM array and with example observations displayed in *Song et al.* [2004]. In general, the spacing of the temporary arrays is about 20 km or less and the spacing of the TriNet is on average 50 km and sometimes less than 30 km. Detailed descriptions about the purpose of these passive source seismic experiments (PASSCAL) are summarized in *Sheehan et al.* [2005]. We will illustrate how to model such dense arrays to retrieve information about the velocity structure from the waveform modeling of upper mantle triplications. Note, also, that we use triplications of direct S which give a better spatial constraint than using triplications of multiple S waves, such as SS, SSS and SSSS [*Grand and Helmberger,* 1984a, b; *Helmberger et al.,* 1985; *Graves and Helmberger,* 1988]. In addition, we benefit from the dense networks with station spacing less than 20 km, which is an order of magnitude better than that in previous studies. We review some of the 1D modeling results supporting the existence of the LVZ [*Song et al.,* 2004]) (Plate. 2) followed by 2D simulations from the 3D tomographic study by *Godey et al.* [2004] of paths passing through the anomalous region. Next, we introduce a new 3D hybrid method for waveform modeling and model the TriNet data producing a 3D image of the western edge of the proposed LVZ beneath the western Basin and Range.

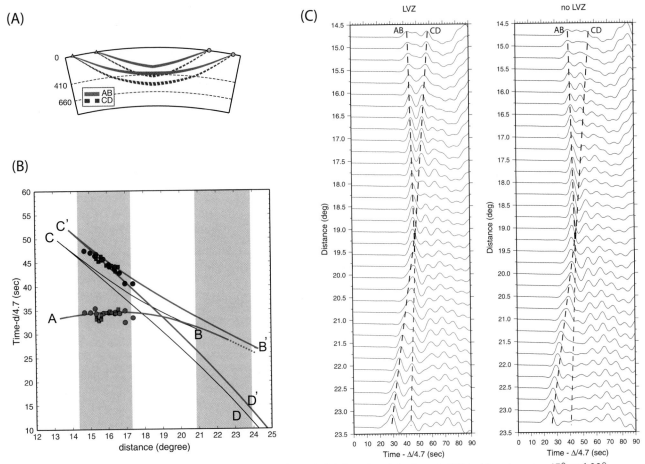

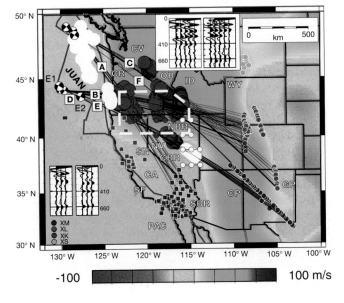

Plate 1. (A) Display of ray paths representing the branch AB in red and the branch CD in blue at ranges 17° and 22°. (B) An example of triplication of AB and CD branch. Red line indicates predicted triplication from a low velocity layer directly above the 410 (dVS = -5%, HLVZ = 40 km). The data points are picked from path A and path B (Plate. 2). (C) Synthetics displacement waveforms from models with (left)/without LVZ (right). The LVZ contains a dVS = -5%, HLVZ = 40 km layer just above the 410. Dashed lines indicate the arrival of AB and CD branches predicted by LVZ model.

Plate 2. Low velocity region atop the transition zone in the western US [*Song et al.*, 2004]. Velocity anomaly imaged by *VanderLee and Frederiksen* [2005] at 400 km is shown in color. Red patches indicated the sampled low velocity zone and their size corresponds to the Fresnel zone of 6 secs S wave. Small red dots are locations where anomalous negative pulses are observed in receiver function profiles [*Gilbert et al.*, 2003]. White patches and white dots show no sign of LVZ in triplication study and receiver functions.

2. DATA AND WAVEFORM ANALYSIS

2.1. Waveform Modeling With 1D Layered Models

We first compute full 1D (ωk) frequency-wavenumber waveform synthetics [*Saikia*,1994], and model the direct S wave triplications at epicentral distances of 14°–17° and explain the timing and amplitude of multiple arrivals coming from fine structures near the 410. We apply the 1D attenuation structure suited for the Western United States and East Pacific Rise [*Ding and Grand*, 1993]. At these distances, we first modify the shallow mantle structure from the 1D model to fit the timing and amplitude of S waves and Love waves [*Grand and Helmberger*, 1984b]. By adding a 20 km lid of 4.55 km/s, we are able to fit most recorded sections collected in the Western United States [*Melbourne and Helmberger*, 2002]. With a better constraint on the shallow mantle structure, we then further model the differences in the CD branch between data and synthetics. To resolve the trade offs between discontinuity topography and mantle velocity directly above or below the discontinuity, we model S wave triplications at epicentral distances of 21°–24°. A LVZ atop the 410 produces a secondary pulse not normally seen at these distances as discussed earlier. Using this secondary pulse as a proxy for the existence of a LVZ atop the 410, we examine waveforms recorded by the TriNet broadband network and several temporary PASSCAL broadband arrays from evens offshore of Washington-Oregon, United States. Perturbed velocity models near the turning point of CD branch are generated as a sensitivity study. The size of the perturbed features are estimated as the size of the Fresnel zone in both along-path and cross-path directions. These zones are represented by the colored circles in the mapping (Plate. 2). This issue will be addressed later in 3D simulations. Here we discuss modeling results along path A, B and C (Plate. 2).

Record section A (event 010914, Table. 1) shows typical waveform characteristics sampling this area at epicentral distances of 14°–17° (Plate. 3). It samples the region beneath the California-Oregon border. Relative to the 1D reference model (on the left), we observe a considerable increase of separation in timing between the AB branch and the CD branch of about 5 seconds. At these ranges, the branch AB represents the S wave propagating through the upper mantle and turning a depths between 200-300 km and the CD branch represent the S wave bottoming near the 410. Record section B (event 970711, Table. 1) trends more east-west, samples the same regions and shows similar waveforms to those in record section A.

Synthetics calculated from a perturbed 1D model with the 410 depressed by 60 km (model Topo) improves the differential time between the AB and CD branches considerably but disagrees with the receiver functions addressed earlier. Model gradient B produces reasonable fits except the CD branch is too strong near 16° as is the case for model LVZ. Model LVZ+Topo fits data well both in timing and amplitude.

These two paths sample regions close together in space and appear to have a common explanation, namely, a low velocity zone. Although it is clear that the velocity is slow just above the 410, it is not easy to distinguish which model explains the waveform data better with only data in this limited range. The synthetic predictions at larger ranges sampling the extended AB branch is effective for this purpose (Fig. 1). Here we illustrate how the long range data can be used to distinguish different models. As an example, we compute the synthetics along path C and the detailed comparison between data and synthetics has been presented by *Song et al.* [2004] earlier. In this example, Model Topo does not produce the observed secondary AB branch along record section C (event 990702, Table 1) at the distance range. Though a gradient model such as model Gradient B does produce the secondary AB branch, the amplitude and timing are different depending on which model we use. After crossover of AB and CD branch, the gradient model often predicts earlier emergence of the AB branch near 21.5°. The differences between Model LVZ and Model LVZ+Topo are that the AB branch, emerging from the crossover with the CD branch, appears at shorter ranges while the CD branch is delayed near 21° due to a deeper 410.

2.2. Modeling Complex Structures

To characterize the origin of the velocity anomaly, it is useful to examine the geometry of the anomaly testing for lateral sharpness. Smooth features are usually attributed to variations in temperature. A velocity anomaly with a sharp edge is often considered to be chemically distinct or due to partial melt [*Ni et al.*, 2002]. There are several cases where the velocity structure is shown to be relatively 1D within the Pacific plate [*Melbourne and Helmberger*, 2002; *Tan and Helmberger*, 2005]. However, in this section, we will discuss the structure underneath the plate boundary in the western Basin and Range, where the velocity structure changes rapidly both in the shallow upper mantle and above the transition zone.

In this section, we have increased the time window to include the beginning of the Love wave as displayed in Fig. 2. Note that the later arriving Love wave is delayed by about 10 to 12 secs along inland paths (Az = 141°) compared to coastal paths (Az = 150°). This observation is consistent with a model where the lid increases its thickness from inland

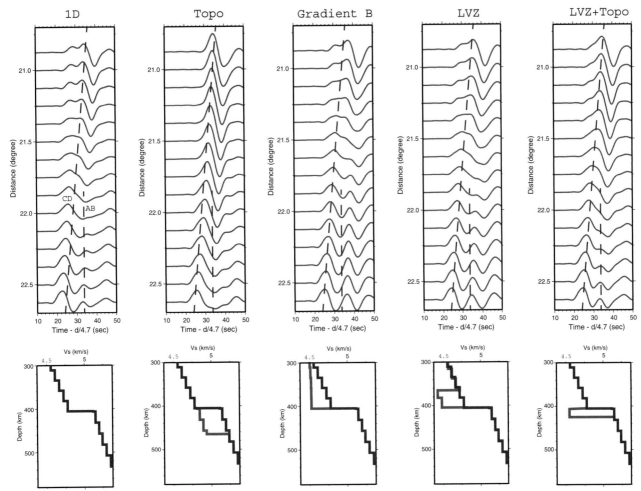

Figure 1. Sensitivity test at 20.7°–22.7° along path C. The top panels show the displacement synthetics and the lower panels show the corresponding velocity models. From left to right: 1D (dep410 = 405 km), Topo (dep410 = 465 km), Gradient B, LVZ (dVS = -5%, HLVZ = 40 km) and LVZ+Topo (dVS = -5%, HLVZ = 20 km, dep410 = 425 km). Dashed lines are the timings of the observed AB and CD branches.

California to coastal California [*Melbourne and Helmberger*, 2001]. At 15°–17°, a slower, thinner lithosphere should slow down AB more than CD and make the phases closer together. However, the data indicates that CD is delayed more than the AB at inland paths, requiring variations in deep upper mantle structure.

We introduce some TriNet data sampling the western edge of this interesting structure (Plate 4). The blue and green traces are observations along record sections of constant

azimuth while the red traces are along constant distance. The latter style of plotting data is commonly referred to as a "fan shot", a technique used to detect low velocity salt plugs in exploration. These recordings contain two pulses associated with an earlier shallow upper mantle arrival with a slow move-out (labeled AB) and a secondary arrival (CD) with a fast move-out, bottoming below the 410 discontinuity.

To model these records requires accounting for significant shallow structure and including 2D and 3D models. We

Table 1. Source parameters

Path	Event date	Lon	Lat	Depth	Strike	Dip	Rake	Mw
A	01/09/14	-128.59	48.82	6.0	330	84	-176	5.9
B	97/07/11	-129.65	44.11	6.0	301	87	-172	5.4
C	99/07/02	-129.38	49.13	6.0	315	81	176	5.9

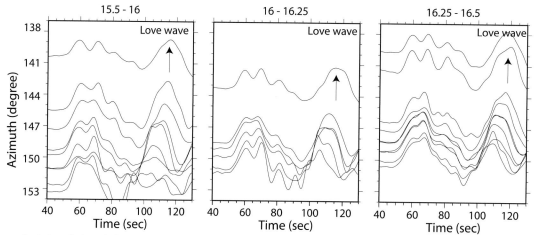

Figure 2. Azimuthal variations of tangential displacements (path A) recorded at the TriNet array. Azimuth sections are shown at three distance ranges: 15.5°–16°, 16°–16.25°, 16.25°–16.5°. Arrows indicate the Love wave.

found the 2D finite-difference (FD) code with embedded earthquake sources quite useful for this purpose [*Vidale et al.*, 1985; *Helmberger and Vidale,* 1988]. If one applies the moving window approach introduced by *Vidale* [1988], we can generate FD synthetics with computation times faster than 1D ωk synthetics. This approach allows us to efficiently propagate waves through 2D sections of tomographic models to generate synthetics. We note that a particular choice of a tomography model is likely to introduce some bias in constructing our model. But the modeled velocity contrast across different azimuths should be rather robust. The recent tomographic model by *Godey et al.* [2004] predicts the changes in Love waves across the array well, and thus provides a reasonable correction for shallow structure.

Below this depth, we used TNA for coastal stations and the model displayed in Plate 5 for the easternmost stations. We note that there is a trade-off between the spatial location of LVZ and the depth of the 410. When the 410 is deeper by 10 km, the LVZ structure moves further to the source side by 100 km. When the 410 is shallower by 10 km, the LVZ structure moves further to the receiver side by 100 km. In this case, we set the 410 at the depth of 420 km, which is very similar to other estimates near the region [*Gilbert et al.*, 2003].

Our preferred model has been slightly modified from the LVZ model discussed earlier to better explain the southernmost TriNet stations. In particular, stations in the Owens Valley are less anomalous although they are late relative to coastal stations probably caused by shallow mantle structure beneath the Sierra Nevada [*Savage and Helmberger,* 2003]. These two models fit the data along the east and west pure paths as displayed in record sections given in Plate 6. We have included a comparison of the more inland data with the model lacking the LVZ to show the obvious mismatch.

The observations at mid-azimuths (Plate 4, red traces) fit neither model although they appear to be a combination of the two models or essentially involving 3D propagation. To correct for these effects, we generate four 2D synthetics which are weighted according to their position in the extended Fresnel zone (Appendix A, Fig. A2–Fig. A4). The method is similar to that introduced in *Helmberger and Ni* [2005] except it uses numerical synthetics instead of analytical. Thus, the contribution from the lit region is broadband while those outside sample the diffracted zone and have their high frequencies removed. We refer to these approximate 3D synthetics as DiFD for convenience. This technique has been validated against SEM [*Komatitsch and Tromp,* 1999] in a similar application by *Ni et al.* [2005] and produces satisfactory results. Here we generate DiFD synthetics at a constant distance, 1960 km and compare with data recorded at different azimuths (Fig. 3).

The data show systematic changes in the timing and amplitude of the CD branch. The westernmost record (DEV) can be explained with the tomographic model without the LVZ while the easternmost station (DAN) can be explained with the low velocity zone model. The middle trace (station MCT), however, can not be explained by either model since the CD branch arrives slightly earlier with smaller amplitude. Instead of generating more 2D models, we explore the synthetics generated by combining the two models into a 3D structure using the DiFD. The 3D model has a sharp edge which separates the western side (normal velocity structure) from eastern side (low velocity zone model). For a station sitting directly above the boundary, each side contributes about half of the response and this synthetic explains the middle trace quite well (Fig. 3). We display typical comparisons against observations with the middle traces in Plate 7. Though few recordings are shown in this swath, we

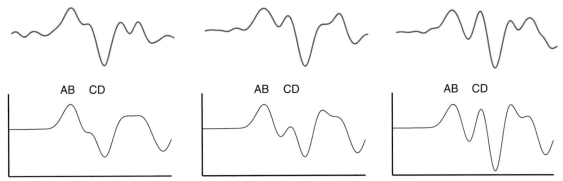

Figure 3. Data (top panel) vs. Diffracted finite difference synthetics (DiFD (bottom panel). From left to right: DEV (Δ=17.3°, Az=144.5°), MCT (Δ=17.3°, Az=142.7°) and DAN (Δ=17.2°, Az=140.5°). Each trace is 40 secs long.

find the DiFD synthetics predict the separation of the AB and CD branches better near 15.5° and generate a shoulder near 17°–18° comparable to the data. The boundary separating these two structures is obviously more complicated than modeled here but the general behavior appears to be explained. To systematically locate the boundary of such low velocity zone structure, we compute five synthetics for each station according to their position relative to the boundary (Fig. A4) and cross-correlate with the data (Fig. 4). Following this modeling strategy, we are able to delineate the likely location of the boundary since the total contribution from the lit zone and the diffraction zone is different depending on the location of the boundary. Also, we assume the boundary is simply vertical in this analysis.

The best fitting synthetic is given in the middle and the poorest fitting synthetic on the right. We have included a cross-correlation measure as a goodness-of-fit criterion (Table 2). In general, the first 25 secs fits quite well which is controlled by the AB- CD interference which is our primary interest. Also, we perform Kruskal-Wallis non-parametric one-way analysis to test the significance of the differences among different models (Table 3). Clearly, the 2D model with LVZ fits the inland data (Az = 137°–143°) significantly better than the 2D model at 95% confidence level (Table. 3). However, it fits the coast data (Az = 145°–147°) much worse at 95% confidence level (Table. 2). Overall, the 3D LVZ model with a vertical boundary fits the whole dataset significantly better than 2D model with LVZ at 99% confidence level (Table. 3). The 2D model with no LVZ seems to fit the

whole data almost equally well simply because large number of stations (24) are located within the coastal swath. We have plotted the stations with a color code to indicate their preference to the weighting (Plate 8).

In general, data recorded on the eastern side of the Californa shear zone can be explained with a boundary of low velocity zone located north of Lake Tahoe. Some of the stations north of the Garlock fault plot as blue suggesting they might sample the structure to the north of the anomalous zone. These paths primarily sample regions beneath southern Oregon, northern California and the western end of the Basin and Range. We conclude that, though the 2-D LVZ model we constructed is not unique, this sharp velocity contrast in the deep upper mantle between easternmost and central northern California is very robust and the velocity decreases rapidly from the east to the west at the scale of less than 100 km.

3. DISCUSSIONS

Recently, *Fee and Dueker* [2004] analyzed receiver function profiles across Nevada- Idaho-Montana and found negative polarity pulse approximately 30 km above the converted pulse from the 410. This section is very close to the area sampled by path C discussed earlier. *Jasbinsek* [2005] also presents similar results beneath NW Colorado suggesting a 5% velocity reduction 30 km atop the 410. As we have shown previously (Fig. 1), a low velocity gradient in the deep upper mantle can also produce secondary AB branch at the distance of 21°–23°. However, the onset of the AB branch appears

Table 2. Correlation coefficient between data and synthetics

Model Type	az 137–143 (11)	az 143–145 (15)	az 145–147 (24)	az 137–147 (50)
2D	0.58±0.09	0.79±0.14	0.82±0.10	0.76±0.15
2D+LVZ	0.73±0.16	0.69±0.19	0.63±0.16	0.67±0.17
3D	0.69±0.12	0.76±0.14	0.74±0.11	0.80±0.12

(): No. of stations

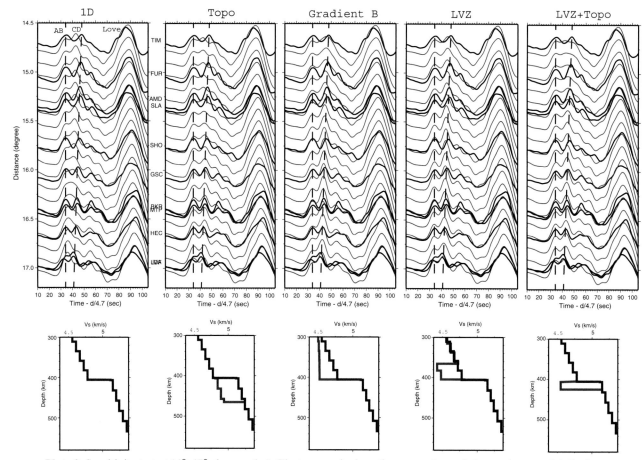

Plate 3. Sensitivity test at 14°–17° along path A. The top panels show the comparison of observed tangential displacement records (in black) and synthetics (in red). The lower panels show the corresponding velocity models. From left to right: 1D (dep410 = 405 km), Topo (dep410 = 465 km), Gradient B, LVZ (dVS = -5%, H*LV Z* = 40 km) and LVZ+Topo (dVS = -5%, H*LV Z* = 20 km, dep410 = 425 km). Dashed lines are the timings of the observed AB and CD branches.

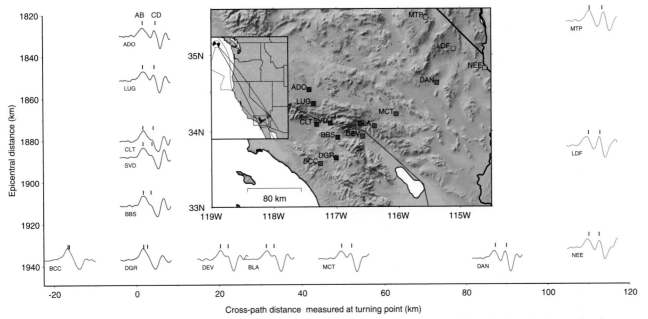

Plate 4. 3D variations in upper mantle triplications recorded at TriNet (path A). A subset of data is plotted against azimuth (138°–155°) and epicentral distance (16.3°–17.5°, 1820-1950 km). Blue traces indicate a distance section with normal moveout of AB and CD branches. Green traces indicate a distance section with anomalously delayed CD branches. Red trances show an azimuth section where the interferences of AB and CD branches are very similar to a distance section but changing in azimuth. Each trace is 40 secs and vertical bars mark the arrivals of AB and CD branches. The locations of these stations are shown in the topographic map of Southern California, with a inset showing great-circle-paths from the source to the east and west California.

DISTANCE = 1870 ~ 1890 km

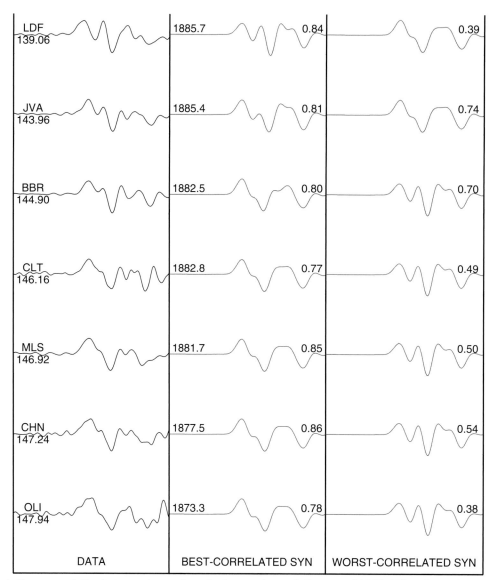

Figure 4. Cross-correlation between data and synthetics at epicentral distance of 16.8°–17°. Left panel: Data. Middle panel: Best correlated DiFD synthetics. Right Panel: Worst correlated DiFD synthetics. Azimuth of each station is shown below the trace. Cross-Correlation coefficient is shown above each synthetic.

earlier than the LVZ model and its amplitude is too large. Also, the gradient models are not consistent with the receiver function profiles in the western US ([*Gilbert et al.*, 2003], Plate. 2, where a negative pulse appears before the converted phase from the 410. Our analysis indicates that the western margin of the LVZ is sharp, less than 100 km. An alternative interpretation of our findings is that the deeper upper mantle in fact is faster along the coastal path due to the existence of dipping Gorda slab [*Bijwaard et al.*, 1998] beneath northern

California, although this would not produce negative peaks observed in receiver function profiles [*Gilbert et al.*, 2003]. If it is true, the dipping fast slab will advance CD branch more than AB branch and decrease their separation, which has been modeled well with current tomographic model [*Godey et al.*, 2004]. Therefore, we prefer the LVZ model but do not rule out this alternative interpretation. However, low velocity zones atop the 410 have been suggested in many other regions. *Revenaugh and Sipkin*[1994] constructed

Table 3. Confidence level of model differences

Model Type	az 137–143	az 143–145	az 145–147	az 137–147
2D, 2D+LVZ	95%	94%	95%	97.5%
2D+LVZ, 3D	30%	69%	95%	99%
2D, 3D	80%	30%	84%	62%

Kruskal-Wallis nonparametric one-way analysis of variance

reflectivity profiles in China and found 5% impedance decrease undeneath northeast China and the Sea of Japan, where the flattened slab is stagnant in the transition zone [*Fukao et al.*, 2001]. *Nolet and Zielhuis* [1994] produced a tomography image beneath Tornquist-Teissyre zone in the western boundary of the Russian Platform. They found a low velocity zone exists beneath this ancient subduction of a wide Tornquist ocean plate between the Baltica and Avalonia shields. These observations are consistent with the idea that these low velocity zones are related to subduction zones, including Farallon plate subduction underneath the western US, which may have horizontal component in the transition zone [*VanderLee and Nolet*, 1997]. If such low velocity zones represent molten layers atop the transition zone, dehydration melting could play a very important role in generating these melts. It is possible that some water is transported by downgoing slabs into the transition zone [*Komabayashi*, this volume]. In particular, a cold slab can be the likely candidate keeping the hydrous silicates stable and carrying water into the transition zone [*Zhao et al.*, 1997; *Komabayashi*, this volume]. *Komabayashi* [this volume] shows that at cold mantle temperaure of less than 1473 K, hydrous phase E is a stable phase. Above 1473 K, that is, at normal mantle temperature in the transition zone, the dehydration of phase E can produce hydrous wadsleyite, which is stable in the transition zone. Because the water-storage capacity of olivine is less than that of wadsleyite, dehydration melting might occur as wadsleyite transforms into olivine during an upwards pass through the 410-km discontinuity, which could affect the chemical differentiation of the upper mantle [*Karato et al.*, this volume]. As suggested by *Komabayashi* [this volume], a stagnant slab in the transition zone is a good reservoir of dense hydrous magnesium silicate, phase E, at temperatures less than 1473 K. Phase E would break down and release water to wadsleyite as the stagnant slab warms up. *Schmid et al.* [2002] estimated that the edge of the Farallon plate that subducted beneath the western US could currently be only 200 K colder than the surrounding mantle, allowing the Farallon slab to dehydrate in the transition zone, consistent with the existence of an LVZ atop the 410-km discontinuity.

4. CONCLUSIONS

We model the upper mantle triplication associated with the 410 seismic discontinuity and find a low velocity zone (5%) above the transition zone consistent with observations and receiver function profiles in parts of the western US. This zone varies in thickness from 20 to 90 km and covers an area from southwest Oregon to central Nevada and extending into northernmost California, roughly a 10° square. Our analysis indicates that the low velocity zone pinches out beneath the western edge with a relatively sharp boundary. Such a structural model produces synthetics with AB vs. CD interferences equally strong in distance and azimuth as observed. In mapping out this LVZ we found that the existence of the AB branch at distances beyond 21° is a particularly useful criterion. Thus, a global search of this portion of the triplication could prove valuable in a mapping exercise and provide constraints on the spatial extent of low velocity zones in the deep upper mantle.

Acknowledgments. The authors would like to thank the editors Suzan Van der Lee and Steven Jacobsen and two reviewers for their constructive reviews and comments. PASSCAL data used in this chapter are requested from IRIS Data Management Center. The work is supported by Tectonic Observatory, Caltech and NSF grant EAR-0337491.

5. APPENDIX

5.1. Finite Difference

To further constrain the lateral extent of the LVZ, it is necessary to extend and go beyond 1-D modeling. Here we adopt the 2-D finite difference method developed by *Vidale et al.* [1985] and *Helmberger and Vidale* [1988]. No attenuation is implemented in the 2D calculation, which might cause over-prediction in the amplitude of AB branches. In the case with no attenuation, the FK wavenumber synthetics produce a very good match with finite difference synthetics (Fig. A1). In addition, the finite difference synthetics will over-predict the higher mode associated with the existence

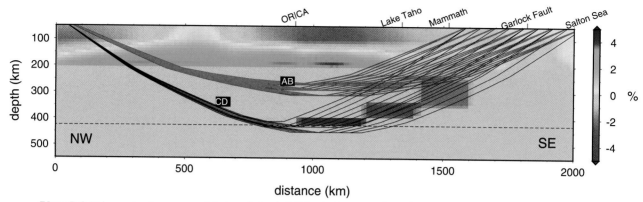

Plate 5. 2-D low velocity zone model along inland California (Az = 137°–143°). Velocity anomaly at teach block from the bottom to the top: dVs = -5%, dVs = -4%, dVs = -3%. Raypaths of AB and CD branches are shown for references.

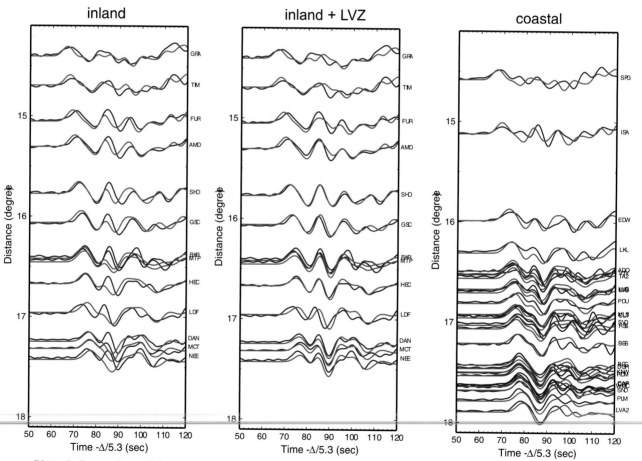

Plate 6. Comparison of data sections and FD synthetics along inland paths (Az = 137°–143°) and coastal paths (Az = 145°–147°). Left panel: synthetics from 2D slices without LVZ. Middle panel: synthetics from 2D slice with LVZ. Right panel: synthetics from 2D slices without LVZ. Data are plotted in blue and synthetics in red.

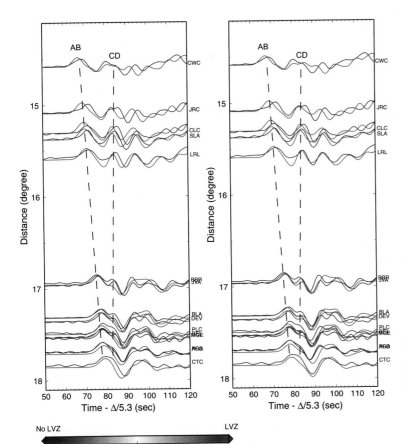

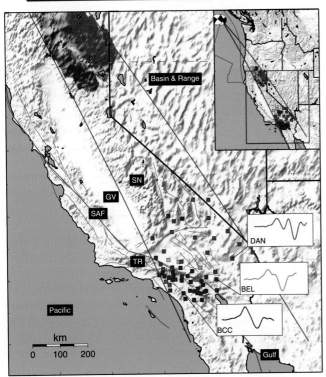

Plate 7. Comparison of data sections and DiFD synthetics along complex middle paths (Az = 143°–145°). Left panel: synthetics from 2D slice without LVZ. Right panel: LVZ/no LVZ boundary directly below the section. Data are plotted in blue and synthetics in red.

Plate 8. Map of low velocity zone atop the 410 beneath the western Basin and Ranges. Typical waveform recored across TriNet are shown for station DAN (Δ=17.3°, Az=140.5°), BEL (Δ=17.5°, Az=143°), BCC (Δ=17.4, Az=146.7°). Data best fitted by the 3D LVZ model where the edge of LVZ is on the eastern/western side of the station beyond the lit zone are plotted in dark blue/dark red. Data best fitted by the 3-D LVZ model where the edge of LVZ is on the eastern/western side of the station within the lit zone are plotted in light blue/light red. Inset on the upper right includes the source-receiver geometry.

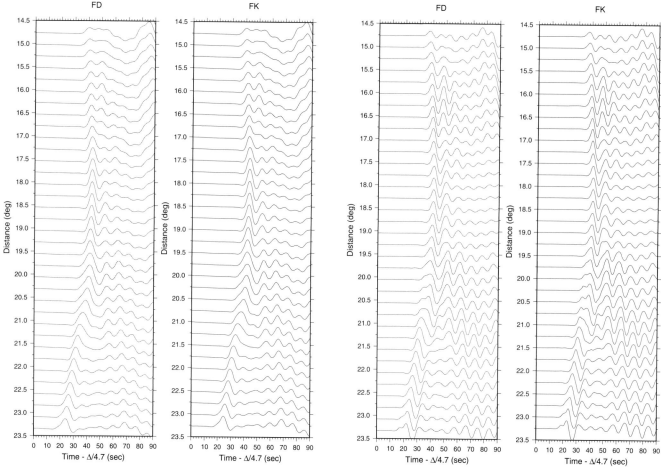

Figure A1. Comparison of Finite difference (FD) and frequency-wavenumber (FK) synthetics at epicentral distance of 14.5°–23.5°. Left panel: FD tangential displacement and velocity. Right: FK tangential displacement and velocity.

of low velocity zone beneath the lithosphere because of the approximate spreading correction. However, we find it is quite useful for our purposes to compute proposed tomographic models and examine their influences on the beginning of the Love waves and the interferences of mantle triplications.

5.2. Diffracted Finite Difference (DiFD)

While 2-D modeling has shown its usefulness [*Stead and Helmberger,* 1988], it can not take into account possible 3-D effects, such as out-of-plane multipathing. To overcome such problems, *Helmberger and Ni* [2005] proposed an approximate algorithm to include responses off the great circle path using 2-D WKM [*Ni et al.,* 2000]. Here we illustrate some of the approximations implemented in the 2-D finite difference calculation. We apply the well-known 2D to 3D mapping approximation

$$\Psi(r, z, t)_P = \sqrt{\frac{2}{r}\frac{1}{\pi}} \left[\frac{1}{\sqrt{t}} * \Psi(r, z, t)_L\right] \qquad (1)$$

where $\Psi(r, z, t)_L$ is essentially a line source response containing a double-couple source (Helmberger and Vidale, 1988) and can be computed with the Finite Difference Method in a 2-D velocity model. To obtain a point source response $\Psi(r, z, t)_P$, simply convolve with an $1/\sqrt{t}$ operator. However, this only accounts for variations in velocities along the Great Circle Path (GC P), where the structure is assumed to be uniform, 2D, at right angles to the plane of propagation.

To estimate the 3D contributions we invoke Huygens' principle, which states the disturbance at some later time can be obtained by summing secondary sources on the reference surface S. In this case, responses off GCP are included. It can be shown that the 2-D Kirchhoff integral over the 3-rd dimension simply represents a $1/\sqrt{t}$ operator if the structure is truly 2D. The procedure is outlined in Fig. A2 with the geometry displayed in (A). To add energy coming from paths outsides the GCP, we simply sum-up contributions along the other paths, essentially a line integration (step B). This response can be written as

(a)

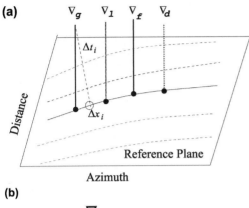

(b)

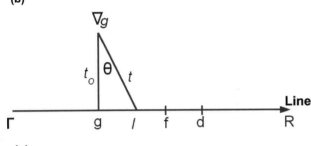

(c)

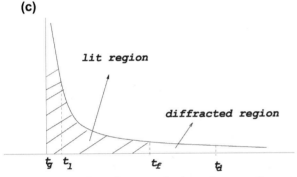

Figure A2. A surface of constant depth serving as a reference plane for interfacing Kirchoff solutions is presented in (a). We assume 2D geometry and integrate along the arc (heavy line) which corresponds to constant epicentral distance with variable azimuth. The other dotted lines correspond to other neighboring distances. Four locations are indicated corresponding to the surface of geometric arrival ∇g, the Fresnel boundary ∇f, the two virtual stations ∇l, ∇d representing the lit-region and the diffracted-region. The middle column (b) illustrates how the energy is added along line Γ, which is at right angle to the GCP. The lower diagram (c) displays the time corresponds to the above four positions relative to the square root weighting.

$$\phi(t) = \frac{1}{2}\frac{d\theta}{dt} + \frac{1}{2}\frac{d\theta}{dt}, \qquad (2)$$

where the angle θ sweeps out paths to the left and to the right. The strongest response comes from $\theta = 0$, or the geometric arrival time (t_0), namely

$$\frac{d\theta}{dt} = \frac{H(t-t_0)}{\sqrt{t-t_0}}, \qquad (3)$$

where t equals the travel time from any point on the line to the receiver. If the field along the line Γ is uniform (independent of azimuth), this line integration produce a $1/\sqrt{t}$ convolution which is that given by equation (1). However, if the left is delayed relative to the right, we obtain a split response as discussed in *Helmberger and Ni* [2005].

Thus, to take into account variations in velocities in the 3-rd dimension, 2-D Kirchhoff line integral is performed to sum all secondary sources at a reference plane S along the line Γ (Fig. A3). In short, we replace $\left[\frac{1}{\sqrt{t}} * \Psi(r,z,t)\right]$ in (1) with

$$\frac{1}{2}\sum_{i=1}^{n}\left[\frac{H(t-t_i)}{\sqrt{t+\Delta t_i}} * \Psi_i(t) - \frac{H(t-t_{i+1})}{\sqrt{t+\Delta t_{i+1}}} * \Psi_{i+1}(t)\right], \quad (4)$$

where ti is the arrival time of path i; Δti is $ti - to$; to is the arrival time along GCP and $\Psi_i(t)$ is the response computed for path i; n is the number of secondary sources considered. $1/\sqrt{t+\Delta t_i}$ becomes a frequency dependent weighting factor. Several assumptions are made:

(a) Reference plane S is placed above the anomalous region.

(b) Velocity variations in the domain are effectively 2-D except the anomaly of interest.

(c) The receiver is sufficiently far away from the diffractor ($T \gg \tau$, T is travel time and τ is period of signal).

5.3. Estimate Zone of Influence

Define the Fresnel zone limit as

$$X_f \approx \sqrt{h\alpha T}, \qquad (5)$$

where h is the depth of reference plane; α is mean velocity between the reference plane and surface and T is the duration of signal. The region inside X_f is defined as the Lit region and the region outside as the Diffraction region. Define the lit operator as

$$O_L(t) \equiv \frac{d}{dt}\left[\left(\frac{H(t)}{\sqrt{t}} - \frac{H(t-t_f)}{\sqrt{t+\Delta t_f}}\right) * \dot{D}(t)\right], \quad (6)$$

where t_f is the arrival time corresponding to the Fresnel limit; Δt_f is $t_f - t_o$; t_o is the arrival time for GCP. Define the diffraction operator as

$$O_D(t) \equiv \frac{d}{dt}\left[\frac{H(t-t_f)}{\sqrt{t}} * \dot{D}(t)\right]. \qquad (7)$$

In this case, the diffraction operator produces the long period signal coming from the secondary sources off GCP along Γ.

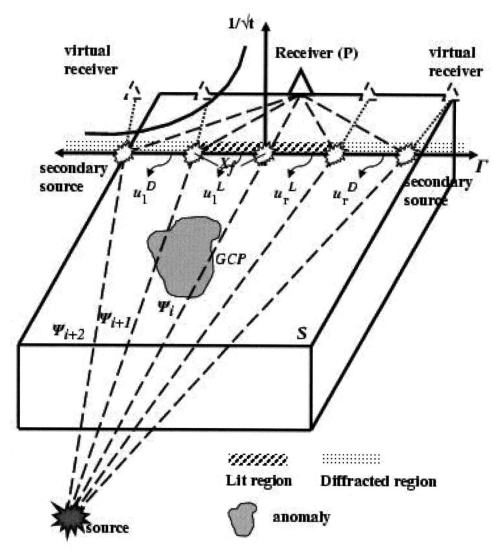

Figure A3. Schematic map showing the calculation of Diffracted finite difference.

To find sharp velocity contrast in the mantle, given a seismic array at the surface, we can use this method to delineate where is the sharp boundary relative to a particular observation. Define

$$u_l^D \equiv O_D(t) * \Psi_l^D \qquad (8)$$

$$u_r^L \equiv O_L(t) * \Psi_r^L \qquad (9)$$

$$u_l^D \equiv O_D(t) * \Psi_l^D \qquad (10)$$

$$u_r^L \equiv O_L(t) * \Psi_r^L, \qquad (11)$$

where u_l^D, u_l^L the responses from diffraction and lit region along Γ of the left side; u_r^L, u_r^D the response from the lit and diffraction region along Γ of the right side. For an exact 2-D structure along GCP, there is no variations along Γ and $\Psi_l^D = \Psi_r^D = \Psi_l^L = \Psi_r^L$ Displacement at receiver P can be expressed as

$$u(P) = u_l^D + u_l^L + u_r^L + u_r^D. \qquad (12)$$

If velocity changes abruptly along Γ, line source responses Ψ_l^D, Ψ_r^D, Ψ_l^L, Ψ_r^L are calculated based upon the relative location between the assumed velocity boundary and the receiver (Fig. A4). With multiple receivers and multiple events, it is possible to determine the location of the boundary in parallel to locating the epicenter of earthquakes. If the

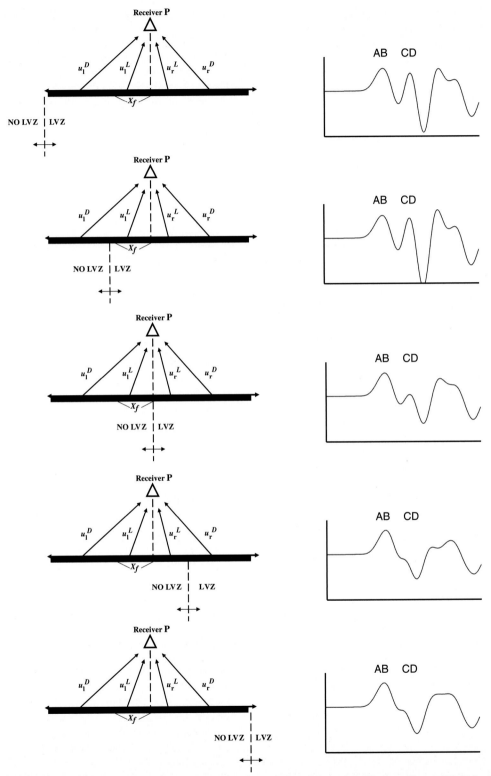

Figure A4. DiFD synthetics. Left panels: Cross section view of 3D LVZ model. The boundary of LVZ is shown as vertical bar. Lit region and diffracted region are shown in hatched and grey color. Right panels: DiFD synthetics corresponds to 3D LVZ model in the left panels.

waveforms changes within X_f then the diffraction surface S has to be shallower to account for the rapid variation, which is similar to determine the hypocenter of earthquakes.

REFERENCES

Agee, C. B., Crystal-liquid density inversion in terrestrial and luner magmas, *Phys. Earth Planet. Inter.*, *107*, 63–74, 1998.

Bercovici, and Karato, Transition zone water filter, *Nature*, *425*, 39–44, 2003.

Bijwaard, H., W. Spakman, and E. R. Engdahl, Closing the gap between regional and global travel time tomography, *J. Geophys. Res.*, *103*, 30,055–30,078, 1998.

Chen, G. Q., T. J. Ahrens, and E. M. Stolper, Shock-wave equation of state of molten and solid faylite, *Phys. Earth Planet. Inter.*, *134*, 35–52, 2002. Ding, X., and S. Grand, Upper mantle q structure beneath the east pacific rise, *J. Geophys. Res.*, *98*, 1973–1985, 1993.

Fee, D., and K. Dueker, Mantle transition zone topography and structure beneath the yellowstone hotspot, *Geophys. Res. Lett.*, *31*, 2004GL020,636, 2004.

Fukao, Y., S. Wdiyantoro, and M. Obayasha, Stagnant slab in the upper and lower mantle transition region, *Rev. Geophysics*, *39*, 291–323, 2001.

Gao, W., S. Grand, W. S. Baldridge, D. Wilson, M. West, J. Ni, and R. Aster, Upper mantle convection beneath the central rio grande rift imaged by p and s wave tomography, *J. Geophys. Res.*, *109*, 2003JB002,743, 2004.

Gilbert, H. J., A. F. Sheehan, K. G. Dueker, and P. Molnar, Receiver functions in the western united states, with implications for upper mantle structure and dynamics, *J. Geophys. Res.*, *108*, 2001JB001,194, 2003.

Godey, S., F. Deschamps, J. Trampert, and R. Snieder, Thermal and compressional anomalies beneath the north american continent, *J. Geophys. Res.*, *109*, 2002JB002,263, 2004.

Grand, S., and D. V. Helmberger, Upper mantle shear structure of north america, *Geophys. J. R. Astro. Soc.*, *76*, 399–438, 1984a.

Grand, S., and D. V. Helmberger, Upper mantle shear structure beneath the northwest atlantic ocean, *J. Geophys. Res.*, *89*, 11,465–11,475, 1984b.

Graves, R., and D. V. Helmberger, Upper mantle cross section from tonga to newfoundland, *J. Geophys. Res.*, *93*, 4701–4711, 1988.

Helmberger, D., and S. Ni, Approximate 3d body-wave synthetics for tomographic models, *Bull. Seism. Soc. Am.*, *95*, 212–224, 2005.

Helmberger, D., and J. Vidale, Modeling strong motioned produced by earthquakes with two-dimensional numerical codes, *Bull. Seism. Soc. Am.*, *78*, 109–121, 1988.

Helmberger, D., and R. A. Wiggins, Upper mantle structure of midwestern united states, *J. Geophys. Res.*, *76*, 3229, 1971.

Helmberger, D. V., G. Engen, and S. Grand, Note on wave propagation in laterally varying structure, *58*, 82–91, 1985.

Hirschmann, M. M., C. Aubaud, and A. C. Withers, Storage capacity of h2o in nominally anhydrous minerals in the upper mantle, *Earth Planet. Sci. Lett.*, *236*, 167–181, 2005.

Jasbinsek, J. J., Strong evidence for 380 and 580 km negative velocity gradients beneath the lodore array in nw colorado, *Eos Trans. AGU*, *86*, 52, 2005.

Johnson, L. R., Array measurements of p velocities in upper mantle, *J. Geophys. Res.*, *38*, 6309–6325, 1967.

Karato, S., D. Bercovici, G. Leary, G. Richard, and Z. Jing, The transition-zone water filter model for global material circulation: Where do we stand?, this volume.

Kawamoto, T., R. L. Hervig, and J. R. Holloway, Experimental evidence for a hydrous transition zone in the early earth's mantle, *Earth Planet. Sci. Lett.*, *142*, 587–592, 1996.

Kohlstedt, D. L., H. Keppler, and D. C. Rubie, Solubility of water in the phases of $(mg, fe)2sio4$, *Contrib. Mineral. Petrol.*, *123*, 345–357, 1996.

Komabayashi, T., Phase relations of hydrous peridotite: Implications for water circulation in the mantle, this volume.

Komatitsch, D., and J. Tromp, Introduction to the spectral-element method for 3-d seismic wave propagation, *Geophys. J. Int.*, *139*, 806–822, 1999.

Matsukage, K. N., Z. Jing, and S. I. Karato, Density of hydrous silicate melt at the conditions of earth's deep upper mantle, *Nature*, *438*, 488–491, 2005.

Melbourne, T., and D. V. Helmberger, Whole mantle shear structure beneath the east pacific rise, *J. Geophys. Res.*, *107*, 2001JB000,555, 2002.

Melbourne, T. I., and D. V. Helmberger, Mantle control of plate boundary deformation, *Geophys. Res. Lett.*, *28*, 4003–4006, 2001.

Ni, S., X. M. Ding, and D. V. Helmberger, Constructing synthetics from deep earth tomographic models, *Geophys. J. Int.*, *140*, 71–82, 2000.

Ni, S., E. Tan, D. V. Helmberger, and M. Gurnis, Sharp sides to the african superplume, *Science*, *296*, 1850–1852, 2002.

Ni, S., D. V. Helmberger, and J. Tromp, Three-dimensional structure of the african superplume from waveform modeling, *Geophys. J. Int.*, *161*, 283–294, 2005.

Nolet, G., and A. Zielhuis, Low s velocities under the tornquist-teisseyre zone: Evidence for water injection into the transition zone by subduction, *J. Geophys. Res.*, *99*, 15,813–15,821, 1994.

Revenaugh, J., and S. A. Sipkin, Seismic evidence for silicate melt atop the 410 km mantle discontinuity, *Nature*, *369*, 474–476, 1994.

Rigden, S. M., T. J. Ahrens, and E. M. Stolper, Densities of liquid silicates at high pressure, *Science*, *226*, 1071–1074, 1984.

Saikia, C. K., Modified frequency-wavenumber algorithm for regional seismograms using filons quadrature-modeling of l(g) waves in eastern north-america, *Geophys. J. Int.*, *118*, 142–158, 1994.

Savage, B., and D. Helmberger, Velocity variaions in the uppermost mantle beneath the southern sierra nevada and walker lane, *J. Geophys. Res.*, *108*, 2001JB001,393, 2003.

Schmid, C., S. Goes, S. Vanderlee, and D. Giardini, Fate of the cenozoic farallon slab from a comparison of kinematic thermal mdoelling with tomographic images, *Earth Planet. Sci. Lett.*, *204*, 17–32, 2002.

Sheehan, A., V. Schulte-Pelkum, O. Boyd, and C. Wilson, The rocky mountain region: evolving lithosphere: tectonics, geochemistry, geophysics, *154*, 309–315, 2005.

Song, T.-R. A., *PhD Thesis, California Institute of Technology*, pp. 1–176, 2006.

Song, T.-R. A., D. V. Helmberger, and S. Grand, Low velocity zone atop the 410 seismic discontonuity beneath the northwestern us, *Nature*, *427*, 530–533, 2004.

Stead, R., and D. V. Helmberger, Numerical-analytical interfacing in 2 dimensions with applications to modeling seismograms, *Pure Appl. Geophys.*, pp. 157–193, 1988.

Stolper, E. M., D. Walker, B. H. Hager, and J. F. Hays, Melt segregation from partially molten source regions: the importance of melt density and source region size, *J. Geophys. Res.*, *86*, 6261–6271, 1981.

Tan, Y., and D. V. Helmberger, Trans-pacific upper mantle shear velocity structure, *to be submited*, 2005.

VanderLee, S., and A. Frederiksen, *Surface wave tomography applied to the North American upper mantle*, Geophysical Monograph, 67-80 pp., American Geophysical Union, 2005.

VanderLee, S., and G. Nolet, Seismic images of the subducting trailing fragments of the farallon plate, *Nature*, *386*, 266–269, 1997.

Vidale, J., Finite-difference calculation of travel-times, *Bull. Seism. Soc. Am.*, *78*, 2062–2076, 1988.

Vidale, J., D. V. Helmberger, and R. W. Clayton, Finite difference seismograms for sh waves, *Bull. Seism. Soc. Am.*, *75*, 1765–1782, 1985.

Walck, M. C., The p-wave upper mantle structure beneath an active spreading center: the gulf of california, *Geophys. J. R. Astro. Soc.*, *76*, 697–723, 1984a.

Walck, M. C., The upper mantle beneath the north-east pacific rim: a comparison with the gulf of california, *Geophys. J. R. Astro. Soc.*, *81*, 243–276, 1984b.

West, M., W. Gao, and S. Grand, A simple approach to the joint inversion of seismic body and surface waves applied to the southwest u.s., *Geophys. Res. Lett.*, *31*, 2004GL020,373, 2004.

Wilson, D., R. Aster, J. Ni, S. Grand, M. West, W. Gao, W. S. Baldridge, and S. Semken, Imaging the seismic structure of the crust and upper mantle beneath the great plains, rio grande rift, and colorado plateau using receiver functions, *J. Geophys. Res.*, *110*, 2004JB003,492, 2005.

Zhao, D., Y. Xu, D. A. Wiens, L. Dorman, J. Hildebrand, and S. Webb, Depth extent of the lau back-arc spreading center and its relation to subduction zone process, *Science*, *278*, 254–257, 1997.

Teh-Ru Alex Song, Seismo Lab, Division of Geological and Planetary Sciences, South Mudd, Pasadena, CA 91125, USA. (alex@gps.caltech.edu)

Mantle Transition Zone Thickness in the Central South-American Subduction Zone

Jochen Braunmiller*

Institute of Geophysics, ETH Zurich, Switzerland

Suzan van der Lee

Department of Geological Sciences, Northwestern University, Evanston, Illinois, USA

Lindsey Doermann

Brown University

We used receiver functions to determine lateral variations in mantle transition zone thickness and sharpness of the 410- and 660-km discontinuities in the presence of subducting lithosphere. The mantle beneath the central Andes of South America provides an ideal study site owing to its long-lived subduction history and the availability of broadband seismic data from the dense BANJO/SEDA temporary networks and the permanent station LPAZ. For LPAZ, we analyzed 26 earthquakes between 1993–2003 and stacked the depth-migrated receiver functions. For temporary stations operating for only about one year (1994–1995), station stacks were not robust. We thus stacked receiver functions for close-by stations forming five groups that span the subduction zone from west to east, each containing 12 to 25 events. We found signal significant at the 2σ level for several station groups from P to S conversions that originate near 520– and 850–900 km depth, but most prominently from the 410- and 660-km discontinuities. For the latter, the P to S converted signal is clear in stacks for western groups and LPAZ, lack of coherent signal for two eastern groups is possibly due to incoherent stacking and does not necessitate the absence of converted energy. The thickness of the mantle transition zone increases progressively from a near-normal 255 km at the Pacific coast to about 295 km beneath station LPAZ in the Eastern Cordillera. Beneath LPAZ, the 410-km discontinuity appears elevated by nearly 40 km, thus thickening the transition zone. We compared signal amplitudes from receiver function stacks calculated at different low-pass frequencies to study frequency dependence and possibly associated discontinuity sharpness of the P to S converted signals. We found that both the 410- and 660-km discontinuities exhibit amplitude increase with decreasing frequency. Synthetic receiver function calculations for discontinuity topography mimicking observed topography show that the observed steep topography can adequately explain the observed frequency dependence. The large transition zone thickness beneath LPAZ can be explained by a ~300 K cooler slab or a mantle saturated with water (~1 wt %).

*Now at College of Oceanic and Atmospheric Sciences, Oregon State University, Corvallis, OR.

Earth's Deep Water Cycle
Geophysical Monograph Series 168
Copyright 2006 by the American Geophysical Union.
10.1029/168GM16

INTRODUCTION

The transition from the upper to the lower mantle in the Earth is bound by two major, globally observed seismic velocity contrasts that occur near 410 km and 660 km depth [e.g., *Dziewonski and Anderson*, 1981]. Mineral physics suggests the discontinuities are due to phase changes from olivine to wadsleyite and from ringwoodite to perovskite and magnesiowüstite at pressures equivalent to these depths [e.g., *Ringwood*, 1975]. The associated increase in S wave velocity for PREM [*Dziewonski and Anderson*, 1981] is 3% and 7%, respectively. Both phase transitions are reported to occur over a fairly narrow (~8 km) depth interval [*Yamazaki and Hirahara*, 1994; *Benz and Vidale*, 1993; *Paulssen*, 1988]. Reasonable temperature variations [*Wood*, 1990] and increasing water content in mantle minerals [*Higo et al.*, 2001] have little seismically detectable effect on the sharpness of the 660-km phase transition. The 410-km phase transition interval, however, might reach widths of up to 20–40 km under hydrous conditions [*Helffrich and Wood*, 1996; *Smyth and Frost*, 2002]. Lower temperatures also widen the interval, but the effect of a cold slab, for example, is much smaller than that of a few tenths of wt % water [*Helffrich and Bina*, 1994; *Helffrich and Wood*, 1996]. Based on observed frequency dependence of P to S converted wave amplitudes, *Van der Meijde et al.* [2003] suggested the 410-km phase transition interval is at least 20 km thick (and in one location 35 km) beneath the Mediterranean region where subduction has been occurring for the past 190 m.y.. Because this thickening is more than plausible from temperatures consistent with tomographically imaged high S velocities [*Helffrich and Bina*, 1994; *Marone et al.*, 2004], the thickening was interpreted as evidence for up to 0.1 wt % of water in the transition zone, brought to depth through subduction.

The overall mantle transition zone thickness also depends on temperature because the 410- and 660-km phase transitions exhibit opposite pressure-temperature Clapeyron slopes [e.g., *Katsura and Ito*, 1989; *Bina and Helffrich*, 1994]. The 410-km transition is exothermic with lower temperatures resulting in decreased pressure (elevated depth) of the phase transition, while the 660-km transition is endothermic and would be depressed under cool conditions. Variations in the depths of the 410- and 660-km discontinuities are thus expected to be anti-correlated near cold subducted slabs resulting in a thickened transition zone. Results from global SS precursor studies [e.g., *Flanagan and Shearer*, 1998; *Gu et al.*, 1998] agree with predicted anti-correlation.

In this study, we calculate receiver functions for stations above the central South American subduction zone to estimate temperature anomalies and water content of the deep upper mantle. To this end we investigate the thickness of the transition zone from the delay time between the P660s and P410s arrivals and the thickness of the olivine to wadsleyite phase transition from analysis of frequency-dependence of receiver functions.

DATA AND METHOD

We are interested in detecting variations of overall mantle transition zone thickness and in characterizing widths of phase transition intervals in well-established subduction zone environments. We thus study the central Andean part of the South American subduction zone because convergence at 50–150 mm/yr is nearly orthogonal since about 50 Ma [*Pardo-Casas and Molnar*, 1987]. In addition, broadband seismic data are available from a dense seismic network. Stations from the temporary BANJO and SEDA networks [*Beck et al.*, 1994], operational from April 1994 to September 1995, and the permanent station LPAZ (Table 1) provide an E-W transect perpendicular and a roughly N-S oriented transect parallel to the strike of the subduction zone (Figure 1). Seismicity defines the subducting slab down to ~350 km depth and between 490 to 640 km depth (Figure 1); slab continuity through the aseismic zone is indicated by seismic tomography [*Bijwaard et al.*, 1998].

We calculated receiver functions from teleseismic earthquakes in the 30°–95° distance range with magnitude M ≥ 6.0 (Figure 2). Most events producing quality receiver functions (Table 2) occurred in the Central American and Mexican subduction zones with back-azimuth 310°–325° and some in the South Sandwich Islands region with back-azimuth 145°–180°. Receiver functions were obtained through frequency domain deconvolution [*Langston*, 1979; *Ammon*, 1991]. We deconvolved 90 s long vertical from 250 s long radial (and transverse) component data, each window starting 20 s before the P wave arrival. We filled spectral gaps to stabilize the deconvolution such that amplitudes on the vertical component were at least 10% of its peak value (0.1 water level). Seismograms were high-pass filtered at 0.05 Hz prior to deconvolution and receiver functions were computed for nine low-pass corner frequencies between approximately 0.15 and 0.75 Hz. We inspected seismograms and radial and transverse receiver functions visually to select quality receiver functions for depth conversion and later stacking. We used the *iasp91* velocity model [*Kennett and Engdahl*, 1991] to perform a pre-stack time-to-depth conversion. Resulting absolute depths are thus probably biased, given that we ignore lateral heterogeneity associated with subduction zones. Our main interest is to determine the transition zone thickness, and since this derives from relative rather than absolute timing of two converted phases it is virtually insensitive to lateral heterogeneity in the top 400 km. Our

Table 1. Station list. #: Station number. EU: number of events used; about 50% of all events could be used. Thickness, 410, 660: transition zone thickness, nominal depth of 410- and 660-km discontinuity with 1σ uncertainty from averaging over 9 frequencies for each station group (Figure 3); 'central-east' group values are from amplitudes not significant at 2σ-level. 410F and 660F indicates probability for frequency dependence of amplitudes based on Gaussian distribution test. Stations 1–3, 4–6, 7–11, 13–17, and 18–20 are 'west', 'central', 'north', 'central-east', and 'far-east' station groups.

#	Name	Latitude [°]	Longitude [°]	EU	Thickness [km]	410 [km]	660 [km]	410F [%]	660F [%]
1	PICH	-19.869	-69.420	3					
2	LIRI	-19.852	-68.849	3	254±1	417±1	671±2	79	24
3	HIZO	-19.607	-68.326	6					
4	SALI	-19.621	-67.726	9					
5	DOOR	-19.354	-67.223	9	274±3	430±2	704±1	84	75
6	CHIT	-20.077	-66.886	5					
7	POOP	-18.387	-67.018	4					
8	CHUQ	-17.945	-67.818	5					
9	LAJO	-17.776	-67.479	3	276±1	425±3	700±2	91	71
10	SICA	-17.292	-67.749	5					
11	COLL	-16.922	-68.314	2					
12	LPAZ	-16.173	-68.078	26	296±6	369±1	666±5	81	67
13	TACA	-18.828	-66.734	7					
14	CRUZ	-19.103	-66.221	6					
15	CRIS	-19.375	-65.932	2	(290)	(400)	(690)		
16	BATO	-19.626	-65.437	4					
17	YUNZ	-19.158	-65.069	6					
18	SCHO	-19.148	-64.643	9					
19	ROSL	-19.486	-64.178	8					
20	PICH	-19.811	-63.721	3					

procedures follow *Van der Meijde et al.* [2003] such that inferences based on depth-converted receiver functions can be directly compared with their interpretations.

RESULTS

The pre-stack depth-converted receiver function stacks are shown for each station group in Figure 3 for the nine investigated low-pass filters (0.75 to 0.15 Hz from left to right); the shaded area corresponds to amplitudes larger than zero at the 95% confidence level (mean minus 2 standard deviations). Station groups and number of events analyzed are given in Table 1. Table 1 also lists results for transition zone thickness and nominal depths of the 410- and 660-km discontinuity derived for the *iasp91* velocity model.

For permanent station LPAZ, we obtained 26 quality receiver functions (Table 2) from 62 analyzed events from 1993 to 2003. The receiver function stack reveals significant signals from the upper mantle phase transitions.

For the 19 BANJO/SEDA stations, we could analyze maximally 17 events (SCHO) per station due to the relatively short-term deployment resulting between only 2 (CRIS and COLL) and 9 (DOOR, SALI, and SCHO) quality receiver functions. Individual station stacks usually did not yield stable results and we grouped stations located at similar

positions relative to the subducting slab (Figure 1). These multi-station stacks (Figure 3) show signals from mantle converted phases. We tested signal robustness by repeating the stacking process for the 'west' group, which has fewest events (12), each time omitting one different receiver function. The resulting 12 stacks composed of the 11 remaining receiver functions showed no significant differences and we conclude that our stacks are stable and indeed reflect subsurface structure.

For the LPAZ and 'central-east'-group stacks, containing the largest number of events (26 and 25, respectively) we investigated the effect of event back-azimuth by stacking only events from northwest and south, respectively. The northwest sub-stacks, which contain most events, are stable and more closely resemble the overall stacks than south sub-stacks. Differences between the sub-stacks are presumably due to rays sampling different mantle parts above, below or inside the slab. The differences also result in incoherent stacking, with incoherency decreasing with decreasing signal frequency.

Observed Phase Transitions and Mantle Transition Zone Thickness

The 'west' stack is composed of three stations located ocean-ward where the slab is above the mantle transition zone

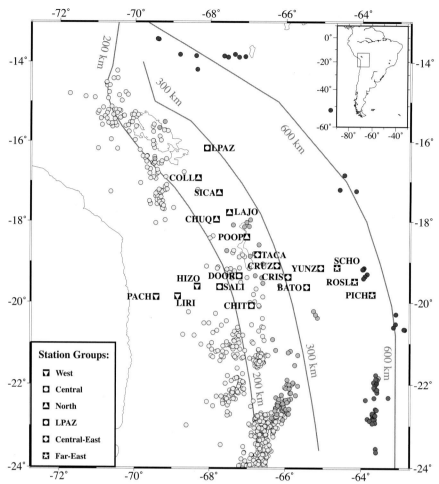

Figure 1. Map of stations used from the BANJO/SEDA temporary networks and permanent station LPAZ. Legend (lower left) indicates station groups for receiver function stacks. Earthquake epicenters are from the Engdahl et al. [1998] catalog with darker gray shading according to four depth intervals (150–250, 250–350, 450–550, and 550+ km depth, no events occurred between 350 and 490 km depth). 200-, 300-, and 600-km Wadati-Benioff zone contours are from Hasegawa and Sacks [1981] and Cahill and Isacks [1992]. Inset (upper right) shows study area.

(Figure 1, Table 1) and we expect to find undisturbed 410- and 660-km discontinuities. The stack shows clear conversions near 417 km and 671 km depth (Figure 3). We also detect conversions near 740 km and 870 km depth (and possibly a weak signal from ~500 km depth). These depths (and others cited in Table 1) are averages determined at maximum amplitudes for 2σ-significant signals at nine frequency bands; uncertainties are simply one standard deviation from averaging and do not include picking errors or systematic errors due to differences between real velocity and *iasp91*. Absolute depths could be overestimated because *iasp91* has a thinner crust than the Andes, which includes a 40–70 km thick slow felsic crust [*Beck and Zandt*, 2002], but could be underestimated because *iasp91* has lower velocities than the seismic velocities in cold subducting lithosphere. Accounting for differences in the upper 70 km alone could systematically raise all conversion

depths by roughly 5–15 km (larger values for thick crust). However, the transition zone thickness of 254 km (Table 1) is normal compared to the 250 km of 1D Earth model *iasp91* and slightly thickened compared to a global average of about 242 km [*Flanagan and Shearer*, 1998]; our absolute thickness uncertainties are probably less than ±10 km.

The 'central' and 'north' groups are located at similar positions relative to the down-going slab (Figure 1). Both groups show clear conversions from the 410- and 660-km discontinuities at nominal depths near 425–430 km and 700–705 km. The transition zone thus thickens from west to east to 274 km and 276 km, respectively. Additional conversions are detected from about 530–535 km and from 845 and 875 km depth, respectively.

The LPAZ stack shows a conversion near 369 km depth indicating a significant 40-km elevation of the 410-km phase transition. *Collier and Helffrich* [2001, their Fig. 6) also found a similarly elevated phase transition depth near the LPAZ 410-km conversion points (Figure 4). Conversions for northwestern events, which dominate the stack, occur where the slab top is ~300 km deep and the slab dips steeply (Figure 4). The 410-km conversion thus probably occurs inside the cold slab and is elevated. We found a fairly normal value of 666 km depth for the 660-km phase transition, which may or may not be biased because of ignored lateral heterogeneity. Overall, the transition zone thickness is widened to about 296 km. We also found signal originating near 915 km depth; but found no clear signal from near 500 km depth.

The 'central-east' and 'far-east' stacks are composed of stations where the Benioff zone beneath dips steeply and its top is between 300 and 600 km depth. Both groups show no clear signal from the 410- or the 660-km discontinuity. Weak signals, not significant at 2σ, possibly indicate a transition zone thickness of ~290 km for the 'central-east' stack. For the 'central-east' stack, we found significant signal amplitudes only at 500 km and 860 km depth. The group's stations are located at a similar position relative to the down-going slab as LPAZ, which shows both phase conversions. We attribute the absence of coherent signal for the 'central-east' group to possible differences in sampling relative to the slab. At LPAZ, signal stacks coherently because most conversion points occur at similar slab depth and thus similar discontinuity depth (Figure 4). This is likewise further west (Figure 4) where the slab is above the 410-km discontinuity and dips relatively gently ('western' and 'central' groups) or, for steeper dip, conversions occur at similar position (depth) relative to the slab ('north' group). In contrast, conversions for eastern groups occur at different positions relative to the steeply dipping slab thus sampling the discontinuities at varying depths such that signal stacks incoherently. This effect is strongest for the 'far-east' group where we found no significant signal from any depth across the frequencies investigated. For the eastern groups, stacking all events thus does not enhance the signal; we suspect stacking events from small source regions would provide coherent signal for specific small conversion regions. However, data quantity is insufficient to obtain reliable stacks from subsets of events for each eastern station group.

Frequency Dependence of Phase Transition Intervals

We calculated receiver function stacks for nine different frequency bands (Figure 3). Amplitudes (and widths) of the 410- and 660-km phase transition signals clearly change with frequency. The amplitudes in Figure 3 are relative to the direct P wave, set to 1, and reach a maximum of 0.133 for the lowest frequency (0.15 Hz) of the 'west'-groups' 410-km

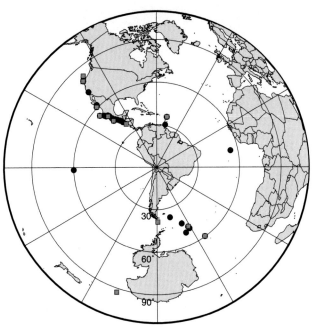

Figure 2. Teleseismic earthquakes used for analysis in equidistant azimuthal projection (thin lines are 30°-distance steps). Dark circles are events used for LPAZ; light squares for BANJO/SEDA stations. Most events are located in Central and North America northwest of the study area, dominating our receiver function stacks.

signal. We observe three trends. First, amplitudes generally increase form high to low frequencies (from left to right) for each phase in each stack. Second, amplitudes vary by a factor of about 2 for the 410-km signal, but only by about 1.5 for the 660-km signal. This is consistent with a wider transition interval for the 410- than for the 660-km phase transition. Signal near 520-km depth, visible for the 'north', 'central', and 'central-east' stacks, is marginal at high frequencies (0.75 Hz) particularly for the otherwise normal 'north' and 'central' stacks, indicating an even wider transition interval for the wadsleyite to ringwoodite phase transition. And third, low frequency amplitudes are larger for the 410- than for the 660-km transition indicating that the shallower phase transition has a stronger impedance contrast, which is in contrast to global velocity models (e.g. *iasp91*).

Following *Van der Meijde et al.* [2003], we estimated frequency dependence of signals from the 410- and the 660-km discontinuity by comparing Gaussian distributions given by the stacks' mean amplitude and standard deviation for highest and lowest frequency receiver function. The non-overlapping area of the two Gaussians then represents the probability of frequency dependence. We read mean amplitudes and standard deviations from the stacks for each group where both phases were present and calculated the probability of frequency dependence (Table 1).

Table 2. Events used for receiver function analysis. Locations and depths are from the PDE catalog, Mw from Harvard CMT, b: body wave magnitude.

Date	Latitude [°]	Longitude [°]	Depth [km]	Magnitude M_w	Station Used
1993/09/10	14.72	-92.65	34	7.3	12
1993/09/27	-53.65	-51.62	33	6.6	12
1993/09/30	15.42	-94.70	19	6.5	12
1993/10/24	16.76	-98.72	21	6.6	12
1994/01/17	34.21	-118.54	18	6.7	12
1994/02/12	-10.79	-128.80	15	6.7	12
1994/05/23	18.17	-100.53	55	6.3	12
1994/07/04	14.89	-97.32	15	6.5	3–15, 17–20
1994/07/25	-56.36	-27.37	81	6.3[b]	3–11,13,17–20
1994/09/01	40.40	-125.68	10	7.0	3–4,6–8,10,13–14,17–19
1994/10/27	43.52	-127.43	20	6.3	5
1994/12/10	18.14	-101.38	48	6.4	1–2,4–10,13,17,19
1995/01/03	-57.70	-65.88	14	5.9	4–6,15,17–19
1995/02/19	40.56	-125.54	10	6.6	8,10,12–13,18–19
1995/03/08	16.56	-59.56	8	6.2	12–13,16–20
1995/05/23	-55.95	-3.36	10	6.8	4–5,12,14,18
1995/05/31	18.96	-107.42	33	6.3	1–5,13–14,18–19
1995/06/14	12.13	-88.36	25	6.6	3–5,8,16
1995/06/21	-61.67	154.77	10	6.7	16
1995/06/30	24.69	-110.23	10	6.2	1–5,12,14,16,18
1995/08/28	26.09	-110.28	12	6.6	12
1995/09/14	16.78	-98.60	23	7.4	12
1996/01/22	-60.61	-25.90	10	6.2	12
1996/02/18	-1.27	-14.27	10	6.6	12
1997/01/11	18.22	-102.76	33	7.2	12
1997/04/22	11.11	-60.89	5	6.7	12
1997/05/01	18.99	-107.35	33	6.9	12
1997/05/22	18.68	-101.60	70	6.5	12
1998/02/03	15.88	-96.30	33	6.4	12
2002/11/12	-56.55	-27.54	120	6.2	12
2002/11/15	-56.05	-36.40	10	6.7	12
2002/12/18	-57.09	-24.98	10	6.1	12
2003/01/21	13.63	-90.77	24	6.5	12
2003/01/22	18.77	-104.10	24	7.5	12

The statistical test confirms our visual observation that signal amplitude for both discontinuities increases with decreasing frequency. Probabilities are generally higher for the 410-km (all >75%) than for the 660-km discontinuity (all ≤75%) and, except for the 660-km discontinuity of the 'west' group, are always significant (>66%). A conventional two-sample t-test confirmed the high statistical likelihood for frequency dependence. Part of the frequency dependence may be the result of data processing as we also found some (30 %) frequency dependence in P410s from stacks of synthetic receiver functions for a flat discontinuity. The apparent frequency dependence in synthetics for a flat interface is due to negligible standard deviations of P410s amplitudes, making small variations in amplitude seem significant; for observed data, the standard deviations are not negligible.

DISCUSSION

Frequency dependence of the P to S converted signal amplitude had been found earlier for some stations in the Mediterranean region [*Van der Meijde et al.*, 2003]. There, only the 410-km signal showed strong dependence while the 660-km signal varied little with frequency. *Van der Meijde et al.* [2003] interpreted their observations as indicating a sharp 660-km phase transition and a thickened 410-km phase transition interval, implying the presence of as much as 500–1000 ppm by weight of water in olivine. They suggested the water is brought into the mantle by subduction processes from above. Here we have studied P to S converted signals from mantle discontinuities in the long-lived, stable South American subduction zone to test their inferences.

We found strong frequency dependence for the 410-km, but contrary to *Van der Meijde et al.* [2003] also significant dependence for the 660-km discontinuity. Our results could imply that both phase transitions occur over a wider than usual depth interval; however, significant widening of the 660-km transition due to temperature variations or increased water content is not expected [*Wood*, 1990; *Higo et al.*, 2001]. Furthermore, we stack spatially distributed converted sig-

nals for station groups in a steep-relief environment while *Van der Meijde et al.* [2003] stacked spatially concentrated converted signals for single stations atop gentler discontinuity topography, which yields the present study more susceptible to incoherent stacking. Discontinuity relief not only affects timing of the converted phases in seismograms but also scatters their energy, distorting the waveforms. An interpretation in terms of scattering by discontinuity relief is consistent with stacks from LPAZ and the 'central-east' group. Conversions for LPAZ are closely spaced resulting in coherent stacks; conversion points of the group are widely distributed (Figure 4), sample the slab at different depths

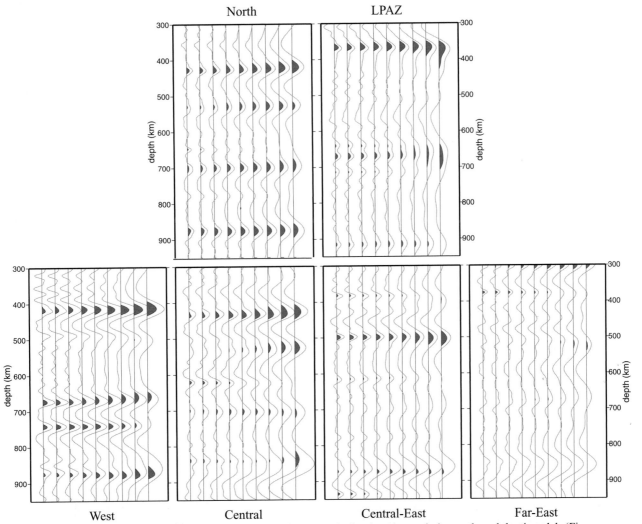

Figure 3. Stacked receiver functions recorded at stations at similar locations relative to the subducting slab (Figure 1 and Table 1). Each panel shows receiver function stacks for nine different low-pass filters with corner frequencies (from left to right) at 0.75, 0.63, 0.5, 0.4, 0.35, 0.3, 0.25, 0.2, and 0.15 Hz, respectively. Solid lines are average values; shaded areas indicate signal is greater than zero at 95% confidence level (2σ). Absolute depths, based on *iasp91* velocity model, ignore velocity heterogeneity and could thus be biased; depth difference estimates (mantle transition zone thickness), however, are robust.

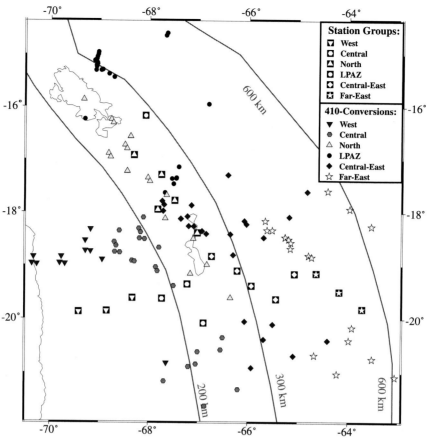

Figure 4. Spatial distribution of P to S conversion points at 410 km depth, with Wadati-Benioff zone contours. Note that groups with coherent stacks of P410s have conversion points predominantly under the slab (west, central) or in the slab and roughly parallel to the slab-depth contours (north, LPAZ), whereas eastern group conversions, resulting in incoherent stacks, occur near or above a steeply dipping slab. Position of conversion points are approximate and do not account for lateral heterogeneity.

and signal stacks incoherently. Multiple-station stacks are coherent further west where phase transitions occur below the slab and the slab has no or only slight influence on the transition depths.

A thickened transition zone can be explained by lowered temperatures or by elevated water content. The observed thickening of the transition zone from 255 km below the slab to 295 km where the slab enters the transition zone (Figure 3) over a lateral distance of about 200 km implies a steep relief on one of or both the 410- and 660-km discontinuities with a maximum amplitude of ~40 km. Absolute depths, although not as well constrained, imply almost all of this relief is on the 410-km discontinuity. However, the P660s signal is less coherent at higher than at lower frequencies suggesting the 660-km discontinuity also has relief.

Most 410-km Ps conversion points for LPAZ (Figure 4) occurred within the slab. Assuming the 660-km discontinuity beneath LPAZ is truly at 666 km depth (Table 1)

requires a 34-km uplift of the 410-km discontinuity. This uplift implies the slab would be about 250–350 K cooler, using a range of Clapeyron slopes [*Bina and Helffrich*, 1994; *Katsura et al.*, 2004], than the surrounding mantle. If this thermal anomaly extends to the 660-km discontinuity, its small (6 km) associated depression would be consistent with the gentle Clapeyron slopes recently inferred for the ringwoodite to perovskite and ferropericlase transition [*Litasov et al.*, this volume], but inconsistent with relief causing the frequency-dependence of P660s. Using the Clapeyron slopes of *Bina and Helffrich* [1994] of 9 km uplift and 6 km depression per 100 K cooling for the 410-km and 660-km discontinuity, respectively, we obtain a vertically homogeneous temperature anomaly of -267 K, ascribing 24 km to uplift of the 410-km discontinuity and 16 km of downwarp to the 660-km discontinuity. However, the transition-zone can also thicken in the presence of water because of the great water storage capacity of transition zone minerals [*Smyth and Jacobsen*, this volume; *Higo et al.*, 2001; *Wood*,

1995; *Smyth and Frost*, 2002; *Hirschmann et al.*, this volume]. If the slab or nearby mantle is hydrous (up to a few tenths of wt % water), it could explain up to about 20 km [*Smyth and Frost*, 2002; *Hirschmann et al.*, this volume] of the uplifted 410-km signal, leaving the slab only 50–150 K cooler. However, such hydration would also thicken the 410-km discontinuity itself by about 20 km and cause a stronger frequency dependence of converted signal from the discontinuity unless the mantle was entirely saturated with water [*Chen et al.*, 2002; *Hirschmann et al.*, this volume]. In the latter case the discontinuity would be sharp and produce frequency-independent seismic conversions, but could have a thin layer of melt atop (*Karato et al.*, this volume; *Hirschmann et al.*, this volume].

We also examined whether such extreme discontinuity relief (a 15° slope] affects frequency dependence by calculating synthetic receiver function stacks for a discontinuity with relief [following *Van der Lee et al.*, 1994]. The relief is a steep Gaussian-shaped hill or dip in the direction perpendicular to the strike of the slab and is constant parallel to it. The Gaussian hill or dip has a half width of 60 km and amplitude of 50 km. We used the same event-station geometry and processing steps in these calculations as for the data. A 50-km up-warp, simulating an elevated 410-km discontinuity within the slab, results in a high (70%) probability for frequency dependence while a 50-km down-warp, mimicking a depressed 660-km discontinuity, has a 50% probability. Frequency dependence is caused by signal focusing and scattering because of the curved interfaces. Our tests show that steep relief can explain most of the observed frequency dependence. However, we performed the tests for the eastern station groups; to reconcile the discrepancy between clear synthetics and lack of observed signal probably implies that the actual relief or shape of the discontinuity is more complicated than in our model, or that discontinuity thickening plays a more important role east of the slab. 3-D velocity heterogeneity would cause further differences in Ps arrival time, even for a flat discontinuity, and contribute to incoherent stacking.

In addition to signals from the 410- and 660-km discontinuities, we found significant amplitudes near 520-km depth for the 'north', 'central' and 'central-east' groups. The exothermic wadsleyite to ringwoodite phase transition occurs around this depth, but the transition interval is broad [e.g., *Helffrich*, 2000] and a 520-km discontinuity is mainly seen using long-period seismic data [*Shearer*, 1990]. This is consistent with our stacks that show increased significance for signal amplitudes with lower frequencies. We also observe coherent signals from a depth of around 850–900 km with signal strength apparently decreasing from west to east (where the signal is absent). Discontinuities in the upper part

of the lower mantle with limited spatial extent have been reported in the literature [e.g., *Kawakatsu and Niu*, 1994], but their origins are currently unknown.

CONCLUSIONS

We conclude that subduction of the Nazca plate is associated with extreme transition-zone thickening over short lateral distances. Such thickening can be caused by cooler temperatures, increased water content in the transition zone, or both. The observed frequency dependence of P to S signals generated at the 410- and 660-km phase transitions in the South American subduction zone is probably predominantly due to this steep discontinuity relief rather than thickened phase transition intervals. Normal transition intervals occur in a water-saturated mantle as well as in a dry mantle. In the absence of water, our results imply a slab up to 350 K cooler than the mantle trenchwards. The presence of close to 1 wt % of water (up to saturation) in or near the slab can explain some of the transition-zone thickening but not all, leaving an inferred temperature anomaly of 150 K at most. Our study documents characteristics of the transition zone in a subduction environment, most notably extreme thickening of the transition zone where the slab is thought to penetrate it. However, our data provide insufficient resolution, particularly because of the limited duration of seismic experiments, to establish the presence or absence of water.

Acknowledgements. BANJO/SEDA and LPAZ seismic data were obtained from the Incorporated Research Institutions for Seismology (IRIS) Data Management Center. This study is based upon work performed by LD during an IRIS Summer Undergraduate Internship at the ETH Zurich, Switzerland supported by IRIS under their Cooperative Agreement No. EAR-0004370 with the National Science Foundation. We thank Steve Jacobsen for valuable suggestions and two anonymous reviewers for critically reading the manuscript and prompt comments.

REFERENCES

Ammon, C. J., The isolation of receiver effects from teleseismic P waveforms, *Bull. Seis. Soc. Am.*, **81**, 2504-2510, 1991.
Beck, S. L., and G. Zandt, The nature of orogenic crust in the central Andes, *J. Geophys. Res.*, **107**, 2230, doi:10.1029/2000JB000124, 2002.
Beck, S. L., et al., Across the Andes and along the Altiplano: a passive seismic experiment, *IRIS Newsletter*, **13**, 1-3, 1994.
Benz, H., and J. Vidale, Sharpness of upper mantle discontinuities determined from high-frequency reflections, *Nature*, **365**, 147-150, 1993.
Bijwaard, H., W. Spakman, and E. R. Engdahl, Closing the gap between regional and global travel time tomography, *J. Geophys. Res.*, **103**, 30055-30078, 1998.

Bina, C., and G. Helffrich, Phase transition Clapeyron slopes and transition zone seismic discontinuity topography, *J. Geophys. Res.*, **99**, 15853-15860, 1994.

Cahill, T., and B. L. Isacks, Seismicity and shape of the subducted Nazca plate, J. Geophys. Res., **97**, 17503-17529, 1992.

Chen, J., T. Inoue, H. Yurimoto, and D. J. Weidner, Effect of water on olivine-wadsleyite phase boundary in the $(Mg,Fe)_2SiO_4$ system, Geophys. Res. Lett., 29, doi:1810.1029/2001GL014429, 2002.

Collier, J. D., and G. R. Helffrich, The thermal influence of the subducting slab beneath South America from 410 and 660 km discontinuity observations, *Geophys. J. Int.*, **147**, 319-329, 2001.

Dziewonski, A., and D. L. Anderson, Preliminary reference Earth model, *Phys. Earth Planet. Inter.*, **25**, 297-356, 1981.

Engdahl, E. R., R. van der Hilst, and R. Buland, Global teleseismic earthquake relocation with improved travel times and procedures for depth determination, *Bull. Seis. Soc. Am.*, **88**, 722-743, 1998.

Flanagan, M. P., and P. M. Shearer, Global mapping of topography on transition zone velocity discontinuities by stacking SS precursors, *J. Geophys. Res.*, **103**, 2673-2692, 1998.

Gu, Y. J., A. M. Dziewonski, and C. B. Agee, Global de-correlation of the topography of transition zone discontinuities, *Earth Planet. Sci. Lett.*, **157**, 57-67, 1998.

Hasegawa, A. and I. S. Sacks, Subduction of the Nazca plate beneath Peru as determined from seismic observations, J. Geophys. Res., **86**, 4971-4980, 1981.

Helffrich, G., Topography of the transition zone seismic discontinuities, *Reviews of Geophysics*, **38**, 141-158, 2000.

Helffrich, G., and C.R. Bina, Frequency dependence of the visibility and depths of mantle seismic discontinuities, Geophys. Res. Lett., 21, 2613-2616, 1994.

Helffrich, G., and B. Wood, 410-km discontinuity sharpness and the form of the olivine α-β phase diagram: Resolution of apparent seismic contradictions, *Geophys. J. Int.*, **126**, F7-F12, 1996.

Higo, Y., T. Inoue, T. Irifune, and H. Yurimoto, Effect of water on the spinel-postspinel transformation in Mg_2SiO_4, *Geophys. Res. Lett.*, **28**, 3505-3508, 2001.

Hirschmann, M. M., A. C. Withers, and C. Aubaud, Petrologic Structure of a Hydrous 410 km Discontinuity, this volume, 2006.

Karato, S., D. Bercovici, G. Leahy, G. Richard, and Z, Jing, The transition-zone water filter model for global material circulation: Where do we stand?, this volume, 2006.

Katsura, T., S. Yokoshi, M. Song, K. Kawabe, T. Tsujimura, A. Kubo, E. Ito, Y. Tange, N. Tomioka, K. Saito, A. Nozawa, and K. Funakoshi, Thermal expansion of Mg_2SiO_4 ringwoodite at high pressures, *J. Geophys. Res.*, **109**, B12209, doi:10.1029/2004JB003094, 2004.

Katsura, T., and E. Ito, The system Mg_2SiO_4-Fe_2SiO_4 at high pressures and temperatures; precise determination of stabilities

of olivine, modified spinel, and spinel, *J. Geophys. Res.*, **94**, 15663-15670, 1989.

Kawakatsu, H., and F. Niu, Seismic evidence for a 920-km discontinuity in the mantle, *Nature*, **371**, 301-305, 1994.

Kennett, B., and E. Engdahl, Traveltimes for global earthquake location and phase identification, *Geophys. J. Int.*, **105**, 429-465, 1991.

Langston, C., Structure under Mount Rainier, Washington, inferred from teleseismic body waves, *J. Geophys. Res.*, **84**, 4749-4762, 1979.

Litasov, K. D., E. Ohtani, and A. Sano, Influence of water on major phase transitions in the earth's mantle, this volume.

Marone, F., S. van der Lee, and D. Giardini, 3D upper mantle S-velocity model for the Eurasia-Africa plate boundary region, Geophys. J. Int., **158**, 109-130, 2004.

Pardo-Casas, F., and P. Molnar, Relative motion of the Nazca (Farallon) and South Ameriacan plates since Late Cretaceous time, *Tectonics*, **6**, 233-248, 1987.

Paulssen, H., Evidence for a sharp 670-km discontinuity as inferred from P-to-S converted waves, *J. Geophys. Res.*, **93**, 10489-10500, 1988.

Ringwood, A., Composition and petrology of the Earth's mantle, *McGraw-Hill, New York*, p. 618, 1975.

Shearer, P., Seismic imaging of upper-mantle structure with new evidence for a 520-km discontinuity, *Nature*, **344**, 121-126, 1990.

Smyth, J., and D. Frost, The effect of water on the 410-km discontinuity: An experimental study, *Geophys. Res. Lett.*, **29**, 10.1029/2001GL014418, 2002.

Smyth, J. R. and S. Jacobsen, Nominally anhydrous minerals and earth's deep water cycle, this volume, 2006.

Van der Meijde, M., F. Marone, D. Giardini, and S. van der Lee, Seismic evidence for water deep in Earth's upper mantle, *Science*, **300**, 1556-1558, 2003.

Van der Lee, S., H. Paulssen, and G. Nolet, Variability of P660s phases as a consequence of topography of the 660 km discontinuity, *Phys. Earth Planet. Inter.*, **86**, 147-164, 1994.

Wood, B., The effect of H_2O on the 410-kilometer seismic discontinuity, *Science*, **268**, 74-76, 1995

Wood, B., Postspinel transformations and the width of the 670-km discontinuity: A comment on "Postspinel transformations in the system Mg_2SiO_4-Fe_2SiO_4 and some geophysical implications" by E. Ito and E. Takahashi, *J. Geophys. Res.*, **95**, 12681-12688, 1990.

Yamazaki, A., and K. Hirahara, The thickness of upper mantle discontinuities, as inferred from short-period J-array data, *Geophys. Res. Lett.*, **21**, 1811-1814, 1994.

Jochen Braunmiller, Institute of Gephysics, ETH Zurich, 8093 Zurich, Switzerland, jochen@sed.ethz.ch

Towards Mapping the Three-Dimensional Distribution of Water in the Upper Mantle From Velocity and Attenuation Tomography

Azusa Shito[1], Shun-ichiro Karato[2], Kyoko N. Matsukage[3], and Yu Nishihara[4]

A new method is developed to determine the three-dimensional variation in water content, temperature, and other parameters such as major element chemistry or the melt fraction from anomalies in seismic wave velocities and attenuation. The key to this method is mineral physics observations indicating different sensitivity of seismic wave velocities and attenuation to temperature, water content and other parameters such as major element chemistry, melt fraction or grain-size. Our analysis shows that among these parameters, temperature and water content generally have a more important influence on seismic wave velocities and attenuation than other factors such as major element chemistry, which are important only in limited regions. The method is applied to the upper mantle beneath northern Philippine Sea including the Izu-Bonin subduction zone, where high-resolution velocity and attenuation tomographic models are available down to a depth of ~400 km. We show that the tomographic images of this region can be explained by lateral variations in temperature and water content, with only little influence of major element chemistry. A broad region of high attenuation with modestly low velocities at 300–400 km depth away from the slab in this region is interpreted as region of high water contents. We speculate that this water-rich region may have been formed by the efficient transport of water to deeper mantle by a fast (and cold) subducting slab in this region or water may come from the transition zone.

1. INTRODUCTION

The distribution of water (hydrogen) in Earth's mantle holds key information about the dynamics and evolution of this planet for two different reasons. First, the way in which water is currently distributed reflects the processes of material circulation including the location and degree of melting. Consequently, from the current distribution of water, one can place some constraints on the evolutionary processes of Earth [e.g., *Thompson,* 1992]. Second, the spatial variation in water content has an important influence on material properties such as viscosity that in turn controls dynamic processes in Earth [e.g., *Karato and Bercovici,* 2006]. For example, a water-rich region will have lower viscosity, which will have a strong influence on the material circulation and the gravitational signal of mantle flow [e.g., *Billen and Gurnis,* 2001; *Ito et al.,* 1999].

The distribution of water is usually inferred directly from rock samples on Earth's surface [e.g., *Bell and Rossman,* 1992]. This direct, petrological (or geochemical) approach has some limitations. First, the regions for which one can

[1]Earthquake Research Institute, University of Tokyo, Tokyo, Japan

[2]Department of Geology and Geophysics, Yale University, New Haven, Connecticut, USA

[3]Department of Earth Sciences, Ibaraki University, Mito, Japan

[4]Department of Earth and Planetary Sciences, Tokyo Institute of Technology, Tokyo, Japan

Earth's Deep Water Cycle
Geophysical Monograph Series 168
Copyright 2006 by the American Geophysical Union.
10.1029/168GM17

obtain data are limited by the availability of samples, and the maximum depth is typically ~200 km. Second, even if one has a sample, the degree to which the water content in the sample reflects the water content in the mantle can be questioned [e.g., *Mackwell and Kohlstedt*, 1990]. Some of the water may have been removed or added during the transport of these samples to Earth's surface. Consequently, alternative approaches using geophysical observations deserve special consideration.

Among the various geophysical observables, we will use seismological data to infer water content in the upper mantle. Electrical conductivity can also provide us with the information on water content [e.g., *Huang et al.*, 2005; *Karato*, 1990], however the spatial resolution of mapping electrical conductivity is limited compared to that of seismological observations. Previous studies in which seismological observations were converted to water content [e.g., *Blum and Shen*, 2004; *Nolet and Zielhuis*, 1994; *Van der Meijde et al.*, 2003; *Wood*, 1995] have some limitations. *Van der Meijde et al.* [2003] used the thickness of the 410 km discontinuity to infer the water content. The method of inferring water content is subject to uncertainties due to the non-uniqueness: the thickness of the 410 km discontinuity can be attributed to not only water content but also Fe content, Al content, and temperature [*Weidner and Wang.*, 2000]. This is the same for the other previous studies inferring water content from only one observation. The situation is also the same for studies trying to interpret one observation to temperature anomaly [*Sobolev et al.*, 1996; *Goes et al.*, 2002; *Cammarano et al.*, 2003]. Although they used multiple data sets to correct for the effects of other parameters (e.g., chemical composition and partial melt), the validity of the assumptions on the other parameters is not clear and, more importantly, they did not consider the influence of water, which can be large in some regions.

In this study, we apply new mineral physics-based formulations for the influence of various parameters, including water content, to retrieve anomalies in these parameters from seismological observations. Challenges with this approach include evaluating of the relative importance of various parameters (from mineral physics and geological considerations) and to obtain multiple high-resolution, mutually-consistent tomographic data for multiple parameters. This type of approach was first proposed by *Karato* [2003] and applied to the Philippine Sea upper mantle by *Shito and Shibutani* [2003b] where high-quality tomographic results are available both for velocity and attenuation. In the present paper, we go one step beyond *Shito and Shibutani* [2003b] in two respects. First, the study by *Shito and Shibutani* [2003b] presented only qualitative results and did not provide the exact locations of water-rich regions nor

the magnitude of water content. Second, the formulation used by *Shito and Shibutani* [2003b] did not include the influence of factors other than temperature and water, such as major element chemistry, partial melting, and grain-size. The purposes of this paper are (i) to present a new formulation of simultaneous inversion for multiple geophysical and geochemical parameters, and (ii) to apply this method to subduction zones to obtain new insights into deep-mantle water circulation.

2. CONVERSION OF SEISMOLOGICAL DATA TO GEOPHYSICAL PARAMETERS

Basic Considerations

In order to develop a method to infer water content from seismological observations, we will provide a brief review of the possible factors that may influence seismic wave velocities and attenuation.

The importance of temperature on seismic wave velocities and attenuation is well known. The influence of temperature on seismic wave velocities and attenuation can be formulated using a model of anharmonicity and anelasticity [*Karato*, 1993].

The influence of water on seismic wave velocities is well known only for high-frequency, unrelaxed velocities (for a review, see [*Jacobsen*, 2006; *Jacobsen and Smyth*, this volume]). These studies showed that the addition of water reduces unrelaxed seismic wave velocities, but the magnitude of this direct effect is small for low water contents (velocity reduction is ~0.3 % for a water content of ~0.1 wt%. *Karato* [1995] proposed that a more important effect of water is its effects through the enhancement of anelasticity. Here "water" means hydrogen incorporated as point defects in anhydrous minerals. This model is based on the well-known relation $Q^{-1} \propto f^{-\alpha} \propto \tau^{-\alpha}$ with $\alpha \sim 0.25$ (f: frequency, τ: characteristic time for relaxation [*Jackson*, 2000]), and the experimental observation that all kinetic processes in silicate minerals are enhanced by water. In all cases so far investigated, the characteristic time for a kinetic process in silicate minerals is a function of water content as $\tau \propto W^{-r}$ where W is the water content with $r = 1–2$ [e.g., *Karato*, 2006]. If water enhances the kinetics of relaxation, then τ will be reduced by the addition of water and hence anelasticity will be enhanced. The results of a reconnaissance study by *Jackson et al.* [1992] and the preliminary results by *Aizawa* (private communication, 2006) are consistent with this relation. Therefore we assume a relation $Q^{-1} \propto W^{\alpha r}$ (parameters α and r need to be known for our inversion and are estimated in the Appendix). Because the water content in the upper mantle can change dramatically due to partial melting [e.g., *Hirth and Kohlstedt*, 1996; *Karato*, 1986; 2006], the influence of water on seismic wave velocities and attenuation can be large. A change in water content by a factor of 100 will have the same effect of a temperature change of ~200–400 K (a change in Q^{-1} by a factor of ~3–8, and a change in velocity of ~1–2 %).

The influence of major element chemistry can be evaluated based on elasticity data and on the phase diagrams of upper mantle minerals [*Jordan*, 1979; *Matsukage et al.*, 2005; *Schutt and Lesher*, 2006]. In particular, *Matsukage et al.* [2005] noted that the compositional dependence is relatively large in the spinel lherzolite field (shallower than ~50–120 km) but it is small in the garnet lherzolite field (deeper than ~50–120 km). However, in most cases, the magnitude of this effect is small (less than ~1 % in velocity). The influence of major element chemistry on seismic wave attenuation is suggested to be small from creep experiments [*Darot and Gueguen*, 1981; *Darham and Goetze*, 1977].

Partial melting is often invoked to interpret low velocities (and high attenuation) [e.g., *Shankland et al.*, 1981]. However, a comparison of recent laboratory data on the influence of partial melting on seismic wave velocities and attenuation with seismological observations suggests that the influence of partial melting is not well documented in the majority of the upper mantle [e.g., *Faul and Jackson*, 2005; *Shito et al.*, 2004]. In particular, the influence of partial melting on seismic wave attenuation is questioned based on model calculations [*Hammond and Humphreys*, 2000a; b] and on experimental and seismological observations [*Faul and Jackson*, 2005; *Shito et al.*, 2004; *Wolfe and Solomon*, 1998]). This is due either to the absence of a significant fraction of melt due to efficient compaction or to melt-caused anelasticity having higher characteristic frequencies than seismic frequencies (or both). However, the potential influence of partial melting on (relaxed) seismic wave velocities cannot be ignored if there are regions with a significant melt fraction.

Influence of grain-size on anelasticity in olivine has been suggested [*Faul and Jackson*, 2005]. However, a compari-

son of the laboratory results ($Q^{-1} \propto d^{-S}$ where $Q^{-1} \propto d^{S}$ is the grain-size with $s{\sim}0.25$ [*Jackson et al.*, 2002]) and a plausible range of grain-size in a typical upper mantle (3–10 mm [*Mercier*, 1980]) indicates that the influence of grain-size is minor, a change in seismic wave velocity of less than ~0.2 % and attenuation by ~20%. This is much smaller than the influence of other factors such as temperature and water content. Furthermore, the effect of grain-size on seismic wave attenuation (and velocity through anelasticity) decreases significantly with pressure. Therefore, the effect appears not important in the deep upper mantle. Consequently, we will not consider the influence of grain-size in this paper.

Anisotropy can influence on seismic wave velocity and attenuation. However, if the ray coverage of the seismic tomography is sufficiently close to uniform, the anisotropic signal can be cancelled out [*Schmid et al*, 2004].

The above discussion is summarized in Table 1. We conclude that temperature and water content are potentially important parameters that control the lateral variation in seismic wave velocity and attenuation. The influence of temperature on seismic wave velocities is through anharmonicity and anelasticity. Water content affects anelasticity and hence indirectly seismic wave velocities. Since water content has only a small effect on unrelaxed velocities (velocities at infinite frequency), the influence of water content is larger for attenuation than for velocities. Consequently, a region with high water content will be seen, in the tomographic image, as a region of high attenuation with modest low velocities. In addition, the influence of major element chemistry and/or of partial melting may also have some effects on seismic wave velocity but not on seismic wave attenuation. These two factors are important only in the shallower mantle. For simplicity, we assume that there is no influence of major element chemistry and/or partial melting deeper than 150 km. The degree to which the major element chemistry and partial melting may affect the results will also be considered when necessary.

Table 1. The influence of various factors on seismic wave velocities and attenuation

	velocity	*attenuation*
temperature	large*	large
water	modest*	large
major element chemistry	small[1]	very small
partial melting	potentially large[2]	small*
grain-size	very small	small

* "large", "modest" or "small" is relative to the amplitude of anomalies (velocity anomalies are typically ~1-5%, and attenuation anomalies are a factor of ~1–3 in the upper mantle).
[1] influence of major element chemistry on seismic wave velocities is small in most cases except for the influence of hydrous minerals
[2] influence of partial melting depends on the fraction and the geometry

Formulation

Lateral variation of seismic wave velocities and attenuation could be caused by lateral variation of various parameters. In general, for a small variation of the parameters Y_j, one can write

$$\delta \ln X_i = A_{ij} \cdot \delta Y_j \qquad i = 1, \cdots, n, j = 1, \cdots, m \qquad (1)$$

where $\delta \ln X_i$ are the data $(\delta \ln V_{P,K}, \delta \ln V_S, \delta \ln Q_{P,S}^{-1})$, δY_j are small changes in parameters Y_j that cause the lateral variation of X_i, and $A_{ij} \equiv \frac{\partial \ln X_i}{\partial Y_j}$. If one has n data types at each spatial position, then one has n uch relationships, so that one can determine n unknown parameters, $Y_j (j = 1, \ldots, n)$, out of m unknowns.

When both velocity and attenuation are measured, we use $(\delta \ln V_{P,K}, \delta \ln V_S, \delta \ln Q_{P,S}^{-1})$ as observed data, so we have a maximum of three independent data ($n = 3$) at each grid point (note that Q_P^{-1} and Q_S^{-1} are not independent in our data set). In such a case, the number of variables that can be determined from seismological data is three ($m = 3$). Consequently, the equation (1) can be written as

$$
\begin{pmatrix} \delta \ln V_{P,K} \\ \delta \ln V_S \\ \delta \ln Q_{P,S}^{-1} \end{pmatrix} =
\begin{pmatrix} A_{V_{P,K}T} & A_{V_{P,K}W} & A_{V_{P,K}Z} \\ A_{V_S T} & A_{V_S W} & A_{V_S Z} \\ A_{Q_{P,S}^{-1}T} & A_{Q_{P,S}^{-1}W} & A_{Q_{P,S}^{-1}Z} \end{pmatrix}
\begin{pmatrix} \delta T \\ \delta \ln W \\ \delta Z \end{pmatrix} \qquad (2)
$$

where $(\delta \ln V_P, \delta \ln V_S, \delta \ln Q_P^{-1})$ or $(\delta \ln V_K, \delta \ln V_S, \delta \ln Q_S^{-1})$ are observed anomalies in seismic wave velocities and attenuation and $(\delta T, \delta \ln W, \delta Z)$ is a set of three unknowns (δT: temperature anomaly, $\delta \ln W$: anomaly in water content, δZ: anomaly in major element chemistry δC or melt fraction $\delta \phi$).

Each coefficient A_{ij} represents a partial derivative of a seismological observation with respect to one unknown physical/chemical parameter (e.g., $A_{V_P T} \equiv \left(\frac{\partial \ln V_P}{\partial T} \right)_{W,Z}$). The details of the mineral physics of these coefficients are discussed in the Appendix.

3. INVERSION

We applied the method developed in the previous section to the upper mantle beneath the Philippine Sea, including the Izu-Bonin subduction zone, where high resolution velocity and attenuation tomographic models are available. The Philippine Sea is located to the south of Japan and the upper mantle in this region has been subjected to long lasting subduction of old plates (Figure 1). As a consequence, there have been several back-arc opening events in this region during the last 40–20 Mys [e.g., *Seno and Maruyama, 1984*]. There is high seismicity in the Izu-Bonin slabs down to deep

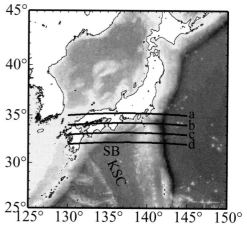

Figure 1. A bathymetry map showing the study area of the northern Philippine Sea region. Four lines a, b, c, and d indicate locations of cross-sections. SB: Shikoku Basin, KSC: Kinan Seamount Chain.

mantle (~660 km). Furthermore, there are dense seismic networks in Japan. The combination of high deep seismicity and an appropriate source-receiver distance makes this region one of the best subduction zones for which detailed seismological studies can be performed. This is distinct from other subduction zones.

We use the results of velocity tomography by *Gorvatov and Kennett* [2003] and attenuation tomography by *Shito and Shibutani* [2003a] (Plate 1). In *Gorvatov and Kennett* [2003], P and S wave travel times (around 1 Hz) from common sources and receivers are used in the joint inversion for bulk sound and shear wave velocities. We calculated P wave velocity by using bulk sound and shear wave velocities. For attenuation tomography, we used the results from *Shito and Shibutani* [2003a]. In *Shito and Shibutani* [2003a], attenuation was determined by the phase pair method which measures the relative high frequency decay of P wave and S wave amplitude spectra. This technique was used because using the relative amplitude decay of P and S waves minimizes the influence of source processes and station bias. In this technique, P wave attenuation Q_P^{-1} and S wave attenuation Q_S^{-1} are not independent, because Q_P / Q_S must be assumed. Therefore, Q_S^{-1} is calculated using the best fit value of $Q_P / Q_S = 2.15$. In both velocity and attenuation tomographic studies, a similar frequency range (0.5–1.5 Hz) was used so that there is no need for corrections for the frequency dependence of seismic wave velocities and attenuation. The spatial resolutions of the velocity and the attenuation tomographic models are also comparable (the grid spacing is 0.5 degree in horizontal (longitude) direction and 50 km in vertical direction). However, the exact position of the grid points is different. Therefore, in order to arrange the grids at the same spatial location for both velocity and attenuation, we produced new grid data by resampling the original data through linear interpolation.

Anomalies in the seismic wave velocity and attenuation ($\delta \ln V_P, \delta \ln V_S, \delta \ln Q_P^{-1}$) are defined by the normalized difference between the observed data and the one-dimensional reference models. For one-dimensional reference velocity and attenuation models we used *ak135* [*Kennett et al.,* 1995] and *PREM* [*Dziewonski and Anderson,* 1981] respectively. The amplitudes of perturbations are relatively insensitive to the choice of the reference model except shallow regions of attenuation model. The average standard error (derived from spectral analysis) of the attenuation model is less than 10% of the estimated values [*Shito and Shibutani,* 2003a]. The standard errors of the velocity model are unknown, but, the average error in regional travel time tomography is estimated to be 10% of the velocity anomalies [*Bijwaard et al.,* 1982].

We invert the velocity and attenuation tomographic models for the three-dimensional distribution of anomaly in temperature, water content, and major element chemistry or partial melting, relative to the standard models ($\delta T, \delta \ln W,$ δZ). Here and in equation (2), the unknown parameter δZ can be an anomaly in major element chemistry, δC or an anomaly in melt fraction, $\delta \phi$. We will first summarize the results of inversion in which δZ is the anomaly in major element chemistry δC. The results for a case of partial melt will be discussed in connection to the error analysis in section 4.

Equation (2) is generally nonlinear because the matrix elements, A_{ij}, depend on the solution ($\delta T, \delta \ln W, \delta C$). Consequently, we solve it iteratively modifying the matrix elements until the difference of solution for the k-th step and for the (k+1)-th step becomes small (i.e. $\delta T < 1$ K). This typically takes 2–7 steps. The unknown parameters $\delta T, \delta \ln W,$ and δC are defined as perturbations from the reference models. The temperature anomaly is denoted by $\delta T \equiv T - T_0$(K). The $\delta \ln W \equiv \ln W - \ln W_0$ is the water content anomaly. The anomaly in major element chemistry is denoted by $\delta C \equiv C - C_0$, where C is the molar concentration ratio of magnesium to magnesium plus iron, Mg/(Mg+Fe) = Mg#. The subscripts "0" indicate the values of the reference models. The reference model for temperature is a one-dimensional oceanic geotherm corresponding to the age of 60 Ma and with a potential temperature of 1,600 K for the sub-lithospheric portion. We assume a reference, homogeneous distribution of water of 800 ppm H/Si $\approx$ 0.005 wt% H_2O [*Hirth and Kohlstedt,* 1996] for the whole upper mantle. We assume the pyrolite model as a reference model of composition with Mg# = 0.89.

The results of the inversion for three unknown parameters ($\delta T, \delta \ln W, \delta C$) are shown in Plate 2. Low temperature anomalies near the slabs and high temperature anomalies

in the wedge are well resolved. Temperature in the wedge mantle is generally higher than the standard model we used (60 Ma oceanic geotherm). This is consistent with relatively young ages in this region of Philippine Sea [*Seno and Maruyama,* 1984]. More specifically, there are distinct high temperature anomalies beneath active Izu-Bonin volcanic chain. The amplitude of this temperature anomaly beneath the Izu-Bonin arc is ~200–300 K (actual temperature at 100 km there is ~1,750 K), which is in reasonable agreement with some modeling results [*Kelemen et al.,* 2003; *Van Keken et al.,* 2002].

A broad region of anomalous water content is found in the deep upper mantle (~300–400 km) away from the slab. It occurs in a region with high attenuation and modestly low velocities (see Plate 1). The anomaly in that region of $\delta \ln W$ = 2–3 corresponds to a water content of ~10,000–20,000 ppm H/Si or 0.06–0.13 wt% H_2O which is close to the saturation limit of water in upper mantle minerals [e.g., *Bolfan-Casanova,* 2005]. The value of r (see Appendix) has some influence on the estimated value of the water content. The amplitude of anomalies in W increases with the decrease of r. We tested a range of values for r, from 1.0 to 2.0. For the value of r= 1.0, the estimated value of $\delta \ln W$ is about twice that for r= 2.0 (which results in extremely large water contents exceeding 100,000 ppm H/Si ppm $\approx$ 0.6 wt% H_2O). Plate 2 shows the results for r = 2.0. In the results on major element chemistry (anomaly in Mg#), we note that there is some indication for a high Mg# in the upper mantle beneath the Shikoku Basin, which may correspond to regions with high depletion after the formation of a marginal basin. However we limit further discussion of major element chemistry because of the large uncertainty.

4. DISCUSSION

Error Analysis and Trade-off: How Much Can We Resolve the Water Content Anomalies?

Because our inversion method is non-linear, it is impossible to calculate the resolution matrix. Instead, we employed a grid search to estimate the resolution of the unknown parameters. We calculated the normalized RMS error by changing the values of the unknown parameters ($\delta T, \delta \ln W, \delta C$). The normalized RMS error is described as follows.

$$E = \sqrt{\frac{\sum_{i}^{n}\left(\frac{\delta \ln X_{i\_obs} - \delta \ln X_{i\_cal}}{\delta \ln X_{i\_obs}}\right)^2}{n}} \quad (3)$$

The grid spacing explored of the unknown parameters is from -300 K to 300 K for temperature, from -0.1 to 0.1 in

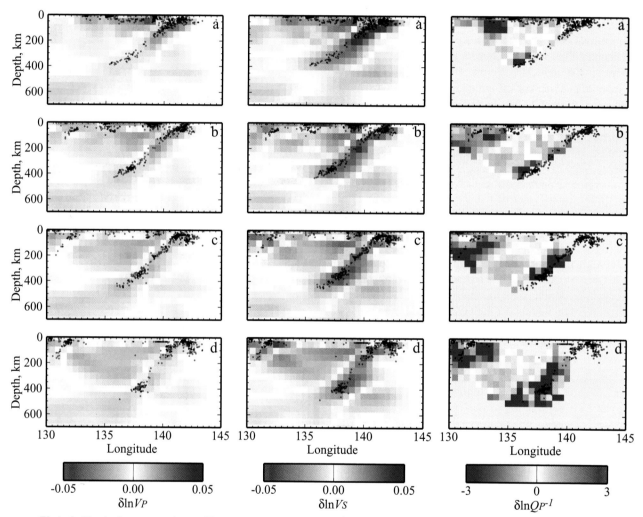

Plate 1. Vertical cross-sections of P wave velocity anomaly, $\delta \ln V_P$ (left column), S wave velocity anomaly, $\delta \ln V_S$ (middle column), and P wave attenuation anomaly, $\delta \ln Q_P^{-1}$ (right column) models used in the Philippine Sea region. The cross-sections are oriented in east-west directions and the locations are shown in Figure 1. Small circles indicate the locations of recent seismicity.

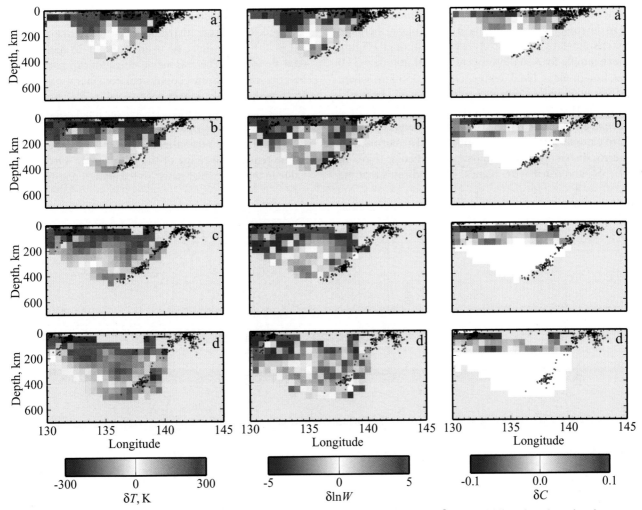

Plate 2. Distribution of anomalies in temperature, δT (left column), water content $\delta \ln W$ (middle column), and major element chemistry δC (right column). The parameter $r = 2$ was used in the inversion (the results for $r = 1$ are similar in pattern with a larger amplitude). The locations of grid points in the cross-sections are the same as those in the tomographic data shown in Plate 1. Small circles indicate the locations of recent seismicity. The top layer (0–50km) corresponds to the crust, and is excluded from the discussion.

Mg# for chemical composition, and -5.0 to 5.0 for water content. Figure 2 shows examples of normalized RMS error contoured against the unknown parameters. Figure 2a is an example for a deeper grid point (Latitude = 33.0 degree, Longitude = 134.0 degree, Depth = 400 km) in which we assume there is no influence of major element chemistry. The ellipsoidal distribution of normalized RMS error contours indicates that there is some trade off between temperature and water content. However a significant minimum can be found in the reasonable range of the $\delta T - \delta C$ plane, therefore the two parameters are relatively well constrained. Figure 2b is an example for a shallower grid point (Latitude = 33.0 degree, Longitude = 134.0 degree, Depth = 150 km) where the influence of major element chemistry may not be negligible. The normalized RMS error contours against the three parameters are projected on two parameter planes on which the third parameter takes the best-fit value. The normalized RMS error distributions are almost perpendicular to δC in the $\delta T - \delta C$ and $\delta \ln W - \delta C$ planes. This indicates that the

parameter δC is only weakly constrained. However, we find a significant minimum in the $\delta T - \delta \ln W$ plane. We can conclude that the seismic wave velocity and attenuation are not very sensitive to major element chemistry and it is weakly constrained. However, temperature and water content are well constrained in this inversion. The resultant best fit values of the parameters (indicated by crosses in Figure 2) are consistent with those from the solution of the inversion.

We also conducted a grid search to calculate the normalized RMS error distribution for unknown parameters $(\delta T, \delta \ln W, \delta \phi)$, where $\delta \phi$ is an anomaly in melt fraction; $\delta \phi \equiv \phi - \phi_0 \equiv \phi$. The reasonable range for this parameter can be constrained from geophysical considerations; the melt fraction should take zero or positive value ($\delta \phi \geq 0$), and if melt fraction has a positive value $\delta \phi > 0$, the temperature anomaly should also be positive ($\delta T > 0$). However, the significant minimum of the normalized RMS error cannot be found in the reasonable range of the parameters. The minimum value in the area which satisfies the above condition is much

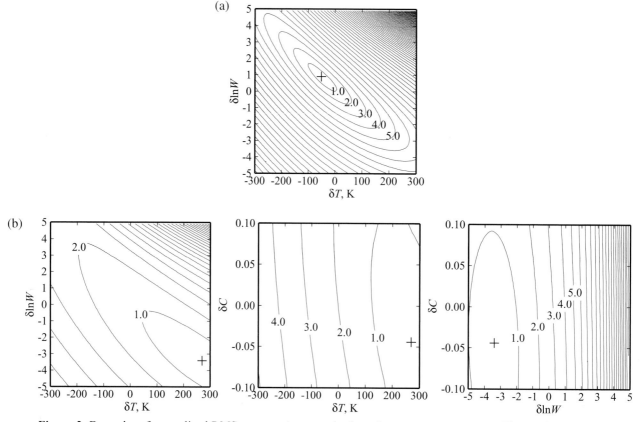

Figure 2. Examples of normalized RMS error contour map in the unknown parameter space. The numbers on the contours indicate the error defined by equation (3) in the text. (**a**) A contour map against the parameters $(\delta T, \delta \ln W)$ for a deep grid point (Latitude = 33.0 degree, Longitude = 134.0 degree, Depth = 400 km). (**b**) A contour map against the parameters $(\delta T, \delta \ln W, \delta C)$ for a shallow grid point (Latitude = 33.0 degree, Longitude = 134.0 degree, Depth = 150 km). Crosses indicate the points corresponding to the minimum values of normalized RMS error.

larger than that in the case of $(\delta T, \delta \ln W, \delta C)$. Therefore, we conclude that anomalous major element chemistry is more suitable than melt fraction as the third unknown parameter. This conclusion is consistent with the tectonic setting of Shikoku Basin where there is no presently active volcanism and thus significant amounts of melt are unlikely.

We note that chemical composition and melt content are inferred by the difference in the results of observed anomalies in P and S wave velocities using the difference in the partial derivatives with respect to these parameters. Neither of them are well constrained (in particular the partial derivatives of velocities with respect to melt fraction is highly sensitive to melt geometry, which is not well constrained [*Takei, 2002*]). We deduce that this leads to the large uncertainties in inferred anomalies in major element chemistry and melt fraction. In contrast, the anomalies in temperature and water content are inferred largely from the difference in observed velocity and attenuation anomalies using the partial derivatives estimated from mineral physics considerations. The partial derivatives of seismic wave velocities with respect to temperature and water content are markedly different from those of seismic wave attenuation, and consequently, when results of high-resolution velocity and attenuation tomography are available, both temperature and water content can be inferred with some confidence.

The combination of the parameters; anomaly in temperature, water content, and major element chemistry can explain the observed seismic wave velocity and attenuation in the Philippine Sea. We note that the success in resolving water content variation in the Philippine Sea upper mantle is due mainly to suitable choice of the parameters. However, this choice of the parameters may not be appropriate for another region. Tomographic images in other regions such as Tohoku Japan [*Nakajima et al., 2001; Tsumura, 2000*] are not easy to interpret by the anomalous temperature, water content, and major element chemistry. In Tohoku Japan, there is a region of very low velocities with modest anomalies in attenuation. A localized region which has significant amount of partial melt may be required to explain such an anomaly.

Geodynamic Implications

We have identified a water-rich region in the deep upper mantle (~300–400 km) beneath the Philippine Sea. This is a region of high attenuation with only modestly low seismic wave velocities. The amount of water inferred depends on the value of r, which is not well constrained. With $r = 2$, we obtain ~0.05–0.1 wt%, which may be close to the solubility limit in upper mantle minerals [e.g., *Bolfan-Casanova, 2005*]. A larger amount of water is inferred if a smaller value of r is used. We note that in this region a cold plate has been

subducting that can carry water to a great depth [e.g., *Rüpke et al., 2004*]. An alternative mechanism is the supply of water from the upwelling current across the transition zone. *Karato et al.* [this volume] suggests that water-rich materials can be transported upward across the 410 km if the water content in the transition zone is high (~1 wt%). *Suetsugu et al.* [this volume] infer such high water content in the transition zone from seismological studies of the topography of the 660 km discontinuity and velocity anomalies in the transition zone.

Our analysis illustrates the importance of using high resolution seismological data combined with mineral physics that provides us with a three-dimensional view of water content in Earth's upper mantle. Water transported to the deep mantle wedge should have a number of potential geodynamic and geochemical consequences. Regions with high water content should have low viscosities. The low viscosity regions have some geodynamic effects such as the topography and gravity field [*Billen and Gurnis, 2001*]. Our results could provide profound implications for a broad range of Earth sciences including geodynamics, geochemistry and evolution of the subduction zones and continents. However, we should also point out major limitations in this approach. First, the resolution of seismological observation is limited and the resolution of different types of seismological observations is not always comparable. Second, constraints on major element chemistry and/or the amount of melt are rather poor in this region. Finally, many mineral physics issues must be understood better. Especially, the dependence of attenuation on water content should be better constrained. With these improvements, this type of interdisciplinary approach will help understand the evolution and dynamics of this planet.

Acknowledgements. We are grateful to Suzan van der Lee and Steven Jacobsen for constructive comments and continuous encouragement. Suzan van der Lee also helped us to grammatically improve the manuscript. We also thank Douglas Wiens and Ian Jackson for constructive and thoughtful reviews. Many discussions with Junichi Nakajima, Yasuko Takei, Noriko Tsumura, Takao Koyama, and Daisuke Suetsugu were helpful to improve the manuscript. Dr. Gorbatov provided us digital data of high-resolution three-dimensional velocity structure. This research is partly supported by the fellowship by JSPS (for AS) and by NSF (for SK).

APPENDIX: SOME NOTES ON PARTIAL DERIVATIVES A_{ij}

(1) Influence of Temperature

Seismic wave velocities depend on temperature, including both anharmonic and anelastic effects [*Minster and Anderson, 1980; Karato, 1993*], as follows:

$$A_{V_{P,K,S}T} = \frac{\partial \ln V_{P,K,S}}{\partial T} = \frac{\partial \ln V_{P,K,S}^{\infty}}{\partial T} - F(\alpha)Q_{P,K,S}^{-1}\frac{\alpha H^*}{RT^2} \quad (A\text{-}1)$$

with $F(\alpha) = \frac{1}{2}\cot\left(\frac{1}{2}\alpha\pi\right)$.

The first term, $\frac{\partial \ln V_{P,K,S}^{\infty}}{\partial T}$ represents the anharmonic effect and the second term the anelastic effect. We use a power-law relation for the anelastic effect, viz., $Q^{-1} \propto \omega^{-\alpha}\exp\left(-\frac{\alpha H^*}{RT}\right)$ where ω is frequency and H^* is the activation enthalpy (~400 kJ/mol for olivine). Values for $\frac{\partial \ln V_{P,K,S}^{\infty}}{\partial T}$ are available for upper mantle minerals [*Ita and Stixrude*, 1992]. Parameters related to anelasticity (α and pressure dependent H^*) are from *Jackson* [2000] and *Jackson et al.* [1992]. Note that these coefficients contain the values of attenuation (Q^{-1}), and consequently the values of these coefficients change from one place to another. Theoretically, the bulk attenuation Q_K^{-1} equals zero.

(2) Influence of Water Content

Although there are no direct experimental data, we formulate the effect of water on seismic wave attenuation extending the experimental result that water enhances plastic deformation [*Karato*, 2003; 2005]. Seismic wave attenuation in minerals occurs either by the motion of defects such as dislocations or grain-boundaries [e.g., *Gueguen et al.*, 1989; *Jackson et al.*, 2002; *Karato and Spetzler*, 1990]. Hydrogen is known to have dramatic effects on all known defect-related properties of silicates [e.g., *Karato*, 2005]. Therefore it is highly likely that water (hydrogen) has an important influence on seismic wave attenuation. In fact, an earlier study by *Jackson et al.* [1992] showed the decrease in attenuation after drying dunite. Their results are similar to those by *Chopra and Paterson* [1984] on the same dunite (Åheim dunite), who observed a significant increase in creep resistance after drying. Results of *Chopra and Paterson* [1984] are interpreted as due to the effects of water and this has been confirmed by a number of follow-up studies [e.g., *Karato et al.*, 1986; *Mei and Kohlstedt*, 2000a; b]. Therefore we conclude that water enhances anelasticity of olivine aggregates. Noting that solid-state attenuation assumes the form, $Q^{-1} \propto (f\tau)^{-\alpha}$, a plausible form for the dependence of attenuation on water content is

$$Q^{-1} \propto W^{\alpha r} \quad (A\text{-}2)$$

where we used a commonly observed relation $\tau \propto W^{-r}$ [e.g., *Karato*, 2005]. The value of r depends on the particular mechanism of energy dissipation. Experimental observations and theoretical models on related processes, i.e., creep and grain-growth kinetics, suggest that r will be between 1 and

2 [e.g., *Karato*, 2005]. Considering the uncertainty in such a relation, we explore a range of r in the present paper.

$$A_{Q_{P,S}^{-1}W} = \frac{\partial \ln Q_{P,S}^{-1}}{\partial \ln W} = \alpha r \quad (A\text{-}3)$$

The influence of water on seismic wave velocity is often analyzed based on its effects on unrelaxed elastic moduli [e.g., *Inoue et al.*, 1998; *Jacobsen et al.*, 2004], see also *Jacobsen and Smyth* [this volume]. However, *Karato* [1995] noted that a more important effect of water on seismic wave velocity is through anelasticity and the change in anisotropy (since we focus only on anomalies in velocities averaged over all direction at each point, we will not consider the influence of anisotropy). This notion is supported by later experimental studies that show that the direct influence of water on unrelaxed seismic wave velocities is less than ~0.1–0.2% if the water content is less than ~0.1 % (see *Jacobsen and Smyth* [this volume]). Consequently, we will ignore the influence of water on unrelaxed seismic velocities (this is justifiable for most of the upper mantle, but not for the transition zone where water content can be as high as ~3 %). Therefore, we describe the effect of water on seismic wave velocity as

$$A_{V_{P,K,S}W} = \frac{\partial \ln V_{P,K,S}}{\partial \ln W} - \alpha r F(\alpha)Q_{P,K,S}^{-1} \quad (A\text{-}4)$$

(3) Influence of Major Element Chemistry

The influence of major element chemistry on anelasticity has not been studied in detail, but is likely small (for example, the creep strength of forsterite (Mg_2SiO_4) is indistinguishable from that of San Carlos olivine ($(Mg_{0.9},Fe_{0.1})_2SiO_4$ [e.g., *Darot and Gueguen*, 1981; *Durham and Goetze*, 1977] and therefore only the influence on seismic wave velocities is considered, i.e.,

$$A_{Q_{P,S}^{-1}C} = \frac{\partial \ln Q_{P,S}^{-1}}{\partial C} = 0 \quad (A\text{-}5)$$

Therefore we focus on the influence of major element chemistry on seismic wave velocities. In principle, the major element chemistry involves a large number of parameters and cannot be constrained from a limited number of seismological observations. However, the recent analysis by *Matsukage et al.* [2005] showed that under some conditions, one can use a single parameter to characterize the influence of major element chemistry on seismic wave velocities. This is a consequence of the fact that the major element chemistry of upper mantle rocks is controlled primarily by a single process: partial melting in the shallow mantle. We use this approach to simplify the inversion, viz.,

$$A_{V_{P,K,S}C} = \frac{\partial \ln V_{P,K,S}^{\infty}}{\partial C} \quad (A\text{-}6)$$

where C is a magnesium number Mg# (the molar concentration ratio of magnesium to magnesium plus iron). The values of coefficients $A_{V_{P,K,S}C}$ are taken from study of *Matsukage et al.* [2005]. An important complication is the influence of the spinel lherzolite to garnet lherzolite transition. In the spinel lherzolite region, the influence of Mg# is through the change of Mg# in each mineral and hence $A_{V_{P,K,S}C} = \frac{\partial \ln V_{P,K,S}}{\partial C} > 0$. In contrast, when garnet is present, the variation in Mg# changes the fraction of garnet which has much higher seismic wave velocities than other minerals. An increase in Mg# in such a case is associated with a decrease in garnet content and hence $A_{V_{P,K,S}C} = \frac{\partial \ln V_{P,K,S}^{\infty}}{\partial C} < 0$. The transition pressure from spinel lherzolite to garnet lherzolite depends on chemical composition and to a lesser extent to temperature. This can cause the regional variation in $A_{V_{P,K,S}C}$ in the depth region (~80–120 km) where this phase transformation occurs. The effect are limited to the shallow mantle. Therefore, the partial derivatives $A_{V_{P,K,S}C} = \frac{\partial \ln V_{P,K,S}^{\infty}}{\partial C} = 0$ below 150 km depth.

REFERENCES

Bell, D. R., and G. R. Rossman, Water in Earth's mantle: The role of nominally anhydrous minerals, *Science, 255,* 1391–1397, 1992.

Bijwaard, H., W. Spakman, and E.R. Engdahl, Closing the gap between regional and global travel time tomography, *J. Geophys. Res., 103,* 30,055–30,078, 1998.

Billen, M. I., and M. Gurnis, A low viscosity wedge in subduction zones, *Earth Planet. Sci. Lett., 193,* 227–236, 2001.

Blum J., and Y. Shen, Thermal, hydrous, and mechanical states of the mantle transition zone beneath southern Africa, *Earth Planet. Sci. Lett., 217,* 367–378, 2004.

Bolfan-Casanova N., Water in the Earth's mantle, *Mineralogical Magazine, 69,* 229–257, 2005.

Cammarano F., S. Goes, P. Vacher, and D. Giardini, Inferring upper-mantle temperatures from seismic velocities, *Phys. Earth Planet. Int., 138,* 197–222, 2003.

Chopra P. N., and M. S. Paterson, The role of water in the deformation of dunite, *J. Geophys. Res., 89,* 7861–7876,1984.

Darot M., and Y. Gueguen, High-temperature creep of forsterite single crystals, *J. Geophys. Res., 86,* 6219–6234, 1981.

Durham W. B., and C. Goetze, Plastic flow of oriented single crystals of olivine 1. Mechanical data, *J. Geophys. Res., 82,* 5737–5353, 1977.

Dziewonski, A.M., and D. L. Anderson, Preliminary Reference Earth Model, *Phys. Earth Planet. Int., 25,* 297–356, 1981.

Faul, U. H., and I. Jackson, The seismological signature of temperature and grain size variations in the upper mantle, *Earth Planet. Sci. Lett., 234,* 119–134, 2005.

Gorbatov, A., and B. L. N. Kennett, Joint bulk–sound and shear tomography for Western Pacific subduction zones, *Earth Planet. Sci. Lett., 210,* 527–543, 2003.

Goes, S., and S. Van der Lee, Thermal structure of the North American uppermost mantle inferred from seismic tomography, *J. Geophys. Res., 107,* 10.1029/2000JB000049, 2002.

Gueguen Y., M. Darot, P. Mazot, and J. Woirgard, Q^{-1} of forsterite single crystals, *Phys. Earth Planet. Int., 55,* 254–258, 1989.

Hammond W. C., and E. D. Humphreys, Upper mantle seismic wave velocity: Effects of realistic partial melt geometries, *J. Geophys. Res., 105,* 10975–10986, 2000a.

Hammond W. C., and E. D. Humphreys, Upper mantle seismic wave attenuation: Effects of realistic partial melt distribution, *J. Geophys. Res., 105,* 10987–10999, 2000b.

Hirth, G., and D. L. Kohlstedt, Water in the oceanic upper mantle: Implications for rheology, melt extraction and the evolution of the lithosphere, *Earth Planet. Sci. Lett., 144,* 93–108, 1996.

Huang X., Y. Xu, and S. Karato, Water content in the transition zone from electrical conductivity of wadsleyite and ringwoodite, *Nature, 434,* 746–749, 2005.

Inoue T., D. J. Weidner, P. A. Northrup, and J. B. Parise, Elastic properties of hydrous ringwoodite (γ-phase) of Mg_2SiO_4, *Earth Planet. Sci. Lett., 160,* 107–113,1998.

Ita, J., and L. Stixrude, Petrology, elasticity, and composition of the mantle transition zone, *J. Geophys. Res., 97,* 6849–6866, 1992.

Ito G, Y. Shen, G. Hirth, and G. J. Wolfe, Mantle flow, melting, and dehydration of the Iceland mantle plume, *Earth Planet. Sci. Lett., 165,* 81–96, 1999.

Jackson, I., Laboratory measurement of seismic wave dispersion and attenuation: Recent progress, in *Earth's Deep Interior Mineral Physics and Tomography from the Atomic to the Global Scale, Geophysical Monograph,* edited by S. Karato, A.M. Forte, R. C. Liebermann, G. Masters, L. Stixrude, pp. 265–289, Am. Geophys. Union, Washington DC, 2000.

Jackson, I., J. D. Fitz Gerald, U.H. Faul, and B. H. Tan, Grain-size sensitive seismic wave attenuation in polycrystalline olivine, *J. Geophys. Res., 107,* doi: 10.1029/2001JB001225, 2002.

Jackson I., M. S. Paterson, and J. D. Fitz Gerald, Seismic wave dispersion and attenuation in Åheim dunite, *Geophys. J. Int. 108,* 517–534, 1992.

Jacobsen S. D., Effect of water on the equation of state of nominally anhydrous minerals, *Rev. Mineral. Geochem., 62,* in press.

Jacobsen S. D., and J. R. Smyth, Effect of water on the sound velocities of ringwoodite in the transition zone, this volume.

Jacobsen S. D., J. R. Smyth, H. A. Spetzler, C. M. Holl, and D. J. Frost, Sound velocities and elastic constants of iron-bearing hydrous ringwoodite, *Phys. Earth Planet. Int., 143,* 47–56, 2004.

Jordan T. H., Mineralogies, densities, and seismic velocities of garnet lerzolite and their geophysical implications, in *The mantle sample: Inclusions in kimberlites and other volcanics,* edited by F.R. Boyd, and H.O.A. Meyer, pp. 1–14, Am. Geophys. Union, Washington DC, 1979.

Karato S., Does partial melting reduce the creep strength of the upper mantle? *Nature, 319,* 309–310, 1986.

Karato S., The role of hydrogen in the electrical conductivity of the upper mantle, *Nature, 347,* 272–273, 1990.

Karato S., Importance of anelasticity in the interpretation of seismic tomography, *Geophys. Res. Lett., 20,* 1623–1626, 1993.

Karato S., Mapping water content in Earth's upper mantle, in *Inside the Subduction Factory, Geophysical Monograph,* edited by J. Eiler, pp. 135–152, Am.Geophys. Union, Washington DC, 2004.

Karato S., Microscopic models for the influence of hydrogen on physical and chemical properties of minerals, in *Superplume: Beyond Plate Tectonics*, edited by D.A. Yuen, S. Maruyama, S. Karato, and B.F. Windley, pp. in press. Springer, 2005.

Karato S., Remote sensing of hydrogen in Earth's mantle, in *Deep Earth Water Circulation,* edited by H. Keppler and J. R. Smyth, Mineralogical Society of America, 2006.

Karato, S., and D. Bercovici, Transition zone water filter model for global material circulation: Where do we stand?, this volume.

Karato S., M. S., Paterson, and J. D. Fitz Gerald, Rheology of synthetic olivine aggregates: influence of grain-size and water, *J. Geophys. Res., 91,* 8151–8176, 1986.

Karato, S., and H. A. Spetzler, Defect microdynamics in minerals and solid-state mechanisms of seismic-wave attenuation and velocity dispersion in the mantle, *Rev. Geophys., 28,* 399–421, 1990.

Kelemen, P. B., J. L. Rilling, E. M. Parmentier, L. Mehl, and B.R. Hacker, Thermal structure due to solid-state flow in the mantle wedge beneath arcs, in *Inside the Subduction Factory, Geophysical Monograph,* edited by J. Eiler, pp. 293–311 Am. Geophys. Union, Washington DC, 2004.

Kennett, B.L.N., E.R. Engdahl, and R. Buland, Constraints on seismic velocities in the Earth from travel times, *Geophys. J. Int., 122,* 108–124, 1995.

Mackwell, S. J., and D. L. Kohlstedt, Diffusion of hydrogen in olivine: implications for water in the mantle, *J. Geophys. Res., 95,* 5079–5088, 1990.

Matsukage K. N., Y. Nishihara, and S. Karato, Seismological signature of chemical differentiation of Earth's upper mantle. *J. Geophys. Res., 110,* 10.1029/2004JB003504, 2005.

Mei S., and D. L. Kohlstedt, Influence of water on plastic deformation of olivine aggregates 1. Diffusion creep regime, *J. Geophys. Res., 105,* 21457–21469, 2000a.

Mei S., and D. L. Kohlstedt, Influence of water on plastic deformation of olivine aggregates 2. Dislocation creep regime, *J. Geophys. Res., 105,* 21471–21481 2000b.

Mercier J.-C. C., Magnitude of the continental lithospheric stresses inferred from rheomorphic petrology, *J. Geophys. Res., 85,* 6293–6303, 1980.

Minster J. B., and D. L., Anderson, Dislocations and nonelastic processes in the mantle, *J. Geophys. Res., 85,* 6347–6352, 1980.

Nakajima, J., T. Matsuzawa, A. Hasegawa, and D. Zhao, Three-dimensional structure of Vp, Vs, and Vp/Vs beneath northeastern Japan: Implications for arc magmatism and fluids, *J. Geophys. Res., 106,* 21843–21857, 2001.

Nolet G., and A. Zielhuis, Low S velocities under the Tornquist-Teisseyre zone: evidence for water injection into the transition zone by subduction, *J. Geophys. Res., 99,* 15813–15820, 1994.

Rüpke L. H., J. P. Morgan, M. Hort, and J. A. D. Connolly, Serpentine and the subduction zone water cycle, *Earth Planet. Sci. Lett., 223,* 17–34, 2004.

Schmid, C., S. Van der Lee, and D. Giardini, Delay times and shear wave splitting in the Mediterranean region, *Geophys. J. Int., 159,* 275–290, 2004.

Schutt, D. L. and C. E. Lesher, Effect of melt depletion on the density and seismic velocity of garnet and spinel Lherzolite, *J. Geophys. Res., 111,* 10.1029/2003JB002950, 2006.

Seno, T., and S. Maruyama, Paleogeographic reconstruction and origin of the Philippine Sea, *Tectonophysics, 102,* 53–84, 1984.

Shankland T. J., R. J. Oconnell, and H. S. Waff, Geophysical Constraints on partial melt in the upper mantle, *Rev. Geophys., 19,* 394–406, 1981.

Shito, A., S. Karato, and J. Park, Frequency dependence of Q in Earth's upper mantle inferred from continuous spectra of body waves, *Geophys. Res. Lett., 31,* doi: 10.1029/2004GL019582, 2004.

Shito, A., and T. Shibutani, Anelastic structure of the upper mantle beneath the northern Philippine Sea, *Phys. Earth Planet. Int., 140,* 319–329, 2003a.

Shito, A., and T. Shibutani, Nature of heterogeneity of the upper mantle beneath the northern Philippine Sea as inferred from attenuation and velocity tomography, *Phys. Earth Planet. Int., 140,* 331–341, 2003b.

Sobolev, S. V., H. Zeyen, G. Stoll, F. Werling, R. Altherr, and K. Fuchs, Upper mantle temperatures from teleseismic tomography of French Massif Central including effects of composition, mineral reactions, anharmonicity, anelasticity and partial melt, *Earth Planet. Sci. Lett., 139,* 147–163, 1996.

Suetsugu, D., H. Shiobara, M. Obayashi, and T. Inoue, Towards mapping the three-dimensional distribution of water in the upper mantle from P-velocity tomography and 660-km discontinuity depth, this volume.

Takei, Y., Effect of pore geometry on V_P/V_S: From equilibrium geometry to crack, *J. Geophys. Res., 107,* doi:10.1029/2001JB000522, 2002.

Thompson, A. B., Water in the Earth's upper mantle, *Nature, 358,* 295–302, 1992.

Tsumura, N., S. Matsumoto, S. Horiuchi, and A. Hasegawa, Three-dimensional attenuation structure beneath the northeastern Japan arc estimated from spectra of small earthquakes, *Tectonophysics, 319,* 241–260, 2000.

Van der Meijde, M., F. Marone, D. Giardini, and S. Van der Lee, Seismic evidence for water deep in Earth's upper mantle, *Science, 300,* 1556–1558, 2003.

Van Keken P. E., B. Kiefer, and S. M. Peacock, High-resolution models of subduction zones: Implications for mineral dehydration reactions and the transport of water into the deep mantle, *Geochemistry, Geophysics, Geosystems, 3,* doi:10.1029/2001GC000256, 2002.

Weidner, D. J. and Wang Y., Phase Transformations: Implications for Mantle Structure, in *Earth's Deep Interior Mineral Physics and Tomography From the Atomic to the Global Scale, Geophysical Monograph,* edited by S. Karato, A. M, Forte, R.C. Liebermann, G. Masters, and L. Stixrude, pp. 215–235. Am. Geophys. Union, 2000.

Wood B. J., The effect of H_2O on the 410-kilometer seismic discontinuity, *Science, 268,* 74–76, 1995.

Towards Mapping the Three-dimensional Distribution of Water in the Transition Zone From P-Velocity Tomography and 660-km Discontinuity Depths

Daisuke Suetsugu[1], Toru Inoue[2], Akira Yamada[2],
Dapeng Zhao[2], Masayuki Obayashi[1]

We estimate temperature anomalies and water content in the mantle transition zone from the depth of the "660-km discontinuity" and tomographically determined P-velocity anomalies. We assume a linear dependence of the discontinuity depths and P-velocity on temperature anomaly and water content. Beneath the Philippine Sea, where the Pacific plate is subducted, temperature anomalies are as low as -500K to -700 K within and near the stagnant Pacific slab and the water content is estimated to be in the range of 1–1.4 wt.%H_2O. The west Philippine basin, away from the Pacific slab, does not have a significant temperature anomaly or water content. Beneath western Japan, where the Pacific slab is subducted, we obtain temperature anomalies up to -300 to -600 K and water content up to 1–1.5 wt.%H_2O. Many problems remain to be solved for obtaining a definitive conclusion on the presence of water and quantitative estimates of temperature and water content. Estimates of the temperature anomaly and water content are highly sensitive to the input seismic parameters (the discontinuity depths and P-velocities) and the assumed dependence of the seismic parameters on temperature and water content determined from experimental studies. More accurate estimates of the seismic parameters and the experimental data measured under the pressure and temperature conditions of the mantle transition zone are necessary. Anelastic attenuation is probably enhanced by water, which might break down the linear dependence of the discontinuity depths and P-velocity on temperature anomaly and water content. A non-linear optimization approach using better seismic and experimental data may be required for obtaining more conclusive evidence for the presence of water.

1. INTRODUCTION

Subduction of oceanic crust transports water, in the form of hydrous minerals, from the surface into the upper mantle and may carry H_2O as deep as the mantle transition zone (hereafter called the MTZ) at 410–660 km depth, and possibly deeper [*e.g., Ohtani et al.,* 2004; *Omori et al.,* 2004; see also *Komabayashi,* this volume]. Recent experimental studies indicate that wadsleyite and ringwoodite can store water up to 2–3 weight % in their crystal structures [*e.g., Smyth, 1987; Inoue et al.,* 1995; *Kohlstedt et al.,* 1996]. Such water capacity is much greater

[1] Institute for Research on Earth Evolution, Japan Agency for Marine-Earth Science and Technology, Yokosuka, Kanagawa 237-0061, Japan

[2] Geodynamics Research Center, Ehime University, Matsuyama 790-8577, Japan

Earth's Deep Water Cycle
Geophysical Monograph Series 167
Copyright 2006 by the American Geophysical Union.
10.1029/167GM18

than that of olivine in the upper mantle or silicate perovskite and magnesiowüstite in the lower mantle [*e.g., Bolfan-Casanova et al.*, 2003; *Ohtani et al.*, 2004; *Mosenfelder et al.*, 2006], suggesting that the MTZ can be a significant water reservoir. *Inoue et al.* [2004] suggested that hydrous wadsleyite may be widely distributed by comparing the globally estimated P-velocity jump across the "410-km discontinuity" to experimental estimates of hydrous and anhydrous wadsleyite. The MTZ beneath subduction zones is thus one of the most likely candidates for a water-rich site in the mantle, since water may be supplied from the subducted slab.

The existence of a large amount of water should in principle be detectable by a seismological analysis [*e.g., Inoue et al*, 1995; *Jacobsen et al.*, 2004]. If we could observe seismic parameters which are influenced strongly by water, we could estimate the water content from such parameters. The sharpness of the "410-km discontinuity" and "660-km discontinuity" (hereafter called the '660' and '410', respectively) are examples of such parameters, because the sharpness of the phase transition at these boundaries is strongly influenced by the presence of hydrogen in the solid phases and also by the presence of hydrous melts [*e.g., Wood*, 1995; *Chen et al.*, 2002; *Smyth and Frost*, 2002; *Hirschmann et al.*, 2005; also see *Hirschmann et al.*, this volume]. On this basis *Van der Meijde et al.* [2003] estimated the water content of olivine at the '410' to be 0.05–0.1 wt.%H_2O beneath the Mediterranean region. However, most of other types of data, e.g., seismic velocity, depths of mantle discontinuities, seismic attenuation, and electrical conductivity are sensitive both to temperatures and water and so it is difficult to separate the effects of temperature and water from any single type of data. Table 1 summarizes the dependences of seismic velocity and the discontinuity depths on temperature and water content. A slower velocity can be caused either by higher temperature or by higher amounts of water. The deeper '660' and shallower '410' can be explained either by lower temperatures or by higher amounts of water [*Wood*, 1995; *Smyth and Frost*, 2002; *Chen et al.*, 2002; *Higo et al.*, 2001]. In principle, a simultaneous use of two or more types of data should enable us to separate thermal and compositional effects, although errors and uncertainties may limit the extent to which this is possible in practice. Recently *Blum and Shen* [2004] determined temperature anomalies and water content in the MTZ beneath the Archaean cratons in southern Africa to be -100 K and 0.3–0.7 wt.%H_2O, respectively, using the MTZ thickness and velocities obtained by seismic tomography. *Shito and Shibutani* [2003] determined temperature anomalies and iron and water contents at depths shallower than 400 km beneath the northern Philippine Sea using tomographically determined P and S velocity and attenuation anomalies.

Table 1. Dependence of seismic velocity and the mantle discontinuity depths on temperature and water

	Seismic velocity	"410" depth	"660" depth
Low T	High V	Elevated	Depressed
High T	Low V	Depressed	Elevated
water	Low V	Elevated	Depressed

In the present study, we present a simple method similar to that used by *Blum and Shen* [2004] for separately determining temperature anomalies and water content from seismic velocities and the mantle discontinuity depths. We apply the method to existing velocity and discontinuity depth data beneath the Philippine Sea and Japan to estimate the water content and temperature anomalies in the MTZ beneath several subduction zones, where the presence of water is most likely.

2. METHODS

In this study we assume a linear relationship between the seismic parameters (velocity perturbation, discontinuity depths) and the target parameters (temperature anomalies and water content) and ignore the effects of anelasticity on seismic velocity. For simplicity, we ignore other compositional factors, e.g., Mg/(Mg+Fe). The effect of iron content on the pressure of a phase change is small for the post-spinel phase change [*Ito and Takahashi*, 1989] and significant for the olivine-wadsleyite phase change: An increase of the iron content by 1mol% results in a 3–4 km elevation of the '410' [*Akaogi et al.*, 1989]. The harzburgite in subducted slabs is lower in iron content than the surrounding mantle peridotite, which may depress the '410' by 10–15 km. We use the '660' depth and P-velocity perturbation to avoid the iron effect. The methods of the present study could be extended to include compositional factors other than water by adding other types of data, e.g., electrical conductivity and attenuation, which may enable us to use the '410' depth data.

We express the velocity perturbation and the discontinuity depths in terms of temperature and water content as:

$$\delta V = \frac{\partial V}{\partial T}\delta T + \frac{\partial V}{\partial w}\delta w$$
$$\delta d = \frac{\partial d}{\partial T}\delta T + \frac{\partial d}{\partial w}\delta w, \tag{1}$$

where δV and δd are the P-wave velocity perturbation (in km/s) from the one-dimensional laterally averaged earth model obtained by tomographic studies [*Fukao et al.*, 2001; *Zhao*, 2001] and deviation of the discontinuity depth from the global average (in km), respectively. We used 660 km as the average '660' depth [*Flanagan and Shearer*, 1998]. The temperature anomaly δT (in K) and the water content δw (in weight percent) at the discontinuities are the unknowns in the inversion. Note that the estimates of the water content, as well as that of temperatures, are relative values with respect to a global average (while it has not been determined), because we used deviations of velocity and discontinuity depths from the global averages to determine the water content. The global averages of velocity and discontinuity depth may reflect the presence of water in the MTZ, e.g., beneath subduction zones [*e.g., Inoue.et al.*, 2004; *Ohtani et al.*, 2004].

The values of the coefficients $\partial V/\partial T$, $\partial d/\partial T$, $\partial V/\partial w$, and $\partial d/\partial w$ were chosen by referring to experimentally obtained values (Table 2). The temperature dependence of elastic moduli was obtained by, e.g., *Sinogeikin et al.* [2003] for ringwoodite, while their result was not obtained under the pressure-temperature condition of the MTZ. From their results, we adopted $\partial V/\partial T$ of -5.5 x 10^{-4} km/s/K in the present study.

The values of $\partial d/\partial T$ are obtained from the Clapeyron slopes of the post-spinel phase change [*e.g., Ito and Takahashi*, 1989; *Akaogi and Ito*, 1993; *Bina and Helffrich*, 1994; *Irifune et al.*, 1998; *Katsura et al.*, 2003; *Fei et al.*, 2004]. The Clapeyron slope of the post-spinel phase change from the experimental studies ranges from about -3 MPa/K (-2.8 MPa/K by *Ito and Takahashi* [1989]; -2.9 by *Irifune et al.* 1998]; -3.0MPa/K by *Akaogi and Ito* [1993]) to about -2MPa/K or smaller (-2.0 MPa/K by *Bina and Helffrich* [1994]; -1.3 MPa/K by *Fei et al.* [2004], -0.4 ~ -2.0 MPa/K by *Katsura et al.* [2003]). Recently *Litasov et al.* [2005] estimated the Clapeyron slope to be -2.0 MPa/K in hydrous peridotite. We tentatively adopted -2.0 MPa/K for the Clapeyron slope in the present study (-0.06 km/K for $\partial d/\partial T$) and change the value from -1.0 MPa/K to -3.0 MPa/K to examine the sensitivity of obtained temperature and water content to the Clapeyron slope.

The dependence of P-velocity on water content $\partial V/\partial w$ for ringwoodite was estimated from difference in the velocity of anhydrous and hydrous ringwoodite which were measured in ambient condition. It was estimated to be -0.23 km/s/wt.%H_2O from *Inoue et al.* [1998], -0.15 km/s/wt.%H_2O from *Wang et al.* [2003], and -0.4 km/s/wt.%H_2O from *Jacobsen et al.* [2004]. We tentatively adopted -0.2 km/s/wt.%H_2O as $\partial V/\partial w$ in the present study, but note that this value is highly speculative for the mantle because it is based only upon the available data obtained at one atmosphere pressure and 300

K. It is hopeful that pressure and temperature derivatives of the velocities will soon be available for hydrous olivine, wadsleyite, and ringwoodite [see, *e.g., Jacobsen and Smyth* of this volume].

The dependences of the '660' depth on water content $\partial d/\partial w$ was estimated on the basis of high-pressure and high-temperature experiments [*e.g., Higo et al.*, 2001; *Litasov et al.*, 2005]. The post-spinel phase boundary is depressed by 8–15 km for the water-saturation condition. We use the coefficient $\partial d/\partial w$ of 2.7 km/wt.%H_2O in the present study. The discontinuity depths are dominantly controlled by temperature in equation (1), since the water-dependence of the discontinuity depths is rather small as compared with the temperature dependence of the depths.

We used values of δV and δd obtained by previous studies. We took P-wave velocity perturbations at around the '660' from three-dimensional P-velocity models obtained by P-wave delay time tomography [*Fukao et al.*, 2001; *Zhao*, 2001]. We used δd determined using a receiver function method for the Philippine Sea basin [*Suetsugu et al.*, 2003, 2005] and those determined with ScS reverberation method [*Yamada et al.*, 2003] and a receiver function method [*Niu et al.*, 2005] for the Japanese Islands. The uncertainties in the depth estimates are about 10–15 km for all the cases.

Multiple phase transitions at around the '660' have been proposed [*e.g, Vacher et al.*, 1998; *Akaogi et al.*, 2002]. Under a cold condition such as subducted slabs, there may be transitions from garnet to ilmenite, from ringwoodite to perovskite + magnesiowüstite (the post-spinel phase transition), and from ilmenite to perovskite. The two garnet-related transitions other than the post-spinel transition are rather gradual: The transition intervals are 60–100 km and 20–60km for the garnet-ilmenite and ilmenite-perovskite transitions [*Vacher et al.*, 1998; *Akaogi et al.*, 2002], respectively, while the post-spinel transition is as sharp as a few km. The garnet-related transitions should not significantly affect the Ps-converted waves at the dominant periods of 4–8 sec used in the present study. The sScS reverberated waves analyzed in the present study have a dominant period of 50 sec, and thus may sense the garnet-related transitions. If that is the case, the '660' depths estimated from the sScS reverberation should be systematically greater than those from the Ps waves around subducted slabs. However, as shown in the next section, the

Table 2. Coefficients used in equation (1).

	Dependence on temperature	Dependence on water content
P-velocity	-5.5×10^{-4} km/s/K ($\partial V/\partial T$)	-0.2 km/s/wt.%H_2O ($\partial V/\partial w$)*
The '660' depth	-0.06 km/K ($\partial d/\partial T$)	2.7 km/ wt.%H_2O ($\partial V/\partial w$)

*Based on values measured at ambient conditions [*Inoue et al.*,1998; *Wang et al.*,2003; *Jacobsen et al.*, 2004]

'660' depths from the sScS reverberation are systematically shallower than those from the Ps waves. Seismic detectability of the multiple transitions is still controversial. Recent studies using Ps converted waves failed to find definitive evidence for multiple discontinuities beneath the Japanese islands (e.g., Tonegawa et al., 2005; Niu et al., 2005). We assume that the '660' depths used for the present study represent the post-spinel phase transition.

It is often the case that velocity perturbations from tomographic inversions are damped for regularization and underestimated due to limited ray coverage. While it is difficult to estimate the true velocity perturbations from a finite amount of data, we tentatively estimated a degree of the under-estimation by "point-spreading function" which was computed as follows. A unit velocity perturbation was given to a target block of a velocity model with null perturbations given to the rest of the blocks. Synthetic delay times were computed using the same event-station pairs as real data. The velocity perturbation was then reconstructed using the synthetic delay times in the same way as the real inversion. The perturbation δV_u at the target block is usually less than input value (unity), which qualitatively represents the degree of the under-estimation. We used δV_u (called the velocity correction factor hereafter) to account for the underestimation of velocity anomalies obtained by tomography and used $\partial V/\partial V_u$ instead of δV in solving equation (1) throughout the present paper. Beside the systematic underestimates, we must also account for the standard deviation in the P-velocity anomalies. The standard deviation was not given by Fukao et al. [2001] or Zhao [2001]. Inoue et al. [1990] estimated the standard deviation of their global P-velocity tomographic model to be 0.05–0.2 % using ISC data similar to those used by Fukao et al. [2001]. In the present study we adopt a conservative error estimate of 0.05 km/s (about 0.5 %) for P-velocity anomalies of Fukao et al. [2001] and Zhao [2001].

3. APPLICATION TO SUBDUCTION ZONES

3.1. Philippine Sea

The Philippine Sea is a site beneath which the oldest oceanic plate (older than 150 Ma) is being subducted, where the cold Pacific slab could store a large amount of water in hydrous minerals. We took P-velocity anomalies from the WEPP2 model [Fukao et al., 2001]. The WEPP2 model was obtained using 2 million P first arrival times reported to Bulletin of the International Seismological Centre (ISC). The model comprises the crust and whole mantle of the globe with a block size of 1.4° x 1.4° beneath the western Pacific with high resolution in the MTZ. We used the '660' depth

obtained by a receiver function method using broadband ocean bottom seismographs and deep-sea broadband bore hole seismograph [Suetsugu et al., 2003, 2005].

The '660' depths were estimated from Ps-P differential times using velocity models beneath stations obtained by data including surface waves [Ritsema and Heijst, 2000; Nakamura and Shibutani, 1998], since velocity models from body wave data, e.g., the WEPP2 model, do not have sufficient resolution in the upper mantle above the MTZ because of scarcity of seismic stations in the Philippine Sea. The P and Ps waves are low-pass filtered at a period of 8 sec to reduce oceanic noises. The Fresnel zone of the Ps wave is about 200–300 km in diameter. The '410' depths could not be determined reliably, since Ps waves from the '410' are faint. The faint '410' signal may be attributed to the low S/N ratio of the waveform data under conditions at the ocean bottom or to a broadening of the olivine-wadsleyite phase transition in a wet condition [Wood, 1995; Smyth and Frost, 2002; Van der Meijde,et al., 2003]. Suetsugu et al. [2005] divided the data into four regions according to geographical distribution of Ps conversion points and determined the '660' depths in each region: the Daito ridge, the Parece Vela basin, the Mariana trough, and the west Philippine basin (Plate 1a). The estimated '660' is deepest beneath the Daito ridge, where the WEPP2 model exhibits fast velocity anomalies (Plate 1a). These observations are qualitatively consistent with the presence of the cold temperature anomalies, which may be associated with the cold Pacific slab stagnant in the MTZ [e.g., Fukao et al., 2001; Zhao, 2004]. Significant anomalies are not apparent for the '660' depth and P-velocity beneath the west Philippine Basin [Suetsugu et al., 2005], which can be explained by absence of temperature anomalies there, because the west Philippine basin is far from the subducted Pacific slab. Beneath the Mariana trough and the Parece Vela Basin, however, the '660' is as deep as 690–700 km even though pronounced P-velocity anomalies are absent in the WEPP2 model. Non-thermal factors are thus required to reconcile the P-velocity anomalies and the '660 depth.

We applied equation (1) to the '660' depths and P-velocity anomalies with a velocity correction factor δV_u of 0.7-0.8. The velocity anomalies used for equation (1) are computed by averaging data of the WEPP2 model in each region denoted by the ellipses (Plate 1). We obtained the temperature anomalies and water content as shown in Plate 1b and Table 3. Uncertainties in the estimate of the temperature and water content were computed from those of the '660' depths and P-velocity anomalies. The uncertainty in the '660' depth was obtained from a bootstrap method [Efron and Tibshirani, 1986].

Table 3 shows that the bottom of the MTZ beneath the Daito ridge, the Parece Vela basin, and the Mariana trough

Table 3. The '660' depth anomaly, P-velocity anomaly, temperature anomaly, and water content determined for the MTZ beneath the Philippine Sea

	'660' depth anomaly (km)	P-velocity anomaly (km/s)	Temperature anomaly (K)	Water content (wt.%H_2O)
Mariana Trough	35±14	0.06±0.05	-537±208	1.0±0.6
Parece Vela Basin	45±9	0.06±0.05	-685±134	1.4±0.4
Daito Ridge	48±16	0.13±0.05	-744±238	1.2±0.7
West Philippine Basin	9±9	0.02±0.05	-139±134	0.2±0.4

could be 500–750 K colder than the global average. The Daito ridge is coldest, although the uncertainty level is large. The water content could range from 1.0 to 1.4 wt.%H_2O (± 0.4–0.7 wt.%H_2O) beneath the three regions. On the other hand, the temperature beneath the west Philippine basin is close to the global average and the water content is negligibly small. The difference between the first three regions and the west Philippine basin may be attributed to distance from the subducted Pacific slab. The MTZ beneath the Daito ridge is coldest possibly because of the stagnant cold Pacific slab existing there.

3.2. Japanese Islands

In the present study, we used the '660' depths determined by *Yamada* [2003] who analyzed sScS reverberated waves and those by *Niu et al.* [2005] who analyzed Ps waves. *Yamada* [2003] determined the depth variations of the '660' beneath the Japan Islands using data from deep earthquakes which occurred around Japan recorded by the broadband networks F-net [*Fukuyama et al.*, 1996] and J-Array [*J-Array group*, 1993] in Japan. A postcursor to sScS, which is identified as a top-side reflection from the '660' (called sScS$_{660}$ hereafter), was analyzed to determine the '660' depth. Relative travel times of the sScS$_{660}$ and sScS waves involve information on two reflections from the '660' beneath the events and the receivers. Using sScS$_{660}$-sScS times from several different events, the source-side and receiver-side depths of the '660' were determined (Plate 2a). The waveform data are low-pass filtered at a period of 50 sec, which corresponds to the Fresnel zone of 500–600 km in diameter. The deeper '660' is found beneath western Japan and can be caused by cold materials of the stagnant Pacific slab. Beneath central Japan, depressions of the '660' are also indicated, but this may not be due to the Pacific slab itself because this region is located to the trench side of the stagnant slab. Elevations of the '660' are found beneath northern Japan. *Niu et al.* [2005] obtained the '660' depths using receiver functions for the Hi-net (High sensitivity seismograph network) data. Hi-net is a dense short-period seismic network in Japan operated by the National Institute for Earth Science and Disaster Prevention. *Niu et al.* [2005] used Ps-waves with a predominant period of 3–4

sec, which corresponds to a Fresnel zone of about 150 km. The depths were determined using the WEPP2 model. The topography of the '660' determined by *Niu et al.* [2005] shows a depressed '660' except for northern Japan and the deepest '660' at westernmost Japan (Plate 3a), which is consistent with *Yamada* [2003]. Amplitudes of the topography of *Niu et al.* [2005] are greater than that of *Yamada* [2003] by a factor of two. This may be because the sScS$_{660}$-sScS times are measured at longer periods (> 50 sec) than the Ps-P time data from the receiver functions (> 4sec). The long-wavelength sScS and sScS$_{660}$ waves, in effect, spatially average the '660' topography, which may result in smaller topography than that obtained by Ps-P times. Besides the above mentioned studies, the '660' topography has been studied extensively using different types of seismic waves and analysis techniques [*Li et al.*, 2000; *Tono et al.*, 2005; *Tonegawa et al.*, 2005]. The depression of the '660' beneath western Japan is a common feature in the previous studies and is considered to be robust, while the topography beneath central and northeastern Japan found by *Yamada* [2003] differs among the previous studies. In the present study we determined temperature anomalies and water content only in western Japan, where the direct effects of the Pacific slab are expected and the '660' topography data are robust.

We used P-velocity anomalies obtained by *Zhao* [2001] and the WEPP2 model [*Fukao et al.*, 2001]. *Zhao* [2001] determined the three-dimensional P-velocity model of the whole mantle with a block size of 5° × 5°, using arrival times of P, pP, PP, and PcP waves from the ISC data reprocessed by *Engdahl et al.* [1998]. The seismic rays were computed by three-dimensional ray tracing in *Zhao* [2001]. The obtained P-velocity anomalies around the '660' are characterized by fast velocity anomalies in western Japan (130°E–136°E) and slower anomalies in central and northeastern Japan (35°N–40°N) (Plate 2b). Based on their locations, the former anomalies can be attributed to the subducted Pacific slab. It is not clear what the latter slow anomalies represent. *Zhao* [2001] suggested that the slow anomalies may represent a small mantle upwelling caused by the subduction of the Pacific slab or a small hot plume rising from the lower mantle. Plate 3b shows velocity anomalies of the WEPP2 model near the '660' beneath Japan, which is composed of blocks of 1.4°×1.4°.

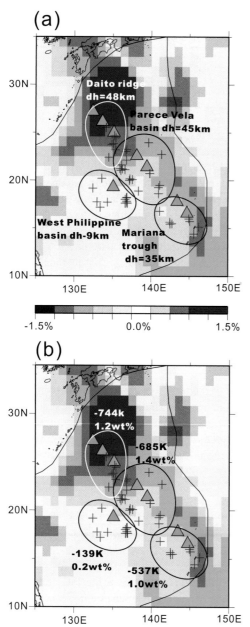

Plate 1. (a) P-velocity anomalies at around 660 km [*Fukao et al.,* 2001] and the '660' depth anomalies [*Suetsugu et al.,* 2003, 2005] beneath the northern Philippine Sea. The four ellipses and numbers contained therein indicate regions where Ps waves sample the '660' and the determined '660' depths. The ocean bottom stations are denoted by triangle and Ps conversion points at the '660' are indicated by crosses, which are grouped by taking into account the Fresnel zone of the Ps waves at a period of 8 sec. (b) Temperature anomalies and water content in the regions.

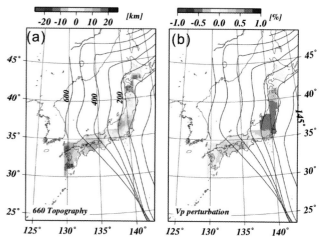

Plate 2. Data set 1 for the Japanese islands. (a) Topography on the '660' [*Yamada,* 2003] and (b) P-velocity anomalies [*Zhao, 2001*] at around 660 km beneath the Japanese Islands. Deep seismic zones are indicated by isodepth curves.

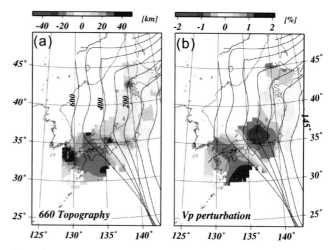

Plate 3. Data set 2 for the Japanese islands. (a) Topography on the '660' [*Niu et al.,* 2005] and (b) P-velocity anomalies of the WEPP2 model [Fukao et al., 2001] at around 660 km beneath the Japanese Islands.

Both of the models were obtained using similar data sources and delay time tomography, while the parameterization, ray tracing technique and regularization (damping) methods are different. The WEPP2 model has a similar velocity pattern at the bottom of the MTZ to *Zhao* [2001]: fast velocities in western Japan and slow velocities in central Japan, while velocity anomalies are not similar in northern Japan. The amplitudes of velocity anomalies are greater for the WEPP2 model than for *Zhao* [2001] by a factor of two. The different amplitudes may be attributed to finer parameterization of the WEPP2 model than Zhao's model and different regularization methods used for the two models.

We computed temperature anomalies and water content beneath western Japan (Plate 4) from two sets of '660' topography data and P-velocity anomaly data. One data set is composed of *Yamada* [2003] for the '660' depth and from *Zhao* [2001] for velocity anomaly (data set 1); the other is composed of *Niu et al.* [2005] for the '660' and the WEPP2 model for velocity anomalies (data set 2). The velocity correction factors δV_u are 0.8 and 0.5 for the Zhao and WEPP2 models, respectively. The former data set has a spatial wavelength of 500–600 km and the latter has that of 100–200 km. Both data sets result in a cold and wet MTZ in western Japan. The westernmost part is the coldest, where the cold Pacific slab is expected to be stagnant. There is a peak of water content found near the Pacific slab beneath westernmost Japan, unlike the cold temperatures beneath all of western Japan. This may be because water is not released to the trench side from the bottom of the Pacific slab. The cold Pacific slab is stagnant in the MTZ beneath western Japan and lowers the temperature of both sides of the surrounding mantle due to thermal conduction. The amplitudes of temperature anomalies and water content are smaller for the result from data set 1 (Plates 4a, 4b) than that for data set 2 (Plates 4c, 4d) by a factor of 1.5~2. From data set 1, the temperature anomalies and water content in westernmost Japan could be -150~-300K and 0.5~1.0wt.%H_2O, respectively. From data set 2, the temperature anomalies and water content could be -400 ~ -700K and 0.5~1.5 wt.%H_2O, respectively. We estimated the statistical uncertainties to be 150 K and 0.5 wt.%H_2O for temperature anomaly and water content, respectively. The uncertainties were obtained only from uncertainties in the seismic data without taking into consideration the uncertainties in the experimentally derived coefficients in equation (1). There is another area of high water content in central Japan (36°N, 136°E) as inferred from data set 2 (Plate 4d), while data set 1 does not indicate a significant water content there. The high water content is caused by low velocity anomalies of the WEPP2 model, which is also seen in Zhao's model, but located 300 km east of the low velocity anomalies in the WEPP2 model (Plates 2b

and 3b). The different location of the low velocity anomalies is mainly due to the different block sizes in the two models. The lowest temperature anomalies are -300 K and -600 K from data sets 1 and 2, respectively, both in westernmost Japan. Westernmost Japan could have water content of 0.5–1.0 wt.%H_2O and 1.0–1.5 wt.%H_2O for data sets 1 and 2, respectively. The differences may be due to the different wavelengths of spatial averaging in the two data sets and different methods of regularization used for tomography.

4. DISCUSSION AND CONCLUSION

The results of the present study suggest the presence of water and low temperature anomalies in the MTZ beneath subduction zones around Japan. The estimates may however be subject to uncertainties of experimental data used for the linear coefficients used, those of the input seismic data, the '660' depth and P-velocity anomaly, and even those in the linear formulation used in the present study. We now discuss how much these uncertainties could affect estimates of the water content and temperature anomaly.

4.1. Uncertainties due to Experimental Data

The coefficients in equation (1) were obtained by experimental studies which show rather large scatter. The Clapeyron slope of the post-spinel phase transformation, which is expressed as $\partial d / \partial T$ in equation (1), was estimated to be about -3 MPa/K [*Ito and Takahashi*, 1989; *Akaogi and Ito*, 1993; *Irifune et al.*, 1998], but recent studies by in situ X-ray diffraction suggest much lower values from -0.4 to -2.0 MPa/K [*e.g., Katsura et al.*, 2003; *Fei et al.*, 2004]. The dependence of seismic velocity on temperature $\partial V / \partial T$ [*Sinogeikin et al.*, 2003] and water content $\partial V / \partial w$ [*Inoue et al.*, 1998; *Wang et al.*, 2003; *Jacobsen et al.*, 2004] was not measured under pressure and temperature conditions of the MTZ. Figure 1 shows temperatures and water content estimated using different values of the coefficients. We used the '660' depth and the P-velocity anomaly obtained for the Daito ridge in the Philippine Sea (708 km and 0.16 km/s, respectively) in Fig. 1 while the tendency described below can apply to other regions. Temperature estimates are nearly inversely proportional to the absolute value of the Clapeyron slope (Fig. 1a). Greater absolute values of the slope result in the smaller absolute values of the temperature anomaly, and vice versa. This is because the temperature anomaly is controlled mostly by the '660' depth in equation (1). A Clapeyron slope of -1.0 MPa/K gives very large negative temperature anomaly of -1300 K and large water content (2.9 wt.%H_2O), which is almost the maximum possible amount of water that can be stored in ringwoodite. Temperature anomalies are not much

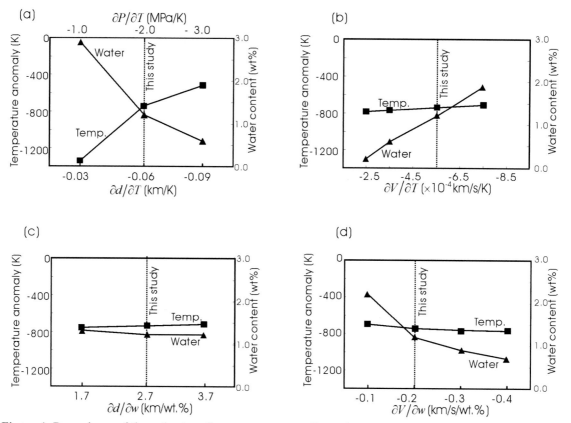

Figure 1. Dependence of the estimates of temperature anomalies and water content on the linear coefficients in equation (1). Dependence on (a) $\partial d/\partial T$; (b) $\partial V/\partial T$: (c) $\partial V/\partial w$; (d) $\partial V/\partial w$, where d, V, T, w denote the '660' depth (in km), P-velocity (km/s), temperature (K), and water content (wt.% H_2O), respectively.

affected by the other coefficients (Figs. 1b, c, d). Estimates of the water content are substantially influenced by changing the coefficients $\partial d/\partial T$, $\partial V/\partial T$, and $\partial V/\partial w$ (Figs. 1a, b, and d, respectively). The estimate becomes greater for the smaller absolute values of the Clapeyron slope (equivalent to $\partial d/\partial T$), the greater values of $\partial V/\partial T$, and the smaller values of $\partial V/\partial w$. Nevertheless the estimated water content is always positive, suggesting that some amount of water is required beneath the Daito ridge. The coefficients $\partial d/\partial w$ affect little the estimates of temperature or water content.

4.2. Uncertainty due to Seismic Data

We examine the sensitivity of the estimates of temperature anomalies and water content to the input '660' depths and P-velocity anomalies. Besides random measurement errors, there are sources of systematic errors in the P-velocity anomaly and the '660' depth. P-velocity anomalies could be underestimated by seismic tomography due to regularization (damping and smoothing). The '660' depth could be biased by an incorrect velocity model used to translate Ps-P or

sScS$_{660}$-sScS times to the '660' depth, where the average velocity along ray path above the '660' is important. Another potential source of the systematic error is the trade-off between the P-velocity anomaly and the '660' depth which were separately determined. The trade-off can be accounted for by jointly inverting P and S wave travel times and Ps-P or sScS$_{660}$-sScS times for three-dimensional P and S velocity models and topography on the '660', but such models are still few. On a global scale, *Gu et al.* [2003] conducted joint inversion of S$_{660}$S and S$_{410}$S waves with travel times of various body waves and surface waves for the long-wavelength S-velocity model in the mantle and the topography on the '410' and '660'. Comparing their results to results from separate inversions, they indicated that the S-velocities did not change significantly due to the joint inversion, but the topography changed by ~10 km. *Zhao* [2004] inverted P-wave travel times for a whole mantle three-dimensional velocity model with topography of the Moho, '410', and '660' discontinuities constrained to those of existing models [*Mooney et al.*, 1998; *Flanagan and Shearer*, 1998], indicating that the effect of the topography on the

P-wave velocities is only 0.2–0.3 %, within the error of 0.5 % assigned in the present study. It is difficult to evaluate the error of the '660' depth due to the separate determination of the '660' depth and velocity. The trade-off effect on the '660' depth should arise via average P and S velocities along rays above the '660'. Here we evaluate the effect of the presumably underestimated velocity anomalies due to regularization in tomography and the error of the '660' depth caused by the incorrect average velocity on the estimates of water content and temperature anomalies.

Regarding the P-velocity anomalies, while we have taken the under-estimation of the P-wave velocity anomalies (at a depth of 660 km) by tomography into consideration by the velocity correction factor δV_u, it is possible that the corrected velocity anomalies are still underestimated. The WEPP2 model of data set 2 (section 3.2) has fast P-velocity anomalies of 1–1.5 % in the MTZ beneath western Japan (Plate 3b) and we applied δV_u of 0.5 to the velocity anomalies. Waveform modeling studies around Japan have suggested fast P velocity anomalies of 2–3 % [e.g., Tajima and Grand, 1998; Shito and Shibutani, 2001], which is consistent with a value of 0.5 for δV_u. While a δV_u value of 0.5 seems reasonable, we show the influence of using a smaller value of δV_u (0.33) in Plate 5. The water content becomes nearly zero in most of western Japan, but the water content is still more than 1 % beneath westernmost Japan, where data set 1 also indicates a water-richest site. The amplitudes of velocity anomalies influence the estimate of the water content significantly and do not affect the estimate of temperature anomalies.

We estimated the effect of a change in the average velocity in the upper mantle velocity model by -1 % to be -8 km for the '660' depth determined by the receiver function analysis (the Philippine Sea and the Japanese islands) [e.g., Suetsugu et al., 2005] and -10 km for the '660' depth determined from the $sScS_{660}$-sScS times [Yamada, 2003] (the Japan islands). We demonstrated the influence on the temperature anomalies and water content by shifting the '660' by ±10 km (Plate 6), which should be caused by changing the average velocity above the '660' by ±1 %, respectively. The deepened '660' provides colder temperatures (Plate 6a) and a greater amount of water (Plate 6b) than in Plate 4, while an elevated '660' gives warmer temperatures (Plate 6c) and a lesser amount of water (Plate 6d). The influence of elevating the '660' by 10 km on estimates of the temperature and water content is -100 ~ -200 K and -0.5 wt.%H_2O, respectively. In summary, the estimates of water content are sensitive both to the '660' depth and P-velocity anomaly, estimates of the temperature anomaly are sensitive only to the '660' depth. For all of the cases tried in the present study (Plates 4, 5, 6), the MTZ has low temperature anomalies beneath western Japan and positive water content in westernmost Japan. Magnifying

the P-velocity anomalies of the seismic tomography and elevating the '660' diminishes the estimate of water content (Plates 5b and 6d) in most of western Japan, but an area of water content greater than 1.0 wt.%H_2O still remains in westernmost Japan, where the Pacific slab encounters the '660'. While the influence of separate determination of the '660' depth and seismic velocity is difficult to quantify, a full joint inversion of P, S, and Ps-P or $sScS_{660}$-sScS times is desirable to minimize the systematic errors due to the trade-off of the '660' depth and velocities.

4.3. Influence of Anelasticity

The presence of water could enhance anelastic attenuation (i.e., lower value of Q parameter) of minerals in the upper mantle [e.g., Jackson et al., 1992; Karato, 1995], which should reduce seismic velocities through physical dispersion [Kanamori and Anderson, 1977] and increase the dependence of seismic velocities on temperature and water content [e.g., Karato, 1993, 2004; Shito and Shibutani, 2003]. The anelasticity effects make temperature and water effects coupled and introduce a non-linear relationship among velocity, temperature, and water content, which needs to be solved by non-linear techniques, e.g., a global optimization. To evaluate the anelasticity effect quantitatively, we require three-dimensional anelasticity data for the MTZ and experimental data for anelasticity of wadsleyite and ringwoodite, which are unavailable at present. Instead, we made a rough estimate of the anelasticity effect in the linear framework of equation (1) by correcting $\partial V/\partial w$ and $\partial V/\partial T$ for anelasticity using the formulation of Karato [1993].

For $\partial V/\partial w$, we evaluated the velocity perturbation due to anelasticity (physical dispersion) [Minster and Anderson, 1981] with a weak frequency dependence given by

$$V_{anelastic}/V_{elastic} = 1 - \cot(\gamma\pi/2)/2Q \quad (0 \leq \gamma < 1) \quad (2)$$

where $V_{elastic}$ and $V_{anelastic}$ are velocities in elastic and anelastic cases, respectively, and Q is the quality factor. We adopted $\gamma = 0.1$, but the choice of γ is not significant in the following discussion. We corrected Vp values of dry and wet ringwoodite under elastic condition [Inoue et al., 1998] for anelasticity with equation (2), then we computed $\partial V/\partial w$ for anelastic ringwoodite. According to Inoue et al. [1998], Vp values under dry and wet (2.2 wt.%H_2O) conditions are 9.79 km/s and 9.27 km/s, respectively, and $\partial V/\partial w = -0.23$ km/s/ wt.%H_2O. For the anelastic correction, we used $Q = 362$, taken from PREM [Dziewonski and Anderson, 1981], as the quality factor Q_{dry} of dry ringwoodite. We estimated the anelasticity of wet ringwoodite to be $Q_{wet} = Q_{dry}/2$ assuming the water dependence of anelasticity for ringwoodite is the

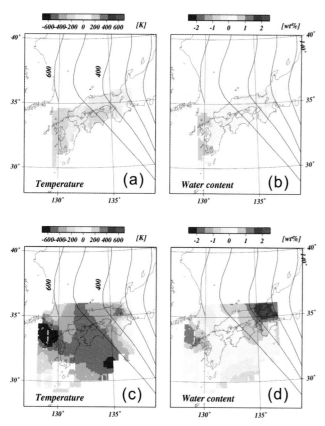

Plate 4. (a) Temperature anomalies and (b) water content beneath the Japanese Islands determined from data set 1. (c) Temperature anomalies and (d) water content from data set 2.

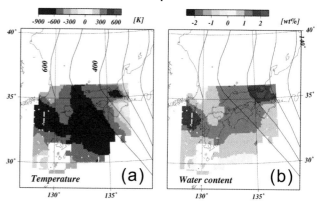

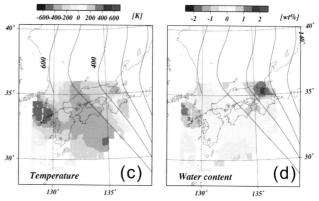

Plate 6. Influences by deepening (a, b) and elevating (c, d) the '660' by 10 km for data set 2. Temperature anomalies are shown in (a) and (c) and water content is shown in (b) and (d).

Velocity correction factor = 0.33

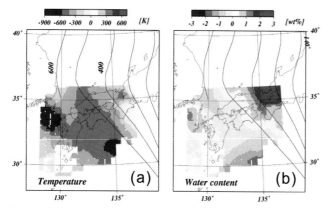

Plate 5. Influence by velocity correction factor of 0.33 for data set 2. (a) Temperature anomalies and (b) water content beneath the Japanese Islands.

same as that of olivine obtained experimentally in a seismic frequency band [*Jackson et al.,* 1992]. *Vp* values corrected for anelasticity with equation (2) are 9.67 km/s and 9.04 km/s for the dry and wet conditions, respectively, and $\partial V/\partial w = -0.29$ km/s/ wt.%H_2O.

To correct the value of $\partial V/\partial T$ for anelasticity, we employed tabulated values of $\partial V/\partial T$ as a function of Q for olivine (Table 1 of *Karato* [1993]) because of the lack of data for ringwoodite. For olivine, the $\partial V/\partial T$ value is larger for $Q = 362$ (PREM) than that for $Q = \infty$ by a factor of 1.4. Assuming the rate of increase of $\partial V/\partial T$ with decreasing Q is same for olivine and ringwoodite, and using -5.5×10^{-4} km/s/K for $\partial V/\partial T$ of the elastic ($Q = \infty$) ringwoodite [*Sinogeikin et al.,* 2001], we have $\partial V/\partial T = -7.8 \times 10^{-4}$ km/s/K. Thus the absolute values of both of the derivatives become greater under wet condition.

Using a '660' depth of 708 km and δVp value of 0.16 km/s of the Daito ridge as sample input data, we estimated the influence of anelasticity on temperature and water content using the above values of $\partial V/\partial w$ and $\partial V/\partial T$ in equation (1). Correcting for anelasticity, the water content is estimated to be 1.4 wt.%H_2O, greater than the value of 1.2 wt.%H_2O computed without anelasticity, and the temperature anomaly is estimated to be -736 K, close to -744 K determined without anelasticity. The influence of Q on estimate of water content is modest and it also does not change our estimate of temperature and water content significantly, considering the large standard errors, which are 150K and 0.5wt.%H_2O for temperature anomaly and water content, respectively. The anelastic effects of $\partial V/\partial T$ and $\partial V/\partial w$ work in opposite ways to the estimate of water content and are partially cancelled: the anelasticity increases via $\partial V/\partial T$ and decreases via $\partial V/\partial w$ the estimate of the water content. The above estimate of the anelastic effect is, however, highly sensitive to experimental data of anelasticity for ringwoodite, though most of the data is lacking. It is desirable to obtain more experimental data on elasticity and anelasticity and use a fully non-linear treatment to solve for temperature and water content.

The seismic data used in the present study suggest that the water may be present in the MTZ beneath the Philippine Sea and the Japanese islands where the subducted Pacific slab may transport water deep into the mantle. The greatest water content is estimated to be 1–1.5 wt.%H_2O in westernmost Japan and Philippine Sea, where the Pacific slab is stagnant. However, large uncertainties in the seismic and experimental data prevent us from obtaining definitive evidence for the presence of water and determining temperature and water content quantitatively. The estimate of water content is very sensitive to the '660' depth, seismic velocity, and the values of the linear coefficients in equation (1) determined experimentally. The temperature anomaly is sensitive to the '660' depth and the Clapeyron slope of the post-spinel phase

change. Besides the above-mentioned problems of uncertainties in data, many problems remain to be clarified. Most of the linear coefficients in equation (1) have not been obtained under the pressure-temperature condition of the MTZ. The linearity assumed in equation (1) can also be broken by effects other than anelasticity. For instance, the Clapeyron slope may be also dependent on water content [*Litasov et al.,* 2005]. More precise determination of seismic velocity anomalies and the discontinuity depths is obviously required. Accurate determination of the mantle discontinuity depths is crucial for reliable estimates of both temperature anomalies and water content. Absolute values of seismic velocity anomalies are also important, particularly for estimating water content. Joint analysis of other types of data, e.g., seismic attenuation and electrical conductivity, would provide more constraint on the water content and temperature anomaly, since their sensitivities to water and temperature are different from those in the present study [*Shito and Shibutani,* 2003; *Huang et al.,* 2005; *Koyama et al.* in this volume]. A joint inversion of the seismic attenuation is particularly important and useful, since the attenuation data are sensitive to temperature and water content and are required to correct seismic velocities for the physical dispersion as discussed above.

Acknowledgments. We thank M. Ichiki, S. Ono, and T. Koyama for stimulating discussion and thank F. Niu for providing their '660' depth data. Critical reading by R. J. Geller was valuable. Comments by two anonymous reviewers were helpful to improve the manuscript. The research was supported by the MUD project of IFREE/JAMSTEC and Grant-in-Aid for Scientific Research [KAKENHI 16075208] from MEXT.

REFERENCES

Akaogi, M. and E. Ito, Olivine-modified spinel-spinel transitions in the system Mg_2SiO_4-Fe_2SiO_4: Calorimetric measurements, thermochemical calculation, and geophysical application, *J. Geophys. Res., 94,* 15671–15685, 1989.

Akaogi, M and E. Ito, Refinement of enthalpy measurement of $MgSiO_3$ perovskite and negative pressure-temperature slopes for perovskite-forming reaction, *Geophys. Res. Lett., 12,* 1839–1842, 1993.

Akaogi, M., A. Tanaka, and, E. Ito, Garnet-ilmenite-perovskite transitions in the system $Mg_4Si_4O_{12}$-$Mg_3Al_2Si_3O_{12}$ at high pressures and high temperatures: phase equilibria, calorimetry and implications for the mantle structure, *Phys. Earth Planet. Inter., 132,* 303–324, 2002.

Bina, C. R. and G. Helffrich, Phase transition Clapeyron slopes and transition zone seismic discontinuity topography, *J. Geophys. Res., 99,* 15853–15860, 1994.

Blum, J. and Shen, Y., Thermal, hydrous, and mechanical states of the mantle transition zone beneath southern Africa, *Earth Planet. Sci. Lett., 217,* 367–378, 2004.

Bolfan-Casanova N., H. Keppler, and D. C. Rubie, Water partitioning at 660 km depth and evidence for very low water solubility in magnesium silicate perovskite, *Geophys. Res. Lett., 30,* doi:10.1029/2003GL017182, 2003.

Chen, J., T. Inoue, H. Yurimoto, and D. J. Weidner, Effect of water on olivine-wadsleyite phase bondary in the $(Mg, Fe)_2SiO_4$ system, *Geophys. Res. Lett., 29,* 1875, doi:10.1029/2001GL014429, 2002.

Dziewonski, A.M. and D. L. Anderson, Preliminary reference Earth model, *Phys. Earth Planet. Inter., 25,* 297–356, 1981.

Efron, B. and R. Tibshirani, Bootstrap methods for standard errors, confidence intervals, and other measures of statistical accuracy, *Stat. Sci., 1,* 54–77, 1986.

Engdahl, E., R. van der Hilst, R. Buland, Global teleseismic earthquake relocation with improved travel times and procedures for depth determination, *Bull. Seismol. Soc. Am., 88,* 722–743, 1998.

Fei, Y., J. V. Orman, J. Li, W. van Westrenen, C. Sanloup, W. Minarik, K. Hirose, T. Komabayashi, M. Walter and K. Funakoshi, Experimentally determined postspinel transformation boundary in Mg_2SiO_4 using MgO as an internal pressure standard and its geophysical implication, *J. Geophys. Res., 109,* B02305, doi:10.1029/2003JB002562, 2004.

Flanagan, M. and P. Shearer, Global mapping of topography on transition zone velocity discontinuities by stacking SS precursors, *J. Geophys. Res., 103,* 2673–2692, 1998.

Fukao, Y., S. Widiyantoro, and M. Obayashi, Stagnant slabs in the upper and lower mantle transition zone, *Rev. Geophys., 39,* 291–323, 2001.

Fukuyama, E., M. Ishida, S. Hori, S. Sekiguchi, and S. Watada, Broadband seismic observation conducted under the FREESIA Project, *Rep. Nat'l Res. Inst. Earth Sci. Dsas. Prev., 57,* 23–31, 1996.

Gu, Y. J., A. M. Dziewonski, and G. Ekström, Simultaneous inversion for mantle shear velocity and topography of transition zone discontinuities, *J. Geophys. Res., 154,* 559–583, 2003.

Higo, Y., T. Inoue, T. Irifune and H. Yurimoto, Effect of water on the spinel-postspinel transformation in Mg_2SiO_4. *Geophys. Res. Lett., 28,* 3505–3508, 2001.

Hirschmann, M. M., C., Aubaud, and A. C. Withers, Storage capacity of H_2O in nominally anhydrous minerals in the upper mantle, *Earth and Planetary Science Letters., 236,* 167–181, 2005.

Huang, X., Y. Xu, and S. Karato, Water content of the mantle transition zone from the electrical conductivity of wadsleyite and ringwoodite, *Nature, 434,* 746–749, 2005.

Inoue, H., Y. Fukao, K. Tanabe and Y. Ogata, Whole mantle P-wave travel time tomography, *Phys. Earth Planet. Inter., 59,* 294–328, 1990.

Inoue, T., H. Yurimoto and Y. Kudoh, Hydrous modified spinel, $Mg_{1.75}SiH_{0.5}O_4$: a new water reservoir in the mantle transition region, *Geophys. Res. Lett., 22,* 117–120, 1995.

Inoue, T., D. J. Weidner, P. A. Northrup and J. B. Parise, Elastic properties of hydrous ringwoodite (γ-phase) on Mg_2SiO_4, *Earth Planet. Sci. Lett., 160,* 107–113, 1998.

Inoue T., Y. Tanimoto, T. Irifune, T. Suzuki H. Fukui and O. Ohtaka, Thermal expansion of wadsleyite, ringwoodite, hydrous

wadsleyite and hydrous ringwoodite, *Phys. Earth Planet. Inter., 143–144,* 279–290, 2004.

Irifune, T., N. Nishiyama, K. Kuroda, T. Inoue, M. Isshiki, W. Utsumi, K. Funakoshi, S. Urakawa, T. Uchida, T. Katsura and O. Ohtaka, The post spinel phase boundary in Mg_2SiO_4 determined by in situ x-ray diffraction. *Science, 279,* 1698–1700, 1998.

Ito, E. and E. Takahashi, Postspinel transformations in the system Mg_2SiO_4-Fe_2SiO_4 and some geophysical implications, *J. Geophys. Res., 94,* 10,637–10,646, 1989.

Jackson, I., M. S. Paterson, J. D. Fitz Gerald, Seismic wave dispersion and attenuation in Aheim dunite, *Gephys. J. int., 108,* 517–534, 1992.

Jacobsen, S. D., J. R. Smyth, H. Spetzler, C. M. Holl, and D. J. Frost, Sound velocity and elastic constants of iron-bearing hydrous ringwoodite, *Phys. Earth Planet. Inter., 143–144,* 47–56, 2004.

J-Array Group, The J-Array Program: System and Present Status, *J. Geomag. Geoelectr. 45,* 1265–1274, 1993.

Kanamori, H. and D. L. Anderson, Importance of physical dispersion in surface-wave and free-oscillation problems, review, *Rev. Geophys. Space Phys., 15,* 105–112, 1977.

Karato, S., Importance of anelasticity in the interpretation of seismic tomography, *Geophys. Res. Lett., 20,* 1623–1626, 1993.

Karato, S., Effects of water on seismic wave velocities in the upper mantle, *Proc. Japan Acad., 71,* 61–66, 1995.

Karato, S., Mapping water content in upper mantle, in *Inside the Subduction Factory, edited by J. M. Eiler,* AGU Geophysical Monograph, 138, 135–152, 2004

Katsura, T., H. Yamada, T. Shimei, A. Kubo, S. Ono, M. Kanzaki, A. Yoneda, M. J. Walter, , E. Ito, S. Urakawa, K. Funakoshi and W. Utsumi, Post-spinel transition in Mg_2SiO_4 determined by high P-T in situ X-ray diffractometry, *Phys. Earth Planet. Inter., 136,* 11–24, 2003.

Kohlstedt, D. L., H. Keppler, D. C. Rubie, Solubility of water in the α, β, and γ phases of $(Mg, Fe)_2SiO_4$. *Contrib. Miner. Petrol., 123,* 345–357, 1996.

Li, X., S. V. Sobolev, R. Kind, X. Yuan and A. Estabrook, A detailed receiver function image of the upper mantle discontinuities in the Japan subduction zone, *Earth Planet. Sci. Lett., 183,* 527–541, 2000.

Litasov, K. D., E. Ohtani, A. Sano and A. Suzuki, Wet subduction versus cold subduction, *Geophys. Res. Lett., 32,* L13312, doi:10.1029/2005GL022921, 2005.

Minster, J. B. and D. L. Anderson, A model of dislocation-controlled rheology for the mantle, *Phil. Trans. R. Soc. London, 299,* 319–356, 1981.

Mooney, W., G. Laske, and G. Masters, A new global crustal model at 5 x 5 degrees: CRUST-5.1, *J. Geophys. Res., 103,* 727–747, 1998.

Mosenfelder J. L., N. I. Deligne, P. D. Asimow, and G. R. Rossman, Hydrogen incorporation in olivine from 2–12 GPa, *Am. Mineral. 91,* 285–294, 2006.

Nakamura, Y. and T. Shibutani, Three-dimensional shear wave velocity structure in the upper mantle beneath the Philippine Sea region, *Earth Planets Space, 50,* 939–952, 1998.

Niu, F. A. Levander, S. Ham, M. Obayashi, Mapping the subducting Pacific slab beneath southwest Japan with Hi-net receiver functions, *Earth Planet. Sci. Lett., 239,* 9–17, 2005.

Ohtani, E., K. Litasov, T. Hosoya, T. Kubo and T. Kondo, Water transport into deep mantle and formation of a hydrous transition zone, *Phys. Earth Planet. Inter., 143–144,* 255–269, 2004.

Omori, S., T. Komabayashi and S. Maruyama, Dehydration and earthquakes in the subducting slab: empirical link in intermediate and deep seismic zones, *Phys. Earth Planet. Inter., 146,* 297–311, 2004.

Ritsema, J. and H. J. van Heijst, Seismic imaging of structural heterogeneity in Earth's mantle: Evidence for large-scale mantle flow, *Science Progress, 83,* 243–259, 2000.

Shito, A. and T. Shibutani, Upper mantle transition zone structure beneath the Philippine Sea region, *Geophys. Res. Lett., 28,* 871–874, 2001.

Shito, A. and T. Shibutani, Nature of heterogeneity of the upper mantle beneath the northern Philippine Sea as inferred from attenuation and velocity tomography, *Phys. Earth Planet. Inter., 140,* 331–341, 2003.

Sinogeikin, S. V., J. D. Bass, and T. Katsura, Single-crystal elasticity of ringwoodite to high pressures and high temperatures: implications for 520 km seismic discontinuity, *Phys. Earth Plane. Inter., 136,* 41–66, 2003.

Smyth, J.R., β-Mg_2SiO_4: A potential host for water in the mantle? *Am. Mineral. 72,* 1051–1055, 1987.

Smyth, J. R. and D. J. Frost, The effect of water on the 410-km discontinuity: an experimental study, *Geophys. Res. Lett., 29,* 1485, doi:10.1029/2001GL014418, 2002.

Suetsugu, D., H. Shiobara, H. Sugioka, S. Kodaira, Y. Fukao, K. Mochizuki, T. Kanazawa, R. Hino and T. Saita, Thick mantle transition zone beneath the Philippine Sea as inferred using the receiver function method for data from the long term broadband ocean bottom seismograph array, *abstract of IUGG2003,* Sapporo, Japan, JSS06/07P/A04-005,2003.

Suetsugu, D., M. Shinohara, E. Araki, T. Kanazawa, K. Suyehiro, T. Yamada, K. Nakahigashi, H. Shiobara, H. Sugioka, K. Kawai and Y. Fukao, Determination of mantle discontinuity depths by a receiver function analysis of broadband seismic records from deep-sea borehole observatory and ocean bottom seismograph in the west Philippine Sea basin, *Bull. Seism. Soc. Am., 95,* 1947–1956, 2005.

Tajima, F. and S. P. Grand, Variation of transition zone high velocity anomalies and depression of 660km discontinuity associated with subduction zones from the southern Kuriles to Izu-Bonin and Ryukyu, *J. Geophys. Res., 103,* 15015–15036, 1998.

Tonegawa, T., K. Hirahara and T. Shibutani, Detailed structure of the upper mantle discontinuities around the Japan subduction zone imaged by receiver function analyses, *Earth Planets Space, 57,* 5–14, 2005.

Tono, Y., T. Kunugi, Y. Fukao, S. Tsuboi, K. Kanjo and K., Kasahara, Mapping of the 410- and 660-km boundaries beneath the Japanese islands, *J. Geophys. Res., 110,* B3, B03307, 10.1029/2004JB003266, 2005.

Vacher, P., A. Mocquet, and C. Sotin, Computation of seismic profiles from mineral physics: the importance of the non-olivine components for explaining the 660 km depth discontinuity, *Phys. Earth Planet. Inter., 106,* 275–298, 1998.

Van der Meijde, M., F. Marone, D. Giardini and S. Van der Lee, Seismic evidence for water deep in Earth's upper mantle, *Science, 300,* 1556–1558, 2003.

Wang, J., S. V. Sinogeikin, T. Inoue and J. D. Bass, Elastic properties of hydrous ringwoodite, *Am. Mineral., 88,* 1608–1611, 2003.

Weidner, D. J. and Y. Wang, Phase transformations: Implications for mantle structure, in *Earth's Deep Interior: Mineral Physics and Tomography From the Atomic to the Global Scale, edited by S. Karato et al.,* AGU Geophysical Monograph, 117, 215–235, 2000.

Wood, B. J., The effect of H_2O on the 410-kilometer seismic discontinuity, *Science, 268,* 74–76, 1995.

Yamada, A., Depth variation of the 660 km discontinuity beneath Japan Islands inferredfrom a postcursor to sScS, *Earth Monthly, 289,* 561–566, 2003 (in Japanese).

Zhao, D, Seismic structure and origin of hotspots and mantle plumes, *Earth Planet. Sci. Lett., 192,* 251–265, 2001.

Zhao, D., Global tomographic images of mantle plumes and subducting slabs: insight into deep Earth dynamics, *Phys. Earth Planet. Inter., 146,* 3–34, 2004.

Seismic Evidence for Subduction-Transported Water in the Lower Mantle

Jesse F. Lawrence

IGPP, Scripps Institute of Oceanography, La Jolla, California, USA

Michael E. Wysession

Department of Earth and Planetary Sciences, Washington University, St. Louis, Missouri, USA

We use seismic attenuation tomography to identify a region at the top of the lower mantle that displays very high attenuation consistent with an elevated water content. Tomography inversions with >80,000 differential travel-time and attenuation measurements yield 3D whole-mantle models of shear velocity (V_S) and shear quality factor (Q_μ). The global attenuation pattern is dominated by the location of subducting lithosphere. The lowest Q_μ anomaly in the whole mantle is observed at the top of the lower mantle (660–1400 km depth) beneath eastern Asia. The anomaly occupies a large region overlying the high-Q_μ sheet-like features interpreted as subducted oceanic lithosphere. Seismic velocities decrease only slightly in this region, suggesting that water content best explains the anomaly. The subducting of Pacific oceanic lithosphere beneath eastern Asia likely remains cold enough to transport stable dense hydrous mineral phase D well into the lower mantle. We propose that the eventual decomposition of phase D due to increased temperature or pressure within the lower mantle floods the mantle with water, yielding a large low-Q_μ anomaly.

INTRODUCTION

The lower mantle accounts for 62% of Earth's volume, so even if lower mantle minerals have low H_2O solubility (< 0.1 wt%), the lower mantle may contain more water than the Earth's oceans. How much water the lower mantle can hold is a subject of some debate, and considerable discussion in this volume. Estimates for the water solubility in silicate perovskite range from a few ppm [*Bolfan-Casanova et al.*, 2003] up to 0.2–0.4 wt% (i.e., 2000–4000 ppm H_2O by weight) [*Litasov et al.*, 2003; *Murakami et al.*, 2002]. The water solubility for magnesiowüstite (~16% of the lower mantle) is more uncertain, with measurements from zero [*Bolfan-Casanova et al.*, 2002] to 0.2 wt% H_2O [*Murakami et al.*, 2002]. For lower mantle solubilities of 0.05 to 0.2 wt%, the lower mantle may hold 1 to 5 times the water in the Earth's oceans.

While solubility provides an upper limit on possible water concentration, the actual concentration of the lower mantle remains uncertain. For water to reside in the lower mantle, water must either exist in primitive reservoirs or circulate back into the lower mantle. Because partial melts likely transport H_2O from the lower mantle into the transition zone [e.g., *Bercovici and Karato*, 2003] the lower mantle should be relatively dry without replenishment. The subduction of cold lithosphere containing high-pressure hydrous phases, like phase D, provides a mechanism to re-hydrate the lower mantle.

Earth's Deep Water Cycle
Geophysical Monograph Series 168
Copyright 2006 by the American Geophysical Union.
10.1029/168GM19

Current tomographic models of the mantle show narrow sheets of high velocity material descending continuously from subduction zones down deep into the lower mantle, with no indication of a thermal or chemical boundary layer at the 660-km discontinuity [*Antolik et al.*, 2004; *Grand*, 2002; *Karason and van der Hilst*, 2001; *Masters et al.*, 2000; *Megnin and Romanowicz*, 2000; *Ritsema and van Heijst*, 2000]. Whether slabs penetrate to the core-mantle boundary or founder within the lower mantle, subduction clearly transports large volumes of material down into the lower mantle.

Recent mineral physics work suggests that significant volumes of water may also be transported into the lower mantle. Large concentrations of water ($\gg 0.05$ wt%) can exist within the oceanic crust and lithosphere in common low-pressure hydrous silicate phases like serpentine. As the lithosphere subducts into the top of the upper mantle, serpentine becomes unstable, but not all of its water is released: some water continues downward within the slab, locked in a suite of hydrous phases that are stable at different depths [*Shieh et al.*, 1998; *Angel et al.*, 2001]. In the lower mantle, large amounts of water can be stored in the cold portions of subducted lithosphere by the presence of phase D, which can hold 10 wt % water [*Ohtani et al.*, 1997]. When the subducted slab reaches a depth of 1200–1400 km [*Shieh et al.*, 1998], phase D decomposes and H_2O is released into the lower mantle, to be absorbed by Mg-perovskite and magnesiowüstite, potentially filling the top of the lower mantle with vast quantities of water.

Difficulties arise in estimating how much water reaches the lower mantle because we do not know how much water is initially stored in hydrous phases nor how much leaves the subducting slab at shallower depths. We have a fairly good idea of how much water comes out of the slab in shallow subduction environments, because this water comes back out of arc volcanoes. *Peacock* [1990] estimates that 8.7×10^{11} kg/year of water enters subduction zones in sediments and ocean crust. Some amount of water may also be stored in the peridotite layer of subducting lithosphere. *Peacock* [1990] estimates that only 2×10^{11} kg/year of water degasses at arc and mid-ocean ridge volcanoes, suggesting that there is a net flux of water into the mantle at subduction zone regions.

It is unclear how much water is released and reabsorbed by other hydrous phases in a slab as it passes through the instability depths of the various hydrous phases. Water may preferentially leave the warm portions of a slab at the bottom of the transition zone depending on temperature and degree of slab stagnation in this layer [*Komabayashi et al.*, 2004]. In contrast, a cold slab may cool and viscously entrain water-rich transition zone minerals down into the lower mantle, creating another mechanism for bringing H_2O into the lower

mantle [*Bercovici and Karato*, 2003]. The lower mantle can hold oceans of water even if the average water concentration is low (< 0.05 wt%) and there is a good mechanism for water to reach the lower mantle. But we don't know how much water, if any, is actually down there. We try to address these issues here by providing new seismic evidence that sheds some light on the fate of water in the lower mantle.

SEISMIC ATTENUATION

Seismic imaging provides our best means of examining the actual state of Earth's deep interior. This imaging is usually done with seismic velocities, showing lateral variations of a few percent within the mantle. In this study we use seismic attenuation, which ideally is a measure of energy loss per cycle of a seismic wave. Although seismic wave propagation relies on the elastic behavior of rock and fluids, the Earth does not behave perfectly elastically, and seismic energy is continuously lost to friction. This frictional anelastic energy loss, or attenuation, occurs more rapidly for higher-frequency than lower-frequency waves, which gives highly-attenuated seismic signals a smoothed and long-period appearance. The rate at which the seismic energy is damped out is expressed using the seismic quality factor Q (often conceptualized as the number of cycles it takes a dampened oscillating wave to reach $e^{-\pi}$, or ~4 %, of its original amplitude). High values of Q represent highly-elastic materials with slow rates of energy loss, so attenuation is often quantified in terms of the reciprocal of the quality factor, Q^{-1}.

Although seismic attenuation can have several different causes, including reflective scattering, focusing and defocusing, and geometric spreading, the primary cause of seismic attenuation in the mantle is considered to be anelasticity. This frictional energy loss is thought to occur over a wide range of scales from atomic to that of grain-boundary deformational processes, so attenuation is often associated with the same processes as creep, and Q is related to viscosity. This means that attenuation can be used to assess different rheologic properties than those observed with seismic velocities. Lateral variations in Q^{-1} are often considered to result from temperature variations [e.g., *Romanowicz and Durek*, 2000] because Q^{-1} increases exponentially with temperature for most silicate materials [*Jackson*, 2000].

Relevant to this monograph, the presence of water can cause attenuation to increase drastically and viscosity decrease by several orders of magnitude [*Karato*, 2003]. For olivine grain boundary diffusion creep, *Mei and Kohlstedt* [2000] found that strain rate is proportional to water fugacity (i.e. OH concentration) approximately to the first power. It is proposed that the reason that the asthenosphere has unusually high attenuation and low viscosity is because it is *not* partially melted, and is therefore water-rich in comparison to the lithosphere [*Hirth and Kohlstedt*, 1996; *Karato and Jung*, 1998]. The role of water is not a simple one, however. Because an increase in water fugacity can affect relative changes in the resistance to dislocation motion and also increase the mobility of grain boundaries, the addition of water can change the fabric of silicates, even creating aniso-tropic textures [*Katayama et al.*, 2004].

Very little experimental work exists on quantifying the effect of increased water content on the seismic attenuation of significant mantle phases. Work has been done to examine the effect of water on seismic velocities, but the results are complicated. It is generally considered that seismic velocities of mantle silicates decrease slightly when water is added to their structure [*Inoue et al.*, 1998; *Crichton and Ross*, 2000]. For example, altering dry wadsleyite to incorporate 3.4 wt% H_2O results in a bulk sound velocity decrease of 2.8%. In the transition zone, where very high solubilities are pos-sible, the decrease in velocity due to water saturation can be even greater than that expected from temperature increases [*Smyth et al.*, 2004; *Jacobsen et al.*, 2004].

Even less is known about the effects of water on seis-mic *attenuation*. However, it is likely that an increase in water content causes a large increase in seismic attenuation [*Karato*, 2003, 2005]. This has been found experimentally [*Jackson et al.*, 1992], and is expected theoretically [*Karato*, 2005]. In the latter, the attenuation (Q_μ^{-1}) is proportional to $C_w^{\alpha q}$, where C_w is the water concentration, $q = 1$ for disloca-tion mechanisms, $q = 2$ for grain-boundary mechanisms, and $\alpha = 0.3$ for the lower mantle. This suggests that a drop of seismic Q_μ in the lower mantle from a value of 300 to 100 could be explained by an increase in water content on the order of a factor of 10 or so. If ambient water concentrations in the lower mantle were initially small ($<< 0.1$ wt% H_2O), this kind of increase in attenuation could be expected for even modest increases in water content.

EXPECTATIONS FROM RADIAL MODELS

Radial profiles of lower mantle Q_μ, from *Lawrence and Wysession* [2006a] do not show evidence of a significant amount of water in the lower mantle. They found that the transition zone has an average Q_μ of 276, increasing to 325 at the top of the lower mantle. The mean attenuation in the uppermost lower mantle is roughly constant as a function of depth, and decreases slightly in the mid-lower mantle, with a Q_μ value of 287 at a depth of 1500 km. The low attenuation at the top of the lower mantle suggests that the region should be relatively dry. However, the anelastic behavior of lower mantle phases is poorly understood, so there is no baseline by which to determine this.

3D ATTENUATION OBSERVATIONS

Three-dimensional modeling may show otherwise. Plate 1 shows a series of horizontal slices through a global 3D model (VQM3DA) of the mantle shear wave attenuation and veloc-ity from *Lawrence and Wysession* [2006b]. This 12-layer model, with equal-area 5-degree blocks, is an inversion using an LSQR algorithm of over 80,000 seismic shear-wave measurements recorded globally for 898 earthquakes during 1990-2002. For a given seismogram, differential travel-time and attenuation measurements are computed between the *S* wave (the first shear-wave arrival) and later shear-wave arrivals (*S, sS, ScS, sScS, SS, sSS*). As shown in Plate 1, there are enormous lateral variations in lower mantle attenuation, and while the top part of the lower mantle has generally low attenuation, there are regions of very high attenuation. The most significant of these is a broad low-Q_μ anomaly beneath eastern Asia that extends nearly from the equator to the pole. It is striking that this anomaly lies directly above the high-*Q* and high-velocity anomaly associated with the subducted Pacific oceanic lithosphere in the lower mantle.

This low-Q_μ anomaly beneath eastern Asia extends from about 700 km to 1400 km in depth, and reaches its minimum value of $Q_\mu = 95$ at about 1000 km depth (Plate 2). This minimum occurs beneath northern China, just northwest of Beijing. The mean value of Q_μ is 311 at this depth with a maximum Q_μ of 367 within the western Pacific subducted lithosphere. Within the "China anomaly" seismic attenua-tion is almost as high as the mean asthenosphere value, even though the mean attenuation at that depth is much lower. Plate 2 shows a checkerboard retrieval test for this depth, demonstrating that there is ample resolution within the data to retrieve a large anomaly of this size and location.

The presence of subduction totally alters the attenuation structure of the mantle, as seen in the vertical great-circle cross-section of Plate 2, which shows relative variations in attenuation as a function of depth. Along the southern part of the cross-section there are higher-than-average Q_μ val-ues at the top of the lower mantle. We consider these to be representative of typical upper lower mantle. At the bottom of the lower mantle along the southern part of this cross-section we see lower-than-average Q_μ values, and at the top of the mantle, the path beneath young oceanic lithosphere has lower-than-average Q_μ values, which is expected due to relatively high temperatures.

Along the northern part of this great-circle cross-section the Q_μ structure is very different. The largely continental path appears as high-Q_μ regions at the top of the mantle. The base of the mantle, which has received cold subducted lithosphere throughout the Mesozoic and Cenozoic Eras, has anomalously high-Q_μ values. At the top of the lower mantle,

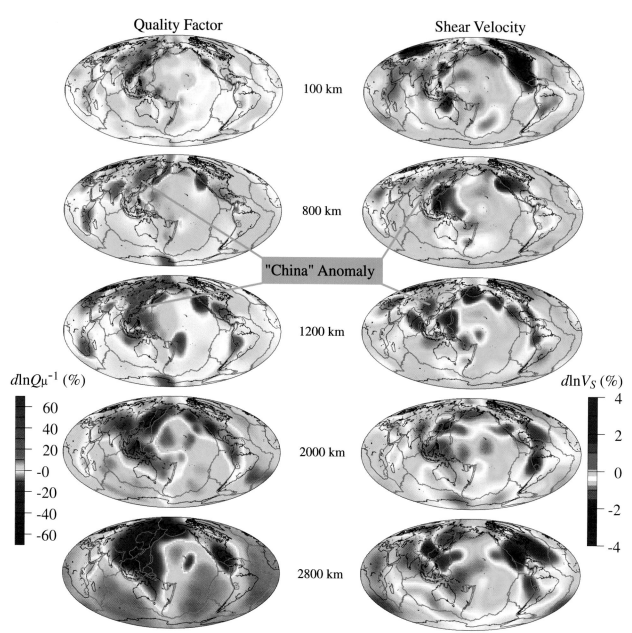

Plate 1. Six horizontal cross-sections through the quality factor (Q_μ) and shear velocity models of *Lawrence and Wysession* [2006b] showing large 3-D variation in the whole mantle. Blue indicates elevated Q_μ and velocity while red indicates low Q_μ and velocity. In the upper mantle the highest Q_μ is associated with dehydration and subduction processes. Deeper in the mantle (700–1400 km) a sizeable attenuation anomaly is visible beneath eastern Asia (a reduction of Q_μ from the mean value of around 300 to a value of around 100). High Q_μ and velocity form a ring around the Pacific in the lower mantle.

above the subducted lithosphere, is a long low-Q_μ anomaly that extends from beneath Southeast Asia around the Pacific Rim to Alaska, and is also seen beneath North America. Using a 5% decrease in Q_μ as the delineation of the China anomaly, its volume is 1.8×10^{10} km^3. If it were to contain 0.1% water, then the mass of water would be slightly greater than the amount of water in the Arctic Ocean.

DISCUSSION AND CONCLUSIONS

The lower-mantle high-attenuation anomaly beneath eastern Asia is very unusual, so it warrants a discussion of what could cause it, as well as if the anomaly is real and required by the data. Seismic attenuation is a difficult parameter to use, and attenuation tomography is in its relative infancy, yet there is good evidence that this signal is real. All of the techniques in the study are well-established, including both the methods used to obtain the differential attenuation measurements [*Flanagan and Wiens*, 1990; *Bhattacharyya et al.*, 1996] and the LSQR tomographic inversion method [*Paige and Saunders*, 1982; *Nolet*, 1987; *Masters et al.*, 1996]. The region beneath Asia also has some of the best data coverage of the mantle, and checkerboard resolution tests (Plate 2) show that we have good resolution for structures smaller-scale than this anomaly.

However, checkerboard resolution tests only show what is computationally possible from the inversion. A complication arises from the fact that seismic velocity anomalies will cause ray paths to anomalously bend (focus and defocus), which changes wave amplitudes and might mimic attenuation anomalies [*Ritsema et al.*, 2002]. However, the effect is hard to isolate because velocity models are highly contaminated by attenuation effects to begin with. The degree to which elastic effects affect attenuation will not be known until extensive modeling with fully 3-D synthetic seismograms can be done, which is computationally prohibitive at this point. Because of these concerns, we keep our model fairly low-order and do not attempt to interpret small-scale anomalies. In this context, we feel that the attenuation observed here is not due to velocity variations because (1) the lower-mantle attenuation anomalies are so much larger than the velocity anomalies, and (2) where there is overlap with existing attenuation models using other data and methods, agreement is very good.

While there are no other global models for the lower mantle that we can compare with, there are upper mantle models. Plate 3 shows our attenuation model compared with the attenuation model QRLW8 of *Gung and Romanowicz* [2002] at a depth of 300 km. The *Gung and Romanowicz* [2002] model is inverted from surface waves, and therefore has its best resolution closer to the surface because only long-period

surface waves sample deeply into the upper mantle. In contrast, our model has better coverage and resolution in the lower mantle, but we still have strong signal retrievability in the upper mantle. It is clear from Plate 3 that these models are very similar, sharing the general trend that old continents have low-attenuation deep keels and ocean regions at this depth have high attenuation. Many of the smaller-scale anomalies are common to both models as well. To have two different models using two very different datasets and inversion methods give similar results is a reassuring measure of resolvability. Note that the effects of elasticity would be very different for both models, suggesting that effects of elastic focusing do not dominate. As 3D heterogeneity is generally weaker in the lower mantle [e.g., *Masters et al.*, 1996] than in the upper mantle, scattering and multipathing have weaker effects here, suggesting that the lower mantle values are even less contaminated than in the upper mantle, where our model is consistent with other results.

Assuming that the China anomaly is real, and not an artifact of elastic focusing effects, the next question is whether it is a measure of anelasticity. As *Cormier* [2000] showed, scattering from small-scale heterogeneities can in some cases cause pulse broadening that has a similar appearance to the effects of anelasticity. Very little is known about the possible existence of very-small-scale heterogeneity in the lower mantle. *Kaneshima and Helffrich* [1998] found evidence of mantle scattering northeast of the Mariana subduction zone from *S*-to-*P* conversions, and interpreted them as chemically-distinct reservoirs. *Kaneshima and Helffrich* [1999] and *Castle and van der Hilst* [2002] postulated that these reflections, at a depth of about 1600 km, were due to fragments of ancient subducted oceanic crust. However, *Helffrich* [2002] estimated that these scatterers may be on the order of 8 km in diameter, and *Brana and Helffrich* [2004] identified a region at the base of the mantle with scatterers they estimated to be less than a kilometer in diameter. The presence of scatterers in the mantle is also consistent with *Hedlin et al.* [1997], who used *PKP* precursors to suggest that small-scale scatterers were distributed throughout the mantle.

The possibility of seismic scatterers causing the China anomaly cannot be ignored. However, we do not think it is likely. The effect is not expected to be so large as to reduce Q_μ from 300 to 100. It would also be difficult for the scattering to mimic the constant-Q_μ behavior over the very wide frequency band of 0.01–0.1 Hz, which is what we observe in the slopes of the spectral quotients of our differential phases. We need also to explain the location of the scatterers, lying above the subducted lithosphere in the lower mantle. One hypothesis is that the former ocean crust could break into many small pieces and delaminate from the rest of the subducting lithosphere. However, this is problematic because the ocean crust is rich in garnet, which is much stiffer than the surrounding mantle [*Karato*, 1997], and the effects of seismic scattering would be countered by a decrease in anelasticity. Advances in theoretical seismology need to be made in order to adequately measure the seismic attenuation of body phase resulting from various distributions of elastic scatterers within the mantle.

If the China anomaly is the result of anelasticity, then there are several different possible causes for it. One is grain size. Experiments show that smaller grain sizes cause increased seismic attenuation [*Jackson et al.*, 2001] due to increased surface-to-volume ratios of smaller grains, which promotes frictional heat loss through grain boundary deformation. Small grains can also align into microfractures, are localizations for high levels of strain accumulation, and result in increased seismic attenuation [*Cooper*, 2002].

The region just above subducted lithosphere at the top of the lower mantle might have smaller grain sizes if rock is brought down through viscous drag along with the subducted lithosphere. Transition zone phases will convert to perovskite and periclase as they pass into the lower mantle—more rapidly for the $(Mg,Fe)_2SiO_4$ system (where spinel converts to perovskite and magnesiowüstite over a narrow pressure range), more slowly for the $(Mg,Fe)SiO_3$ system (where the garnet transition to perovskite and magnesiowüstite is quite gradual). New perovskite and periclase crystals will nucleate, and for some distance into the lower mantle, there might be a smaller mean grain size. Because Q_μ may be proportional to the square root of the grain size [*Cooper*, 2002], reducing the grain size by a factor of 10 could provide the reduction in Q_μ by a factor of 3.

However, if viscous drag were bringing surrounding transition zone material into the lower mantle, this should happen on both sides of the slab. The low-Q anomaly is only observed above the subducted lithosphere, not beneath it. Also, if the anomaly were due to grain size, other slabs entering the lower mantle would show evidence of low-Q anomalies surrounding them, but this is not observed. It is also hard to drag such a large volume of material as this down into the lower mantle with the subducting lithosphere.

While temperature is the usual prime suspect in interpreting seismic attenuation anomalies, especially high-attenuation anomalies, temperature is an unlikely cause in this case. Ongoing subduction for > 200 Ma has transported an enormous volume of cold oceanic lithosphere into the mantle beneath eastern Asia. Even though the thermal conductivity of rock is exceedingly low, and so might not be an efficient way of chilling the lower mantle, it is hard to conceive that adding cold material would warm the lower mantle. A Q_μ value of less than 100 is on par with asthenosphere values, and much lower than the low-Q_μ anomalies associated with

the megaplumes beneath the Pacific and African plates, which appear in our model as only broad slightly-low-Q_μ features. There is no known evidence of plume-like structures in the area of the China anomaly from seismic tomographic studies [*Montelli et al.*, 2004].

If the anomaly were due to increased temperatures, there would also be a very large low-velocity anomaly. This is suggested in Figure 1, which is an extrapolation based upon the calculations of *Karato* [1993] describing variations in Q and V as functions of water and temperature. While these relations are intended for upper mantle minerals and pressures, the general trend should hold true for the lower mantle. An increase in Q^{-1} of 200% here (a decrease in Q from 300 to 100) requires a temperature increase of ~450°C. However, such a large temperature increase would also cause a decrease in seismic shear velocity of ~4%, which is clearly not observed in any seismic velocity model in this area.

The remaining candidate for this low-Q_μ anomaly is an elevated concentration of water. The key factor to the transport of water into the lower mantle via slab subduction is maintaining a low slab temperature (< 1000°C) [*Komabayashi et al.*, 2004]. As the subducting lithosphere sinks into the mantle and heats up, it dehydrates and loses its water into the overlying mantle. If the subducting slab in the lower mantle is cold enough for hydrous phase D to remain stable,

Velocity and Quality Factor Variations

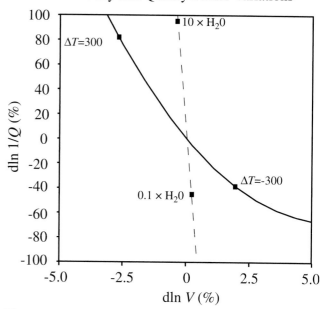

Figure 1. Plot showing the changes in temperature and water concentration that would be needed to give a 200% increase in attenuation, based upon *Karato* [1993]. Note that the increase in temperature is expected to also cause a large velocity decrease, which is not observed in the data.

then water can remain within its cold, central core down to depths of 1200–1400 km according to *Shieh et al.* [1998], who observed the breakdown of phase D above 42 GPA at 1550°C and above 46 GPa at 1320°C. At pressures closer to a depth of 660 km *Kawamoto* [2004] found the breakdown to be at around 1200°C. The slabs sinking beneath the western Pacific are very cold, and have some of the highest thermal parameter values of any subduction zone (product of the slab age, plate velocity, and the sine of the dip angle) [*Kirby et al.*, 1996], so they are efficient at bringing cold material deep into the lower mantle, perhaps down to 1400 km depth.

There need not be a large amount of phase D present in the subducting lithosphere to cause a large concentration of water at the top of the lower mantle. The rapid rate and long duration of subduction beneath the western Pacific would provide sufficient flooding of water in the lower mantle to account for the observed anomaly. The relative difference in water saturation between the anomalous region and the surrounding mantle is more important than the absolute amount. For example, an increase in the amount of water in the lower mantle by a factor of 10–20 might be sufficient to cause the decrease in Q_μ from 300 to 100 [*Karato*, 2005]. Interestingly, this change in water might not cause a correspondingly large change in seismic velocity, as seen in Figure 1. This is important because as Plate 1 shows there is only a slight corresponding negative velocity anomaly in the region of the large attenuation anomaly.

The model we suggest is shown in Plate 4. Water is brought into the lower mantle via the subducting Pacific lithosphere, which is old, cold and sinking fast enough for hydrous phase D to be stable to depths of at least 1400 km. As the slab gradually warms, phase D becomes less stable and the water is released. Because the entire slab does not cool instantaneously, the water is released over a broad range of depths. Our attenuation model suggests that there is a large difference in water concentration between the China anomaly and the ambient upper lower mantle, but this can occur in two ways: 1) It is possible that the top of the lower mantle is normally quite dry (<< 0.1 wt% H_2O), with water contents well below the maximum saturation (on the order of 0.1 wt% H_2O). The regional flooding of the lower mantle might still be below saturation levels, but the elevated levels would be enough to cause the large attenuation increase. 2) The ambient lower mantle might normally contain higher H_2O concentrations, but the lower-mantle region above the subducted Pacific slab is over-saturated.

This second scenario would require that fluids have a low diffusion rate; otherwise they might rapidly escape up into the transition zone, which has a much higher water capacity. Saturation would likely result in partial melting, which could have a variety of effects on seismic velocity and Q

depending on the geotherm, type of melting, and geometry of flow [*Karato and Jung*, 1998]. The lack of a strong velocity anomaly suggests that there is likely not much melt present, so the anomaly is not over-saturated.

In either case, the flow rates of water are an important, and as of yet unconstrained factor. Laboratory experiments suggest that water might be stable in a solid ice state at deeper lower mantle pressures [*Lin et al.*, 2005], but this has not been experimentally observed for upper lower mantle conditions. The low-Q_μ anomaly is distributed over a large volume. So either fluid transport (most likely along grain boundaries) or advection of mantle rock (facilitated by decreased viscosities) would be needed to distribute the water efficiently. Such increased water levels would have enormous, if poorly understood, implications for convection due to the strong dependencies of viscosity upon water. Water content usually lowers a rock's melting point and weakens it, decreasing its viscosity [*Hirth and Kohlstedt*, 1996]. However, there are also suggestions that an increase in water might have the opposite effect, actually causing the rock to stiffen while still displaying an increased seismic attenuation [*Karato*, 2005]. If the rock were to stiffen, high fluid flow rates would be required to explain the broad distribution of water.

Another unanswered question regarding the fate of dehydrated water is whether water in the lower mantle would interact with the transition zone, and if so, then how? The low-Q_μ anomaly has a sharp, roughly flat upper boundary that coincides with the 660-km discontinuity. Either the water concentration would not increase in this region of the transition zone, or the rise in water concentration is small relative to the average transition zone water concentration. The attenuation in the transition zone seems to be low [e.g., *Durek and Ekstrom*, 1996], suggesting that added water there has little effect on attenuation. This makes sense because the attenuation is a measure of the instability of a material, and the transition zone is perfectly stable containing large quantities of water. It is also possible that water may pass up into the transition zone, but that some of that rock may be advected back down into the lower mantle adjacent to the subducting lithosphere, creating a circulation vortex that may trap water at the region above and below the 660-km discontinuity in this region.

As with many geological processes, there may be multiple causes influencing the anomaly. The low-Q_μ anomaly could involve the presence of water combined with small grain sizes due to a phase change to perovskite/periclase and also combined with the presence of scattering chemical heterogeneities associated with fragments of post-eclogitic ocean crust. Yet, as with dehydration in the upper mantle wedge [*Zhao et al.*, 1992], it seems that dehydration could

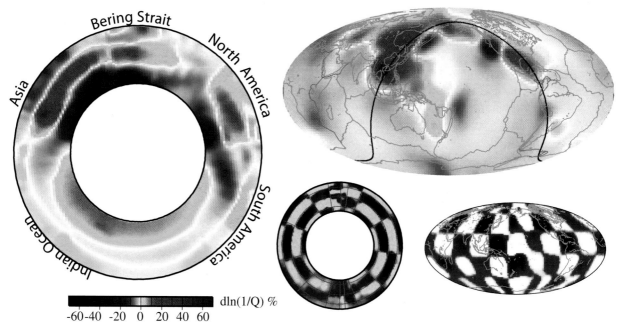

Plate 2. Great-circle vertical cross-section and horizontal cross-section at a depth of 1000 km through the attenuation model of *Lawrence and Wysession* [2006b], with accompanying checkerboard resolution test images. There is a long high-attenuation (low-Q_μ) anomaly at the top of the lower mantle just above the locations of the subducted Pacific ocean lithosphere along the western and northern rim of the Pacific.

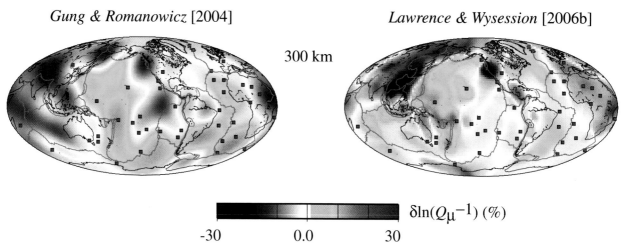

Plate 3. Comparison of the body shear wave attenuation model of *Lawrence and Wysession* [2006b] compared with the surface wave attenuation model QRLW8 of *Gung and Romanowicz* [2002] at a depth of 300 km. These models are constructed with different data types and different inversion methods, which lends support to the reality of the shared anomalies.

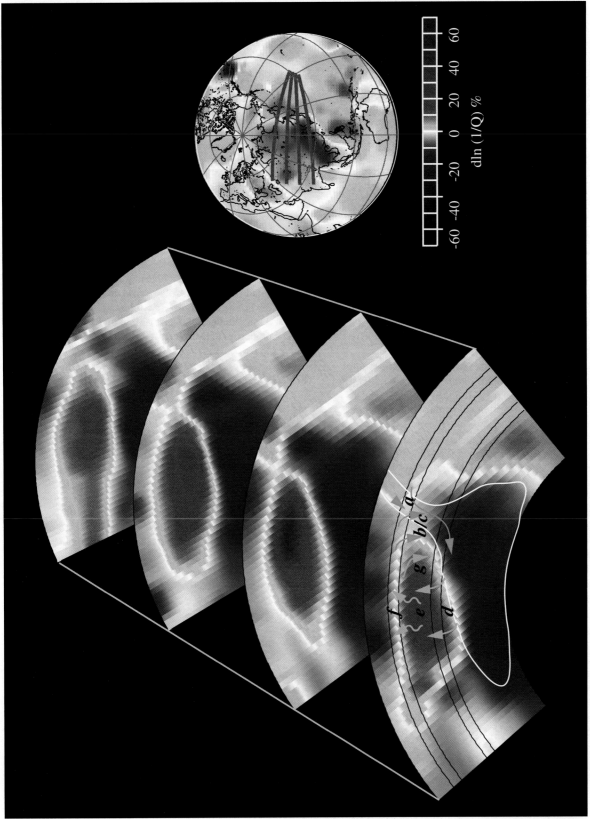

Plate 4. Tomographic slices through the high-attenuation anomaly, with some possible flow paths for the water that could be entering the lower mantle via subduction of oceanic lithosphere. a) Water can enter the deep mantle in hydrous phases contained within a cold slab. b) Hydrous transition zone minerals may be viscously entrained along with c) a cold slab containing hydrous phase D. d) at 1100 to 1400 km depth hydrous phase D may break down releasing water into the lower mantle. e) This water may percolate or diffuse through the lower mantle and possibly f) back up into the transition zone. g) Alternatively, water may continue to circulate through the lower mantle.

play a significant role in the formation of this anomaly. Imaging the presence of water in the deep mantle is a relatively recent endeavor [*van der Meijde et al.*, 2003], and the interpretation of our seismic attenuation anomalies as the result of high water content remains highly speculative, but one that could prove to be very important in understanding the mechanisms of convection within planets.

Acknowledgments. We thank IRIS for providing the digital seismic data. We also thank Guy Masters for supplying subroutines useful for global tomography. We thank S. van der Lee, S. Jacobsen and an anonymous reviewer for very helpful comments. This project was supported through funding by NSF-EAR-0207751 and NSF-EAR-0544731.

REFERENCES

Angel, R. J., D. J. Frost, N. L. Ross, and R. Hemley (2001), Stabilities and equations of state of dense hydrous magnesium silicates, *Phys. Earth Planet Inter,. 127*, 181–196.

Antolik, M., Y. J. Gu, G. Ekstrom, and A. M. Dziewonski (2004), J362D28: a new joint model of compressional and shear velocity in the Earth's mantle, *Geophys. J. Int., 153*, 443–466.

Bercovici, E., and S.-I. Karato (2003), Whole-mantle convection and the transition –zone water filter, *Nature, 425*, 39–44, 2003.

Bhattacharyya, J., G. Masters, and P. M. Shearer (1996), Global lateral variations of shear wave attenuation in the upper mantle, *J. Geophys. Res, 101*, 22,273–22,289.

Bolfan-Casanova, N., S. Mackwell, H. Kappler, C. McCammon, and D. C. Rubie (2002), Pressure dependence of H solubility in magnesiowüstite up to 25 GPa: Implications for the storage of water in the Earth's lower mantle, *Geophys. Res. Lett., 29*, 89–92.

Bolfan-Casanova, N., H. Kappler, and D. C. Rubie (2003), Water partitioning at 660 km depth and evidence for very low water solubility in magnesium silicate perovskite, *Geophys. Res. Lett.*, 1905, doi:10.1029/2003GL017182.

Brana, L., and G. Helffrich (2004), A scattering region near the core-mantle boundary under the North Atlantic, *Geophys. J. Int., 158*, 625–636.

Castle, J. C., and R. D. van der Hilst (2003), Searching for seismic scattering off mantle interfaces between 800 km and 2000 km depth, *J. Geophys. Res., 108*, pp. ESE 13-1, DOI 10.1029/2001JB000286.

Cooper, R. F. (2002), Seismic wave attenuation: Energy dissipation in viscoelastic crystalline solids, in *Plastic Deformation of Mineral and Rocks*, S.-i. Karato and H.-R. Wenk, eds., *Rev. Mineral. Geochem., 51*, 253–290.

Cormier, V. F. (2000), D" as a transition in the heterogeneity spectrum of the lowermost mantle, *J. Geophys. Res., 105*, 16,193–16,205.

Crighton, W. A., and N. L. Ross (2002), Equation of state of dense hydrous magnesium silicate phase A, $Mg_7Si_2O_8(OH)_6$, *Amer. Mineralogist, 87*, 333–338.

Durek, J. J. and G. Ekström (1996), A radial model of anelasticity consistent with long period surface wave attenuation, Bull. Seism. Soc. Am., 86 144–158.

Flanagan, M. P., and D. A. Wiens (1990), Attenuation structure beneath the Lau back-arc spreading center from teleseismic S phases, *Geophys. Res. Lett., 17*, 2117–2120.

Grand, S. P. (2002), Mantle shear-wave tomography and the fate of subducted slabs, *Phil. Trans., Math. Phys. Eng. Sci., 360*, 2475–2491.

Gung, Y. and B. Romanowicz (2002), Q tomography of the upper mantle using three-component long-period waveforms, *Geophys. J. Int., 157(2)*, 813–830.

Hedlin, M. A. H., P. M. Shearer, and P. S. Earle (1997), Seismic evidence for small-scale heterogeneity throughout the Earth's mantle, *Nature, 387*, 145–150.

Helffrich, G. (2002), Chemical and seismological constraints on mantle heterogeneity, *Phil. Trans.: Math., Phys. And Eng. Sci., 360*, 2493 – 2505.

Hirth G. and D. L. Kohlstedt (1996), Water in the oceanic upper mantle; implications for rheology, melt extraction and the evolution of the lithosphere, *Earth Planet. Sci. Lett., 144*, 93–108.

Inoue, T., D. J. Wiedner, P. A. Northrup, and J. B. Parise (1998), Elastic properties of hydrous ringwoodite (γ-phase) in Mg_2SiO_4, *Earth Planet. Sci. Lett., 160*, 107–113.

Jackson, I. (2000), Laboratory Measurements of Seismic Wave dispersion and Attenuation: Recent Progress. In *Earth's Deep Interior: Mineral Physics and Tomography from the Atomic to the Global Scale*, AGU Geophysicsl Monograph Service, vol.117, S. Karato et al (eds), pp. 265–289.

Jackson, I., M. S. Paterson, and J. D. Fitz Gerald (1992), Seismic wave attenuation in Aheim dunite: an experimental study, *Geophys. J. Int., 108*, 517–534.

Jackson, I., J. D. Fitz Gerald, U. H. Faul, and B. H. Tan (2001), Grain-size sensitive seismic wave attenuation in polycrystalline olivine, *J. Geophys. Res., 107*, 2360, doi: 10.1029/2001JB001225.

Jacobsen, S.D., J.R. Smyth, H. Spetzler, C.M. Holl, and D.J. Frost (2004), Sound velocities and elastic constants of iron-bearing hydrous ringwoodite, *Phys. Earth Planet. Inter., 143–144*, 47–56.

Kaneshima, S., and G. Helffrich (1998), Detection of lower mantle scatterers northeast of the Mariana subduction zone using short-period array data, *J. Geophys. Res., 103*, 4825–4838.

Kaneshima, S., and G. Helffrich (1999), Dipping Low-Velocity Layer in the Mid-Lower Mantle: Evidence for Geochemical Heterogeneity, *Science, 283*, 1888–1892.

Karason, H., and R. D. van der Hilst (2001), Tomographic imaging of the lowermost mantle with differential times of refracted and diffracted core phases (*PKP, Pdiff*), *J. Geophys. Res., 106*, 6569–6587.

Karato, S. (1993), Importance of anelasticity in the interpretation of seismic tomography, Geophys. Res. Lett., 20, 1623–1626.

Karato, S.-i (1997). On the separation of crustal component from subducted oceanic lithosphere near the 660-km discontinuity, *Phys. Earth Planet. Inter., 99*, 103–111.

Karato, S.-i. (2003), Mapping water content in the upper mantle, in *Inside the Subduction Factory,* J.M. Eiler ed., *Geophys. Monogr., 138,* 135–152.

Karato, S.-i. (2005), Microscopic models for the effects of hydrogen on physical and chemical properties of earth materials, In *Superplume: Beyond plate tectonics,* ed. by D. A. Yuen, S. Maruyama, S. Karato, and B. F. Windley, in press.

Karato, S.-i., and H. Jung (1998), Water, partial melting and the origin of the seismic low velocity and high attenuation zone in the upper mantle, *Earth. Planet. Sci. Lett., 157,* 193–207.

Katayama, I., Jung, H., Jiang, Z. and Karato, S., 2004. A new olivine fabric under modest water content and low stress, Geology, 32, 1045–1048.

Kawamoto, T. (2004), Hydrous phase stability and partial melt chemistry in H_2O-saturated KLB-1 peridotite up to the uppermost lower mantle conditions, *Phys. Earth Planet. Int., 143,* 387–395.

Kirby, S. H., S. Stein, E. A. Okal, and D. C. Rubie (1996), Metastable mantle phase transformations and deep earthquakes in subducting oceanic lithosphere, *Rev. of Geophys., 34,* 261–306.

Komabayashi, T., S. Omori, and S. Maruyama (2004), Petrogenetic grid in the system MgO-SiO2-H2O up to 30 GPa, 1600{degree sign}C: applications to hydrous peridotite subducting into the Earth's deep interior, *J. Geophys. Res., 109,* B03206, doi:10.1029/2003JB002651.

Lawrence, J. F., and M. E. Wysession (2006a), QLM9: A new radial quality factor (Q) model for the mantle, *Earth Planet. Sci. Lett., 241,* 962–971.

Lawrence, J. F., and M. E. Wysession (2006b), A low-order whole-mantle model of seismic attenuation, *J. Geophys. Res.,* in preparation.

Litasov, K., E. Ohtani, F. Langenhorst, H. Yurimoto, T. Kubo, and T. Kondo (2003), Water solubility in Mg-perovskites and water storage capacity in the lower mantle, *Earth Planet. Sci. Lett., 211,* 189–203.

Lin, J.-F., E. Gregoryanz, V. V. Struzhkin, and M. Somayazulu (2005), Melting behavior of H2O at high pressures and temperatures, *Geophys. Res. Lett., 32,* L11306, doi:10.1029/2005GL022499.

Masters, G., S. Johnson, G. Laske, and H. Bolton (1996), A shear-velocity model of the mantle, *Phil. Trans. R. Soc. Lond. A, 354,* 1385–1411.

Masters, G., G. Laske, H. Bolton, and A. M. Dziewonski (2000), The relative behavior of shear velocity, bulk sound speed, and compressional velocity in the mantle: Implications for chemical and thermal structure, in *Earth's Deep Interior,* S. Karato, A. M. Forte, R. C. Liebermann, G. Masters and L. Stixrude, eds., AGU Monograph 117, AGU, Washington D.C..

Megnin, C., and B. Romanowicz (2000), The shear velocity structure of the mantle from the inversion of body, surface and higher modes waveforms, *Geophys. J. Int., 142,* 709–726.

Mei, S., and D. L. Kohlstedt (2000), Influence of water on plastic deformation of olivine aggregates 1. Diffusion creep regime, *J. Geophys. Res., 105,* 21457–21470.

Montelli, R., G. Nolet, F. A. Dahlen, G. Masters, E. R. Engdahl, and S.-H. Hung (2004), Finite frequency tomography reveals a variety of plumes in the mantle, *Science, 303,* 338–343.

Murakami, M., K. Hirose, H. Yurimoto, S. Nakashima, and N. Takafuji (2002), Water in Earth's lower mantle *Science, 295,* 1885–1887.

Nolet, G. (1987), Seismic wave propagation and seismic tomography, in *Seismic Tomography,* G. Nolet, ed., Reidel, Dordrecht, 1–23.

Ohtani, E. (2005), Water in the mantle, *Elements, 1,* 25–30.

Ohtani, E., H. Mitzobata, Y. Kudoh, and T. Nagase (1997), A new hydrous silicate, a water reservoir, in the upper part of the lower mantle, *Geophys. Res. Lett., 24,* 1047–1050.

Paige, C.C., and M. A. Saunders (1982), LSQR: An algorithm for sparse linear equations and sparse least squares, *TOMS, 8,* 43–71.

Peacock, S. (1990), Fluid processes in subduction zones, *Science, 248,* 329–337.

Ritsema, J., and H. J. Van Heijst (2000), Seismic imaging of structural heterogeneity in Earth's mantle: Evidence for large-scale mantle flow, *Science Progress, 83,* 243–259.

Ritsema, J., L. A. Rivera, D. Komatitsch, and J. Tromp (2002), Effects of crust and mantle heterogeneity on PP/P and SS/S amplitude ratios, *Geophys. Res. Lett., 29,* 72–75.

Romanowicz, B., and J. J. Durek (2000), Seismological constraints on attenuation in the earth: A review, in *Earth's Deep Interior,* AGU Geophysical Monograph, *117,* 265–289.

Shieh, S. R., H.-k. Mao, R. J. Hemley, and L. C. Ming (1998), Decomposition of phase D in the lower mantle and the fate of dense hydrous silicates in subducting slabs, *Earth Planet. Sci. Lett., 159,* 13–23.

Smyth, J. R., C. M. Holl, D. J. Frost, and S. D. Jacobsen (2004), High pressure crystal chemistry of hydrous ringwoodite and water in the Earth's interior, *Phys. Earth Planet. Int., 143,* 271–278.

Van der Meijde, M., F. Marone, D. Giardini, and S. Van der Lee (2003), Seismic evidence for water deep in Earth's upper mantle, *Science, 300,* 1556–1558.

Warren, L. and P. M. Shearer (2002), Mapping lateral variations in upper mantle attenuation by stacking *P* and 4 spectra, *J. Geophys. Res., 107,* doi:10.1029/2001JB001195.

Zhao, D. (2004), Global tomographic images of mantle plumes and subducting slabs: insight into deep Earth dynamics, *Phys. Earth Planet. Inter., 146,* 3–34.

Implications of Subduction Rehydration
for Earth's Deep Water Cycle

Lars Rüpke

*Physics of Geological Processes, University of Oslo, Oslo, Norway, and
SFB 574 Volatiles and Fluids in Subduction Zones, Kiel, Germany*

Jason Phipps Morgan

*Cornell University, Ithaca, New York, USA and SFB 574 Volatiles and
Fluids in Subduction Zones, Kiel, Germany*

Jacqueline Eaby Dixon

RSMAS/MGG, University of Miami, Miami, Florida, USA

The "standard model" for the genesis of the oceans is that they are exhalations from Earth's deep interior continually rinsed through surface rocks by the global hydrologic cycle. No general consensus exists, however, on the water distribution within the deeper mantle of the Earth. Recently *Dixon et al.* [2002] estimated water concentrations for some of the major mantle components and concluded that the most primitive (FOZO) are significantly wetter than the recycling associated EM or HIMU mantle components and the even drier depleted mantle source that melts to form MORB. These findings are in striking agreement with the results of numerical modeling of the global water cycle that are presented here. We find that the *Dixon et al.* [2002] results are consistent with a global water cycle model in which the oceans have formed by efficient outgassing of the mantle. Present-day depleted mantle will contain a small volume fraction of more primitive wet mantle in addition to drier recycling related enriched components. This scenario is consistent with the observation that hotspots with a FOZO-component in their source will make wetter basalts than hotspots whose mantle sources contain a larger fraction of EM and HIMU components.

1. INTRODUCTION

The presence of running water is one of the key distinctions between the Earth and other planets in the solar system. The obvious surface expression of this is the global ocean that covers some 71% of Earth's surface. Although the origin of the oceans is still under debate, current conventional wisdom is that they are exhalations from the Earth's interior (see [*Drake and Righter*, 2002] for a review). The 'hidden oceans' that may exist inside the Earth's deep interior remain, however, elusive and with them the question of water distribution inside the deeper Earth. Mineral physics tells us that upper mantle rocks, especially transition zone rocks, have the chemical ability to incorporate significant

Earth's Deep Water Cycle
Geophysical Monograph Series 168
Copyright 2006 by the American Geophysical Union.
10.1029/168GM20

amounts of water into their crystal structure [*Hirschmann, et al.*, 2005; *Ohtani*, 2005; *Smyth*, 1987; *Williams and Hemley*, 2001]—up to several oceans worth of water. Geochemical evidence seems to show a different and drier picture of the Earth's mantle. *Dixon et al.* [2002] have recently determined water concentrations in various OIB and MORB type basaltic rocks and found a maximum water concentration of only ~750-ppm for the most primitive source components (FOZO)—see Table 1 for an explanation of the used terminology. FOZO appears therefore to be wettest mantle component—wetter than average mantle and the even drier depleted MORB source.

Here we will briefly review the geochemical evidences before presenting the results of a numerical model for the chemical evolution of the Earth. This model puts some constraints on how mantle outgassing and potential regassing through plate subduction may have evolved through time.

1.1. Mantle Geochemistry

The sources of both mid-ocean ridge basalts (MORB) and ocean island basalts (OIB) contain several isotopically distinct source components (e.g. [*Zindler and Hart*, 1986]; see [*Hofmann*, 1997] for a review) which leave their fingerprints on observed basalts to which their melts contribute, with arrays of basalts produced at a given hotspot typically occupying a tubelike subregion in the isotope-space spanned by terrestrial basaltic volcanism [*Hart, et al.*, 1992; *Phipps Morgan*, 1999]. It is now generally agreed that at least several of the isotopically most extreme components contain material that was originally formed by melting and alteration processes near Earth's surface and then recycled by plate subduction back into the mantle to rise and melt beneath present-day hotspots and mid-ocean ridges [*Chase*, 1979; *Hofmann*, 1997; *Hofmann and White*, 1982]. Two

'Enriched mantle' source components EM1 and EM2 and the HIMU source component are believed to have formed in this way through the recycling of sediments and OIB for the EM components, and the recycling of the lower MORB crust for the HIMU component. Another mantle component called 'FOZO' appears to be linked to the melting of the most primitive, undifferentiated, and undegassed component of the mantle mélange. In particular, the partial melting of this component correlates with highest $^3He/^4He$ ratios in basalts erupted at ocean islands and mid-ocean ridges. Other, more incompatible trace element-poor and isotopically depleted components have also been inferred, in particular the prevalent depleted DMM component evident in many MORB.

Recently Dixon and coworkers have measured the water contents of the DMM, FOZO, EM1 mantle components and inferred the value for HIMU [*Dixon, et al.*, 2002]. During mantle melting beneath hotspots and ridges, water partitions from a source component into its melt with roughly the same affinity for the melt phase as the light rare-earth element Ce, i.e. it behaves as a moderately incompatible element during pressure-release melting. Even though both the EM and HIMU mantle components are much richer in incompatible elements than the mantle average, *Dixon et al.* [2002] found that the recycled EM and HIMU components have lower ratios of water to cerium (H_2O/Ce~100) than either typical MORB source (H_2O/Ce ~150–250) [*Michael*, 1995] or FOZO-influenced OIB (e.g. H_2O/Ce ~210 Pacific—South Atlantic, ~250 North Atlantic, ~190 Kilauea source, and up to 350 for the metasomatized source for the peripheral Hawaii alkalic basalts [*Dixon and Clague*, 2001]). This implies in terms of absolute water abundances that the FOZO source has an average of 750±210-ppm water [*Dixon, et al.*, 2002; *Simons, et al.*, 2002]. This value is consistent with the estimate by [*Asimow, et al.*, 2004] who give a value of 700-ppm for the Azores plume end-member and [*Nichols, et al.*, 2002] who calculated that

Table 1. Glossary of the used terminology.

mantle outgassing	water loss from the mantle to the atmosphere during mantle melting at mid-ocean ridges and hotspots.
mantle regassing / water recycling	water transport from the surface back into the mantle by plate subduction.
exosphere	oceans (hydrosphere) and continental crust
ppm	parts per million – by weight – is equivalent to mg/kg
MORB	mid-ocean ridge basalt is made by melting of asthenospheric mantle at mid-ocean ridges
OIB	ocean island basalt is produced by melting of plume mantle at hotspots.
FOZO	focal zone – a common geochemical mixing end-member that probably represents the most primitive mantle component; sometimes also called 'C' (or 'PHEM')
HIMU	mantle end-member that is characterized by a high $\mu=^{238}U/^{204}Pb$ ratio and that is associated with the recycling of subducting plates.
EM-1 / EM-2	enriched mantle – mantle end-members that are thought to result from the subduction of sediments (and sometimes from the recycling of OIB).
DMM	depleted MORB mantle

the enriched mantle end-member under Iceland has 770±150-ppm H_2O. All these estimates depend strongly on the assumed degrees of melting but regardless of assumptions it is safe to say that all recent estimates for water concentrations in the FOZO component are below 1000-ppm and that the estimate used here from [*Simons, et al.*, 2002] of 750-ppm is representative. The average source for mid-ocean ridge basalts has less water and contains about ~100-ppm while the source of the more depleted MORB has a water abundance of ~60-ppm. *Dixon et al.* [2002] concluded that the high H_2O/Ce in the FOZO source component cannot be derived from recycled oceanic crust, and that instead a significant amount of the water in the FOZO component must be juvenile, left over from planetary accretion. They also concluded that the EM and HIMU components lost most of their chemicallybound water, but not their incompatible rare-earth elements by subduction-induced slab-dehydration. These absolute water concentrations in mantle rocks and conclusions on preferential slab-dehydration of recycled sediments and ocean crust put some constraints on possible mantle evolution scenarios. Here we present results of numerical modeling of the Earth's global water cycle and explore which evolution scenarios for the mantle's hydration state are consistent with the geochemical data.

2. EARTH'S DEEP WATER CYCLE

We model the Earth's deep water cycle by assuming that the mantle's present hydration state represents the long-term product of the competing processes water outgassing at mid-ocean ridges and hotspots and mantle regassing at subduction zones. This scenario is illustrated in Fig. 1: oceanic plates become hydrated during their stay at the ocean floor. Upon subduction, most of this water is released and brought back to the surface by arc magmatism and fore-arc fluid venting. Some chemically bound water subducts past the region of sub-arc desiccation to rehydrate the deeper mantle. This deeply subducted water mixes efficiently with the convecting mantle. During mantle upwelling and melting at mid-ocean ridges and hotspots, water is outgassed from the mantle into the hydrosphere thereby 'closing' the global geologic water cycle.

This type of mantle evolution scenario depends on four principle input parameters: (1) the amount of subducted water, (2) the efficiency of water outgassing during mantle melting, (3) mantle overturn rates, and (4) the present-day amount of 'non-mantle' exosphere water, i.e. water in the oceans and in the continental crust.

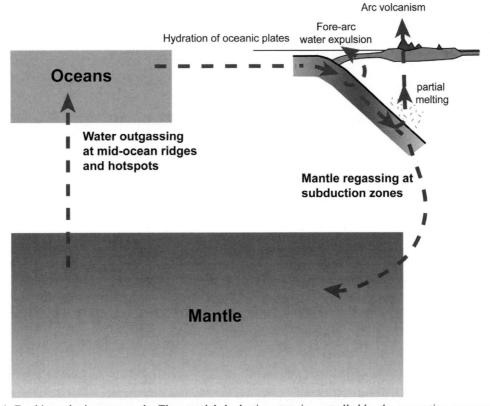

Figure 1. Earth's geologic water cycle. The mantle's hydration state is controlled by the competing processes water outgassing at ridges and hotspots and water regassing through plate subduction.

2.1. Hydration and Dehydration of Oceanic Plates

Oceanic plates undergo a variety of alteration processes during their lifetime on the ocean floor. Hydration starts during ocean crust formation at mid-ocean ridges. Fractures within the young ocean crust lead to seawater infiltration and hydrothermal alteration. After this initial hydration phase the oceanic crust experiences continuous low temperature alteration during its lifetime at the ocean floor. *Staudigel, et al.*, [1996] give an estimate of about 2.7wt.% water within the highly altered upper part of mature oceanic crust. The deeper parts of the crust contain less water and estimates range from almost zero to 1.0wt%. In addition to crustal hydration, sediments containing both chemically bound and pore water are deposited onto oceanic lithospheric plates. *Plank and Langmuir*, [1998] formulated an average sediment composition that they call GLOSS. GLOSS contains about 7wt% of chemically bound water. If we assume an average sediment thickness of 500m and that the 8km thick crust is made of one kilometer of highly altered basalt (2.7wt% H_2O), two kilometers of less hydrated (1wt% H_2O) and 5km of almost dry rocks we obtain a total water content of about 2.5×10^5kg below each square meter of ocean floor (Plate 1). Note that this number gives the total amount of water in crustal and sedimentary rocks in a column of ocean floor and is a conservative estimate; *Schmidt and Poli*, [1998] give an estimate of up to 6.0×10^5kg/m^2 but assume almost water saturated conditions and *Peacock's* [1990] estimate for water subduction rates translates into 3.35×10^5kg/m^2.

These two hydration processes, sediment deposition and crustal alteration, are now well established. However, recent geophysical observations indicate that in addition the cold lithospheric mantle may become hydrated during plate bending. *Ranero, et al.*, [2003] showed that normal faulting during plate bending may provide the pathways for seawater to reach and react with cold lithospheric mantle to make serpentine, i.e. to hydrate it. Heat flow studies [*Grevemeyer, et al.*, 2005] and refraction seismic experiments offshore Nicaragua seem to confirm this finding by showing reduced sub-Moho p-wave velocities consistent with 10–20% serpentinization [*Grevemeyer, et al.*, 2006]. Furthermore, oceanic lithospheric mantle may also become serpentinized during seawater infiltration during plate formation at mid-ocean ridges. In fact, *Marone et al.*, [2003] may have imaged a serpentinized sub-Moho mantle in the NE Atlantic Ocean. If we assume that slabs contain a 5km thick layer that is 10% serpentinized [*Schmidt and Poli*, 1998] we end up with another 2.1×10^5kg/m^2 H_2O in a column of lithospheric mantle. The total water content of slabs is therefore on the order of 4.6×10^5kg/m^2 chemically bound within (1) sediments, (2) crust, and (3) serpentinized mantle. This value can be seen as an average value for present-day water contents of subducting oceanic plates.

Key to understanding the role of plate subduction in the global geologic water cycle is the amount of water that may survive sub-arc water release to be recycled back into the deeper mantle. We have recently explored deep water recycling by plate subduction using a chemo-thermo-dynamical model [*Rüpke, et al.*, 2002; *Rüpke, et al.*, 2004]. We found that at the coldest present-day subduction zones up to one third of the subducted water may be transported into the deeper mantle. Young and hot subduction zones are characterized by very efficient plate dehydration so that almost no water is transported into the deeper mantle. An upper bound for the amount of water that is presently subducted into the deeper mantle is therefore about 1.6×10^5kg/m^2.

In order to understand which role subduction rehydration played for the chemical evolution of the mantle we have to constrain water recycling throughout Earth's history. As no consensus exists on when plate tectonics emerged and what past-subduction might have looked like, we have to explore a wide parameter range to constrain water recycling over geologic time scales. We assume that plate subduction has been sufficiently cold since 2Ga to recycle water into the deeper mantle and that the amount of water retained in slabs increased linearly from zero at the onset of 'cold' subduction to a given present-day value. This present-day value is assumed to vary by factors between 0 and 2 around the average value of 1.6×10^5kg/m^2 given above. As these estimates give total water retention per square meter of subducted ocean floor, the spreading (or subduction) rates must also be known though time in order to determine absolute recycling rates.

2.2. Convection, Mixing and Outgassing

To constrain the mantle flow pattern throughout Earth's history we assume whole mantle flow and follow *Phipps Morgan's*, [1998] approach of inferring mantle overturn rates from the changing radiogenic heat production during Earth's history. From this it follows that mantle overturn rates have exponentially decreased by a factor of about 25 during Earth's history (Fig. 2).

Furthermore, we assume that any water recycled back into the mantle at subduction zones is instantaneously mixed into the convecting mantle. The 'true' characteristic mixing time of the mantle is certainly higher—maybe even on the order of Ga [*Ten, et al.*, 1998]. We, however, neglect the complexity of mantle mixing and assume instantaneous mixing. Furthermore, we neglect possible preferential water storage in the mantle (i.e. the transition zone)—an assumption that is supported by numerical modeling. *Richard, et al.*, [2002]

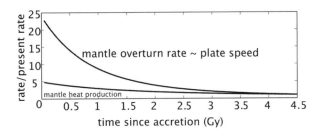

Figure 2. Mantle overturn rates through time. We assume a parameterized mantle convection scenario in which radiogenic heat production is balanced by convective cooling. This assumption yields that mantle overturn rates have decreased by a factor of about 25 through time (see [*Phipps Morgan*, 1998] for details).

showed that mantle water follows the convective flow pattern and that preferential water storage inside the mantle is unlikely.

Mantle water is outgassed during mantle melting at hotspots and mid-ocean ridges. The efficiency of water outgassing is controlled by the 'incompatibility' of water during mantle melting and the solubility of water in the produced melts. While it is clear that water partitions readily into the melt phase (see e.g. [*Asimow, et al.*, 2004] for D-values), the solubility of water is somewhat less well constraint. Water solubility in near-surface melts is very low so that it is efficiently outgassed during sub-aerial hotspot volcanism. At mid-ocean ridges, at 2500m water depth, water solubility in melts is higher. However, the MORB source is dry and depleted and may have experienced a primary melting event during plume melting [*Phipps Morgan and Morgan*, 1999]. It thus appears that 50% is a lower bound for water outgassing efficiency during mantle melting, i.e. half of the primary mantle water during melting at hostspots, mid-ocean ridges, and arcs is outgassed into the atmospheres. Our preferred values would be higher (>90%). However, since we do not have tight constraints on mantle outgassing efficiencies we systematically vary it between 30% and 99% in the model runs. Absolute water outgassing rates thus depend on the outgassing efficiency, the melting mantle's water content and on the assumed spreading or mantle overturn rates.

2.3. Water on the Earth's Surface

The global ocean is the primary liquid water reservoir on Earth's surface. Together with ground water, glaciers, atmospheric water, etc. we end up with a total of 1.46×10^{21} kg of surface water. While the amount of water in the oceans is relatively well constrained the amount of water chemically bound in the continental crust is not. Estimates of the present-day volume of continental crust give a value of about 7.8×10^9 km^3 [*Mooney, et al.*, 1998; *Nataf and Ricard*, 1996].

Assuming an average density of 2700kg/m^3 [*Stacey*, 1992] and an average water concentration of 1–3 wt.% [*Henderson*, 1986], we end up with another $(2.1–6.3) \times 10^{20}$kg of water in the continental crust. The total amount of 'non-mantle' exosphere water is therefore between 1.67×10^{21}kg and 2.09×10^{21}kg. As these numbers are again only estimates we systematically vary the Earth's present-day exosphere water content in the models in order to quantify which effects it has on the modeling results.

Now we have assembled all the ingredients needed to formulate the governing equations of the model:

$$\frac{\partial H_2O_{man}}{\partial t} = -f_{H_2O} * R(t) * H_2O_{man}(t) + S(t)$$

$$\frac{\partial H_2O_{exo}}{\partial t} = f_{H_2O} * R(t) * H_2O_{man}(t) - S(t)$$

$$R(t) = \frac{Q^2}{R_0}$$

$$S(t) = s(t) * s_0 * Q^2$$

$$H_2O_{man}(0) = H_2O_{initial}$$

$$H_2O_{exo}(end) = H_2O_{ocean}$$

Here $R(t) = Q^2(t)/R_0$ is the 'paleo-mantle-processing-rate' of the mantle, $R_0 = 9.8Ga$ is the present-day mantle overturn rate, $Q(t)$ is radiogenic heating through time normalized to the present-day value, and f_{H2O} is the water outgassing efficiency during mantle melting. $S(t)$ is the water recycling (mantle regassing) rate through plate subduction, $s(t)$ the recycling pre-factor, and $s_0 = 1.6 \times 10^5 kg/m^2 \times 2.7 \times 10^{15} m^2/Ga = 4.16 \times 10^{20} kg/Ga$ is the present-day water subduction rate with the second number being the present-day spreading rate. Please refer to [*Phipps Morgan*, 1998] for a discussion of the values chosen for the present-day spreading and mantle overturn rate. H_2O_{ocean} is the present-day water content of the exosphere as discussed above. Solving these equations yields a first order approximation of the Earth's deep water cycle.

3. A MANTLE EVOLUTION SCENARIO

3.1. The Mutual Effect of Model Parameters

Figure 3 shows six example calculations for the geologic water cycle through time. Every model is tuned to reproduce the present day conditions, i.e. the model reproduces the surficial water mass currently found in Earth's continental crust and oceans. As model parameters we assume that water recycling starts 2Ga ago and that the continental crust contains 2wt.% of water. To exemplify the effect of subduction rehydration in (a) a recycling pre-factor of 0.5 is assumed

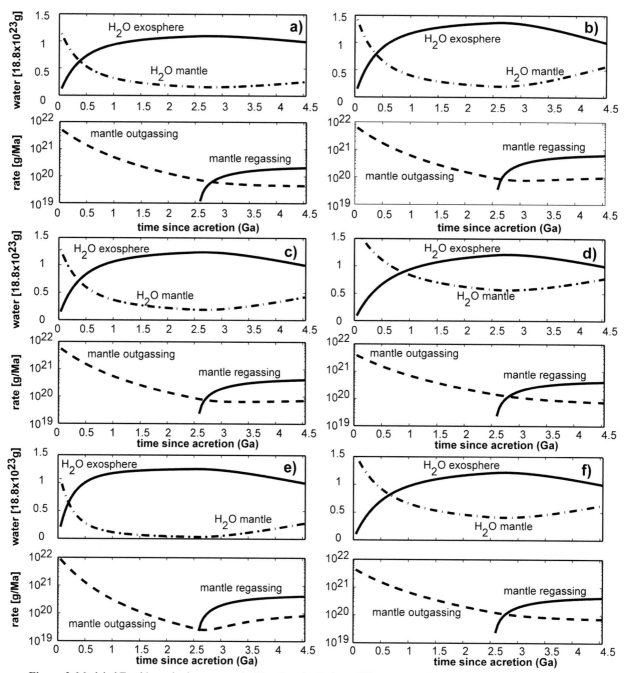

Figure 3. Modeled Earth's geologic water cycle. The plots (a–f) show different evolution scenarios for the Earth's mantle and oceans assuming different water recycling functions and water outgassing efficiencies through time (see Table 2 and text for details). In (a–b) the present-day mantle overturn rate is 9.8Ga while in (e) it is 5.7Ga and in (f) 14.7Ga.

so that 2.08×10^{20} g Ma^{-1} of water are presently recycled into the mantle in (b) this pre-factor is 1.5 and 6.24×10^{20} g Ma^{-1} of water are recycled; outgassing efficiency is 90% in both models. The starting conditions that reproduce the present-day conditions are 1.26 in (a) and 1.57 exospheres in (b) of water initially dissolved in the mantle. In both models high

mantle water contents and high overturn rates in the beginning of Earth history lead to efficient water ougassing and formation of the oceans while the mantle progressively desiccates. Peak sea-level conditions are reached about 2Ga ago when mantle regassing through cold plate subduction starts to dominate over water ougassing through mantle melting.

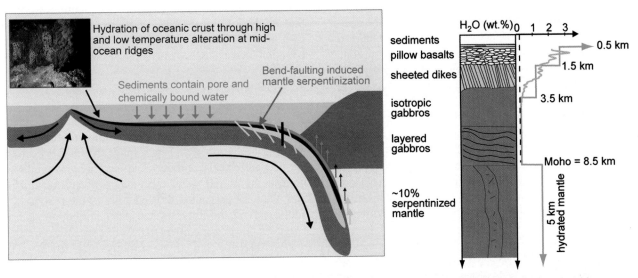

Plate 1. Hydration of oceanic plates. Ocean crust becomes hydrated through hydrothermal alteration at mid-ocean ridges and through low-temperature alteration at the sea floor. Sediments contain both chemically bound and pore water. Bend-faulting at the trench outer rise may lead to additional hydration of the cold lithospheric mantle.

As a consequence the mantle slowly rehydrates. Depending on the assumed water recycling rate the present-day conditions are characterized by a mantle that has lost 79% and 64%, respectively, of its initial water and presently contains 26% and 57%, respectively, of the water in the exosphere. In the absence of subduction rehydration, the mantle would have lost 90% of its initial water. This shows that in both cases the present-day mantle has lost a significant part of its initial water to form the oceans and that plate subduction plays an important role in rehydrating the mantle.

In (c) and (d) we assume the same present-day water regassing rates of 4.16×10^{20} g Ma^{-1} and vary water outgassing efficiency between 90% (c) and 50% (d). As expected reducing the mantle ougassing efficiency increases the amount of initial water in the mantle that is necessary to reproduce the present-day conditions (1.41 in (c) and 1.78 in (d)). The present-day conditions are characterized by a mantle that has lost 71% in (c) and 56% in (d) of its initial water and presently contains 42% to 78% of the water presently stored in the exosphere. In the absence of subduction the mantle would have lost 90% and 72% of its water, respectively.

In (e) and (f) we have varied the present-day mantle processing rate. In all other model runs we assume a mantle processing rate of 9.8Ga, which is a relatively robust estimate that is based on the present-day spreading rate—see [*Phipps Morgan*, 1998] for details. However, to illustrate the effect of changing mantle processing rates we assume a faster rate of 5.7Ga in (e) and 14.7Ga in (f). Note, that the faster rate has been suggested by [*Kellogg and Wasserburg*, 1990]. Faster, processing rates result in very rapid and efficient mantle outgassing early during Earth's history. Slower rates prolong the time to form the oceans. In order to put these values into perspective we have summarized the results in Table 2 and compared them to previous studies.

These six example model runs give some initial insights into how subduction rehydration and water outgassing affect the evolution of the Earth: raising water recycling by plate subduction has a similar effect as lowering water outgassing efficiency—both lead to higher present-day and initial water contents of the Earth's mantle. A somewhat surprising effect of lowering the outgassing efficiency is that this does not lead to lower absolute water outgassing rates (see Table 2, run c and d). In fact, less effective water outgassing leads to a wetter mantle, which increases again the water outgassing rates. This implies that the outgassing efficiency is directly related to the present-day water content of the mantle—the lower the outgassing efficiency the more water is in the mantle. Another effect of lowering the water outgassing efficiency is that the time to form the oceans is prolonged.

In order to give a more complete picture of how the different model parameters affect the results we have performed some 1500 model runs and plotted the results in phase diagrams in Plate 2. Inspecting these results shows that for a wide parameter range the mantle will have lost most of its water to the exosphere and contains presently much less water than the exosphere. Only for very high rehydration pre-factors, i.e. mantle regassing rates, and very low water outgassing efficiencies can the mantle reach equal or higher water contents as the exosphere. This shows that the model for the global geologic water cycle that is presented here favors a 'dry' mantle over a 'wet' mantle. Note, that these numbers imply whole mantle flow, so that eventually most of the mantle participates in near-surface upwelling, melting, and differentiation processes. If a strictly layered Earth is assumed to have always existed, it is possible that a reservoir of incommunicative 'primitive' mantle water exists that would still not yet have 'participated' in the surface + communicative mantle geologic water cycle and water distribution between the oceans and the mantle.

3.2. Comparison to Geochemical Constraints

The above results show that for every combination of model parameters there is one starting condition that re-produces the present-day oceans. This implies that more constraints are needed to make somewhat less ambiguous predictions on how the global water cycle has evolved through time and what the present hydration state of the mantle is. As additional constraints we use the water concentrations in different mantle end-members that were determined by *Dixon et al.,* [2002]. To compare our findings to geochemical data we translate our findings on water distribution between the exosphere and the mantle into average water concentrations in different mantle components (Plate 3). To make predictions for the most primitive mantle components (FOZO) we assume that all the initial water in the mantle is dissolved in the 4.0×10^{24}kg [*Stacey*, 1992] of mantle rocks. The predicted water concentration in average mantle can be determined from the modeled present-day water content of the mantle.

Dixon et al., [2002] proposed an average hydration of the most primitive mantle components of 750±250-ppm. Unfortunately the large uncertainty in this value prevents us from putting tighter constraints on the likely model parameters. If we, however, assume that 750-ppm is the most likely value for the FOZO component we can conclude from Plate 2&3 that only intermediate model parameters are consistent with this value: low water outgassing efficiencies together with high subduction induced regassing of the mantle results in too high water concentrations; high outgassing efficiencies and low regassing rates lead to an underestimation of the water content. *Dixon et al.,* [2002] further proposed that slab influenced mantle components are drier containing

Table 2. Summarized modeling results from the six example runs shown in Fig. 3. To put these values into perspective we compare our findings and assumptions to previous studies: [1][*Peacock, 1990*]; [2][*Schmidt and Poli, 1998*]; [3][*Wallmann, 2001*]; [4][*Hilton, et al., 2002*]; [5][*Bebout, 1996*]; [6][*Hallam, 1992*];[7][*Dixon, et al., 2002*]. Our estimates for water input into subduction zones is higher than the one by Peacock, (1990) who neglected water in serpentinized mantle and lower than the one by *Schmidt and Poli*, [1998] who assume water saturated conditions in the crust. The range of possible water subduction rates given by *Bebout*, [1996] is consistent with our estimate. The estimate by *Hilton et al.*, [2002] is significantly lower than all other estimates. Our estimates for deep water recycling (mantle regassing) are comparable to the ones by *Wallmann*, [2001] and also our calculated values for mantle outgassing are close to the estimate by *Wallmann*, [2001], which is based on the total magma production rate of sub-aerial volcanism [*Schmincke*, 2000]. Note, that mantle outgassing refers to water loss from the mantle to the atmosphere and does not include water released from subducting slabs, i.e. slab derived water at arc volcanoes. Phanerozoic sea-level data indicates a maximum drop in sea-level of 400–600m [*Hallam*, 1992].

present-day values	model run a	model run b	model run c	model run d	model run e	model run f	other studies
H_2O exosphere	18.8×10^{23} g	18.8×10^{23} g	18.8×10^{23} g	18.8×10^{23} g	18.8×10^{23} g	18.8×10^{23} g	
water subducted	12.0×10^{20} g Ma^{-1}	12.0×10^{20} g Ma^{-1}	12.0×10^{20} g Ma^{-1}	12.0×10^{20} g Ma^{-1}	12.0×10^{20} g Ma^{-1}	12.0×10^{20} g Ma^{-1}	[1]8.7×10^{20} g Ma^{-1} [2]$(18.5 - 28.0) \times 10^{20}$ g Ma^{-1} [3]6.7×10^{20} g Ma^{-1} [4]1.6×10^{20} g Ma^{-1} [5]$(9.1 - 19.4) \times 10^{20}$ g Ma^{-1}
outgassing efficiency	90%	90%	90%	50%	90%	90%	
rehydration prefactor	0.5	1.5	1.0	1.0	1.0	1.0	
mantle overturn rate	9.8Gy	9.8Gy	9.8Gy	9.8Gy	5.7Gy	14.7Gy	[8]9.8 Gy [9]5.7 Gy
deep recycling/ mantle regassing	2.08×10^{20} g Ma^{-1}	6.24×10^{20} g Ma^{-1}	4.16×10^{20} g Ma^{-1}	4.16×10^{20} g Ma^{-1}	4.16×10^{20} g Ma^{-1}	4.16×10^{20} g Ma^{-1}	[3]$(1.62 - 4.14) \times 10^{20}$ g Ma^{-1}
mantle outgassing	4.41×10^{19} g Ma^{-1}	9.64×10^{19} g Ma^{-1}	7.02×10^{19} g Ma^{-1}	7.38×10^{19} g Ma^{-1}	8.25×10^{19} g Ma^{-1}	7.18×19^{19} g Ma^{-1}	[3]$(3.6 - 5.4) \times 10^{19}$ g Ma^{-1}
H_2O mantle	4.88×10^{23} g	10.7×10^{23} g	7.80×10^{23} g	14.6×10^{23} g	5.38×10^{23} g	11.9×10^{23} g	
H_2O total	23.6×10^{23} g	29.5×10^{23} g	26.5×10^{23} g	33.4×10^{23} g	24.2×10^{23} g	$30.6 \; 10^{23}$ g	
% of exosphere water in mantle	26%	57%	42%	78%	29%	63%	
mantle water loss	79%	64%	71%	56%	78%	61%	
mantle degassing	90%	90%	90%	72%	96%	79%	
H_2O in FOZO	590-ppm	738-ppm	663-ppm	834-ppm	604-ppm	766-ppm	[7]~750-ppm
Ø H_2O in mantle	122-ppm	268-ppm	195-ppm	365-ppm	134-ppm	297-ppm	[7]<400-ppm
Phanerozoic sea-level drop	186m	605m	397m	386m	386m	389m	[6]<400-600m

less than 400-ppm water and that the MORB source contains only about 100-ppm. If we assume that 400-ppm is an upper bound for the average mantle value we can again rule out combined low outgassing and high recycling efficiencies and thereby a very 'wet' mantle.

As an additional constraint we use observed sea-level changes. The *Hallam*, [1992] sea-level curve suggests a maximum sea-level drop of 400–600m during the Phanerozoic while *Artyushkov and Chekhovich*, [2001] argue against significant sea-level changes. Given the 'dynamics' of water outgassing and water recycling it appears to us almost inevitable that the global ocean volume has changed through time

and with it sea-level. Nevertheless, a conservative upper bound for maximum sea-level drop over the past 600Ma is 400m. In order to convert our findings on variations in ocean volume into sea-level changes we assume that the continent-ocean distribution and the average ocean depth have not changed through time. These are, of course, 'strong' (and hard to constrain) assumptions but since we use predicted sea-level changes only as a 'control' parameter such a rough estimate is sufficient. Changes in sea-level, h, can therefore be expressed as $h = V_{past}/V_{now} * d_{now} - d_{now}$, where V is the past and present ocean volume, respectively, and $d_{now} = 3800m$ is the present average depth of the oceans. The

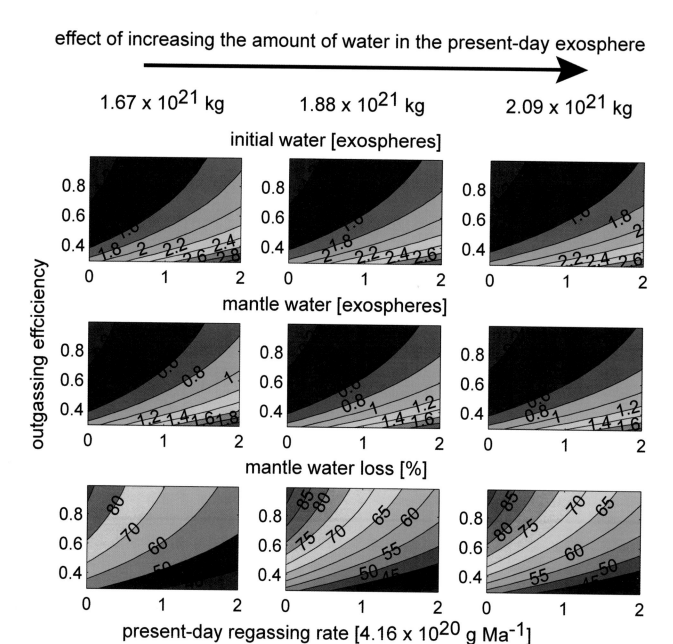

Plate 2. Phase diagram of all model runs. The y-axis of all plots describe the assumed outgassing efficiency which varies between 30% and 99%. The x-axis in all plots describe mantel regassing rates through plate subduction as multiples of the present-day rate of 4.16×10^{20} g H_2O Ma^{-1}. Multiplication of the numbers for water content of the mantle with the assumed water content of the exosphere (see top of each column) gives mantle water contents in kg.

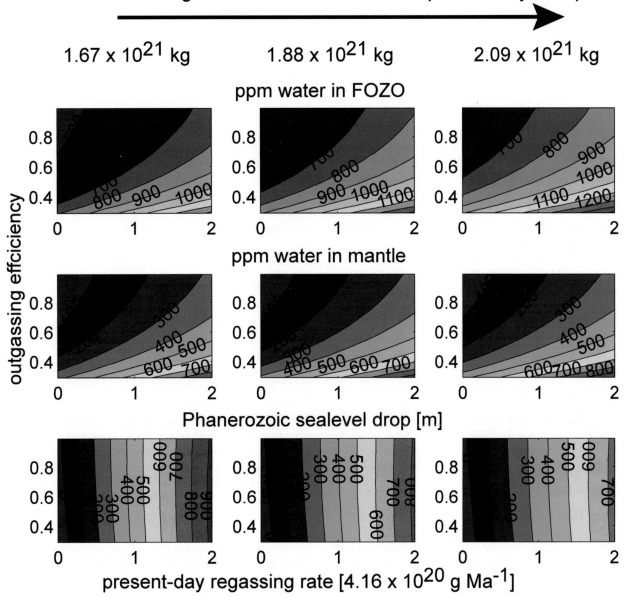

Plate 3. Phase diagram of all model runs. The axes are the same as in Plate 2. See text for details.

last panel plots in Plate 3 show the predicted drop in sea-level over the past 600Ma for all the model runs. These plots show that it is mainly the intensity of subduction rehydration that controls this number: strong water recycling leads to larger sea-level drops through time. Using the upper bound of 400m for Phanerozoic sea-level changes we can rule out present-day water recycling rates greater than about 1x 4.16×10^{20} g Ma^{-1}—a result that is consistent with box-modeling of the global water cycle [*Wallmann*, 2001].

We can now infer mantle evolution scenarios that are consistent with the geochemical data and the sea-level constraints. Two end-member evolution models appear plausible. The fist scenario is what we call the 'water-world' model. Mantle regassing rates through plate subduction have increased over the past 2Ga to the present-day value of about 4.16×10^{20} g H_2O Ma^{-1} leading to significant sea-level changes throughout Earth history. All outgassing efficiencies between the upper bound of 99% and the lower bound of 50% are consistent with the geochemical constraints (Plate 2&3). In this scenario plate subduction plays an important role in connecting the Earth's surface and deep water cycles. A major drop in ocean volume over the past ~2Ga resulted in significant changes in land-sea distribution, and possibly in the emergence of large parts of the continents.

In the second end-member scenario sea-level has not changed significantly during the last 2Ga, which requires that subducting plates dehydrate efficiently so that only little or no water is recycled into the deeper mantle. This scenario requires much lower (~50%) outgassing efficiencies for the modeling results to be consistent with the geochemical constraints.

At this point we favor the 'water-world' scenario but additional data will be required to discriminate between these two end-member models. We can, however, conclude that none of the scenarios explored results in a very wet mantle, i.e. one that has several oceans of water dissolved in it.

4. DISCUSSION AND CONCLUSIONS

These findings are somewhat contrary to some recent studies that proposed a transition zone that is much wetter than the average mantle (e.g. [*Bercovici and Karato*, 2003; *Huang, et al.*, 2005; *Ohtani*, 2005; *Smyth*, 1987]). In fact, *Bercovici and Karato*, [2003] formulated the hypothesis of the 'transition zone water filter' to explain the observed chemical differences between OIB and MORB. In this study we neglect potential preferential water storage in a 'mantle acquifer' such as the transition zone. This model assumption is based on our results on efficient plate dehydration during subduction and the chemical evidence presented by *Dixon et al.*, [2002]. Our recent work on slab dehydration presented

here and in [*Rüpke, et al.*, 2004] implies that it is unlikely that slabs remain sufficiently cold and hydrated during subduction to eventually exceed the water storage capacity of deeper mantle rocks. Such a super-saturation of subducting slabs is, however, critical for forming a fluid phase deep inside the mantle that could possibly lead to preferential water storage inside the transition zone. Furthermore, preferential water storage or water stripping from hydrated mantle material flowing through the transition zone should fractionate water from its magmatic sister element Ce. However, *Dixon et al.,'s* [2002] results do not show such a fractionation. We therefore feel that this is a valid approximation and that the here presented model incorporates the major geologic processes affecting Earth's deep water cycle. If, however, a highly hydrated transition zone does exist, then our prediction of the mantle's hydration state will need to be revised.

In conclusion, it appears likely that Earth's mantle contains less water than the exosphere (oceans plus crust) and we suggest that the equivalent of one exosphere is an upper bound for the total water content of the mantle (see Plate 2&3). Earth's mantle appears therefore to be rather 'dry' or 'damp' than 'wet' as a result of water outgassing at mid-ocean ridges and hotspots, which mitigates mantle rehydration by plate subduction. This finding is supported by the fact that our model predicts realistic water concentrations for the different mantle components—concentrations consistent with the independent estimates by *Dixon et al.* [2002].

One interesting aspect of our findings is that practically all models predict water concentrations in 'average' mantle that are higher than the 'normal' water concentration of 100-ppm in the MORB source. In fact, the strikingly homogeneous geochemistry of MORB relative to OIB is one of the most remarkable observations in geochemistry. *Phipps Morgan and Morgan*, [1999] proposed the two-stage melting model to explain this observation, in which the MORB source represents the 'left-overs' of plume melting beneath hotspots. Applying this model to the water concentrations determined here for the average mantle leads to the following concept: if mantle melting preferentially occurs first in its most volatile-rich or incompatible element-rich components (plausible since both volatiles and incompatible elements tend to reduce the solidus), then it would be completely consistent that hotspots with a FOZO-component in their source will make wetter basalts than hotspots whose mantle sources contain a larger fraction of EM and HIMU components, and also consistent that progressive upwelling and pressure-release melting beneath a mid-ocean ridge should eventually be able to melt the average mantle to the point where ~100-ppm depleted, slab influenced mantle components are the only remaining water source to contribute to the more depleted MORB. Thus our results

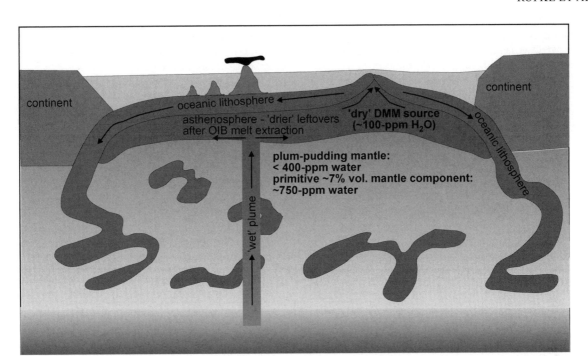

Figure 4. The plume-pudding mantle model. 'Left-overs' from hotspot melting feed the asthenosphere that melts beneath mid-ocean ridges to make MORB. This two-stage melting model may explain the low water content of the MORB source and its remarkable homogeneous chemistry with respect to OIB.

(summarized in Fig. 4) are consistent with a plum-pudding whole mantle convection evolution scenario in which plate subduction plays an important role in partially filtering the water content of the recycled sediments and ocean basalts that form the sources of the more enriched mantle plumes.

Acknowledgments. Many thanks to Terry Plank, Suzan van der Lee, Steve Jacobsen, and an anonymous reviewer for very constructive and helpful comments on the manuscript. Fruitful discussions with Matthias Hort greatly helped to develop the ideas for this study.

REFERENCES

Artyushkov, E. V., and P. A. Chekhovich (2001), The East Siberian basin in the Silurian: evidence for no large-scale sea-level changes, Earth Planet. Sci. Lett., 193, 183–196.

Asimow, P., J. E. Dixon, and C. Langmuir (2004), A hydrous melting and fractionation model for mid-ocean ridge basalts: Application to the Mid-Atlantic Ridge near the Azores, Geochemistry Geophysics Geosystems, 5, doi: 10.1029/2003GC000568.

Bebout, G. E. (1996), Volatile Transfer and Recycling at Convergent Margins: Mass-Balance and Insights from High-P/T Metamorphic Rocks, in Subduction Top to Bottom, edited by G. E. Bebout, et al., pp. 179–193, American Geophysical Union, Washington DC.

Bercovici, D., and S. Karato (2003), Whole-mantle convection and the transition-zone water filter, Nature, 425, 39–44.

Chase, C. G. (1979), Subduction, the geoid, and lower mantle convection, Nature, 282, 464–468.

Dixon, J. E., and D. A. Clague (2001), Volatiles in basaltic glasses from Loihi Seamount, Hawaii: Evidence for a relatively dry plume component, J. Petrol., 42, 627–654.

Dixon, J. E., L. Leist, C. Langmuir, and J.-G. Schilling (2002), Recycled dehydrated lithosphere observed in plume-influenced mid-ocean-ridge basalt, Nature, 420.

Drake, M. J., and K. Righter (2002), Determining the composition of the Earth, Nature, 416, 39–44.

Grevemeyer, I., E. R. Flueh, C. Ranero, D. Klaeschen, and J. Bialas (2006), Passive and active seismological study of bending-related faulting and mantle serpentinization at the Middle America trench, Earth Planet. Sci. Lett., submitted.

Grevemeyer, I., N. Kaul, J. L. Diaz-Naveas, H. W. Villinger, C. R. Ranero, and C. Reichert (2005), Heat flow and bending-related faulting at subduction trenches: Case studies offshore of Nicaragua and Central Chile, Earth Planet. Sci. Lett., 236, 238–248.

Hallam, A. (1992), Phanerozoic Sea-Level Changes, Columbia University Press.

Hart, S. R., L. A. Hauri, L. A. Oschman, and J. A. Whitehead (1992), Mantle plumes and entrainment: isotopic evidence, Science, 256, 517–520.

Henderson, P. (1986), Inorganic Geochemistry, Pergamon Press.

Hilton, D. R., T. P. Fischer, and B. Marty (2002), Noble gases and volatile recycling at subduction zones, in Noble Gases In Geochemistry And Cosmochemistry, edited, pp. 319–370.

Hirschmann, M., C. Aubaud, and A. Withers (2005), Storage capacity of H2O in nominally anhydrous minerals in the upper mantle, Earth Planet. Sci. Lett., 236, 167–181.

Hofmann, A. W. (1997), Mantle geochemistry: the message from oceanic volcanism, Nature, 385, 219–229.

Hofmann, A. W., and W. M. White (1982), Mantle plumes from ancient oceanic crust, Earth Planet. Sci. Lett., 57, 421–436.

Huang, X., Y. Xu, and S. Karato (2005), Water content in the transition zone from electrical conductivity of wadsleyite and ringwoodite, Nature, 434, 746–749.

Kellogg, L. H., and G. J. Wasserburg (1990), The role of plumes in mantle helium fluxes, Earth Planet. Sci. Lett., 99, 276–289.

Marone, M., M. van der Meijde, S. van der Lee, and D. Giardini (2003), Joint inversion of local, regional and teleseismic data for crustal thickness in the Eurasia-Africa plate boundary region, Geophys. J. Int., 154, 499–514.

Michael, P. J. (1995), Regionally distinctive sources of depleted MORB: Evidence from trace elements and H2O, Earth Planet. Sci. Lett., 131, 301–320.

Mooney, W. D., G. Laske, and G. Masters (1998), CRUST 5.1: A global crustal model at 5 X 5, J. Geophys. Res., 103, 727.

Nataf, H.-C., and Y. Ricard (1996), 3SMAC: an a priori tomographic model of the upper mantle based on geophysical modeling, Phys. Earth Planet. Int., 95, 101–122.

Nichols, A. R., M. R. Carroll, and A. Höskuldsson (2002), Is the Iceland hotspot also wet? Evidence from the water contents of undegasssed submarine and subglacial pillow basalts, Earth Planet. Sci. Lett., 202, 77–87.

Ohtani, E. (2005), Water in the mantle, Elements, 1, 25–30.

Peacock, S. M. (1990), Fluid Processes in Subduction Zones, Science, 248, 329–337.

Phipps Morgan, J. (1998), Thermal and rare gas evolution of the mantle, Chem. Geol., 145, 431–445.

Phipps Morgan, J. (1999), The Isotope Topology of Individual Hotspot Basalt Arrays: Mixing Curves or Melt Extraction Trajectories? Geochemistry Geophysics Geosystems, 1.

Phipps Morgan, J., and J. Morgan (1999), Two-stage melting and the geochemical evolution of the mantle: a recipe for mantle plum-pudding, Earth Planet. Sci. Lett., 170, 215–239.

Plank, T., and H. Langmuir (1998), The chemical composition of subducting sediment and its conssequences for the crust and mantle, Chem. Geol., 145, 325–394.

Ranero, C. R., J. Phipps Morgan, K. D. McIntosh, and C. Reichert (2003), Bending, faulting, and mantle serpentinization at the Middle America trench, Nature, 425, 367–373.

Richard, G., M. Monnereau, and J. Ingrin (2002), Is the transition zone an empty water reservoir? Inference from numerical model of mantle dynamics, Earth Planet. Sci. Lett., 205, 37–51.

Rüpke, L. H., J. Phipps Morgan, M. Hort, and J. A. D. Connolly (2002), Are the regional variations in Central American arc lavas due to differing basaltic versus peridotitic slab sources of fluids? Geology, 30, 1035–1038.

Rüpke, L. H., J. Phipps Morgan, M. Hort, and J. A. D. Connolly (2004), Serpentine and the subduction zone wate cycle, Earth Planet. Sci. Lett., 223, 17–34.

Schmidt, M. W., and S. Poli (1998), Experimentally based water budgets for dehydrating slabs and consequences for arc magma generation, Earth Planet. Sci. Lett., 163, 361–379.

Schmincke, H.-U. (2000), Vulkanismus, Wissenschaftliche Buchgesellschaft.

Simons, K., J. E. Dixon, J.-G. Schilling, R. Kingsley, and R. Poreda (2002), Volatiles in basaltic glasses from the Easter-Salas y Gomez Seamount Chain and Easter Microplate: Implications for geochemical cycling of volatile elements, Geochemistry Geophysics Geosystems, 3, 10.1029/2001GC000173.

Smyth, J. R. (1987), beta-Mg2SsO4: A potential host for water in the mantle, Am. Mineral., 72, 1051–1055.

Stacey, F. D. (1992), Physics of the Earth, 3rd ed., Brookfield Press.

Staudigel, H., T. Plank, B. White, and H.-U. Schmincke (1996), Geochemical Fluxes During Seafloor Alteratrion of the Basaltic Upper Ocean Crust: DSDP Sites 417 and 418, in Subduction Top to Bottom, edited by G. E. Bebout, et al., pp. 19–38, American Geophysical Union.

Ten, A. A., Y. Y. Podladchikov, D. A. Yuen, T. B. Larsen, and A. V. Malevsky (1998), Comparison of mixing properties in convection with the particle-line method, Geophys. Res. Lett., 25, 3205–3208.

Wallmann, K. (2001), The geological water cycle and the volution of marine d18O values, Geochim. Cosmochim. Acta, 65, 2469–2485.

Williams, Q., and R. Hemley (2001), Hydrogen in the deep Earth, Ann. Rev. Earth Planet. Sci., 29, 365–418.

Zindler, A., and S. R. Hart (1986), Chemical Geodynamics, Ann. Rev. Earth Planet. Sci., 14, 493–571.

Petrologic Structure of a Hydrous 410 km Discontinuity

Marc M. Hirschmann, Anthony C. Withers, and Cyril Aubaud

Department of Geology and Geophysics, University of Minnesota, Minneapolis, Minnesota, USA

We examine a simple thermodynamic model relating the effect of pressure and H_2O content on the relative stabilities of olivine, wadsleyite, and hydrous melt through the upper mantle/transition zone boundary. As noted previously, the strong preference of H_2O for wadsleyite relative to olivine means that H_2O increases the depth interval over which the two polymorphs coexist and displaces the boundary to shallower depths. Olivine and wadsleyite become increasingly hydrous with decreasing depth through the transformation interval, and if the available H_2O exceeds the storage capacity of olivine, then partial melting will occur at the shallowest portions of the interval. Increases in H_2O beyond that necessary to generate incipient melt result in thinning of the transformation interval, but additional shoaling. In the extreme case where melting occurs throughout the melting interval, the transformation interval will be as thin as for the dry case (i.e., ~7 km). These calculations may serve as a starting point for dynamical calculations of possible melting at the 410 km discontinuity and as a guide for interpretation of observational geophysical characteristics of putative hydrous features at the transition zone/upper mantle boundary.

INTRODUCTION

The recognition that there is a significant contrast in H_2O storage capacity across the 410 km discontinuity has lead to extensive investigation of the effect of H_2O on this boundary [*Gasparik*, 1993; *Young, et al.*, 1993; *Wood*, 1995; *Chen, et al.*, 2002; *Smyth and Frost*, 2002; *Bercovici and Karato*, 2003; *Litasov and Ohtani*, 2003a; *Hirschmann, et al.*, 2005]. The discontinuity is attributed to the transformation between olivine and wadsleyite [e.g., *Ringwood*, 1975]. The region below this discontinuity is generally termed the transition zone and for the purposes of this paper, the region above it is termed the upper mantle. The strong contrast in H_2O storage capacities between olivine and wadsleyite [*Kohlstedt, et al.*, 1996] can influence both the absolute depth and depth interval over which the transformation occurs [*Wood*, 1995]. Additionally, these differences in storage capacity could

Earth's Deep Water Cycle
Geophysical Monograph Series 168
Copyright 2006 by the American Geophysical Union.
10.1029/168GM21

lead to hydrous melting associated with the transformation [*Gasparik*, 1993; *Young, et al.*, 1993; *Bercovici and Karato*, 2003; *Litasov and Ohtani*, 2003a]. Hydrous transition zone material advected into the upper mantle will partially melt if the resulting upper mantle assemblage is incapable of sequestering all of the H_2O formerly held in transition zone phases. This could occur locally, associated with spatially restricted masses of unusually hydrous transition zone material or globally, if significant hydration is a widespread feature of the transition zone [*Bercovici and Karato*, 2003].

Transport of H_2O between the transition zone and the upper mantle is a critical element of the global deep H_2O cycle. The water filter hypothesis of *Bercovici and Karato* [2003] posits that widespread melting at this interface, followed by eventual return of dense hydrous melt to the transition zone produces a global H_2O-rich reservoir in the transition zone and controls the H_2O content of the upper mantle. If this hypothesis is correct, melting at the transition zone/upper mantle interface modulates the supply of H_2O to the upper mantle and hence the flux of H_2O through ridges to the exosphere. If it is not correct, then local melting at the interface may still have an important influence on the flux of H_2O to the upper mantle [*Hirschmann*, 2006a].

It is well-established that the 410 km discontinuity is caused by a phase change in $(Mg,Fe)_2SiO_4$ [*Ringwood*, 1975]. Because the transformation between α (olivine) and β (wadsleyite) polymorphs is divariant, involving equilibration between two phases with differing Fe/Mg ratios, it must occur over a finite pressure interval Katsura et al. 2004; [e.g., *Akaogi, et al.*, 1989; *Katsura and Ito*, 1989; *Wood*, 1995; *Helffrich and Wood*, 1996; *Frost*, 2003; *Katsura, et al.*, 2004], corresponding to a finite depth interval in the mantle.

Water can have notable effects on the transformation between olivine and wadsleyite. *Wood* [1995] used simple thermodynamic theory to demonstrate that preferential partitioning of H_2O into wadsleyite should stabilize the transition zone assemblage to lower pressures, producing a shallower seismic discontinuity, and broadens the depth interval over which the transformation occurs. *Chen et al.* [2002] and *Smyth and Frost* [2002] verified experimentally that H_2O diminishes the pressure of the olivine/wadsleyite transformation, but reached contrary conclusions regarding the predicted thickening effect. *Smyth and Frost* [2002] found that H_2O leads to a thicker transformation depth interval, in agreement with the predictions of *Wood* [1995], whilst *Chen et al.* [2002] found that H_2O reduces the transformation interval.

Seismological studies of the 410 km discontinuity reveal a number of features that may be attributable to the effects of H_2O. To date, all of these features are found in regional studies, meaning there is as yet no evidence that H_2O enrichment has a global effect on the character of the 410 km discontinuity. In most localities, the transformation interval is rather narrow (5–10 km) [*Helffrich*, 2000]. However, *Van der Meijde et al.* [2003] observed broadening of the 410 discontinuity to 20–35 km beneath some portions of the Mediterranean Sea. If this broadening is attributable to H_2O, then the calculations of *Wood* [1995] suggest approximately 700 ppm (by weight) H_2O. However, revised calculation from *Hirschmann, et al.* [2005], based on more recent experimental constraints on the relative preferences of olivine and wadsleyite for H_2O [*Chen, et al.*, 2002], indicate that the amount of H_2O required to account for this thickening would be between 1000 and 2000 ppm. A subsequent receiver function study suggests that the depth of the 410 discontinuity in this region may be locally variable, with depths ranging from 385 to 420 km[*van der Meijde, et al.*, 2005].

Shoaling of the top of the transition zone beneath the Kapvaal craton in southern Africa has also been attributed to the influence of H_2O by *Blum and Shen* [2004], who observed an overall shallowing of the transition zone of ~20 km. Attribution of 410 km discontinuity shoaling to

H_2O must be made with care, however, as variations in temperature and other compositional effects may have similar effects. Indeed *Blum and Shen* [2004] note that the effects beneath southern Africa cannot be attributed solely to H_2O and are perhaps best explained by mantle that is both unusually hydrous and anomalously cold. It is also notable that *Blum and Shen* [2004] did not observe significant thickening of the 410 km transformation interval, as might be expected from enrichments in H_2O.

A range of studies have also revealed seismological features that may be best explained by layers of melt residing atop the 410 km discontinuity [*Revenaugh and Sipkin*, 1994; *Vinnik, et al.*, 2003; *Song, et al.*, 2004; *Chambers, et al.*, 2005]. For example, *Song et al.* [2004] inferred ~5% slowing of shear wave velocities beneath western North America in a layer about 20 km thick. The fraction of melt that would be required to account for 5% slowing of S waves may be in the range of 2–3 volume % [*Yoshino, et al.*, 2005].

In this chapter, we apply a simple thermodynamic model of the 410 km discontinuity to examine the effects of H_2O and hydrous melting on the depth and thickness of the transformation. The quantitative constraints from the calculations may be limited by the simplicity of the model, but nevertheless they provide a framework for understanding interrelationships between the amount of H_2O available, melting, and the petrologic structure of the upper mantle/transition zone boundary. These provide the starting point for discussions of possible dynamic processes at the boundary, as well as for possible geophysically observable characteristics.

MODEL DESCRIPTION AND CONSTRAINTS

Following *Wood* [1995], the equilibrium pressure, P, at which olivine and wadsleyite of a given composition coexist is given by

$$(P - P^0)\Delta V^0 = -RT\ln[(X_{Mg}^{wd})^2 \gamma_{Mg}^{wd}(1 - X_{OH}^{wd})^{0.5}] + RT\ln[(X_{Mg}^{ol})^2 \gamma_{Mg}^{ol}(1 - X_{OH}^{ol})^4 /(1 - 0.5X_{OH}^{ol})] \quad (1)$$

In Eqn. (1), P^0 and ΔV^0 are the pressure and the volume change of the olivine-wadsleyite transition for pure Mg_2SiO_4 at the temperature (T) of interest. For depths near 410 km in mantle away from plumes and subduction zones, this temperature is likely between 1500 and 1600 °C [*Ita and Stixrude*, 1992; *Katsura, et al.*, 2004]. The parameters X_{Mg}^{wd} and X_{Mg}^{ol} in Eqn. 1 are the mole fractions of Mg_2SiO_4 in wadsleyite and olivine, respectively, and γ_{Mg}^{wd} and γ_{Mg}^{ol} are activity coefficients of Mg_2SiO_4 in each phase. Values of ΔV^0, γ_{Mg}^{wd} and γ_{Mg}^{ol} are calculated from the parameters given by *Wood* [1995]. X_{OH}^{wd} is the mole fraction of O1 crystallographic sites in wadsleyite that are occupied by OH, and X_{OH}^{ol} is the mole fraction of OH in olivine. It is important to note that there remains some uncertainty as to whether these thermodynamic models for the substitution of H into wadsleyite and olivine are most appropriate, and alternative models should produce different calculated results [*Karato*, 2006a]. Therefore, the quantitative thicknesses of the transformation interval calculated in this chapter have considerable uncertainty. However, the qualitative relationships between those thicknesses and the effect of melting are likely to be robust.

Application of Eqn. 1 to calculation of the wadsleyite to olivine transformation requires parameterization of the composition of the two phases through the transformation interval. The pressure interval of the olivine to wadsleyite transition depends in large part on the difference in Fe-Mg composition between the two coexisting phases [*Katsura and Ito*, 1989; *Wood*, 1995; *Frost*, 2003; *Katsura, et al.*, 2004], which in turn depends on the Fe-Mg exchange equilibria between the polymorphs. Experimental determinations of Fe-Mg exchange between wadsleyite, expressed in terms of an equilibrium constant K_{DFe-Mg}^{ol-wd}:

$$K_{DFe-Mg}^{ol-wd} = \frac{X_{Fe}^{ol} X_{Mg}^{wd}}{X_{Mg}^{ol} X_{Fe}^{wd}} \qquad (2)$$

are shown in Fig. 1a. Values of K_{DFe-Mg}^{ol-wd} observed from nominally dry experiments cluster around 0.5, although there is considerable scatter (Fig. 1a). One reason for this scatter may be temperature variations in K_{DFe-Mg}^{ol-wd} (Fig. 1b), however present data are not yet accurate enough to assess this dependence.

It is important to note that the thermodynamic analysis of *Wood* [1995] overestimates the value of K_{DFe-Mg}^{ol-wd} (Fig. 1a). Other things being equal, this should result in an underestimate of the thickness of the transition under dry conditions. On the other hand, experimentally-determined phase diagrams for $(Mg,Fe)_2SiO_4$ generally overestimate the thickness of the transformation relative to the ~5 km that is normally observed seismologically [e.g., *Agee*, 1998]. This is likely because the thickness of the phase transformation across the 410 discontinuity in peridotite is influenced by the presence of additional phases such as pyroxene and garnet, owing to the phenomenon of compositional frustration [*Stixrude*, 1997; *Helffrich*, 2000; *Frost*, 2003]. For the purposes of this contribution, the key requirement of the thermodynamic model is that it give a reasonable approximation of the thickness of the dry transition. The model of *Wood* [1995] predicts a thickness of 7 km for dry $(Mg_{0.9}Fe_{0.1})_2SiO_4$, meaning inaccuracies in calculation of K_{DFe-Mg}^{ol-wd} largely compensate for not accounting for the effect of other phases.

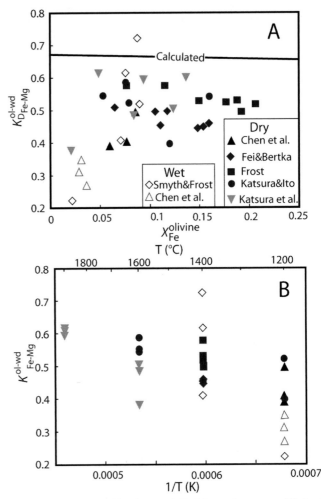

Fig. 1. Experimentally observed Fe-Mg exchange coefficient between olivine and wadsleyite,

$$K_{DFe-Mg}^{ol-wd} = \frac{X_{Fe}^{ol} X_{Mg}^{wd}}{X_{Mg}^{ol} X_{Fe}^{wd}}$$

versus (a) mole fraction of Fe_2SiO_4 in olivine and (b) reciprocal temperature. Filled symbols are nominally anhydrous experiments and open symbols are from hydrous experiments. Solid line, derived from thermodynamic parameters given by *Wood* [1995], is model used for calculations in this paper. Although the calculated values of K_{DFe-Mg}^{ol-wd} that are systematically high, this does not cause significant errors in predictions of the transformation interval (see text). Values of K_{DFe-Mg}^{ol-wd} observed from hydrous experiments show considerable scatter and are not systematically higher or lower than those from nominally anhydrous experiments. Data from *Katsura and Ito* [1989], *Fei and Bertka* [1999], *Chen et al.* [2002], *Smyth and Frost* [2002], *Frost* [2003] and *Katsura et al.* [2004].

One might also infer that partitioning of H_2O between garnet, wadsleyite, and olivine may modulate the influence of H_2O on the thickness of the transformation interval. However, the H_2O storage capacity of majoritic garnet, though not known with great precision, is likely to be small [*Katayama, et al.*, 2003], so this effect should be minimal. The treatment in this paper neglects the role of minerals other than olivine and wadsleyite. Consequently, "bulk" H_2O concentrations are in fact the sum of concentrations in olivine, wadsleyite, and (if present) melt.

An interesting question is whether Fe-Mg exchange between olivine and wadsleyite is affected by the presence of H_2O. Because the H_2O storage capacities of olivine [*Zhao, et al.*, 2004] and wadsleyite [*Hirschmann, et al.*, 2005] increase with Fe content and because wadsleyite preferentially sequesters H_2O, it is to be expected that increased H_2O stabilizes a more FeO-rich wadsleyite in equilibrium with a given composition olivine, thereby reducing K_{DFe-Mg}^{ol-wd}. This should widen the olivine+wadsleyite stability loop and cause an increase in the transformation interval. It also might delay the onset of melting until greater proportions of wadsleyite convert to olivine, as the resulting Fe-rich wadsleyite would have to reach higher H_2O concentrations before melt is stabilized. Experimental data for the magnitude of this effect are inconclusive: some data suggest that K_{DFe-Mg}^{ol-wd} may be smaller for wet compositions than for nominally anhydrous, but other experiments suggest that there is no discernable difference (Fig. 1a). Moreover, some of the observed differences between wet and dry experiments may be attributable to the effects of temperature (Fig. 1b). In the present analysis, we do not account for the effects of H_2O on Fe-Mg partitioning of between olivine and wadsleyite or for the effects of Fe-Mg composition on the H_2O storage capacities of these phases.

Application of Eqn. 1 also requires estimation of the partitioning of H_2O (or OH) between olivine and wadsleyite, D_{OH}^{ol-wd}. Whereas *Wood* [1995] applied a value of 10 for D_{OH}^{ol-wd}, *Hirschmann et al.* [2005] argued that a value closer to 5 may be more appropriate. To our knowledge, the only direct experimental values for D_{OH}^{ol-wd} are those determined by Secondary Ion Mass Spectrometry (SIMS) by *Chen et al.* [2002], who found an average value of 5±0.3 (n=4) at 1200 °C and 13 GPa. These are consistent with the relative solubilities of olivine and wadsleyite determined by Fourier Transform Infrared (FTIR) spectroscopy [*Kohlstedt, et al.*, 1996], once the olivine data are adjusted by a factor of 3 owing to inaccuracies in the FTIR calibration for olivine [*Bell, et al.*, 2003; *Koga, et al.*, 2003; *Hirschmann, et al.*, 2005; *Mosenfelder, et al.*, 2006].

Key variables influencing possible melting across the 410 km discontinuity are the storage capacities of wadsleyite and olivine. The storage capacity is the maximum H_2O that a phase can sequester at a given temperature, pressure, and composition of coexisting fluid phase, and is therefore limited if the coexisting fluid has low H_2O fugacity, as may be the case at high temperature, particularly for complex bulk comostions, which have a propensity to stabilize hydrous

melts [*Hirschmann, et al.*, 2005]. If melting occurs in the transformation interval, olivine and wadsleyite must reach their storage capacity simultaneously, assuming that the two minerals are in equilibrium with each other. Thus the storage capacities of the respective minerals must be linked through the distribution coefficient, D_{OH}^{ol-wd}. The H_2O storage capacity of wadsleyite is ≥2wt.% at low temperatures (<1200 °C) [e.g., *Kohlstedt, et al.*, 1996], but it is reduced significantly at the high temperatures (1500–1600 °C) appropriate for the 410 km discontinuity in the convecting mantle. At these temperatures, the storage capacity for pure Mg_2SiO_4 wadsleyite is approximately 0.3–0.8 wt.% [*Litasov and Ohtani*, 2003b; *Demouchy, et al.*, 2005], but it is likely higher for $(Mg,Fe)_2SiO_4$ wadsleyite [e.g., *Hirschmann, et al.*, 2005]. The storage capacity of olivine at 12–13 GPa and 1100–1250 °C is 0.5±0.2 wt.% (n=10) [*Kohlstedt, et al.*, 1996; *Chen, et al.*, 2002; *Smyth, et al.*, 2005; *Mosenfelder, et al.*, 2006] (with data from *Kohlstedt et al.* [1996] adjusted by a factor of 3, as suggested *Bell et al.* [2003] and *Koga et al.* [2003]). There are at present no data for the storage capacity of olivine at higher temperature, but it is likely to be reduced at 1500–1600 °C.

If a hydrous melt is formed during the transformation from wadsleyite to olivine, the fraction of melt formed will be inversely proportional to the H_2O concentration in the melt. Partial melting experiments of hydrous natural peridotite (KLB-1), performed in the olivine stability field close to the transformation pressure (13–14 GPa) [*Kawamoto, et al.*, 2004] and synthetic CMAS peridotite in the wadsleyite stability field (14–17 GPa) [*Litasov and Ohtani*, 2002], show that the H_2O concentrations of melts depend strongly on temperature (Fig. 2). These individual data have considerable uncertainties, particularly because the estimates of *Kawamoto et al.* [2004] are derived chiefly from microprobe totals on aggregates of minerals±glass quenched from high pressure experiments. However, the trend that they reveal is consistent with the expectation that at these pressures, storage capacity of olivine and wadsleyite will be limited by equilibrium with a fluid that transforms continuously from a H_2O-rich fluid at low temperature to a hydrous melt at high temperature. The data suggest that for peridotite near 14 GPa and 1500–1600 °C, the hydrous melt produced when olivine and wadsleyite reach their storage capacity has ~20 wt.% H_2O. Such large H_2O concentrations are unsurprising, given that these temperatures are ~500 °C cooler than the dry solidus of peridotite under these conditions [*Zhang and Herzberg*, 1994]; ~20 wt.% H_2O may be sufficient to lower the solidus by this amount. Thus, for a given mass of H_2O that cannot be retained in solid minerals at these conditions, the mass of melt formed will be about 5 times greater.

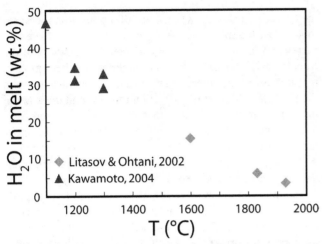

Fig. 2. Concentration of H_2O in melt coexisting with olivine at 13.5–14 GPa [diamonds, *Litasov and Ohtani*, 2002] and wadsleyite at 14–17 GPa [triangles, *Kawamoto, et al.*, 2004] and from partial melting experiments of synthetic [CMAS, *Litasov and Ohtani*, 2002] and natural [*Kawamoto, et al.*, 2004] peridotite. The former are calculated from mass balance constraints and the latter are estimated from electron microprobe totals of quenched melts. The latter may overestimate water concentrations if the water in the liquid was lost from the quenched material during and after quench [*Karato*, 2006b], but much of the H_2O may remain in the charge as quenched amorphous phases and or hydrous minerals, so this effect may not be large. The H_2O concentration required to stabilize melt diminishes with increasing temperature. At the conditions of the transition zone (~1500 °C), it is ~20 wt.%.

RESULTS

We begin by considering the effects of H_2O on the wadsleyite-olivine transformation for the specific case where $D_{OH^-}^{ol-wd}=5$ and the olivine and wadsleyite H_2O storage capacities are equal to 1500 and 7500 ppm, respectively, as may possibly be appropriate at the temperature of the 410 km discontinuity in normal sub-oceanic mantle. Below we examine the quantitative effects of different values for $D_{OH^-}^{ol-wd}$ and storage capacities on the possible structure of the 410 km transformation interval. As will be seen, the quantitative results will vary with the values assumed for these variables, but all calculations show the same general features. Thus, the qualitative structure of the transformation interval is little affected by selection of storage capacities.

Considering the transformation of wadsleyite to olivine during upwelling across the discontinuity, both phases become increasingly hydrous as the relative proportion of olivine to wadsleyite increase (Fig. 3). As originally described by *Wood* [1995], the increasing hydration of wadsleyite with decreasing depth retards its transformation to olivine, leading to a non-linear relationship between the proportions of

the polymorphs with depth (Fig. 4). If, prior to any transformation to olivine, the H_2O concentration in wadsleyite equals the storage capacity of olivine, melting commences at the shallowest portion of the transformation interval, with an infinitesimal quantity of melt forming at the point where wadsleyite disappears. With H_2O contents greater than the olivine storage capacity, melting occurs through more of the transformation interval and in the extreme case where the initial H_2O concentration equals the H_2O storage capacity of wadsleyite, melting occurs throughout the interval.

With increased H_2O content, the thickness of the interval increases from 7 km (dry) to a maximum of 24 km at 1500 ppm H_2O, the storage capacity of olivine. However, further increases in H_2O cause the transformation interval to diminish, as the H_2O concentrations of the solid phases become buffered by the presence of melt over increasing portions of the transformation interval (Fig. 5). Because thickening of the transformation interval is directly related to changes in the H_2O content of wadsleyite during transformation, buffering leads to transformation over a narrow pressure interval. At full wadsleyite saturation (7500 ppm H_2O), the

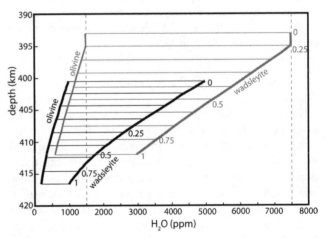

Fig. 3. Calculated water concentration in coexisting olivine and wadsleyite through the transformation interval as a function of depth for two different total H_2O contents (1000 and 3000 ppm H_2O, black and grey lines, respectively), with assumed $D_{OH^-}^{ol-wd} = 5$ and the olivine and wadsleyite storage capacities (indicated by dashed lines) assumed to be 1500 and 7500 ppm, respectively. Concentration of H_2O in each phase (shown in heavy lines) increases as proportion of wadsleyite (indicated by numbers along side "wadsleyite" line for each case) diminishes with decreasing depth. Thinner lines join compositions of phases in equilibrium with each other. In the example with 3000 ppm total H_2O, olivine and wadsleyite reach their storage capacity about 5 km prior to exhaustion of wadsleyite. This results in formation of a small proportion of melt, equal to 0.4% if the H_2O concentration in that melt is assumed to be 20 wt.%.

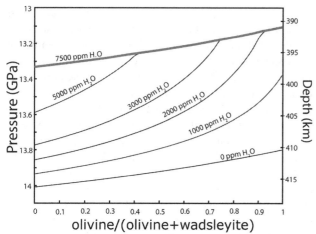

Fig. 4. Variation in proportion of olivine and wadsleyite, expressed as the mass fraction olivine/(olivine+wadsleyite) versus pressure and depth for varying initial wadsleyite concentrations ranging from 1000 ppm to 7500 ppm. Assumptions for $D_{OH^-}^{ol-wd}$ and the olivine and wadsleyite storage capacities are the same as for Fig. 3. Transformation from wadsleyite to olivine per increment of depth diminishes with increasing olivine/(olivine+wadsleyite) ratio, as large increases in H_2O (Fig. 3) enhance the stability of wadsleyite. Once the H_2O storage capacity is reached and the H_2O concentrations of olivine and wadsleyite become buffered (conditions indicated by thick gray line), the proportion of transformation per increment of depth increases greatly.

transformation interval (7 km) is not distinguishable from that of the dry case.

The median transformation depth may be defined as the that where olivine and wadsleyite are equal in proportion. Note that this depth may not correspond directly to the depth of the seismic discontinuity, which may be more sensitive to depth intervals of rapid mineralogic transformation change and which may also depend on the seismic method used to determine the transformation depth. The median transformation depth diminishes with increasing H_2O concentration from 416 km for completely dry phases to 397 km for fully saturated conditions (Fig. 6).

The depth interval over which melting can occur ranges from 0 for relatively dry bulk compositions (≤1500 ppm H_2O) to 7 km for full saturation (7500 ppm H_2O). Thus, even though H_2O can lead to considerable thickening of the wadsleyite to olivine transformation interval, the region of melt stability is relatively limited. Assuming that the melt generated has 20 wt. % H_2O (Fig. 2), the possible extents of melting ranges from 0 to 1.8 wt% for bulk H_2O contents ranging from ≤1500 to 7500 ppm. The region of melt produced by transformation across the 410 km discontinuity is comparatively small; i.e. for 3000 ppm H_2O initially in wadsleyite, the region generated is 2 km thick with a mean

melt fraction of ~0.25 wt.%. For 5000 ppm, this increases to 4 km with a mean melt fraction of ~0.5 wt.%. Note that there is an inverse correlation between the thickness of the transformation interval and the thickness of the region of melting. Similarly, there is an inverse correlation between the thickness of the transformation interval and the amount of melt generated, so long as that amount exceeds zero.

The general effects of progressive hydration on the upper mantle/transition zone boundary are summarized in Fig. 7. A dry boundary occurs deep and over a narrow depth interval (Fig. 7a). With increased H_2O, the boundary becomes thicker and shallower (Fig. 7b). When enough H_2O is present such that melting begins at the top of the transformation interval (Fig. 7c), the interval is yet shallower and at its maximum thickness. Additional H_2O leads to further shallowing, accompanied by thinning of the transformation interval and an increase the proportion of the interval over which melting occurs. In the extreme case where sufficient H_2O is present such that melting occurs throughout the melting interval (Fig. 7d), the thickness of the boundary is the same as for the dry case.

All of the results described thus far pertain strictly to the case where the storage capacities of olivine and wadsleyite are 1500 and 7500 ppm, respectively. The effects of different assumed

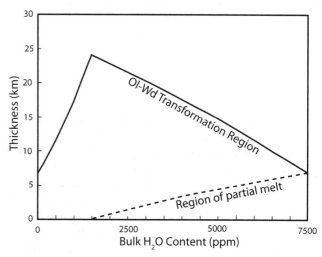

Fig. 5. Depth interval of wadsleyite to olivine transformation as a function of H_2O concentration initially in wadsleyite (i.e., at depths greater than the initial stability of olivine). Assumptions for $D_{OH^-}^{ol-wd}$ and the olivine and wadsleyite storage capacities are the same as for Fig. 3. The thickness of the transformation interval increases with increasing available H_2O until the storage capacities of olivine and wadsleyite are reached. Also shown is the depth interval over which melting may occur, which increases from zero, when incipient melting occurs up to ~7 km, when sufficient H_2O is present such that olivine and wadsleyite are at their storage capacities throughout the melting interval.

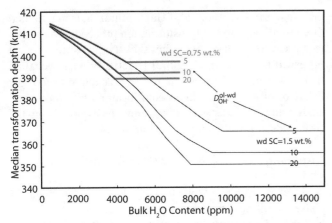

Fig. 6. Median depth of wadsleyite transformation interval for two different wadsleyite storage capacities (wd SC: 0.75 and 1.5 wt.% H_2O) and wadsleyite/olivine partition coefficients ($D_{OH^-}^{ol-wd}$) Trends for 0.75 wt.% are shown in thick grey lines; those for 1.5 wt.% H_2O are shown in narrow black lines. The two trends overlap at low bulk H_2O, but diverge when melting begins for the 0.75 wt.% wd SC case, at which point the median depth no longer increases. The median depth is defined as the depth at which the mantle assemblage consists of equal parts olivine and wadsleyite, and may not correspond to the discontinuity depth inferred seismologically. The figure illustrates that the principal variable influencing the median depth is the wadsleyite storage capacity.

values for $D_{OH^-}^{ol-wd}$ and storage capacities on the thickness of the transformation interval are explored in Fig. 8. For a given bulk H_2O content, increases in $D_{OH^-}^{ol-wd}$ produce greater thickening of the transformation interval. At constant $D_{OH^-}^{ol-wd}$, increases in wadsleyite storage capacity delay melting to greater bulk H_2O contents and hence allow greater thickening. Thus for a wadsleyite storage capacity of 1.5 wt.% and $D_{OH^-}^{ol-wd}=20$, the maximum predicted transformation interval, which occurs when 750 ppm H_2O is present, is 75 km.

Shoaling of the transformation interval depends mainly on the amount of H_2O present and on the assumed storage capacity for wadsleyite (Fig. 6). For fully saturated conditions and a storage capacity of 7500 ppm, the maximum median depth is 390–400 km, depending on the assumed value of $D_{OH^-}^{ol-wd}$. Larger storage capacities delay fluid or melt saturation to greater bulk H_2O concentrations, and hence increase possible shoaling; and because wadsleyite has a much greater storage capacity, it has a much greater influence. For example, if the wadsleyite storage capacity is 1.5 wt.%, large amounts of H_2O can produce shoaling of the upper mantle/transition zone boundary to a median depth of 350–365 km (Fig. 6).

Regardless of the values for olivine or wadsleyite storage capacity or for $D_{OH^-}^{ol-wd}$, addition of H_2O beyond that required to produce melt causes the transformation interval to diminish, reaching 7 km when olivine and wadsleyite are at their storage capacities throughout the transformation interval. Similarly, in all cases the portion of the interval over which melt is present increases from zero, coinciding to conditions of incipient melt generation (i.e., the amount of H_2O present equals the storage capacity of olivine), up to ~7 km for full saturation throughout the melting interval.

DISCUSSION

Comparison to Experimental Studies

The calculated results are similar in some respects to the experimental studies of *Chen et al.* [2002] and *Smyth and Frost* [2002]. *Smyth and Frost* [2002] found that, relative to dry conditions, the presence of H_2O at 1400 °C causes shoaling of the transformation by ~30±15 km and thickening of the transformation interval to about 40 km. *Smyth and Frost* [2002] did not report H_2O concentrations in their phases, so it is difficult to make quantitative comparisons with the calculations presented here. Their experimental charges initially contained 3 wt.% H_2O, and if most of that H_2O was retained, then one would expect saturation with a small fraction of hydrous melt and consequently a narrow transformation interval displaced to a shallow depth. However, *Smyth and*

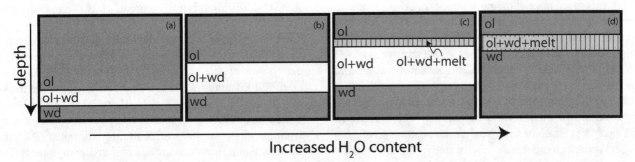

Fig. 7. Cartoon illustrating the general effects of progressive hydration on the wadsleyite/olivine transformation interval. See text for further description.

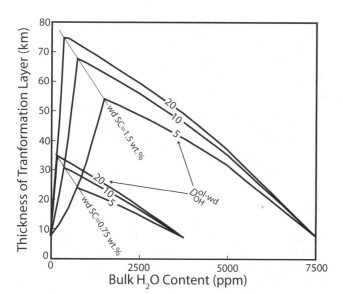

Fig. 8. Thickness of wadsleyite/olivine transformation interval as a function of total H_2O present for a range of wadsleyite storage capacities (wd SC, in wt. % H_2O) and wadsleyite/olivine partition coefficients ($D_{OH^-}^{ol-wd}$). In all cases, the thickness increases with increasing bulk H_2O content until olivine and wadsleyite reach their storage capacity, at which point melt is stabilized and further increases in bulk H_2O diminish the thickness of the transformation interval (see Fig. 5). For a given wadsleyite storage capacity, a smaller value of $D_{OH^-}^{ol-wd}$ delays saturation to a greater bulk H_2O content. A larger wadsleyite storage capacity allows for greater thickening of the transformation interval.

Frost [2002] did not note evidence of hydrous melt in their experimental charges, and the thick transformation interval they observe is more consistent with undersaturated conditions. Their experimental charges may not have retained all of the added H_2O.

Chen et al. [2002] documented about 15 km of shoaling of the transformation from experiments that were fully saturated with hydrous fluid or melt at 1200 °C. Considering that the wadsleyite and olivine from these experiments contained large amounts of H_2O (1.9–3.4 and 0.4–0.7 wt.%, respectively), this amount of shoaling is much less than expected from the calculated results (i.e., Fig. 6). It is also markedly less than *Smyth and Frost* [2002] observed from experiments with minerals that likely contained significantly less H_2O (i.e., the latter were conducted at higher temperature and fluid saturation is not apparent).

Chen et al. [2002] also observed a transformation interval slightly narrower than the dry equivalent. Their failure to find that H_2O produces thickening of the transformation interval is understandable – as illustrated in the calculations presented here and documented in their charges, the olivine and wadsleyite do not change H_2O concentration during the transformation interval if fluid saturation is maintained throughout, and so the transformation proceeds over a narrow pressure interval. The fact that the pressure interval observed by *Chen et al.* [2002] is actually narrower than that observed for the dry case may be related to azeotropic behavior of hydrous melts in equilibrium with olivine and wadsleyite. The latter possibility cannot be explored with the simple ideal model employed in our calculations.

Distribution and Fate of Melts Formed Near 410 km

The analysis applied thus far considers only the effects of phase equilibria. Dynamic processes, such as flow of melt upwards, downwards, or laterally, or kinetic inhibition of phase transformations, may impact the distribution of phases significantly. Although any such dynamic considerations must begin from a description of equilibrium phase relations, as are provided here, melt distributions may be quite different depending on the relative rates of melt generation and melt flow away from (or towards) the transformation interval.

The fate of hydrous melt potentially formed at the boundary between the transition zone and upper mantle depends in part on the buoyancy of the melt. If the melt (or partially molten region) is less dense than surrounding residual mantle, then it may ascend to the shallow mantle and possibly lead to eruptions of kimberlite [*Young, et al.,* 1993] or hydrous komatiite [*Kawamoto, et al.,* 1996; *Litasov and Ohtani,* 2002]. Little is known about the compositions of small-degree hydrous melts that might form at these depths or their compositional evolution during ascent, but they could be highly enriched in incompatible elements and consequently be similar to kimberlites.

Alternatively, the melt could be denser than the bulk upper mantle mineral assemblage [*Bercovici and Karato,* 2003]. Although H_2O has a tendency to reduce the density of silicate melts at pressures appropriate for the 410 km discontinuity, recent studies have suggested partial melts may still be more dense than the upper mantle assemblage if the concentration is not too great [*Matsukage, et al.,* 2005; *Sakamaki, et al.,* 2006]. *Matsukage et al.* [2005] suggested that the crossover occurs at 6 wt.% H_2O, but that uncertainties in density measurements allow for the possibility that it occurs somewhere between 2 and 20 wt.% H_2O. *Sakamaki et al.* [2006] suggested a more precise limit for the density crossover at 6.7±0.6 wt.% H_2O. However the compositions of near-solidus hydrous partial melts that may be produced from peridotite at 400 km depth are still not well known. Depending on the concentration of dense components (FeO, CaO) in those melts, it remains uncertain whether a density crossover occurs at any H_2O concentration. An additional key consideration is that along a normal mantle geotherm, initiation of hydrous melting of peridotite at 410 km may require ~20 wt.% H_2O in the partial melts (Fig. 2), and it seems unlikely that such hydrous melts can be denser than bulk upper mantle at these depths.

If dense hydrous melts are produced by partial melting at the top of a hydrous transition zone-upper mantle transformation region, they will be less dense than the transition zone because the density of wadsleyite is much greater than that of olivine. Presumably the depth of neutral buoyancy would lie within the wadsleyite to olivine transformation interval. Considering that melts are likely formed at the top of the transformation interval, where the mode of wadsleyite is small, the depth of neutral buoyancy could be beneath the level of melt formation, meaning that melts formed at the top of the transformation interval could percolate downwards to depths where the melts are undersaturated. Back reaction would wholly or partially freeze such melts and redistribute the hydration profile across the olivine-wadsleyite transformation interval. This may potentially be a mechanism to trap large concentrations of H_2O in the transformation region, meaning that continued upwelling of solid mantle, melting at the top of the transformation interval, and percolation of dense melts downwards would lead toward a more fully hydrated transformation region. In other words, the condition depicted in Fig. 7C, would evolve towards that depicted in Fig. 7D, though it might not actually go towards a fully hydrated condition, as the melts may not be dense enough to percolate into the lower portions of the region, where the wadsleyite/olivine ratio is high.

If melt collects at a region of neutral buoyancy and if lateral flow away from the region of generation is sluggish compared to the rate of mantle advection through the region, considerable melt may accumulate in the layer. On the other hand, the rate of melt percolation may greatly exceed advection of solid mantle, as is generally true in the shallow mantle [e.g., *McKenzie*, 1984]. In this case, melt would likely drain from the boundary region faster than it may accumulate and dynamic processes would reduce rather than enhance melt accumulation.

Another consideration is that melting may continue in the depth interval above the olivine to wadsleyite transformation. At the end of the transformation process, the upper mantle solid assemblage remains at its storage capacity. If the storage capacity continues to diminish with decreasing depth, small amounts of melt may be generated throughout the upper mantle as advection continues [*Hirschmann, et al.*, 2005]. If the melt formed in this lowermost part of the upper mantle is negatively buoyant compared to the upper mantle matrix, it may pond at its depth of neutral buoyancy at the top of the transition zone. On the other hand, the top of the

transition zone may represent a minimum in the H_2O storage capacity/depth trend, owing to reactions between garnet and pyroxene [*Hirschmann, et al.*, 2005], in which case no additional melting will occur until significantly shallower depths. If there is no such reversal in storage capacity and neutral buoyancy occurs at some depth above the transformation, a modest amount of melt may be added to the layer perched on top of the transition zone. For example, if neutral buoyancy occurs at 300 km depth and if the storage capacity of peridotite diminishes from 1500 ppm at ~400 km down to 750 ppm at 300 km, 0.25% melting may occur in the bottom 100 km of the upper mantle.

Effect of H_2O on Character of Seismologic Characteristics of Upper Mantle-Transition Zone Boundary

The phase equilibria constraints described here may serve as a guide for understanding geophysical observations of the upper mantle-transition zone boundary. If the quantity of H_2O present is modest, then the thickness and the mean depth of the transformation region will be proportional to the concentration of H_2O, as originally envisioned by *Wood* [1995], though the quantitative relationships may differ [*Hirschmann, et al.*, 2005] (Fig. 7b). If sufficient H_2O is present to induce some melting, the observable geophysical signatures may be quite different from this. However, it must be kept in mind that the melt distribution predicted by phase equilibria is almost certainly modified by dynamic processes.

If sufficient H_2O is present to induce incipient melting, as argued by *Huang et al.* [2005] (but see also *Hirschmann* [2006b] and *Huang et al.[2006]*), then the thickness of the transformation interval should be maximal (Fig. 7c). Although the calculated thickness for such conditions may vary with presumed values for olivine and wadsleyite storage capacities (Fig. 8), combination of these calculations with experimental observations of thickening of ~40 km [*Smyth and Frost*, 2002] suggest that incipient melting should correspond to a transformation interval that at least approaches this thickness. To date, the only region for which such thick transformation intervals have been identified is beneath the Mediterranean [*van der Meijde, et al.*, 2003]. Incipient melting should also be accompanied by significant shoaling, as is also observed, at least locally, in this region [*van der Meijde, et al.*, 2005].

If there are any regions on Earth where the transition zone is substantially hydrous (i.e., on the order of ~1 wt.% H_2O), then there could be extensive melting through the transition zone-upper mantle boundary (Fig. 7d). Such melting would not necessarily generate significant thickening of the transformation interval, but would correspond to dramatic shoaling of the boundary. Thus, the observations of *Blum and Shen [2004]* from beneath southern Africa, which suggest that the transition zone-upper mantle boundary is ~20 km shallower than normal but do not indicate significant boundary-broadening may be the best known candidate for such a region.

This amount of shoaling is less than might be expected based on present calculations or the experimental results of *Smyth and Frost* [2002], but are consistent with experiments of *Chen et al.* [2002] for fully saturated conditions.

One important enigma that arises from the calculations presented here is that the thickness of partially molten regions at 410 km is never greater than ~7 km. Thus, hydrous melting may not be capable of explaining seismically slow regions ~20 km thick as inferred beneath western North America by *Song et al.* [2004]. Either these observations suggest a significant dynamic accumulation of hydrous melt at this boundary or the seismological observations may require an additional explanation.

Acknowledgements. We are grateful for the editorial efforts of Steve Jacobsen and Suzan van der Lee in assembling this volume as well as their careful comments and corrections of this ms. The ms. also benefited from the comments of two anonymous referees. We thank Shun Karato and Joe Smyth for access to some of their unpublished work. This work supported by National Science Foundation grant EAR-0456405.

REFERENCES

Agee, C. B. (1998), Phase transformations and seismic structure in the upper mantle and transition zone, *Rev. Mineral.*, *37*, 165–203.

Akaogi, M., E. Ito, and A. Navrotsky (1989), Olivine-modified spinel-spinel transitions in the system Mg_2SiO_4-Fe_2SiO_4: Calorimetric measurements, thermochemical calculation, and geophysical application, *J. Geophys. Res.*, *94*, 15671–15685.

Bell, D. R., G. R. Rossman, J. Maldener, D. Endisch, and F. Rauch (2003), Hydroxide in olivine: a quantitative determination of the absolute amount and calibration of the IR spectrum, *J. Geophys. Res.*, *108*, 2105, doi:10.1029/2001JB000679.

Bercovici, D., and S.-I. Karato (2003), Whole-mantle convection and the transition-zone water filter, *Nature*, *425*, 39–44.

Blum, J., and Y. Shen (2004), Thermal, hydrous, and mechanical states of the mantle transition zone beneath southern Africa, *Earth Planet. Sci. Lett.*, *217*, 367–378.

Chambers, K., A. Deuss, and J. H. Woodhouse (2005), Reflectivity of the 410-km discontinuity from PP and SS precursors, *J. Geophys. Res.*, *110*, B02301, doi:10.1029/2002GL016590.

Chen, J., T. Inoue, H. Yurimoto, and D. J. Weidner (2002), Effect of water on olivine-wadsleyite phase boundary in the $(Mg,Fe)_2SiO_4$ system, *Geophys. Res. Lett.*, *29*, 1875, doi:10.1029/2001GL014429.

Demouchy, S., E. Deloule, D. J. Frost, and H. Keppler (2005), Pressure and temperature-dependence of water solubility in iron-free wadsleyite, *Am. Mineral.*, *90*, 1084–1091.

Fei, Y., and C. M. Bertka (1999), Phase transitions in the Earth's mantle and mantle mineralogy, in *Mantle Petrology: Field Observations and High Pressure Experimentation*, edited by Y. W. Fei, et al., pp. 189–208, Geochemical Society.

Frost, D. J. (2003), The structure and sharpness of $(Mg,Fe)_2SiO_4$ phase transformations in the transition zone, *Earth Planet. Sci. Lett.*, *216*, 313–328.

Gasparik, T. (1993), The role of volatiles in the transition zone, *J. Geophys. Res.*, *98*, 4287–4299.

Helffrich, G. (2000), Topography of the transition zone seismic discontinuities, *Rev. Geophys.*, *38*, 141–158.

Helffrich, G. R., and B. J. Wood (1996), 410 km discontinuity sharpness and the form of the olivine alpha-beta phase diagram: Resolution of apparent seismic contradictions, *Geophys. J. Int.*, *126*, F7–F12.

Hirschmann, M. M. (2006a), Water, Melting, and the Deep Earth H_2O Cycle, *Ann. Rev. Earth Planet. Sci.*, 34: 629–653.

Hirschmann, M. M. (2006b), A wet mantle conductor? *Nature*, *439*, E3–E4, doi:10.1038/nature04528.

Hirschmann, M. M., C. Aubaud, and A. C. Withers (2005), Storage capacity of H_2O in nominally anhydrous minerals in the upper mantle, *Earth Planet. Sci. Lett.*, *236*, 167–181.

Huang, X. G., Y. S. Xu, and S. I. Karato (2005), Water content in the transition zone from electrical conductivity of wadsleyite and ringwoodite, *Nature*, *434*, 746–749.

Huang, X. G., Y. S. Xu, and S. I. Karato (2006), A wet mantle conductor? *Nature*, *439*, E3–E4, doi:10.1038/nature04529.

Ita, J., and L. Stixrude (1992), Petrology, elasticity, and composition of the mantle transition zone, *J. Geophys. Res.*, *97*, 6849–6866.

Karato, S. I. (2006a), Microscopic models for the effects of hydrogen on physical and chemical properties of Earth materials, in *Superplume-Beyond Plate Tectonics*, edited by D. Yuen, et al., Springer Verlag.

Karato, S. I., Bercovici, D., Leahy, G., Richard, G. and Jing, Z. (2006b), The transition-zone water filter model for global material circulation: Where do we stand? in *Earth's Deep Water Cycle*, edited by S. D. Jacobsen and S. van der Lee, American Geophysical Union, Washington D.C.

Katayama, I., K. Hirose, H. Yurimoto, and S. Nakashima (2003), Water solubility in majoritic garnet in subducting oceanic crust, *Geophys. Res. Lett.*, *30*, 2155, doi:10.1029/2003GL018127.

Katsura, T., and E. Ito (1989), The system Mg_2SiO_4-Fe_2SiO_4 at high-pressures and temperatures: precise determination of stabilities of olivine, modified spinel, and spinel, *J. Geophys. Res.*, *94*, 15663–15670.

Katsura, T., et al. (2004), Olivine-wadsleyite transition in the system $(Mg,Fe)_2SiO_4$, *J. Geophys. Res.*, *109*, B02209.

Kawamoto, T., R. L. Hervig, and J. R. Holloway (1996), Experimental evidence for a hydrous transition zone in the early Earth's mantle, *Earth Planet. Sci. Lett.*, *142*, 587–592.

Kawamoto, T., K. N. Matsukage, K. Mibe, M. Isshiki, K. Nishimura, N. Ishimatsu, and S. Ono (2004), Mg/Si ratios of aqueous fluids coexisting with forsterite and enstatite based on the phase relations in the Mg_2SiO_4-SiO_2-H_2O system, *Am. Mineral.*, *89*, 1433–1437.

Koga, K., E. Hauri, M. M. Hirschmann, and D. Bell (2003), Hydrogen concentration analyses using SIMS and FTIR: comparison and calibration for nominally anhydrous minerals, *Geochem. Geophy. Geosys.*, *4*, 1019, doi:10.1029/2002GC000378.

Kohlstedt, D. L., H. Keppler, and D. C. Rubie (1996), Solubility of water in the α, β and γ phases of $(Mg,Fe)_2SiO_4$, *Contrib. Mineral. Petrol.*, *123*, 345–357.

Litasov, K., and E. Ohtani (2002), Phase relations and melt compositions in CMAS-pyrolite-H_2O system up to 25 GPa, *Phys. Earth. Planet. Int.*, *134*, 105–127.

Litasov, K., and E. Ohtani (2003a), Hydrous solidus of CMAS-pyrolite and melting of mantle plumes at the bottom of the upper mantle, *Geophys. Res. Lett.*, *30*, 2143, doi:10.1029/2003GL018318.

Litasov, K., and E. Ohtani (2003b), Stability of various hydrous phases in CMAS pyrolite-H_2O system up to 25 GPa, *Phys. Chem. Mineral.*, *30*, 147–156.

Matsukage, K., Z. Jing, and S. I. Karato (2005), Density of hydrous silicate melt at the conditions of Earth's deep upper mantle, *Nature*, *438*, 488–491.

McKenzie, D. (1984), The generation and compaction of partially molten rock, *J. Petrol.*, *25*, 713–765.

Mosenfelder, J., N. Deligne, P. Asimow, and G. R. Rossman (2006), Hydrogen incorporation in olivine from 2–12 GPa, *Am. Mineral.*, *91*, 285–294.

Revenaugh, J., and S. A. Sipkin (1994), Seismic evidence for silicate melt atop the 410 km mantle discontinuity, *Nature*, *369*, 474–476.

Ringwood, A. E. (1975), *Composition and Petrology of the Earth's Mantle*, 618 pp., McGraw-Hill, New York.

Sakamaki, T., i. A. Suzuk, and E. Ohtani (2006), Stability of hydrous melt at the base of the Earth's upper mantle, *Nature*, *439*, 192–194.

Smyth, J. R., and D. J. Frost (2002), The effect of water on the 410-km discontinuity: an experimental study, *Geophys. Res. Lett.*, *29*, 1485, doi:10.1029/2001GL014418.

Smyth, J. R., D. J. Frost, and F. Nestola (2005), Hydration of olivine and the Earth's deep water cycle, *Geochim. Cosmochim. Acta*, *69*, A746.

Song, T. R. A., D. V. Helmberger, and S. P. Grand (2004), Low-velocity zone atop the 410-km seismic discontinuity in the northwestern United States, *Nature*, *427*, 530–533.

Stixrude, L. (1997), Structure and sharpness of phase transitions and mantle discontinuities, *J. Geophys. Res.*, *102*, 14835–14852.

van der Meijde, M., F. Marone, D. Giardini, and S. van der Lee (2003), Seismic evidence for water deep in earth's upper mantle, *Science*, *300*, 1556–1558.

van der Meijde, M., S. van der Lee, and D. Giardini (2005), Seismic discontinuities in the Mediterranean mantle, *Phys. Earth. Planet. Int.*, *148*, 233–250.

Vinnik, L., M. R. Kumar, R. Kind, and V. Farra (2003), Super-deep low-velocity layer beneath the Arabian plate, *Geophys. Res. Lett.*, *30*, 1415, doi:10.1029/2002GL016590.

Wood, B. J. (1995), The effect of H_2O on the 410-kilometer seismic discontinuity, *Science*, *268*, 74–76.

Yoshino, T., Y. Takei, D. A. Wark, and E. B. Watson (2005), Grain boundary wetness of texturally equilibrated rocks, with implications for seismic properties of the upper mantle, *J. Geophys. Res.*, *110*.

Young, T. E., H. W. Green, A. M. Hofmeister, and D. Walker (1993), Infrared spectroscopic investigation of hydroxyl in β-$(Mg,Fe)_2SiO_4$ and coexisting olivine: implications for mantle evolution and dynamics, *Phys. Chem. Mineral.*, *19*, 409–422.

Zhang, J. Z., and C. Herzberg (1994), Melting experiments on anhydrous peridotite KLB-1 from 5.0 to 22.5 GPa, *J. Geophys. Res.*, *99*, 17729–17742.

Zhao, Y.-H., S. B. Ginsberg, and D. L. Kohlstedt (2004), Solubility of hydrogen in olivine: dependence on temperature and iron content, *Contrib. Mineral. Petrol.*, *147*, 155–161.

The Transition-Zone Water Filter Model for Global Material Circulation: Where Do We Stand?

Shun-ichiro Karato, David Bercovici, Garrett Leahy,
Guillaume Richard and Zhicheng Jing

Yale University, Department of Geology and Geophysics, New Haven, CT 06520

Materials circulation in Earth's mantle will be modified if partial melting occurs in the transition zone. Melting in the transition zone is plausible if a significant amount of incompatible components is present in Earth's mantle. We review the experimental data on melting and melt density and conclude that melting is likely under a broad range of conditions, although conditions for dense melt are more limited. Current geochemical models of Earth suggest the presence of relatively dense incompatible components such as K_2O and we conclude that a dense melt is likely formed when the fraction of water is small. Models have been developed to understand the structure of a melt layer and the circulation of melt and volatiles. The model suggests a relatively thin melt-rich layer that can be entrained by downwelling current to maintain "steady-state" structure. If deep mantle melting occurs with a small melt fraction, highly incompatible elements including hydrogen, helium and argon are sequestered without much effect on more compatible elements. This provides a natural explanation for many paradoxes including (i) the apparent discrepancy between whole mantle convection suggested from geophysical observations and the presence of long-lived large reservoirs suggested by geochemical observations, (ii) the helium/heat flow paradox and (iii) the argon paradox. Geophysical observations are reviewed including electrical conductivity and anomalies in seismic wave velocities to test the model and some future directions to refine the model are discussed.

1. INTRODUCTION

The most well-known on-going chemical differentiation process on Earth is partial melting beneath mid-ocean ridges, which is primarily due to the adiabatic upwelling (*pressure-release melting*) and creates relatively "enriched" oceanic crust and "depleted" residual mantle (e.g., [*Hofmann*, 1997]). If this is the only chemical differentiation process operating at present and in the recent past, then several immediate consequences follow. For example, if one accepts the model

of whole mantle convection, as suggested by seismic tomography ([*Grand*, 1994; *van der Hilst, et al.*, 1997]), then the whole mantle is depleted with only a small volume (~10%) of relatively enriched material continuously replenished by subduction of oceanic crust and sediments. This view is inconsistent with geochemical observations that suggest the entire lower mantle (~70% of the mantle) is relatively enriched [*Albarède and van der Hilst*, 2002; *Allègre, et al.*, 1996]). An alternative model assumes that differentiation at mid-ocean ridges involves material only from the upper mantle (e.g., [*Allègre, et al.*, 1996]). In this model, the deeper mantle is not involved in chemical differentiation at mid-ocean ridges and maintains relatively enriched chemical composition. However, this latter model implies strongly

Earth's Deep Water Cycle
Geophysical Monograph Series 168
Copyright 2006 by the American Geophysical Union.
10.1029/168GM22

layered convection that is not consistent with the results of seismic tomography.

High-pressure experimental studies on properties of minerals suggest that melting may not be limited to the shallow upper mantle, but may occur in the deep upper mantle or the transition-zone (e.g., [*Dasgupta and Hirschmann*, 2006; *Gasparik*, 1993; *Ohtani, et al.*, 2001; *Wang and Takahashi*, 2000; *Young, et al.*, 1993]). *P-T* conditions in the deep mantle are well below the solidus of dry, depleted peridotite so melting under deep mantle conditions would require the addition of "impurities" such as water (*flux melting*). For example, a large contrast in the solubility of water (hydrogen) between minerals in the upper mantle and the transition-zone minerals (e.g., [*Kohlstedt, et al.*, 1996]) could lead to partial melting when the upwelling materials passes the 410-km boundary. [*Bercovici and Karato*, 2003] analyzed possible consequence of this transition-zone melting on the geochemical cycling and mantle convection. They proposed that if the melt produced there is denser than the upper mantle minerals (but lighter than the transition zone minerals) then the melt will remove much of the highly incompatible elements so that the mantle will be chemically layered in terms of highly incompatible elements although the majority of mantle materials undergo the whole mantle wide convection. This transition-zone water filter model (hereafter referred to as the TZWF model) assumes the following processes: (1) The global flow pattern in Earth's mantle is dominated by the slab-related localized downwelling currents and diffuse upwelling flow forced by the slab flux that occurs more-or-less homogeneously in most of the mantle. (2) There is enough water in the transition-zone to cause dehydration-induced partial melting of ambient upwelling material as it goes to from wadsleyite-dominated assemblage to an olivine-dominated one upon crossing the 410-km boundary. (3) The melt formed there is denser than the surrounding materials in the upper mantle but lighter than the transition-zone materials and hence will be trapped at the 410 km discontinuity. (4) Partial melting removes some of the incompatible elements from the original materials and hence the residual materials are depleted with these elements. (5) The residual materials continue upwelling into the upper mantle and become the relatively depleted source materials for MORB. (6) The separation of incompatible elements (filtering) is not effective for hot plumes because hot melts have low density, high ascent velocities, and thus shorter residence times in the transition zone resulting in an "unfiltered" source for OIB. (7) Melt thus formed on top of the 410 km boundary will eventually return to the deeper mantle due to entrainment by downwelling material associated with subducting slabs.

However, the original hypothesis proposed by [*Bercovici and Karato*, 2003] contained a number of loosely defined concepts and was based on then poorly known material properties. Furthermore, some of the geophysical and geochemical consequences of the model were not examined in any detail. During the last a few years, there has been much progress in (i) mineral physics, (ii) geodynamic modeling and (iii) geophysical (geochemical) observations on the issues closely related to this TZWF model. The purpose of this paper is to review the current status of this model in view of these new studies. We first review new experimental observations relevant to melt formation and melt density. Secondly, recent models of water circulation and the influence of re-distribution of radioactive elements on convection pattern will be summarized. This is followed by a discussion of geophysical and geochemical constraints on the model. Consequences of transition-zone melting on electrical conductivity and seismological signatures and the influence of transition-zone melting on trace element distribution will also be reviewed. Finally we will discuss future directions to further test and modify the original model.

2. MINERAL PHYSICS BASIS FOR THE TRANSITION-ZONE WATER FILTER

Melting Relations

In the TZWF model, broadly distributed upwelling materials across the 410-km discontinuity partially melts due to the dehydration of wadsleyite. Flux melting occurs by the presence of impurities (such as hydrogen) and the extent of melting is greater when the solubility of impurities in the solid phases is low. An impurity element such as hydrogen (water) has important influence on melting because it is more dissolvable in a liquid phase than in a solid phase. When an impurity phase is dissolved in a material, its free energy is reduced primarily by the increase in the configurational (mixing) entropy, plus some effects through the change in the intrinsic Gibbs free energy (such as the internal energy and the vibrational entropy). Consequently, when an incompatible phase is present, the Gibbs free energy of a liquid phase is reduced by more than that of a solid phase, resulting in a reduction of the melting temperature. However, since the conditions for melting are controlled by the relative Gibbs free energies of melt and solid, the actual melting conditions also depend on the solubility of impurities in the solid phase. The solubility of hydrogen in wadsleyite is larger than that in olivine by roughly 3-10 times at conditions near 410-km (e.g., [*Kohlstedt, et al.*, 1996]). Consequently, the partial melting of upwelling current across the 410-km discontinuity is a key feature of the TZWF model (e.g., [*Bercovici and Karato*, 2003; *Gasparik*, 1993; *Ohtani, et al.*, 2000; *Young, et al.*, 1993]).

Although the basic physics and chemistry of hydrous silicate melts (and with other incompatible elements) is well known, important details regarding the possibility of melting at ~410-km remain poorly constrained. Experimental difficulties in this area include: (i) melts under these conditions are unquenchable and consequently the water content in melts cannot be determined directly from quenched samples, and (ii) small degrees of melting are hard to identify particularly because the dihedral angle of melt and solid minerals under these conditions becomes zero [*Yoshino, et al.*, 2006]. Also, the influence of other incompatible components such as potassium, sodium and/or carbon is critical in estimating the conditions for formation of a dense melt, have not been considered in the previous transition-zone water filter model (e.g., [*Bercovici and Karato*, 2003; *Hirschmann, et al.*, 2005]). Controversy surrounding the calibration of water content [*Bell, et al.*, 2003; *Koga, et al.*, 2003] (see also [*Aubaud, et al.*, 2006; *Karato*, 2006b]) complicates estimates of the amount of water needed for melting from experimental observations.

Considering these complications, we believe that the best approach is to illustrate the range of behavior using a simplified generic phase diagram (Fig. 1). Let us consider melting of mantle material and the presence of water as a simplified pseudo-binary system composed of "silicate" and "water". The "silicate" may contain incompatible components other than water such as K_2O, CO_2 and Na_2O that will affect the melting behavior. For simplicity, the "melting temperature" of water-free peridotites is shown only by one value, T_2 or T_3 for the upper mantle and transition zone materials respectively. This is a good approximation for the melting of impurity-free peridotites in the deep upper mantle for which the solidus and the liquidus are close (e.g., [*Takahashi*, 1986]). However, in cases where the influence of volatile components such as K_2O, CO_2 and Na_2O other than water is important, these temperatures must be interpreted as the solidi corresponding to appropriate systems (in these cases, a phase diagram in at least three-dimensional space is needed to fully characterize the melting behavior).

Fig. 1 illustrates this point. The mantle rocks without water melt at a certain temperature range (T_2 or T_3). When water is added, then water is distributed between the melt and the solid. Water is dissolved into the melt more than in the solid minerals, and hence the melting temperature will be reduced. Now we have two phases (melt and mineral) coexist at a given P and T each of which has a certain amount of water that is controlled by the partitioning coefficient. Each curve describing the composition and T (P) for each coexisting phase will define the solidus and the liquidus. In Fig. 1, the curve T_2-Q is the solidus, and the curve T_2-X_1 is the liquidus for the upper mantle phase (for the transition zone minerals

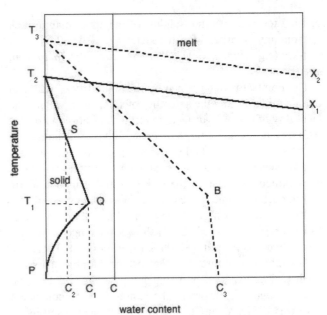

Fig. 1. Schematic phase diagram for a theoretical binary system of silicate + water at ~410 km. Thicker dashed and solid lines represent the solidus for transition zone and upper mantle minerals, respectively. The curve $P-Q$ (thin dashed line) represents the solubility of water in upper mantle minerals as a function of temperature. Above a certain temperature (T_1), a water-rich fluid/melt forms and the amount of hydrogen in solid minerals coexisting with the hydrous melt decreases with temperature (trend $Q-S-T_2$). Given the curves $T_2 - X_1$ (liquidus) and $Q-S-T_2$ (solidus), one can infer the critical water content for melting and the water content in the melt (for details see text).

$T_3 - B$ is the solidus, and $T_3 - X_2$ is the liquidus). The low temperature portions of the phase diagram can be analyzed using the experimental data on the solubility of water in the solid minerals. At low temperatures, a fluid phase that would coexist with the minerals will be free water. Solid minerals have a certain degree of solubility of water. Therefore the phase diagram at low temperature is characterized by a curve $P-Q$ that is essentially the solubility of water in the upper mantle minerals (the solubility of water in the transition zone minerals is much higher and is shown by a curve $C_3 - B$). When water starts to dissolve silicate components, then the fugacity of water in the system will be lowered. As a consequence, the amount of water dissolved in the solid minerals will be reduced. Eventually, the fluid phase becomes indistinguishable from silicate melt, and the solubility curve should become the solidus curve. The inflection point Q defines the temperature at which the solubility of silicates in water becomes large.

Let us first discuss the curve T_2-Q, the solidus of the upper mantle rocks, which defines the conditions for melting. T_2 is the melting temperature (or the solidus) without water. The melting temperature (the solidus) of the upper mantle rock without any volatiles is well constrained. It is ~2400 K at 410-km (e.g., [*Inoue*, 1994; *Ohtani, et al.*, 1995; *Takahashi*, 1986]).

The point Q is determined by the solubility of water in mantle minerals and the solubility of silicates in water. The solubility of water (hydrogen) in olivine is well-constrained ([*Kohlstedt, et al.*, 1996; *Zhao, et al.*, 2004]), and some less detailed data are available for other minerals such as garnet. However, the solubility of silicates in water is not well constrained. Because of the lack of well-defined eutectic melting behavior due to the total miscibility between silicate melts and water at high P, the point Q is not well defined. However, [*Stalder, et al.*, 2001] showed a change in properties of hydrous melt at T~1500 K (at P~14 GPa). Therefore we use this temperature as a rough estimate for the inflection point Q. The experimental studies by [*Inoue*, 1994; *Inoue and Sawamoto*, 1992] show the results that are consistent with this choice of inflection point. The water solubility in olivine at the point Q (T~1500 K and P~14 GPa) is ~0.1 wt%. The total water solubility in upper mantle rocks at this condition depends also on the water solubility in other minerals. Currently there is a large discrepancy on the water solubility in garnets (e.g., [*Lu and Keppler*, 1997; *Withers, et al.*, 1998]). Also the water solubility in pyroxenes is not well constrained under deep upper mantle conditions (e.g., [*Mierdel and Keppler*, 2004; *Rauch and Keppler*, 2002]) (for a review of the experimental data see also [*Bolfan-Casanova*, 2005]). In this paper, we tentatively use the solubility value in olivine C_W~ 0.1 wt% (and T_I~1500 K) to define the conditions at $Q(C_W - T_I)$. The estimated water content at Q has an uncertainty of a factor of ~2 caused by the uncertainties in the water solubility in garnet and pyroxenes and in the FT-IR calibration.

The solidus, $Q - S - T_2$, can be estimated by the interpolation between two fixed points, T_2 and $Q(C_W, T_I)$. The solidus line has a curvature such that it should be concave downward in this plot, but considering the large uncertainties, we use a straight line. If T_2 for volatile free upper mantle (~2400 K) is used, and if the typical mantle temperature of ~1800-2000 K at 410-km is used, then the critical water content is estimated to be ~0.05 wt%. The critical value of water content for partial melting estimated here is much less than the value estimated by [*Hirschmann, et al.*, 2005] (~0.4 wt%) with which one would conclude that melting is highly unlikely in most of the current Earth's mantle at ~410-km. This is due to the fact that [*Hirschmann, et al.*, 2005] essentially identified the solubility limit of water in upper mantle

minerals at the eutectic point, Q, as the solidus. The water content in olivine at the solidus at a higher temperature (S) will be significantly lower than the water content at Q. The solubility of water (hydrogen) in minerals and the water content at the solidus (which defines the minimum water content for melting) are different and should not be confused.

So far, we have only considered the influence of water on partial melting. It is well known that Earth contains other incompatible elements such as carbon, potassium and sodium that partition into the melt (e.g., [*McDonough and Sun*, 1995]). Consequently, these elements should play a similar role in the melting as hydrogen. Experimental studies on melting of upper mantle peridotites have shown that these impurities reduce the melting temperature of dry peridotite (see Table 1). Consequently, for the water-free but otherwise impurity-present mantle, the solidus temperature, T_2, will be ~1800-2000 K at ~410 km as opposed to ~2400 K for volatile-free peridotite. Note that this temperature is close to the estimated temperature at 410-km (~1800-2000 K; e.g., [*Ito and Katsura*, 1989]). In fact, [*Wang and Takahashi*, 2000] suggested a wide-spread (small degree of) partial melting in the transition zone due to the influence of K_2O. The influence of incompatible elements on melting is essentially through their effects on the configurational (mixing) entropy, and hence the conditions for melting at a given temperature and pressure is determined by the net molar abundance of incompatible elements. Consequently, the minimum amount of water to cause partial melting at 410-km is reduced when other incompatible elements are present. We conclude that the minimum amount of water needed for melting is ~0.05 wt% or less for a realistic mantle where other impurities are also present. When the estimated critical value of water content for melting is compared with the recent estimate of water contents in the transition zone (~0.1-0.2 wt% in the Pacific; [*Huang, et al.*, 2005; 2006], up to ~1 wt% in the western Pacific or the Philippine sea region; [*Suetsugu, et al.*, 2006]), we conclude that partial melting at or near 410-km is likely in a broad region of the current Earth's mantle.

The Melt Density

One of the direct consequences of this type of melting (*flux melting*) is that the melt produced should have a large concentration of incompatible elements, and the amount of incompatible elements in the melt increases with a decrease in temperature. To be more specific, "pure" peridotites (peridotites with little or no volatile, incompatible elements)

---

[1] [*Hirschmann et al.*, 2005] began their discussion by distinguishing the solidus and the solubility limit, but in their final estimate of the minimum water content for melting, they used the solubility limit.

Table 1. Amount of reduction in solidus of peridotite by volatiles at ~14 GPa

CO_2	~400 K	eutectic*	[*Dasgupta and Hirschmann*, 2006]
K_2O	~600 K	eutectic*	[*Wang and Takahashi*, 2000]
Na_2O	~150 K	for 0.3 wt%	[*Hirschmann, et al.*, 1998]
H_2O	~600 K	for 2 wt%	[*Litasov and Ohtani*, 2002]

*: For eutectic melting, the solidus reduction is independent of the concentration of impurities. However, the notion of "eutectic" behavior is not well documented in these studies. A small amount of solubilities of these elements in the mineral will change the melting behavior to solid-solution type as shown in **Fig. 1**.

will melt at ~2400 K at ~410-km depth (e.g., [*Inoue*, 1994; *Inoue and Sawamoto*, 1992; *Ohtani, et al.*, 1995; *Takahashi*, 1986]). A plausible temperature at 410-km discontinuity is ~1800-1900 K for typical regions, and ~2000-2100 K for hot regions (i.e., plumes or the Archean mantle). Therefore one needs to have a large enough concentration of impurities to reduce the solidus by ~300-500 K.

The concentration of incompatible elements in the melt depends strongly on the degree of melting temperature reduction needed to cause melting. A simple thermodynamic analysis shows that a decrease in melting point by incompatible elements is due mainly by their effect of increasing the configurational entropy of a melt, thus $\delta S_{config} = \Delta S_m \frac{\delta T}{T_m}$ where δS_{config} is the change in configurational entropy, ΔS_m is the entropy change upon melting, T_m is the melting temperature, and δT is the reduction of melting temperature. Because δS_{config} is a function of concentration of incompatible elements, equation $\delta S_{config} = \Delta S_m \frac{\delta T}{T_m}$ gives a relation between the concentration of incompatible elements and melting temperature reduction. At deep mantle melting, $\Delta V_m \sim 0$ and $\Delta S_m \approx Rlog2 = 0.69$ R and hence $\delta S_{config} \sim 0.15$ R ~ for $T_m = 2400$ K and $\delta T = 500$ K. This can be translated into the (molar) concentration of impurities, x, through $S_{config} = -R\left[xlog\,x + (1-x)\,log(1-x)\right]$ to $x \sim 0.35$. Experimental

studies show that ~10-15 wt% of water is dissolved in the melt at P~14GPa to reduce the melting temperature by ~500 K ([*Litasov and Ohtani*, 2002]). If all hydrogen is in the form of H_2O, this will correspond to $x = 0.15-0.22$, whereas if all hydrogen in the form of OH, $x = 0.30-0.45$. We conclude that the experimental observation by [*Litasov and Ohtani*, 2002] agrees with a simple model and a molar fraction of impurity on the order of ~0.2-0.4 is needed to cause melting at 410-km at a temperature of ~1900 K.

Incompatible elements such as hydrogen enhance melting, but the addition of incompatible elements also influences melt density. Being the lightest element, the addition of hydrogen significantly reduces the density of a melt. [*Matsukage, et al.*, 2005] determined the influence of water on the density of peridotite melt. The density of water in the peridotite melt at P~14 GPa and T~1900 K is ~2.3 × 10^3 kg/m³ (molar volume ~8×10^{-6} m³/mol), and comparable to effects of H2O on the density MORB melt [*Sakamaki et al.*, 2006].

Given the molecular weights and the partial molar volumes of incompatible components, one can estimate the density of melt and examine the critical concentrations of incompatible components needed to produce a melt denser than the surrounding solid mantle at 410-km. *Matsukage et al.* [2005] calculated the critical water contents for a dense peridotite melt at ~410-km conditions (P=14 GPa, T~1900 K) and obtained a value of ~5 (+7, -2) wt% ([*Sakamaki, et al.*, 2006] obtained a similar result for MORB). Although uncertainties are very large, these values are less than the estimated water content in the melt under similar conditions (~10-15 wt%) and we conclude that the conditions for a dense melt are limited if only water is considered.

Water is the lightest impurity component, but in the mantle there are expected to be other impurities with higher density components that have higher densities (Table 2). Based on the above analysis, we now assume that a total amount of impurity is ~0.2-0.4 (molar fraction) through a combi-

Table 2. The abundance and properties of typical "impurity" components

	molecular weight	abundance[1]	density[2]	partial molar volume
H_2O	18	50-1000 (wt ppm)	2.3 (× 10^3 kg/m³)	8 (× 10^{-6} m³/mol)
CO_2	44	1200	2.9	15
Na_2O	62	6200	3.2	19
K_2O	94	300	3.4	28

[1]: water: see papers in this volume, carbon dioxide: [*Dasgupta and Hirschmann*, 2006], Na_2O: [*McDonough and Sun*, 1995], K_2O: [*McDonough and Sun*, 1995].

[2]: The densities and partial molar volumes of these species in silicate melt at P=14 GPa and T=1900 K. The data for water is from [*Matsukage, et al.*, 2005], the density of CO_2 is estimated from the equation of state (e.g., [*Frost and Wood*, 1997]), and those for Na_2O and K_2O are calculated from low P data using the Birch-Murnaghan equation of state for $K' = 6$ with the initial density and K_o summarized by [*Lange and Carmichael*, 1990]. The uncertainty in the density (partial molar volume) estimate is ~±20%.

nation of impurities (K_2O, Na_2O and CO_2). The critical ratio of water to other impurities is $\frac{c_w}{c_x}$ (C_W: molar content of water, C_X: molar content of another volatile impurity such as K_2O). For example, K_2O is heavy and makes a melt denser when K_2O is added (Fig. 2). For a total impurity content of 0.3, the critical value of $\frac{c_w}{c_x}$ is ~0.4 (when C_X is the molar content of K_2O). When the fraction of water exceeds this critical value, then the melt will be light and the water filter leaks. Regions with a very high water content (geochemically "enriched" regions) will undergo melting at ~410-km but the melt will be light and filtering will not occur. These materials will form OIB with enriched geochemical signature upon melting near the surface. Whereas when the fraction of water content is less than this value but the total volatile content is large enough for melting to occur, then the melt is dense and the water filter will work, and a depleted materials after filtering will become the source materials for MORB. Finally, when the total amount of volatiles is too low, then melting does not occur and the water filter will not work. These materials will rise to form MORB or relatively depleted OIB. An important conclusion is that the water filter will work only when the water content is in an intermediate range. We note that the

above estimates of critical conditions for a dense melt have very large uncertainties for two reasons. First, the estimate of the partial molar volumes of volatile components contains large uncertainties (see Table 2). Second, the estimate of the critical molar concentration of volatile from experimental studies has large uncertainties. Consequently, the above values must be considered as rough estimate only.

The new experimental data on melting relationship and the density of hydrous melt indicate that the conditions under which water-filter mechanism operates depend critically on the concentration of incompatible components (as well as the mantle temperature). The previous studies on melting and melt density have been focused on the influence of water only. When only the influence of water is considered, although partial melting likely occurs in most cases (melting occurs if more than ~0.05wt% of water is present), the melt thus produced will probably be lighter than the surrounding solid minerals because the reduction of melting temperature can be achieved only by dissolving a large amount of water in the melt.

However, considering other incompatible impurities such as CO_2, K_2O or Na_2O, flux melting and the formation of denser melt is possible. We conclude that partial melting in the deep mantle at ~410-km is highly likely, and a melt of sufficient density to remain neutrally buoyant at ~410 km will include CO_2, K_2O or Na_2O in addition to H_2O. In a typical oceanic upper mantle, current estimates of water content in the transition zone are small (~0.1 wt%), and the major impurity components in the melt are K_2O or Na_2O. However, in regions of the transition zone where subduction occurs, higher water contents of ~1 wt% are inferred (e.g., [*Suetsugu, et al.*, 2006]). In these cases, melt would contain a large amount of water and be less dense than the surrounding mantle and hence rise and result in volcanism. Some volcanism away from the trench could be due to melting of subducted material with higher water contents. ([*Iwamori*, 1992; *Miyashiro*, 1986]). Some OIBs may be formed by the upwelling of relatively water-rich materials that do not undergo filtering because of the low density of melt.

3. MODELS OF MATERIAL CIRCULATION

Simple Mass-Balance Box Model (Appendix I)

In this section, we review theoretical modeling of dense partial melt at ~410-km and discuss plausible processes by which volatile components such as water are entrained and circulated through the transition zone. In the original TZWF study, [*Bercovici and Karato*, 2003] proposed a simple mass balance or box-model calculation to illustrate the basic predictions of their hypothesis. We briefly review the primary

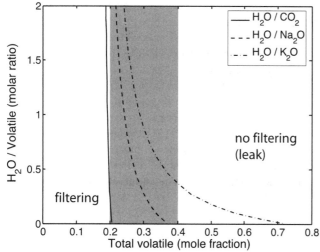

Fig. 2. Critical conditions (in terms of volatile composition) for which water filter works. When the density of melt generated at ~410-km is denser than that of the surrounding upper mantle minerals (but lighter than the transition-zone minerals), then water filter will work. The total volatile content needed for melting at ~410-km is difficult to calculate mainly because of (i) a large uncertainty in the partial molar volume of water (this diagram assumes 8 cm^3/mol) and (ii) a large uncertainty in the experimental determination of volatile content but is estimated to be 0.2–0.4. For a given total volatile content, there is a maximum water content (relative to the concentration of other volatiles) at which the melt is dense and the water filter works. In regions where the water content is very high, the melt will be light and the filter will leak.

ingredients of this model in order to establish the setting for further unanswered geodynamical questions. In particular, we discuss in some detail the assumptions made in constructing this box model, and where recent developments shed new light on these assumptions.

The putative heavy melt-rich layer at the 410-km boundary that is predicted by the TZWF model is assumed to reach a steady thickness once the melt production from the mantle upwelling is balanced by melt entrainment from slab-induced downwelling (see sections below). Since the melt is heavy it will accumulate and the solid matrix will decompact or dilate to accommodate the pooling of heavy melt. However, the matrix will not necessarily completely decompact if the melt is drawn away by slab-associated entrainment is fast enough; i.e., the pooling of heavy melt drives decompaction but the melt could be extracted efficiently enough that the matrix never completely decompacts. Two-phase compaction theory [*Bercovici and Ricard*, 2003; *Bercovici, et al.*, 2001; *McKenzie*, 1984] can be used to infer basic scaling relationships between the peak porosity Φ in the melt-rich layer, the melt-rich layer thickness, H, and the slab-induced entrainment rate U. These relations are shown in Fig. 3 for an input melt fraction of $f = 1\%$. First, we can see that increasing slab entrainment rate U causes both Φ and H to decrease

because as the entrainment is more efficient it both keeps the melt-rich layer thinner and suppresses decompaction of the matrix. H scales with $H^* \equiv \sqrt{\frac{\mu_m W}{|\Delta\rho| g}}$ (μ_m: solid viscosity, W: ambient mantle upwelling velocity ($\sim 10^{-11}$ m/s ([*Bercovici and Karato*, 2003]), $|\Delta\rho|$: density difference between melt and solid). The thickness of the melt-rich layer is sensitive to the solid viscosity and the density contrast. For typical values of $|\Delta\rho| \sim 10$ kg/m^3 and $\mu_m \sim 10^{18}$ Pa s, $H \sim 1$ km (however both parameters can vary by orders of magnitude that leads to a broad range of H: for example, if melt completely wets grain-boundaries under deep mantle conditions as suggested by [*Yoshino, et al.*, 2006], then μ_m will be considerably lower than the above value). Because this melt-rich layer is probably thin (for typical values of parameters), detection of this layer by seismology will require a high-resolution technique that is sensitive to the impedance contrast.

In the essential TZWF hypothesis, water is assumed to be trapped within the transition-zone because of dehydration melting of ambient upwelling at 410 km, and by exolution of water from slabs at 660km [*Bercovici and Karato*, 2003], [*Richard, et al.*, 2006a]. Without the production of liquid phases above and below the transition-zone, water would tend to be largely removed and homogenized throughout the mantle [*Bercovici and Karato*, 2003; *Richard, et al.*,

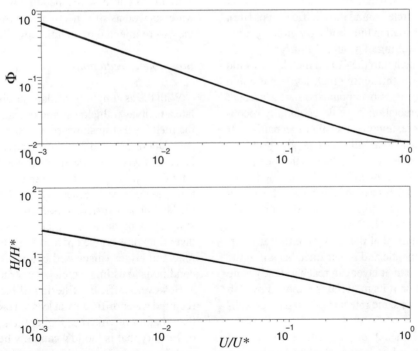

Fig. 3. Steady state scaling relationships for peak porosity Φ and melt layer thickness H versus slab-induced entrainment rate U for the putative heavy melt layerp at 410km (see Appendix I). The melt layer thickness scale is $H^* = \sqrt{\mu_m W/(|\Delta\rho| g)}$, and the entrainment rate scale is U ∗ = scale is $\sqrt{W |\Delta\rho| g/\mu_m}$, where W is the ambient mantle upwelling velocity, μm is the solid matrix viscosity, $\Delta\rho < 0$ is the solid-melt density difference, and g is gravity. Input melt fraction (i.e., of the upwelling mantle) is fixed at f = 1%. See text for discussion and Appendix I for details.

2002]. Of course, the complete entrapment of water in the transition-zone is an end-member case since some leakage is bound to occur (since the upper and lower mantle phases above and below the transition-zone can transport some water, albeit at small concentrations.) With the assumption that recycling of water is primarily closed in the transition-zone (with any leakage largely compensated by input from slabs transporting water from the surface), we posed a simple theoretical mass-balance model with which to understand, for example, the average thickness of the melt layer.

The model system is driven by a background circulation through the transition-zone consisting of downwelling slabs and the resultant passive ambient upwelling. The slab mass flux is in fact the primary driver of the entire system. We assume that the majority of the upwelling mantle rises passively in response to downward slab injection, and that plumes play only a minor role in the upward mass flux. This simple picture of mantle circulation is in keeping with standard models of convection where the heat source is primarily from radiogenic heating and secular cooling, both of which lead to a "top-cooled", downwelling dominated system. The net slab flux is calculated assuming a downwelling rate of order 10 cm/yr (or less), an average slab thickness of approximately 100 km, and a global slab length approximately equal to the circumference of the Earth (since most of the slab flux occurs in a nearly great circle around the Pacific ring of fire). A mass-flux balance calculation thus leads to a globally averaged passive mantle upwelling of around 1 mm/yr.

Continuous influx of melt into the 410km melt layer would eventually lead to an ever thickening melt layer that could possibly reach the density cross-over point and start to ascend; it if ascends through percolation it would essentially recontaminate the upper mantle, undoing the filter mechanism. If it ascends through diapiric instabilities (depending on how compacted and impermeable the solid is above the melt layer) then the melt and solid might remain isolated, allowing for the arrival of depleted MORB-source mantle; however, this would also suggest the arrival of very highly enriched wet melts in anomalous midplate or non-MORB magmatism.

A more likely scenario is that the heavy melt layer interacts with downwelling mantle and is entrained back into the lower mantle. Indeed, the melt layer can reach a steady state thickness and composition if its material is removed by slab induced entrainment at the same rate it is injected by upwelling ambient mantle.

In the original TZWF model, entrainment is assumed to occur solely by freezing of melt onto slabs (or within the slabs' cold thermal halo). Cooling and freezing of the melt near slabs is facilitated by "window-pane" convection (driven by cooling of fluid from the side) and leads to a thick silicate solidification entrainment zone [*Bercovici and Karato*,

2003]. However, the solidified material cannot remove water in sufficient quantities. That is, the silicates crystallize at the wadsleyite saturation value of a few percent (since, at cold near-slab temperatures, the melt solidifies to wadsleyite) which is much less than the melt's water concentration of perhaps tens of percent. Thus solidification of melt leaves large quantities of water, which likely just dissolve in the nearby melt. So water must be removed by additional means. Entrapment of melt and/or exsolved water by inclusions in the solidifying material is unlikely since the solid fraction is probably not large enough and the melt/water pathways not closed (again the dihedral angles are probably nearly zero). In our original model [*Bercovici and Karato*, 2003], the other likely water-removal mechanism was assumed to be chemical diffusion of water. The accumulation of water in the melt would put the melt well beyond chemical equilibrium with the slab and thus diffusion of water into the slab could occur at a sufficient rate to remove the water. In fact, this was one of the main elements of the self-limiting feature of the model that allowed steady state. If water removal was inefficient, then the melt layer would thicken and water would accumulate near the slab; however, the melt layer thickness also represents the slab-melt contact area and the thicker the layer the more water could diffuse into the slab. Thus eventually a thickness could be reached that would allow for the all the water as well as silicates to be removed at the same rate that they were injected by mantle upwelling.

Box-Model Predictions

With the assumptions of slab entrainment, one could calculate a melt layer thickness and overall distribution of water in the melt-layer–transition-zone system. However, such model calculations only indicate the effective thickness of a pure melt layer averaged over the entire globe: they do not account for lateral variations in melt layer thickness, nor do they estimate thickness of a partial melt layer which would conceivably be thicker but have different mechanical properties. Nonetheless, simple box-model results predicted an average effective melt layer thickness on the order of a few tens of kilometers. More recent analyses (discussed below) considers more realistic conditions to estimate melt layer structure.

However, sufficient filtering of the upwelling mantle also required water diffusivity at least an order or magnitude higher than that for olivine (i.e., higher than $10^{-8} m^2/s$). Otherwise, a diffusivity that is too low can only be compensated by thick melt layer, which would take up almost all the water of the system, leaving the transition-zone too dry to produce sufficient melt fractions to allow the filter to work.

In our original work, we argued that water diffusivity in wadsleyite was likely to be as much as 3 orders of magnitude faster than in olivine; no experimental evidence for this was known, although it was thought that the much higher diffusivity of other elements and the higher electrical conductivity in wadsleyite indicated the likelihood of a higher water (i.e., hydrogen) diffusivity. In this case, diffusive entrainment of water by slabs was a plausible mechanism. Yet, recent studies of water diffusivity in transition-zone minerals suggest that it is probably no faster than in olivine [*Hae, et al.*, 2005] which casts doubt on the slab/water-diffusion entrainment assumption. Moreover, preliminary work by our group [*Leahy and Bercovici*, 2006] indicates that the slab entrainment model is a serious oversimplification and that entrainment is a continuous process driven by chemical reaction with dry downwelling mantle and not only by freezing at the so-called interface between the melt and the slab. This new development is discussed below.

Entrainment by Reaction With Downwelling Mantle: Viscous Entrainment Zone

As described above, the original TZWF model stipulated that the melt layer was essentially divided into two regions: a melt production area, where wet upwelling material melts as it passes the 410km boundary, and a slab-entrainment area, where the melted material refreezes as it comes into contact with cold slabs, leading to its entrainment into the deeper mantle. However, this is an oversimplification in many ways, most notably in that it assumes the mantle makes an abrupt transition from upwelling (where melt is produced) to the downwelling slab region (where melt is removed). Being highly viscous, the mantle obviously cannot support such abrupt transitions in flow directions, so clearly there is a continuous change in the upwelling velocity, from its peak value (roughly at some point furthest from slabs, or equidistant from encircling slabs) through zero, across a relatively broad downwelling region being viscously dragged down by slabs, then to the maximum downwelling velocity at the cold slab itself. The region viscously entrained by slabs is roughly at ambient mantle temperatures, as typical for infinite Prandtl number systems where thermal diffusivity is much less than momentum diffusivity. The melt produced in the upwelling region will tend to spread under its own weight into this *viscous entrainment zone* (VEZ) and begin to react with the downwelling ambient mantle in a somewhat complicated mechanism we have proposed [*Leahy and Bercovici*, 2006] and describe below (Plate 1).

Reaction Between Melt and Viscous-Entrainment Zone

In the upwelling melt-production region, the heavy melt and the remaining buoyant solid are at, or nearly at, chemical equi-

librium with respect to water since partial melting is assumed to occur at equilibrium. As the melt spreads into the ambient downwelling, or viscous entrainment zone, it comes into contact with downgoing solid upper mantle that might at best be hydrated enough to still be near equilibrium with the melt (i.e., if the solid has remained unprocessed since passing through the 410km filter), or more likely is relatively dry (assuming further processing by continued melting during upwelling, or further melting at ridges or beneath the lithosphere) and thus well away from equilibrium with the melt. However, the downward traversal of this solid upper mantle material across an ostensibly heavier melt is a nontrivial process. In fact, a key step in the process is that melt is pushed or advected by the downwelling into the wadsleyite stability field, where (because of the higher wet-melting temperature of wadsleyite) the melt freezes (Fig. 4). Although silicates in the melt solidify as water-saturated wadsleyite, the melt's water concentration is much larger than the saturation limit for ambient wadsleyite (of order 1%) and so water is exsolved in the freezing process and, being buoyant, ascends to be reabsorbed by the overlying melt. The melts water concentration thus rises and given the melt's low viscosity, the water concentration of melt also homogenizes relatively quickly. This causes the hydrous melt to be out of equilibrium with the overlying downwelling olivine (even if the olivine had remained unprocessed); the melt thus reacts with the olivine (i.e., hydrates it locally beyond its saturation value) causing the olivine to melt, and likewise acting to dilute the melt with additional silicate (thus reducing its water concentration). However, in this downwelling region the entire process (i.e., the melt layer being pushed and frozen into the transition-zone, the melt becoming subsequently "super-hydrated", and thus olivine reacting and melting into the melt layer) is ongoing and allows the traversal of downgoing material across the melt zone.

In the original simple slab entrainment model, the accumulation of water near the slab purportedly lead to an enhanced diffusion flux of water into the slab. But within the viscous entrainment zone, the accumulation of water is readily damped out by reaction with olivine downwelling from above. A large diffusive flux is probably never necessary to effect melt and water entrainment with this viscous-entrainment mechanism (other than perhaps to hydrate the overlying olivine, but this is also driven by reaction rates and occurs along reaction fronts where local equilibrium is maintained, thus hydrogen diffusion across a thin or infinitesimal reaction front is nearly instantaneous).

Quite significantly, this "viscous-entrainment-zone" process provides an extremely important means for draining the melt layer back into the lower mantle. As with simple slab entrainment, water in the melt can only be drawn back into the transition-zone at wadsleyite saturation concen-

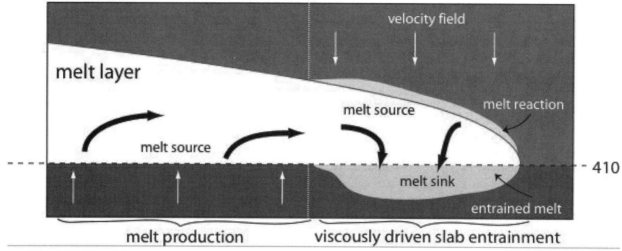

Fig. 4. Model melt layer structure as a spreading gravity current with sources of melt in upwelling regions, and net sink of mass in downwelling regions, where entrainment is accompanied by a melting reaction at the top of the melt layer

trations. However, the melt is entrained over a potentially much broader region (i.e., as much as the entire viscous entrainment zone) rather than just along the melt-slab contact. Since water is essentially only being injected from the transition-zone at roughly 0.1 wt% through the broad upwelling, and taken back down at saturation value of roughly 1 wt%, then water can be drained with a downwelling mass flux roughly 1/10th that of the upwelling mass flux. Indeed, simple calculations indicate this mechanism is extremely efficient at draining the melt layer and could by itself keep the melt zone extremely thin. However, as discussed below [*Leahy and Bercovici*, 2006] the entrainment also affects the background flow which then likely inhibits the entrainment process itself (Fig. 5).

An important implication of the aforementioned "viscous-zone" entrainment mechanism is that material from the melt layer that gets solidified and recycled into the transition-zone has a large, near saturation water content and is thus relatively buoyant [see *Jacobsen*, 2006 for a review of related equations of state]. [*Leahy and Bercovici*, 2006] find that though this hydrated material is embedded in the downwelling, its buoyancy is sufficient to impede the slab-driven flow field (Plate 2). The reduction in downwelling mass flux and melt drainage allows the melt layer to thicken and spread further across the viscous entrainment zone before being drawn away. Earth-like parameter regimes are characterized by competition between buoyant spreading of hydrated wadsleyite and slab-driven downward entrainment. Buoyant spreading of hydrated material tends to spread laterally in the transition zone and drives small scale chemical circulation within the transition-zone, which has a further—albeit more complicated—feedback on melt production and melt drainage.

The drainage of water into the transition zone over a relatively broad viscous entrainment region, and the subsequently induced secondary circulation via chemically driven convection, provides a mechanism for recirculating, mixing and distributing water in the transition zone. In the original TZWF model with simple slab entrainment, water would be returned to the transition-zone in a nearly closed-system cycle. However, the model did not specify how water would be mixed and distributed across broad areas of the transition zone in order to continue to hydrate the broad upwellings rising out of the lower mantle. Drainage of water into the transition-zone through the broad viscous entrainment zones begins to address this important question. Moreover, secondary circulation established by chemical convection would also facilitate mixing in the transition-zone.

Hydration of the Transition-Zone by "Stagnant Slabs"

The water entrainment processes discussed above provides some possible mechanisms for water recycling and distribution in the transition zone. Another important recirculating mechanism concerns slab-induced entrainment and the prevalent horizontal deflection of slabs in the transition zone [*Richard, et al.*, 2006]. The horizontal deflection of many slabs in the transition zone is well documented from seismic tomography [*Fukao, et al.*, 1992; *Fukao, et al.*, 2001; *Zhao*, 2004]. Hydrated material from the oceanic crust and lithosphere, as well as from the putative 410km melt layer that is entrained by slabs either within the slabs cold thermal halo as originally proposed by [*Bercovici and Karato*, 2003], or through the viscous entrainment zone as proposed by [*Leahy and Bercovici*, 2006] and discussed above, would be horizontally deflected

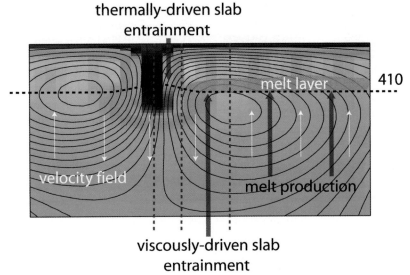

thermally-driven slab
entrainment

melt layer   410

velocity field   melt production

viscously-driven slab
entrainment

Plate 1. The flow field from a convection simulation demonstrates possible water-filter regimes: melt production, where hydrated wadsleyite is advected above the 410 km discontinuity and melting occurs, thermally-driven slab entrainment ([*Bercovici and Karato*, 2003]), where melt is entrained primarily due to diffusion of water into cold slabs, and viscously-driven slab entrainment, where melt is entrained into downgoing wadsleyite at ambient mantle temperatures (after [*Leahy and Bercovici*, 2006]).

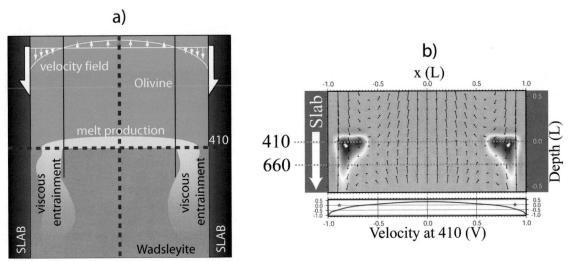

a)

velocity field

Olivine

melt production   410

viscous entrainment   viscous entrainment

SLAB   SLAB

Wadsleyite

b)

x (L)

Slab

410
660

Depth (L)

Velocity at 410 (V)

Plate 2. Buoyancy effect of entrained melt on mantle flow. a) A simple slab-driven velocity model is perturbed due to buoyancy anomalies of entrained melt at the 410. b) Calculations of the new flow field (arrows) and steady-state water distribution for the model. Red areas are enriched in water relative to the background (blue). Entrained melt permits the melt layer to spread further through adjustments to the velocity field, and may contribute to the rehydration of the transition zone.

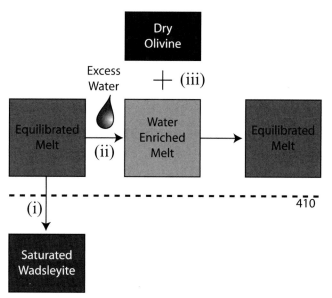

Fig. 5. Flow chart showing the chain of events when a small amount of the melt is dragged below the 410-km discontinuity. (i) The melt is in equilibrium with wet wadsleyite, which crystallizes out of the melt and continues to sink. (ii) The water excess due to crystallization stays in the layer and enriches the melt, which is now out of equilibrium with the overlying olivine solid. (iii) The enriched melt reacts with the olivine, causing melting to return the system to equilibrium. This provides a mechanism for the melt to spread further into the viscous entrainment region.

with the slab, and the upward-facing side of the slab would potentially lose water by diffusive and, if conditions are right, convective transport (Fig. 6).

Diffusive transport of water from the "floating slab" is complicated by the temperature contrast between the cold slab and the overlying "ambient" transition-zone because water solubility is temperature dependent. Even if a system has initially uniform chemical concentration, a contrast in solubility would establish a gradient in chemical potential causing water to diffuse from the low solubility to the higher solubility region until equilibrium is established. Experimental studies [*Demouchy, et al.*, 2005; *Kawamoto et al.*, 1996; *Ohtani, et al.*, 2000; *Zhao et al.*, 2004] indicate that the solubility of water in olivine, wadsleyite, and ringwoodite increase slightly with temperature but only up to temperatures (~1200C) where the system solidus is reached and a melt/fluid forms (see Appendix II). Upon further increase of temperature, as the volume fraction of melt/fluid increases, the amount of water in the coexisting minerals decreases. In the presence of fluid, the storage capacity of water in the minerals decreases rapidly. For example in wadsleyite, the solubility decreases from ~2.5 wt% H_2O at 1400 C to ~1.5 wt% at 1500 C [*Demouchy et al.*, 2005].

Theoretical "double diffusion" studies of coupled diffusive transport of heat into and water out of a floating slab [*Richard, et al.*, 2006b] yield intriguing results. If solubility increases with temperature, thermal gradients enhance chemical transport since the warm mantle overlying the floating slab has higher solubility than does the colder slab, and thus essentially acts to soak up the available water. Alternatively, if solubility decreases with increasing temperature then one nominally expects chemical transport to be inhibited since the warm mantle overlying the slab would have lower solubility than the cold slab. However,

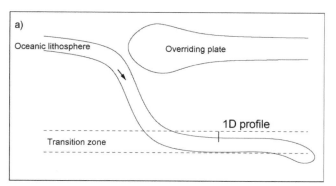

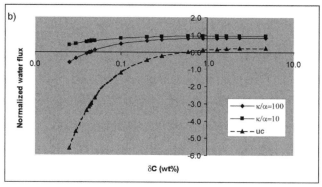

Fig. 6. a) The dehydration of a stagnant slab in the mantle transition zone. The essence of the model hypothesis: The diffusion of water out of the slab and the diffusion of heat into the slab occur simultaneously and are coupled by the temperature dependence of the water solubility. b) Effect of the initial slab water concentration on the outgoing water flux. Normalized water fluxes (NWF: total water flux over pure chemical water flux ratio) as function of the initial water concentration step between slab and mantle (δC (wt%)) for two water diffusivities (α=10^{-8},10^{-7} m^2/s) compared to the NWF if the heat and water diffusion processes were uncoupled (UC, dashed line). The NWF is almost constant and close to one for high slab concentration (δC>0.5wt%) and decreases rapidly for lower values. This plot demonstrates that the coupling strongly mitigates the decrease of water flux induced by the high solubility of the cold slab (E=-100 kJ/mol is the intrinsic chemical potential of the water in wadsleyite). The thermal diffusivity is set to be κ=10^{-6} m^2/s and the mantle water concentration is fixed to 0.1 wt%.

several nonlinear interactions may act to mitigate this effect. First, as the uppermost region of the recumbent, floating and hydrated slab warms, its solubility drops causing it to approach saturation and to be even further from equilibrium with its surroundings. This disequilibrium drives a diffusive flux that goes both up and down; the overlying mantle is warmer (lower solubility) but drier, while the colder deeper parts of the slab are colder (higher solubility) but equally hydrated. However, the back diffusion of water deeper into the slab causes water accumulation in the water concentration, leading a concentration peak which is even further out of equilibrium with overlying warm mantle.

Thus, the back-flux and accumulation of water in the slab drives an enhanced flux into the overlying mantle. Indeed, model calculations suggest that the water flux out of the slab is very weakly dependent on the temperature dependence of the solubility [Richard, et al., 2006]. Moreover, given typical residence times of slabs floating horizontally across the transition-zone, for 50Myrs or more, essentially 50–100% of the water brought by entrainment into the transition-zone is expelled through this double-diffusive mechanism into the transition-zone. That it occurs over a relatively broad region (the extent of slab deflection) allows for enhanced distribution of water through the transition-zone.

Diffusion is important for removing water from the slab itself. However, small-scale chemically driven convection of hydrated material is also likely an important recirculating mechanism [Bercovici and Karato, 2003; Leahy and Bercovici, 2006]. Hydrated material entrained in the vertical downwelling would not only impede the downwelling, but could lead to small-scale instabilities. Moreover, warm mantle material overlying a flat, floating slab would also become buoyant when hydrated by the slab, and thus would likely become unstable. Hydrated material from the viscous-entrainment zone that is advected horizontally by a deflected floating slab would also lead to a relatively broad region of convectively unstable material. Rising chemically buoyant material may exhibit complex circulation patterns if it is stirred by the background mantle flow. Eventually, these buoyant anomalies would pass through the 410km boundary, whereupon they would undergo partial melting and return their water to the water-filter melt layer. The details of this convectively enhance water circulation have, however, yet to be explored.

Convection and Heat Source Distribution

One of the predictions of the water-filter model, even with the most recent adjustments discussed in previous sections, is that the heat-source distribution would be almost entirely confined to the sub-410km mantle. However, [Bercovici and Karato, 2003] conjectured that large-scale, whole mantle flow would be largely unaffected by such a layered, inhomogeneous heat-source distribution as long as there are no internal boundaries impermeable to the bulk flow. This hypothesis was tested in a two-dimensional isoviscous convection model by Leahy and Bercovici [2004] for a suite of possible heat-source distributions ranging from uniform whole-mantle heating, mantle heated uniformly only below 410km, to the extreme case of heating confined entirely to the transition-zone (under the assumption that radiogenic elements follow the water circulation through the transition-zone). These cases were also combined in various proportions with each other as well as with the contribution from secular cooling (which effectively acts as a uniform heat source). Stability and numerical results indicate that in almost every possible heating distribution, the large scale flow is almost completely unaffected by layered heating (Fig. 7). In only one end-member case, in which all possible heating was confined to the transition-zone, was circulation affected; however, even in this situation, while thermal anomalies were concentrated in the upper-most mantle, circulation still extended throughout the entire mantle (Fig. 7, see case 3). The addition of even minimal secular cooling in the mantle or some contribution from broadly distributed heating restores both the thermal anomalies and circulation to a whole-mantle configuration.

4. GEOPHYSICAL AND GEOCHEMICAL CONSEQUENCES

One of the most important predictions of the water-filter model is that there should be a discontinuous jump in the concentrations of incompatible elements at 410-km, although the major (compatible element) abundance should be nearly continuous. The model also predicts large lateral variation in the concentrations of incompatible elements such as hydrogen. Consequently, a useful test is to see if there are any geophysical observables that carry information on the spatial distribution of concentration of incompatible elements. Because of the relatively small concentrations, the concentrations of incompatible elements are difficult to infer from geophysical observations, except for hydrogen. Geophysically observable properties that are potentially sensitive to water (hydrogen) content include electrical conductivity and seismic wave velocities (and attenuation). In these studies, one needs to have high-resolution geophysical data sets and the relationships between water content and physical properties. Particularly critical is to establish the relationships between water content and physical properties based on laboratory data and theoretical models. This topic has recently reviewed by [Karato, 2003; 2006a; c].

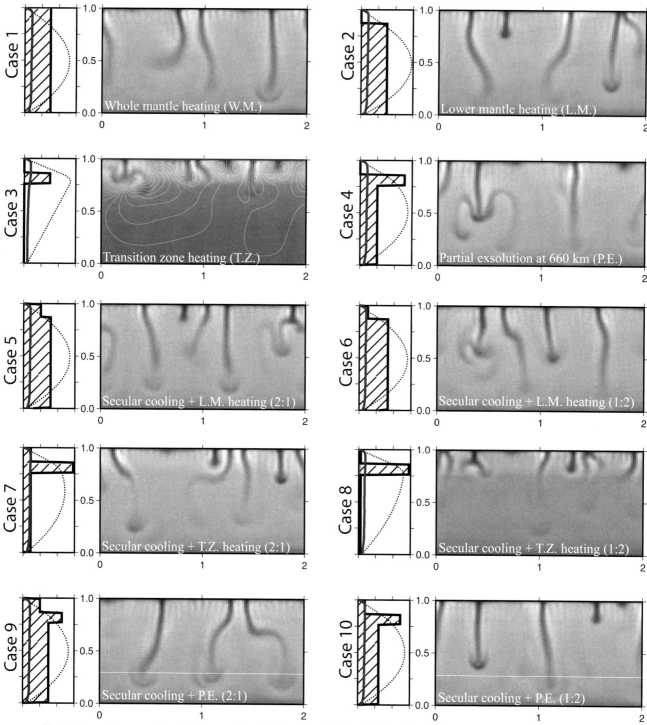

Fig. 7. Convection experiments: Temperature field and heating distributions for the 10 cases labeled. For each case, the left frame gives the heating distribution (shaded area) and the horizontally averaged temperature (dashed line). The right frame for each case is the total temperature field, where white represents hot regions and black represents cold regions. In case 3, stream lines have been added to demonstrate the whole region participates in the convective cycle even though the lower layer appears stable (after [*Leahy and Bercovici*, 2004]).

The TZWF model predicts a contrast in hydrogen (water) and other incompatible element concentrations between the upper mantle and the mantle below 410-km. If material circulation across the 410-km discontinuity is continuous without melting, then the concentrations of incompatible elements such as hydrogen must be the same between these two layers. A comparison of geophysically inferred electrical conductivity profile with the recent experimental data on the influence of water (hydrogen) content on electrical conductivity in wadsleyite and olivine shows that there is a marked contrast in water content across the 410-km. [*Huang, et al.*, 2005] inferred water content from electrical conductivity to conclude ~0.1-0.2 wt%. [*Huang, et al.*, 2006] estimated the ratio of hydrogen content between the transition zone and the upper mantle to be ~10 times more water than a typical upper mantle (with a large regional variation). The results of this analysis are insensitive to the uncertainties in temperature estimate and it provides a strong support for the TZWF model.

Seismological observations have also been used to infer the water content. There are at least two issues that can be addressed from this approach. Studies in this category include those from the thickness of the 410- km ([*van der Meijde, et al.*, 2003; *Wood*, 1995; *Smyth and Frost*, 2002]), and the simultaneous inversion of topography on the '410-km' discontinuity and the velocity anomalies in the transition zone ([*Blum and Shen*, 2004; *Suetsugu, et al.*, 2006]). A difficulty in this approach is the uncertainties in the relationships between water content and seismological observables (for details see [*Karato*, 2006a; c]). The exact relationship between the thickness of the 410- km and the water content is subject to uncertainty due primarily to the uncertainties in the model of dissolution of hydrogen [*Karato*, 2006b, c]. The inference of water content from the joint inversion of the topography of the 410-km discontinuity (or the thickness of the transition zone) and the velocity anomalies in the transition zone can provide a strong constraint. This method utilizes the notion that the anomalies of seismic wave velocity and the topography of the discontinuity depend on anomalies in temperature and water content. Using such an approach, [*Blum and Shen*, 2004] inferred the water content in the transition-zone beneath Africa using the velocity anomalies and the anomalies of the depth of the discontinuity. They inferred ~0.1 wt%. Similarly, [*Suetsugu, et al.*, 2006] used the anomalies of velocity and thickness of the transition-zone beneath western Pacific and inferred that water content varies with geological setting and regions in which old slabs subduct contains ~1 wt% water, but regions in which young slabs subduct has much smaller water content. Recent experimental studies of elastic wave velocities in OH-bearing minerals at high pressure indicates that water has a much

stronger influence on *S*-velocities than *P*-velocities, so that regions with "normal" *P*-velocities associated with high V_P/V_S ratios (resulting from low V_S) may be characteristic of hydration [see *Jacobsen and Smyth*, this volume].

There have been some reports suggesting the presence partial melt around 410-km (e.g., [*Revenaugh and Sipkin*, 1994; *Song, et al.*, 2004]). However, the direct connection between partial melting and the low velocities is not straightforward (e.g., [*Karato*, 2006c]) because regions of high melt fraction are expected to be highly localized, as suggested by theoretical modeling (e.g., see previous section and *Spiegelman and Elliott*, 1993]). A promising method to detect a thin melt-rich layer is to investigate seismological signals that are sensitive to the presence of a sharp impedance contrast (e.g., [*Chambers, et al.*, 2005]).

In addition to partial melting, spatial variation in hydrogen content will also have important effects on seismic wave velocities and attenuation [*Karato*, 2006c]. The TZWF model predicts that a broad region surrounding a subducting slab contains a large amount of enriched (hydrogen-rich) materials [*Leahy and Bercovici*, 2006] and hence will have low seismic wave velocities and high attenuation. *Obayashi et al.* [2006] reported low velocity anomalies in the transition zones ocean-side of the western Pacific subduction zone that is consistent with a model by *Leahy and Bercovici* [2006] (this model predicts hydrogen-rich regions in the *both sides* of a subducting slab. However, in the continental side, velocities will probably not be as low because it is on the colder top side of the slab (e.g., [*Fukao, et al.*, 2001])).

Consequence for the Distribution of Trace Elements

One of the important contrasting geochemical characteristics between MORB and OIB is the generally higher concentrations of incompatible elements in OIB than MORB (Plate 3a). In more detail, the difference between MORB and OIB in terms of trace element abundance is larger for more incompatible elements. Also the degree of difference is large for certain regions of OIBs such as HIMU or EM-1 than OIB from Hawaii.

The TZWF model may explain important characteristics of trace element distribution patterns. Upon partial melting, those elements that have ionic radii and/or electrostatic charge that are largely different from the host ion in minerals tend to go to melt. One of the important conjectures in the TZWF model is that the MORB is produced by the partial melting of materials at shallow mantle that have already undergone "filtering" by partial melting at 410-km, while OIB is formed by partial melting at shallow mantle of materials that have not undergone filtering. Consequently, the MORB source materials are more depleted than the source materials for OIB. Here we show a simple model calculation of trace element abundance based on a model that incorporates TZWF. We assume the following sequence of chemical differentiation. (i) The original chemical composition of Earth is chondritic. (ii) The chondritic material differentiated by the formation of continental crust. The chemical composition of residual materials can be calculated by mass balance by knowing the mass fraction of the continental crust (0.6 %). (iii) The residual materials will undergo transition-zone partial melting. Partial melting removes incompatible elements the degree to which this happens depends on the degree of melting and the partitioning coefficients. (iv) Materials that have undergone these two stages of differentiation will finally partially melt near the surface to form MORB. (v) Materials that have escaped differentiation in the transition-zone will undergo partial melting near the surface to form OIB.

In this calculation, we need to know the degree of melting, the partitioning coefficients of elements between melt and solid, and the mode of melting. The degree of melting for MORB is estimated to be ~10% based on the ratio of oceanic crust and residual lithosphere. The degree of melting for OIB is unknown, and we chose 1-5%. The degree of melting at 410-km depends on the mantle adiabat and the concentration of impurity components. The estimated impurity content in a typical Earth is ~1 wt% or less (Table 2), so we explored a range of degree of melting, 0.5-1.0%. The degree of melting at 410-km is important in assessing the effects of garnet on trace element abundance. Garnet has peculiar partition coefficients due to its flexible crystal structure: garnet has significantly higher partitioning coefficients for modestly compatible elements than other minerals. This garnet signature shows up, however, only for a large degree of melting, larger than a few percent. The partition coefficients are available only at modest pressures for melts without much volatiles [*Hauri, et al.*, 1994]. *Taura et al.* [1998] investigated the influence of pressure on trace element partitioning, but the influence of pressure is only modest. *Wood and Blundy* [2002] estimated the influence of water on trace element partitioning, however, the indirect influence of water (through other physical property changes) in phases such as wadsleyite is unconstrained. Consequently we used the results by *Hauri, et al.* [1994] for element partitioning coefficients under low pressure and anhydrous conditions but assuming 30% of the material is garnet. A gross feature (relative incompatibility of various elements) of trace element distribution pattern is likely represented by the use of this set of partition coefficients, although some refinement will be needed when a better data become available.

The results are shown in Plate 3b. We note that we have reproduced the observed trace element abundance patterns for MORB and OIB (Hawaii) using the published partition-

ing coefficients and plausible degree of melting at 410-km and near the surface. In a classic model by *Hofmann* [1988], chemical evolution in Earth was assumed to have been caused by melting associated with the formation of continental and oceanic crust and subsequent melting at mid-ocean ridges but no melting is assumed at the deep mantle. Since the degree of melting at ~410-km is small, the transition-zone melting does not affect the trace element distribution pattern so much. In particular, the influence of garnet is not seen because the peculiar partition coefficients of trace elements in garnet (and melt) occur for relatively compatible elements such as Yb, Er and Lu for which the partition coefficients are ~10^{-1} and consequently, the "garnet signature" can be seen only after a high degree of melting (~10%). The trace element distribution patterns for more "enriched" OIB such as EM-1 or HIMU cannot be explained by our model. These regions require some additional processes for "enrichment" such as the involvement of continental lower crust and/or sediments.

Hydrogen is one of the highly incompatible elements whose partition coefficient is on the order of 10^{-3}, which is close to (or slightly smaller than) that of Rb. Consequently, if one assumes that Hawaii source regions are representative of deep mantle composition as argued above, then a contrast of hydrogen content between the upper mantle and the deep mantle will be similar to that of Rb, i.e., a factor of ~5 [*Hofmann*, 1997; 2004]. The inferred contrast in hydrogen content by [*Huang, et al.*, 2006] is a factor of ~10 with a large regional variation. In other words, in order to create a factor of ~10 difference in hydrogen content between the upper mantle and the transition zone with the partition coefficient of water of $k=10^{-3}$, one needs ~1% melting for batch melting and ~0.2% for fractional melting. These values are in good agreement with what is expected from the concentration of volatile, incompatible elements and their partitioning coefficients. We therefore conclude that the observed concentrations of water in the upper mantle and the transition zone can be attributed to the filtering processes that occurs at ~410-km. Obviously, the filtering can be leaky as discussed before and consequently, we expect that the water content of the upper mantle is heterogeneous. If water content is significantly less than ~0.05wt%, then no melting occurs, and materials will go through the 410-km without filtering. If a material has large water content, say ~1 wt%, then the melt will be light and again filtering will not occur. These materials will form a localized upwelling to create OIB. If the water content of materials in the transition zone is in between these values, then the filtering will occur to change the water content to ~1/10 of the original value.

In contrast to the present analysis *Hirschmann et al.* [2005] argued that the TZWF model would predict a water content in the upper mantle that is too high compared to the inferred values. They argued that if partial melting occurs at 410-km, then the upwelling materials will continue to melt and the upwelling materials (materials in the upper mantle) will have a water content higher than the inferred value of water in the source region of MORB (~0.01 wt%). The validity of such an argument hinges on the assumed values of partition coefficient of hydrogen and other thermochemical properties of minerals and melt that determine the degree of melting at a given P-T condition. There are large uncertainties in thermochemical properties of melts, but as we discussed above, plausible estimates of the partition coefficient (~10^{-3}) and the degree of melting (~1 %) give a reasonable value of water content in the upper mantle (~0.01 wt%).

Consequence for the Distribution of Isotopes

Contrasting geochemical characteristics between MORB and OIB include not only the abundance of trace elements but also the distribution of isotopes. For example, OIB has higher $\frac{^{87}Sr}{^{86}Sr}$ and $\frac{^{143}Nd}{^{144}Nd}$ than MORB. Also, the variation of these isotopic ratios is larger for OIB than for MORB (e.g., [*Hofmann*, 1997], see however [*Hofmann*, 2004] who emphasized the diversity of isotope and other geochemical data from MORB). Since melting does not change the isotopic compositions, the explanation of these characteristics must be sought for some indirect

mechanisms. One obvious mechanism is the modification of the parent to daughter element abundance ratio by melting. Consider a radioactive decay, $^{\alpha}A \rightarrow {}^{\gamma}B$. If element B has another isotope, $^{\beta}B$, that is not created by radioactive decay, the isotopic ratio $\frac{[^{\gamma}B]}{[^{\beta}B]}$ will increase with time and with increasing concentration of element A, $\frac{[^{\gamma}B]}{[^{\beta}B]} = \left(\frac{[^{\gamma}B]}{[^{\beta}B]}\right)_0 + \frac{[^{\alpha}A]}{[^{\beta}B]}(e^{\frac{t}{\tau}}-1)$ where τ is the half life-time of radioactive decay. If there are two regions (i.e., the OIB source region and the MORB source region) that contain different $\frac{[^{\gamma}B]}{[^{\beta}B]}$ and $\frac{[^{\alpha}A]}{[^{\beta}B]}$ ratios (but the same initial $\frac{[^{\gamma}B]}{[^{\beta}B]}$ then for a period of time t, $\Delta\frac{[^{\gamma}B]}{[^{\beta}B]} = \Delta\frac{[^{\alpha}A]}{[^{\beta}B]}(e^{\frac{t}{\tau}}-1)$, where Δ represents a change in isotopic ratio over time t. We use this relation to calculate $\Delta\frac{[^{\alpha}A]}{[^{\beta}B]}$ from observed $\Delta\frac{[^{\gamma}B]}{[^{\beta}B]}$ and t/τ that is required by the model (τ ~0.5 Gys) and the half-life of the relevant radioactive scheme. Next we compare model Δ ratios with observed Δ ratios inferred from the difference in composition of "depleted" and "undepleted" mantle (e.g., [*Newsom*, 1995]).

Model and observed values of the isotopic ratios for Sr, Nd, Re/Os, and Pb are summarized in Fig. 8. The data on $\Delta\frac{[^{\gamma}B]}{[^{\beta}B]}$ are largely from [*Hofmann*, 1997], but additional data on Os are from [*Hauri and Hart*, 1993]. We conclude that the observed isotopic compositions of MORB and OIB are largely consistent with the TZWF model, although some of the "extreme" compositions such as EM-2 are difficult to explain by this model.

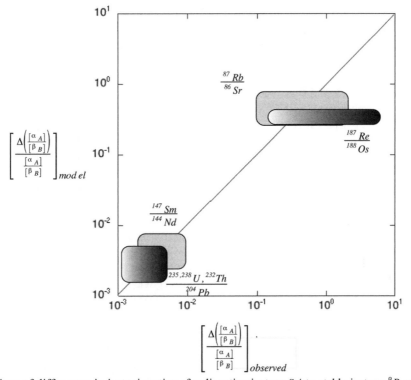

Fig. 8. Comparison of differences in isotopic ratios of radioactive isotope $^{\alpha}A$ to stable isotope $^{\beta}B$ created by partial melting at 410-km required to explain the geochemical observations by the TZWF model and the observed ratios.

It is often argued that the observed isotopic compositions are not consistent with a model in which the whole mantle was depleted at the time of continental crust formation (e.g., [*van Keken, et al.*, 2002]), and that more than 1 billion years are needed to establish the observed differences in isotopic compositions between MORB and OIB (e.g., [*Hofmann*, 1997]). We have shown that ~0.5 Gys of separation is long enough to result in most of the observed differences in isotopic ratios.

The observations on helium isotopic ratio are not straightforward to explain. Briefly, $\frac{[^3He]}{[^4He]}$ for OIB are much more scattered (and in general larger) than those of MORB. ^4He is a radiogenic isotope that is produced by the radioactive decay of ^{238}U or ^{232}Th. Consequently, if the TZWF operates well then the concentration of radioactive parent atoms such as ^{238}U or ^{232}Th will be lower in the upper mantle (MORB source region), and hence MORB would have a *higher* $\frac{[^3He]}{[^4He]}$ than OIB. We note that the observed $\frac{[^3He]}{[^4He]}$ may also be controlled by the degree of depletion of helium. [*Brooker, et al.*, 2003; *Parman, et al.*, 2005] reported that helium behaves like a highly incompatible element during partial melting (and diffusion of helium is very fast and hence distribution of *He* will always be in chemical equilibrium). The small reported partition coefficient (on the order of ~10^{-3}) leads to significant depletion in helium in the source region of MORB in our model, and hence the observed lower $\frac{[^3He]}{[^4He]}$ ratio for MORB can be due to the depletion of helium in the source region of MORB. Similarly, *Ar* and other rare gas elements are likely sequestered into the deep mantle due to their low partition coefficients ([*Brooker, et al.*, 2003]) (and relatively high diffusion coefficients [*Harrison*, 1981]), if TZWF operates. This provides a simple explanation for the *Ar* paradox (i.e., there are too much *Ar* in the atmosphere compared to the amount of *K* in the upper mantle; [*Allègre, et al.*, 1996]).

In examining these models incorporating partial melting and element partitioning, an important implicit assumption is chemical equilibrium. This assumption is not secure for elements with very small diffusion coefficients. The chemical equilibrium will be attained if the time scale of reaction exceeds $t > \frac{d^2}{\pi^2 D}$ where d is grain-size (~ 3 mm) and D is relevant diffusion coefficient. The diffusion coefficients vary from one species to another (see Table 3), but from the Table 3 with t ~10 My, chemical equilibrium is certainly attained for *He, Pb*, and presumably for *Sr* and *Sm*, but the distribution of *U* and *Th* is likely out of equilibrium. *U* and *Th* likely behave like compatible elements, but *He* is a highly incompatible element, and therefore if there is melting at ~410-km, much of *U* and *Th* will be transported to the upper mantle but not much of *He*. This explains the helium paradox, i.e., the amount of *He* flux is too much smaller than expected from the amount of *U* and *Th* [*O'Nions and Oxburgh*, 1983].

Table 3. Diffusion coefficients of some trace elements in typical minerals (cpx, olivine) at $\frac{T}{T_m}$ ~0.7 (T_m is the melting temperature of the mineral)

U, Th	(cpx)	~10^{-20} m^2/s	[*Van Orman, et al.*, 1998]
Pb	(cpx)	~10^{-16}	[*Cherniak*, 1998]
Sr, Sm	(cpx)	~10^{-18}	[*Sneeringer, et al.*, 1984]
He	(ol)	~10^{-11}	[*Hart*, 1984]

The overall processes of geochemical cycling in this model are summarized in Plate 4 in which processes associated with upwelling currents are emphasized. A unique feature of this model, as compared to any previous models, is the role of (a small amount of) melting at ~410-km caused by the presence of volatile components H_2O, CO_2, Na_2O, K_2O. We have shown that melting at this depth is highly likely and that the melt thus formed has a higher density than the surrounding minerals under some conditions. We propose that if the volumetrically dominant part of the deep mantle is made of relatively depleted materials such as the source region of Hawaii or FOZO (after the formation of continental crust), together with volumetrically minor more enriched materials, then almost all geochemical observations can be explained naturally as the result of melting at ~410-km without violating the geophysical constraints suggesting whole mantle wide convection. They include trace element abundance patterns, isotopic ratios, abundance of *He* relative to *U* and *Th* (the helium/heat paradox), and the abundance of *Ar* relative to the abundance of *K, U* and *Th* (the *Ar* paradox). The homogeneity of composition of MORB compared to that of OIB is partly attributed to the dominance of Hawaii (or FOZO) type composition of the deep mantle, and some homogenization due to melting.

Homogeneity of MORB may also be due to the selective filtering: filtering to form MORB source materials occurs only for materials with certain composition as discussed in the previous section. The TZWF acts as a "low-pass" filter through which only materials with relatively depleted character passes into the MORB source region. OIB may also be formed when a hot plume goes through the transition zone that has a material typical of the deep mantle (Hawaii type composition, by assumption). These hot materials will escape TZWF if their upwelling velocity is faster than the velocity of lateral flow of a melt layer and/or when the melt density is lighter than the surrounding materials due to high temperature as discussed by [*Bercovici and Karato*, 2003].

The source region of OIB is considered to occupy a large fraction of mantle (e.g, [*Albarède and van der Hilst*, 2002; *Allègre*, 1997]). In almost all previous models, the source regions for MORB and OIB are assumed to have been isolated for billions of years and the geodynamic difficulty is

(a)

(b)

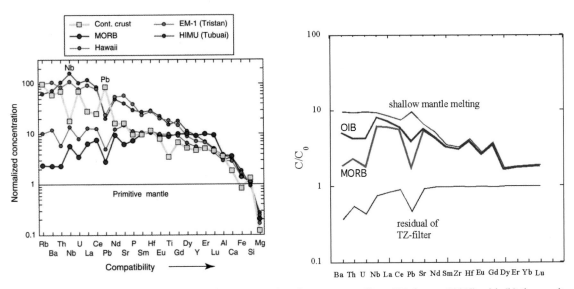

Plate 3. A comparison of (a) the observed trace element abundance pattern (from [*Hofmann*, 1997]) with (b) the results of model calculations. "Residual of TZ-filter" shows the pattern of trace element concentration after the TZWF. The MORB trace element distribution pattern reflects this pattern but involves another high-degree melting at the surface. The OIB distribution pattern does not include the influence of TZWF. The difference between OIB and MORB is either TZWF operates or not.

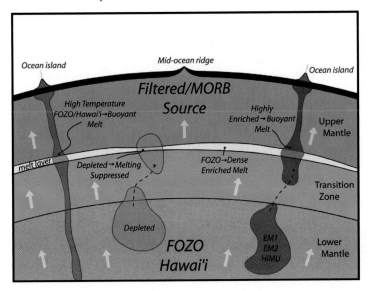

Plate 4. Cartoon showing roles of transition-zone melting on the geochemical cycling in Earth's mantle. The majority of the mantle was assumed to be made of the residual materials after the formation of continental crust (Hawaii or FOZO type composition). These materials are moderately depleted, and will melt, assisted by volatiles such as water, when they move across the 410-km boundary from below to form a dense melt. Much of the incompatible elements will be removed from the upwelling materials and the resultant depleted materials will become the source materials for MORB. Highly enriched materials such as EM-1 will be melted, but the density of these melt is likely low because of high concentration of water (or because of high temperature). These enriched materials will remain enriched as they ascend and will form enriched OIB. Relatively depleted OIB can also be formed when a hot plume goes through the transition zone. In such a case, melting occurs at ~410-km, but the melt will be less dense because of a high temperature, and hence the melt rises to form a OIB with relatively depleted signature. Highly depleted materials in the deep mantle will not undergo melting and will directly be transported to the upper mantle to become a source material for MORB. When a dense melt is formed, then the transition-zone melting will sequester highly incompatible elements into the deeper mantle including *He* and *Ar* (but not *U* and *Th* that have very low diffusion coefficients). This model provides a unified explanation for a broad range of geochemical observations without violating geophysical constraints that mantle convection occurs more-or-less in the whole mantle wide flow.

to reconcile such a model with geodynamic observations suggesting whole mantle convection. Although our model does not exclude the presence of a relatively small amount of materials that have been isolated for billions of years, we propose that there is an additional process that may have an important consequence on geochemical cycling.

We note that the proposed partial melting has an important effect on the distribution of "impurities" such as hydrogen and potassium. All incompatible elements will be sequestered in the deep mantle if the TZWF were to occur. The transition zone and the lower mantle will have a higher hydrogen content than the upper mantle, which will result in a lower viscosity in these regions than in the case of homogeneous composition. Similarly, much of the radioactive elements such as potassium and uranium will also be sequestered in the deep mantle. Inference of the composition of the mantle from upper mantle rocks (e.g., [*McDonough and Sun*, 1995]) will lead some bias for highly incompatible elements.

5. SUMMARY AND PERSPECTIVES

We have presented a review and refinement of the TZWF model originally proposed by *Bercovici and Karato* [2003]. Mineral physics observations suggest that a small degree of partial melting at around 410-km discontinuity is likely under a broad range of conditions, but results in a dense (denser than upper mantle minerals but lighter than transition zone minerals) melt for relatively low concentrations of hydrogen (water), whereas flux melting due to higher concentrations of water results in a light (buoyant) melt. This melting in the transition-zone can modify the distribution of highly incompatible elements without affecting the distribution of relatively compatible elements and thus provides an explanation for a broad range of geochemical observations. conditions for creating a dense melt is difficult at this stage due mainly to the limited knowledge about the role of CO_2, K_2O and Na_2O as well as H_2O in melting and the melt density under transition zone conditions and represents an experimental area ripe for exploration.

We have also reviewed recent observational constraints of this model. A combination of geodynamic modeling and mineral physics suggests several features that are characteristic to the TZWF model. Obviously, any evidence for a melt-rich layer at ~410-km will support the model, but we note formidable challenges in detection of such a layer because even partial melt beneath some mid-ocean ridges has not been well resolved by seismological observations ([*Wolfe and Silver*, 1998]). Because partial melt is probably highly localized, seismological detection of a thin (partially) molten layer will require a high-resolution technique that is sensitive to the impedance contrast (e.g., [*Chambers, et al.*,

2005]). Although indirect, the distribution of hydrogen can provide important data to test the model. Hydrogen distribution can be broad in some cases and hydrogen has strong influence on some geophysically measurable properties such as electrical conductivity and seismic wave velocity and attenuation. One important observation is the jump in electrical conductivity at ~410-km in many regions that suggest a jump in hydrogen content which is consistent with the TZWF model. The TZWF model also predicts that the distribution of hydrogen (and other volatile elements) is laterally heterogeneous. Recent seismological and electrical conductivity observations support this notion (e.g., [*Fukao, et al.*, 2004; *Obayashi, et al.*, 2006; *Suetsugu, et al.*, 2006]). However, much remain poorly constrained and further progress in the following areas is needed: (i) the relationship between geophysical observations (seismic wave velocities and attenuation, electrical conductivity) and water content or partial melting needs to be better constrained, and (ii) more detailed geophysical exploration is needed to map the distribution of hydrogen, and melting in the mantle (for a review see [*Karato*, 2006c]).

In addition, geochemical consequence of the model needs to be explored in more detail. Improved determination of partition coefficients of trace elements between minerals and a volatile-rich melt under the deep mantle conditions is required. The present model has far-reaching implications for the chemical evolution and the volcanism in Earth, but detailed prediction of geochemical consequences is difficult due mainly to the lack of experimental data on element partitioning at deep mantle conditions. The model implies that the mantle is likely layered with respect to highly incompatible elements such as hydrogen, carbon, potassium and sodium.

Acknowledgements. We thank Rajdeep Dasgupta, Marc Hirschmann, Steve Jacobsen, and Joe Smyth for sending us their papers or experimental data prior to publication, and Eiji Ohtani and Jay Ague for discussion. Persistent criticisms on the TZWF model by Marc Hirschmann helped improve the paper. Editorial comments by Steve Jacobsen and Susan van der Lee helped clarify the presentation. This research is supported by National Science Foundation.

APPENDIX I: SIMPLE DECOMPACTION THEORY FOR A HEAVY MELT LAYER

To examine decompaction of the heavy melt layer, we employ the generalized two-phase damage equations [*Bercovici et al.*, 2001a,b; *Ricard et al.*, 2001; *Bercovici and Ricard*, 2003; *Ricard and Bercovici*, 2003; *Bercovici and Ricard*, 2005], which we only briefly summarize in order to develop our decompaction scaling relationships. We note in advance that in the following equations, subscripts f and m refer to fluid (i.e., the melt phase) and matrix (i.e., the solid phase) phases, respectively. The conservation of mass equations in two-phase theories [McKenzie, 1984; Bercovici et al., 2001a] are

$$\frac{\partial \phi}{\partial t} + \boldsymbol{\nabla} \cdot [\phi \mathbf{v}_f] = 0 \qquad (A1\text{-}1)$$

$$\frac{\partial (1-\phi)}{\partial t} + \boldsymbol{\nabla} \cdot [(1-\phi)\mathbf{v}_m] = 0, \qquad (A1\text{-}2)$$

where ϕ is porosity, and $\mathbf{v}_f$ and and $\mathbf{v}_m$ are the fluid and matrix velocities. We assume steady state and that while the solid matrix undergoes 1D vertical de- compaction, the fluid undergoes 2D flow as slab entrainment enforces horizontal drainage of the melt. Thus, we obtain

$$\frac{\partial}{\partial z}[(1-\phi)w_m] = 0 \qquad (A1\text{-}3)$$

$$\frac{\partial \phi w_f}{\partial z} + \frac{\partial \phi u_f}{\partial x} = 0 \qquad (A1\text{-}4)$$

where w_j and u_j are the vertical and horizontal velocity components, respectively, of phase j. We assume that the solid mantle below the 410km boundary is up- welling at constant velocity W and upon crossing the boundary undergoes melting by a fraction f; we assume that melt and solid densities are nearly the same and thus neglect the density difference $\Delta \rho = \rho_m - \rho_f$ unless, as prescribed by the Boussineseq approximation, it is multiplied by gravity g; thus we make no distinction between f being a mass or volume fraction. As solid flow is only one- dimensional the vertical solid flux crossing the 410km boundary never changes and thus integration of (A1-3) yields

$$(1-\phi)w_m = (1-f)W \quad \text{or} \quad w_m = \frac{1-f}{1-\phi}W \quad (A1\text{-}5)$$

We assume for simplicity that the horizontal entrainment rate is given by a constant $\partial u_f /\partial x = U$ which approximates the steady accumulation of melt with horizontal distance from the center of the upwelling; thus (A1-4) becomes

$$\frac{\partial \phi w_f}{\partial z} = -\phi U \qquad (A1\text{-}6)$$

where we assume ϕ is only a function of z. The initial melt fraction or porosity at $z = 0$ is simply f and the input melt flux is fW. Because the melt is heavy it will first accumulate into a high porosity region; however, the horizontal melt entrainment will eventually drain away all the melt so that at some height $z = H$ we get $\phi = 0$. Therefore, if we integrate (A1-6)

from $z = 0$ to $z = H$, with the boundary conditions that $\phi wf = fW$ at $z = 0$, and $wf = 0$ at $z = H$, we obtain

$$fW = U \int_0^H \phi(z)dz = UH\Theta \qquad (A1\text{-}7)$$

where of course $\Theta = \frac{1}{H}\int_0^H \phi dz$.

We will also be concerned with terms in the momentum equation proportional to the velocity difference $\Delta w = w_m - w_f$ and for this we can add (A1-3) and (A1-6) to obtain

$$\frac{\partial \phi \Delta w}{\partial z} = \frac{\partial w_m}{\partial z} + \phi U \qquad (A1\text{-}8)$$

The momentum equations for each phase (assuming the melt is much less viscous than the solid) are

$$0 = -\phi \left[\boldsymbol{\nabla} P_f + \rho_f g \hat{\mathbf{z}} \right] + c\Delta \mathbf{v} \qquad (A1\text{-}9)$$

$$0 = -(1-\phi)\left[\boldsymbol{\nabla} P_m + \rho_m g \hat{\mathbf{z}} \right] + \boldsymbol{\nabla} \cdot [(1-\phi)\underline{\boldsymbol{\tau}}_m] - c\Delta \mathbf{v} + \Delta P \boldsymbol{\nabla} \phi \quad (A1\text{-}10)$$

where g is gravity, $\Delta \mathbf{v} = vm - vf$, and c is the coefficient for viscous drag between phases, also referred to as the Darcy drag coefficient, and as is typically $c = \mu_f/k$ where μ_f is fluid melt viscosity and k is a reference permeability [McKenzie, 1984; Bercovici et al., 2001a]; P_j and ρ_j are the pressure and density, respectively, in phase j and the density of each phase is assumed constant. The deviatoric stress for the solid phase is given by

$$\underline{\boldsymbol{\tau}}_m = \mu_m \left[\boldsymbol{\nabla} \mathbf{v}_m + [\boldsymbol{\nabla} \mathbf{v}_m]^t - \frac{2}{3}(\boldsymbol{\nabla} \cdot \mathbf{v}_m)\underline{\mathbf{I}} \right] \qquad (A1\text{-}11)$$

where μ_m is solid matrix viscosity. The pressure difference $\Delta P = P_m - P_f$ in the absence of surface tension and strongly nonequilibrium effects (such as damage) is

$$\Delta P = -\frac{K\mu_m}{\phi}\boldsymbol{\nabla} \cdot \mathbf{v}_m \qquad (A1\text{-}12)$$

where $K \approx 1$. The momentum equations are combined by taking ϕ times (A1-10) minus $1 - \phi$ times (A1-9); we also substitute (A1-12), and keep only the vertical component, thereby obtaining

$$0 = \mu_m \frac{\partial}{\partial z}\left((1-\phi)\left(\frac{4}{3}+\frac{1}{\phi}\right)\frac{\partial w_m}{\partial z} \right) - (1-\phi)\Delta \rho g - c\Delta w/\phi \quad (A1\text{-}13)$$

which is the standard 1D compaction equation. We further multiply (A1-13) by ϕ^2, take the derivative of the resulting equation with respect to z, and eliminate both w_m, using (A1-5), and Δw, using (A1-8):

$$0 = \mu_m W(1-f)\frac{\partial}{\partial z}\left[\phi^2 \frac{\partial}{\partial z}\left((1-\phi)\left(\frac{4}{3}+\frac{1}{\phi}\right)\frac{\partial}{\partial z}\frac{1}{1-\phi} \right) \right]$$
$$- \Delta \rho g \frac{\partial}{\partial z}(\phi^2(1-\phi)) - c\left(W(1-f)\frac{\partial}{\partial z}\frac{1}{1-\phi} + U\phi \right) \qquad (A1\text{-}14)$$

We can nondimensionalize the above equation by writing $z = \delta z'$, $c = \mu_m \delta^2$, $U = (W/\delta)U'$ and $\Delta\rho g = -(\mu_m W/\delta^2)\beta$ where $\delta = \sqrt{\mu_m/c}$ is the compaction length and $\beta > 0$ is the dimensionless melt "heaviness" (note that with a heavy melt $\Delta\rho = \rho_m - \rho_f < 0$); we thus obtain

$$0 = (1-f)\frac{\partial}{\partial z'}\left[\phi^2\frac{\partial}{\partial z'}\left((1-\phi)\left(\frac{4}{3}+\frac{1}{\phi}\right)\frac{\partial}{\partial z'}\frac{1}{1-\phi}\right)\right]$$
$$+ \beta\frac{\partial}{\partial z'}(\phi^2(1-\phi)) - (1-f)\frac{\partial}{\partial z'}\frac{1}{1-\phi} - U'\phi \quad \text{(A1-15)}$$

We can develop a scaling analysis to (A1-15) by using a trial function for ϕ that has the right symmetry and boundary conditions. To leading order, $\phi(z')$ can be assumed a parabolic function wherein

$$\phi(z') = \Phi - (\Phi - f)\left(\frac{z'}{H'}\left(1+\sqrt{\frac{\Phi}{\Phi - f}}\right) - 1\right)^2 \quad \text{(A1-16)}$$

where we require that $\phi = f$ at $z' = 0$ and $\phi = 0$ at $z' = H'$ where $H' = H/\delta$. Moreover, with this nondimensionalization and the assumed form of $\phi(z')$, (A1-7) yields $f = U' H' \Theta$ where

$$\Theta = \Phi - \frac{1}{3}\frac{\Phi^{3/2}+(\Phi-f)^{3/2}}{\Phi^{1/2}+(\Phi-f)^{1/2}} \quad \text{(A1-17)}$$

For the scaling analysis, we evaluate the amplitude of the terms in (A1-15) by integrating it from $z' = 0$ to H', employing both (A1-16) and (A1-7), and make the reasonable assumption that $f \ll 1$; assuming the resulting viscous resistance terms balance the gravitational settling term, we obtain (after considerable algebra) an implicit scaling relationship for Φ as a function of U (i.e., by writing the function $U(\Phi)$), which we express in dimensional form:

$$U = \sqrt{\frac{W|\Delta\rho|g}{2\mu_m}}\frac{f\sqrt{f}}{\Theta\left(\sqrt{\Phi}+\sqrt{\Phi-f}\right)} \quad \text{(A1-18)}$$

Since $H' = f/(U'\Theta)$ or $H = f W/(U\Theta)$ we also have a relationship for the melt layer thickness (again, in dimensional form)

$$H = \sqrt{\frac{2\mu_m W}{|\Delta\rho|g}}\frac{\left(\sqrt{\Phi}+\sqrt{\Phi-f}\right)}{\sqrt{f}} \quad \text{(A1-19)}$$

Equations (A1-18) and (A1-19) are used to construct Figure 3 for which we consider a full of possible entrainment rates U.

APPENDIX II.

Some Notes on the Temperature (and Pressure) Dependence of Hydrogen Solubility

There is some confusion in the literature on the issues of temperature and pressure dependence of solubility of hydrogen (or another volatile species). The solubility is defined by the reaction between a volatile phase and a solid, viz.,

$$fluid + solid \Leftrightarrow solid \cdot volatile \quad \text{(A2-1)}$$

where fluid could be water, and *solid · volatile* denotes a solid that contains volatile element (such as hydrogen). For example, in case of the reaction between water and wadsleyite (or olivine), the dominant reaction is given by

$$H_2O + mineral \Leftrightarrow (2H)_M^\times + O_O^\times + MgO_{surface} \quad \text{(A2-2)}$$

where $(2H)_M^\times$ is a point defect that contains two protons at M-site, $O_O^\times$ is an oxygen ion at O-site. Applying the law of mass action to reaction (A2-2), one obtains,

$$\left[(2H)_M^\times\right] \propto f_{H2O}(P,T)\cdot\exp\left(-\frac{\Delta u + P\Delta v}{RT}\right)\cdot a_{MgO}^{-1} \quad \text{(A2-3)}$$

where $\left[(2H)_M^\times\right]$ is the concentration of $(2H)_M^\times$, $f_{H2O}(P,T)$ is the fugacity of water, $\Delta u, \Delta v$ are the difference in internal energy and volume associated with reaction (A2-2). The solubility of hydrogen depends on T and P through $f_{H2O}(P,T)$ and $\exp\left(-\frac{\Delta u}{RT}\right)$ terms. Note that the water fugacity is in general a strong function of P and T. This is true for pure water as well as for a case where water forms solution with a silicate melt. For pure water, the fugacity of water at high P and T (P~5-15 GPa, T~1000-2000 K) is approximately an exponential function of P and T, $f_{H2O}(P,T) \propto exp\left(\frac{H_{H2O}}{RT}\right) = exp\left(\frac{E_{H2O}+PV_{H2O}}{RT}\right)$ with $v_{H2O} \approx 11 \times 10^{-6}$ m³/mol (~molar volume of water at P~5-15 GPa) and $E_{H2O} \approx 0$ kJ/mol ($H_{H2O} = E_{H2O} + PV_{H2O} \approx$ 110 kJ/mol). This means that if one observes very small pressure dependence at relatively low temperatures where free water exists (this is what [Demouchy, et al., 2005] reported), then the volume change of the system should be Δv ~11 10⁻⁶ m³/mol rather than ~0 m³/mol as argued by [Demouchy, et al., 2005] (note that the conclusion by [Demouchy, et al., 2005] would be inconsistent with the above defect model (equation (A2-2)) but Δv~11×10⁻⁶ m³/mol is exactly what the model (A2-2) would imply). Similarly, the observed weak temperature dependence of hydrogen solubility at relatively low temperatures implies that $\Delta h \equiv \Delta u + P\Delta v \approx 100$ kJ/mol at P~10-13 GPa. Obviously, when water is present in the silicate melt, then its fugacity will be lower, and as the silicate fraction increases (with temperature), its fugacity decreases.

REFERENCES

Agee, C. B., and D. Walker (1988), Static compression and olivine flotation in ultrabasic silicate liquid, *Journal of Geophysical Research*, *93*, 3437–3449.

Albarède, F., and R. D. van der Hilst (2002), Zoned mantle convection, *Philosophical Transactions of Royal Society of London*, *A360*, 2569–2592.

Allègre, C. J. (1997), Limitation on the mass exchange between the upper and the lower mantle: the evolving convection regime of the Earth, *Earth and Planetary Science Letters*, *150*, 1–6.

Allègre, C. J., et al. (1996), The argon constraints on mantle structure, *Geophysical Research Letters*, *23*, 3555–3557.

Aubaud, C., et al. (2006), Intercalibration of FTIR and SIMS for hydrogen measurements in glasses and nominally anhydrous minerals, *American Mineralogist*, submitted.

Bell, D. R., et al. (2003), Hydroxide in olivine: A quantitative determination of the absolute amount and calibration of the IR spectrum, *Journal of Geophysical Research*, *108*, 10.1029/2001JB000679.

Bercovici, D., and S. Karato (2003), Whole mantle convection and transition-zone water filter, *Nature*, *425*, 39–44.

Bercovici, D., and Y. Ricard (2003), Energetics of two-phase model of lithospheric damage, shear localization and plate-boundary formation, *Geophysical Journal International*, *152*, 1–16.

Bercovici, D., et al. (2001), A two-phase model for compaction and damage 1. General theory, *Journal of Geophysical Research*, *106*, 8887–8906.

Blum, J., and Y. Shen (2004), Thermal, hydrous, and mechanical states of the mantle transition zone beneath southern Africa, *Earth and Planetary Science Letters*, *217*, 367–378.

Bolfan-Casanova, N. (2005), Water in the Earth's mantle, *Mineralogical Magazine*, *69*, 229–257.

Brooker, R. A., et al. (2003), The 'zero charge' partitioning behaviour of noble gases during mantle melting, *Nature*, *423*, 738–741.

Chambers, K., et al. (2005), Reflectivity of the 410-km discontinuity from PP and SS precursors, *Journal of Geophysical Research*, *110*, 10.1029/2004JB003345.

Cherniak, D. J. (1998), Pb diffusion in clinopyroxene, *Chemical Geology*, *150*, 105–117.

Dasgupta, R., and M. M. Hirschmann (2006), Deep melting in the Earth's upper mantle caused by CO_2, *Nature*, *440*, 659–662.

Demouchy, S., et al. (2005), Pressure and temperature-dependence of water solubility in iron-free wadsleyite, *American Mineralogist*, *90*, 1084–1091.

Frost, D. J., and B. J. Wood (1997), Experimental measurements of the properties of H_2O-CO_2 mixtures at high pressures and temperatures, *Geochimica et Cosmochimica Acta*, *61*, 3301–3309.

Fukao, Y., et al. (2004), Trans-Pacific temperature field in the mantle transition region derived from seismic and electromagnetic tomography, *Earth and Planetary Science Letters*, *217*, 425–434.

Fukao, Y., et al. (1992), Subducting slabs stagnant in the mantle transition zone, *Journal of Geophysical Research*, *97*, 4809–4822.

Fukao, Y., et al. (2001), Stagnant slabs in the upper and lower mantle transition zone, *Review of Geophysics*, *39*, 291–323.

Gasparik, T. (1993), The role of volatiles in the transition zone, *Journal of Geophysical Research*, *98*, 4287–4299.

Grand, S. (1994), Mantle shear structure beneath Americas and surrounding oceans, *Journal of Geophysical Research*, *99*, 11591–11621.

Hae, R., et al. (2006), Hydrogen diffusivity in wadelyite and water distribution in the mantle transition zone, *Earth and Planetary Science Letters*, *243*, 141–148.

Harrison, T. M. (1981), Diffusion of [40]Ar in hornblende, *Contributions to Mineralogy and Petrology*, *78*, 324–331.

Hart, S. R. (1984), He diffusion in olivine, *Earth and Planetary Science Letters*, *70*, 297–302.

Hauri, E. H., and S. R. Hart (1993), Re-Os isotope systematics of HIMU and EMII oceanic basalts from the south Pacific Ocean, *Earth and Planetary Science Letters*, *114*, 353–371.

Hauri, E. H., et al. (1994), Experimental and natural partitioning of Th, U, Pb and other trace elements between garnet, clinopyroxene and basaltic melt, *Chemical Geology*, *117*, 149–166.

Hirschmann, M. M. (2006), Petrologic structure of a hydrous 410 km discontinuity, in *Earth's Deep Water Cycle*, edited by S. D. Jacobsen and S. van der Lee, American Geophysical Union, Washington DC.

Hirschmann, M. M., et al. (2005), Storage capacity of H_2O in nominally anhydrous minerals in the upper mantle, *Earth and Planetary Science Letters*, *236*, 167–181.

Hirschmann, M. M., et al. (1998), The effect of alkalis on the silica content of mantle-derived melts, *Geochimica et Cosmochimica Acta*, *62*, 883–902.

Hofmann, A. W. (1988), Chemical differentiation of the Earth: the relationship between mantle, continental crust, and oceanic crust, *Earth and Planetary Science Letters*, *90*, 297–314.

Hofmann, A. W. (1997), Mantle geochemistry: the message from oceanic volcanism, *Nature*, *385*, 219–228.

Ito, E., and T. Katsura (1989), A temperature profile of the mantle transition zone, Hofmann, A. W. (2004), Sampling mantle heterogeneity through oceanic basalts: isotopes and trace elements, in *Treatise on Geochemistry*, edited by H. D. Holland and K. K. Turekian, pp. 61–101, Elsevier, Amsterdam.

Huang, X., et al. (2005), Water content of the mantle transition zone from the electrical conductivity of wadsleyite and ringwoodite, *Nature*, *434*, 746–749.

Huang, X., et al. (2006), A wet mantle conductor? (Reply), *Nature*, *439*, E3–E4.

Inoue, T. (1994), Effect of water on melting phase relations and melt composition in the system Mg_2SiO_4-$MgSiO_3$-H_2O up to 15 GPa, *Physics of Earth and Planetary Interiors*, *85*, 237–263.

Inoue, T., and H. Sawamoto (1992), High pressure melting of pyrolite under hydrous condition and its geophysical implications, in *High-Pressure Research: Application to Earth and Planetary Sciences*, edited by Y. Syono and M. H. Manghnani, pp. 323–331, American Geophysical Union, Washington DC. *Geophysical Research Letters*, *16*, 425–428.

Iwamori, H. (1992), Degree of melting and source composition of Cenozoic basalts in southwestern Japan: evidence for mantle upwelling by flux melting, *Journal of Geophysical Research*, *97*, 10983–10995.

Jacobsen, S. D. (2006), Effect of water on the equation of state of nominally anhydrous minerals, in *Water in Nominally Anhydrous Minerals*, edited by H. Keppler and J. R. Smyth, Mineralogical Society of America, Washington DC.

Karato, S. (2003), Mapping water content in Earth's upper mantle, in *Inside the Subduction Factory*, edited by J. E. Eiler, pp. 135–152, American Geophysical Union, Washington DC.

Karato, S. (2006a), Hydrogen-related defects and their influence on the electrical conductivity and plastic deformation of mantle minerals: A critical review, in *Earth's Deep Water Cycle*, edited by S. D. Jacobsen and S. van der Lee, American Geophysical Union, Washington DC.

Karato, S. (2006b), Microscopic models for the influence of hydrogen on physical and chemical properties of minerals, in *Super-plume: Beyond Plate Tectonics*, edited by D. A. Yuen, et al., p. in press, Springer.

Karato, S. (2006c), Remote sensing of hydrogen in Earth's mantle, in *Deep Earth Water Circulation*, edited by H. Keppler and J. R. Smyth, Mineralogical Society of America, Whagington DC.

Kawamoto, T., and J. R. Holloway (1997), Melting temperature and partial melt chemistry of H_2O-saturated peridotite to 11 gigapascals, *Science*, *276*, 240–243.

Koga, K., et al. (2003), Hydrogen concentration and analyses using SIMS and FTIR: Comparison and calibration for nominally anhydrous minerals, *Geochem. Geophys. Geosyst.*, *4*, 10.1029/2002GC000378.

Kohlstedt, D. L., et al. (1996), Solubility of water in the α, β and γ phases of $(Mg,Fe)_2SiO_4$, *Contributions to Mineralogy and Petrology*, *123*, 345–357.

Lange, R. L., and I. S. E. Carmichael (1990), Thermodynamic properties of silicate liquids with emphasis on density, thermal expansion and compressibility, *Review of Mineralogy*, *24*, 25–64.

Leahy, G., and D. Bercovici (2004), The influence of the transition-zone water filter on convective circulation in the mantle, *Geophysical Research Letters*, *31*, 10.1029/2004GL021206.

Leahy, G., and D. Bercovici (2006), On the entrainment of hydrous melt above the transition zone, *Earth and Planetary Science Letters*, submitted.

Litasov, K., and E. Ohtani (2002), Phase relations and melt compositions in CMAS-pyrolite-H_2O system up to 25 GPa, *Physics of Earth and Planetary Interiors*, *134*, 105–127.

Lu, R., and H. Keppler (1997), Water solubility in pyrope to 100 kbar, *Contributions to Mineralogy and Petrology*, *129*, 35–42.

Matsukage, K. N., et al. (2005), Density of hydrous silicate melt at the conditions of the Earth's deep upper mantle, *Nature*, *438*, 488–491.

McDonough, W. F., and S.-S. Sun (1995), The composition of the Earth, *Chemical Geology*, *120*, 223–253.

McKenzie, D. (1984), The generation and compaction of partially molten rocks, *Journal of Petrology*, *25*, 713–765.

Mierdel, K., and H. Keppler (2004), The temperature dependence of water solubility in enstatite, *Contributions to Mineralogy and Petrology*, *148*, 305–311.

Miyashiro, A. (1986), Hot regions and the origin of marginal basins in the western Pacific, *Tectonophyiscs*, *122*, 195–216.

Newsom, H. E. (1995), Composition of the solar system, planets, meteorites, and major terrestrial reservoirs, in *Global Earth Physics: A Handbook of Physical Constants*, edited by T. H. Ahrens, pp. 159–189, American Geophysical Union, Washington DC.

O'Nions, R. K., and E. R. Oxburgh (1983), Heat and helium in the Earth, *Nature*, *306*, 429–431.

Obayashi, M., et al. (2006), High temperature anomalies oceanward of subducting slabs at the 410-km discontinuity, *Earth and Planetary Science Letters*.

Ohtani, E., et al. (2000), Stability of dense hydrous magnesium silicate phases in the system Mg_2SiO_4-H_2O and $MgSiO_3$-H_2O at pressures up to 27 GPa, *Physics and Chemistry of Minerals*, *27*, 533–544.

Ohtani, E., et al. (1995), Melting relations of peridotite and the density crossover in planetary mantles, *Chemical Geology*, *120*, 207–221.

Ohtani, E., et al. (2001), Stability of dense hydrous magnesium silicate phases and water storage capacity in the transition zone and lower mantle, *Physics of Earth and Planetary Interiors*, *124*, 105–117.

Parman, S. W., et al. (2005), Helium solubility in olivine and implications for high $^3He/^4He$ in ocean island basalts, *Nature*, *437*, 1140–1143.

Rauch, M., and H. Keppler (2002), Water solubility in orthopyroxene, *Contributions to Mineralogy and Petrology*, *143*, 525–536.

Revenaugh, J., and S. A. Sipkin (1994), Seismic evidence for silicate melt atop the 410-km mantle discontinuity, *Nature*, *369*, 474–476.

Richard, G., et al. (2006b), Slab dehydration in the Earth's mantle transition zone, *Earth and Planetary Science Letters*, in press.

Richard, G., et al. (2006a), Slab dehydration and fluid migration at the base of the upper mantle: implication for deep earthquake mechanisms, *Geophysical Journal International*, in press.

Richard, G., et al. (2002), Is the transition zone an empty water reservoir? Inference from numerical model of mantle dynamics, *Earth and Planetary Science Letters*, *205*, 37–51.

Sakamaki, T., et al. (2006), Stability of hydrous melt at the bottom of the Earth's upper mantle, *Nature*, *439*, 192–194.

Sneeringer, M., et al. (1984), Strontium and samarium diffusion in diopside, *Geochimica et Cosmochimica Acta*, *48*, 1589–1608.

Song, T.-R. A., et al. (2004), Low-velocity zone atop the 410-km seismic discontinuity in the northwestern United States, *Nature*, *427*, 530–533.

Spiegelman, M., and T. Elliott (1993), Consequences of melt transport for uranium series disequilibrium in young lavas, *Earth and Planetary Science Letters*, *118*, 1–20.

Stalder, R., et al. (2001), High pressure fluids in the system MgO-SiO_2-H_2O under upper mantle conditions, *Contributions to Mineralogy and Petrology*, *140*, 607–618.

Suetsugu, D., et al. (2006), A study of temperature anomalies and water contents in the mantle transition zone beneath subduction zones as inferred from P-wave velocities and mantle discontinuity depths, in *Earth's Deep Water Cycle*, edited by S. van der Lee and S. D. Jacobsen, American Geophysical Union, Washington DC.

Takahashi, E. (1986), Melting of a dry peridotite KBL-1 up to14 GPa: implications on the origin of peridotitic upper mantle, *Journal of Geophysical Research*, *91*, 9367–9382.

Taura, H., et al. (1998), Pressure dependence on partition coefficients for trace elements between olivine and coexisting melts, *Physics and Chemistry of Minerals*, *25*, 469–484.

van der Hilst, R. D., et al. (1997), Evidence for deep mantle circulation from global tomography, *Nature*, *386*, 578–584.

van der Meijde, M., et al. (2003), Seismic evidence for water deep in Earth's upper mantle, *Science*, *300*, 1556–1558.

van Keken, P. E., et al. (2002), Mantle mixing: the generation, preservation, and destruction of chemical heterogeneity, *Annual Review of Earth and Planetary Sciences*, *30*, 493–525.

Van Orman, J. A., et al. (1998), Uranium and thorium diffusion in diopside, *Earth and Planetary Science Letters*, *160*, 505–519.

Wang, W., and E. Takahashi (2000), Subsolidus and melting experiments of K-doped peridotite KLB-1 to 27 GPa; Its geophysical and geochemical implications, *Journal of Geophysical Research*, *105*, 2855–2868.

Withers, A. C., et al. (1998), The OH content of pyrope at high pressure, *Chemical Geology*, *147*, 161–171.

Wolfe, C. J., and P. G. Silver (1998), Seismic anisotropy of oceanic upper mantle: shear wave splitting methodologies and observations, *Journal of Geophysical Research*, *103*, 749–771.

Wood, B. J. (1995), The effect of H_2O on the 410-kilometer seismic discontinuity, *Science*, *268*, 74–76.

Wood, B. J., and J. D. Blundy (2002), The effect of H_2O on crystal-melt partitioning of trace elements, *Geochimica et Cosmochimica Acta*, *66*, 3647–3656.

Yoshino, T., et al. (2006), Complete wetting of olivine grain-boundaries by a hydrous melt near the mantle transition zone, *Earth and Planetary Science Letters*, submitted.

Young, T. E., et al. (1993), Infrared spectroscopic investigation of hydroxyl in β-$(Mg,Fe)_2SiO_4$ and coexisting olivine: implications for mantle evolution and dynamics, *Physics and Chemistry of Minerals*, *19*, 409–422.

Zhao, D. (2004), Global tomographic images of mantle plumes and subducting slabs: insights into deep mantle dynamics, *Physics of the Earth and Planetary Interiors*, *146*, 3–34.

Zhao, Y.-H., et al. (2004), Solubility of hydrogen in olivine: dependence on temperature and iron content, *Contributions to Mineralogy and Petrology*, *147*, 155–161.